DIGITAL DESIGN

ALSO BY R. K. RICHARDS

Arithmetic Operations in Digital Computers, Van Nostrand, 1955
Digital Computer Components and Circuits, Van Nostrand, 1957
Electronic Digital Systems, Wiley, 1966
Electronic Digital Components and Circuits, Van Nostrand, 1967

Digital Design

R. K. Richards

WILEY-INTERSCIENCE

a Division of John Wiley & Sons, Inc.
New York • London • Sydney • Toronto

Copyright © 1971, by John Wiley & Sons, Inc.

All rights reserved. Published simultaneously in Canada.

No part of this book may be reproduced by any means, nor transmitted, nor translated into a machine language without the written permission of the publisher.

Library of Congress Catalogue Card Number: 73-147235

ISBN 0-471-71945-5

Printed in the United States of America.

10 9 8 7 6 5 4 3 2

1739384

PREFACE

This book is intended to be a text and reference book for persons who plan to be engaged in the design of machines, particularly electronic machines, that employ digital techniques. "Computers" constitute a major category of such machines, but the same design techniques apply to an endless variety of other digital systems ranging from miscellaneous control networks employing only a dozen or so switching devices to very large communications switching systems employing millions of such devices.

Originally (i.e., in the 1940's and early 1950's), relatively little distinction was made between the problems of electrical engineering encountered in the interconnections among resistors, diodes, transistors, and other electronic components and the "logic" problems encountered when large numbers of digitally operating circuits were interconnected. The two types of problems are still not completely separate, because circuits intended for specific digital applications are still being invented, developed, and used. Nevertheless, the vast bulk of present and expected future interest and usage is in the direction of limiting the electrical engineering to the design of switching and storage "modules" that can be interconnected in various ways. The pattern of interconnections is determined by relating the "logic" characteristics of the modules to the task to be performed by the machine as a whole. Although the machine designer must be aware of various engineering-imposed limitations of the modules at hand, the actual design of the digital system can then take place without further attention to electrical engineering considerations—in fact, without even any knowledge or background in electrical engineering at all.

The subject matter of this book also has a historical relationship to mathematics. For one thing, most early digital machines (and many present ones, for that matter) were built for the specific purpose of generating numerical answers to problems in pure and applied mathematics. More pertinently, the determination of the interconnections among the modules is usually found to be sufficiently complex that a scheme of notation (i.e., a mathematical procedure) is greatly helpful and perhaps absolutely necessary to designate the signals and their interrelationships. The origins of the relevant mathematical procedures are traceable to the mid-nineteenth century, but the bulk of the development work for this application has taken place in very recent years. However, the nature of this work is so closely related to problems encountered with physical machines that the tendency has been to associate the work with the machines rather than with mathematics.

In short, although digital design procedures have close relationships both to electrical engineering and to mathematics, the procedures do not really constitute a branch of either. Instead, the subject of digital design is a branch of a new field of knowledge, which has come to be known by names such as "computer science" or "information processing technology."

This book is designed to be used as a text in a course of instruction that will be guided by an instructor, but with reasonable diligence it can be used as a self-study text. The treatment of many of the topics is such as to include more alternative approaches than would probably be of interest from the standpoint of a general introduction to the subject of digital design. The purpose in including the alternatives is to enhance the value of the book as a reference when the student is subsequently faced with the actual task of designing a system or some part of it. An instructor can, however, easily skip over those items that do not fit the objectives of a particular introductory course of instruction. The exercises at the end of each chapter, for the most part, are designed to assist the reader in comprehending the material in the text, although some have the characteristics of "problems" that additionally exercise ingenuity.

The bulk of this book was written so that the only prerequisite is a good knowledge of algebra. A few exceptions proved to be necessary, especially when the application of the machine under design is to be for the solution of more advanced mathematical problems, such as with the digital differential analyzers that are described in a part of Chapter 8. However, skipping over such topics will not impair the understanding of subsequent topics.

I believe that I originated some of the ideas in this book, but the vast bulk of the material originated elsewhere. Assigning credit for each and every design procedure, network type, arithmetic technique, and so on, has proved to be hopelessly impractical. Apart from the very large number of such ideas, the original source of even the major ones is often unknown or at least in doubt. Nevertheless, full recognition is hereby given to the contributions made to the subject of digital design by many, many people. The contributions have reached me primarily through the technical journals and conference proceedings as published by the AIEE, the ACM, and several other organizations and also through other media such as engineering magazines, patents, advertisements, and instruction manuals pertaining to specific digital systems.

In instances where a topic is considered to be pertinent to the general subject of digital design, but too detailed for most expected readers of this book, the topic is summarized and references to other literature are included in the body of the text.

R. K. RICHARDS

Ames, Iowa
January, 1971

CONTENTS

1 Boolean Algebra and Its Relationship to Switching Networks   1
2 AND-OR Networks   24
3 Networks Containing Inverters   76
4 Feedback, Digital Storage, and Time-Dependent Signals in Switching Networks   120
5 Digital Decoding   178
6 Implementing Binary Arithmetic   272
7 Implementing Decimal Arithmetic   376
8 Implementing Other Digital Operations   429
9 An Introduction to the General Purpose Computer and Its Programming   526

Index   573

DIGITAL DESIGN

1
BOOLEAN ALGEBRA AND ITS RELATIONSHIP TO SWITCHING NETWORKS

Boolean algebra derives its name from George Boole who was the author of a book published in 1847 entitled *The Mathematical Analysis of Logic*, a reproduction of which was published by Blackwell (England) in 1948. Although his system of notation, Boolean algebra, became well known to mathematicians working on the theory of logic, the application of this branch of mathematics to switching circuits seems to have not taken place until the comparatively recent date of 1938 when Claude E. Shannon published a paper entitled "A Symbolic Analysis of Relay and Switching Circuits" [*Transactions of the American Institute of Electrical Engineers*, 57 (1938), pp. 713–723].

Boolean algebra was something less than a major influence in the invention and design of early electronic computers, which began to appear in 1946. In fact, the 1950's were well under way before the algebra was used at all at some of the major pioneering electronic computer organizations. To this day, a respectable and competitive electronic computer could probably be designed without recourse to the algebra. On the other hand, virtually all computer engineers have found the algebra so useful in understanding and teaching the workings of digital systems, in recording the relationships between switching elements, and in reducing the number of devices required for a given switching function, that a knowledge of Boolean algebra has become a fundamental requirement for people working in digital technology.

Some Introductory Remarks about Electronic Switching Components

The emphasis in this book will be with relationship to electronic switching components (as opposed to electromagnetic "relays" and other digital components for which the design procedures, although similar in certain basic principles, are very different in detail). In electronic switching systems each wire or "line" is viewed as having a signal "present" or "not present" at any given time, except that a nonzero interval of time is required for a change between the presence and nonpresence of a signal. In nearly all practical applications the presence or nonpresence of a signal is represented by one of two potential levels, where the potential is measured with respect to some reference, usually ground potential. No standards have been established for the potential levels to be used for the presence and nonpresence of signals, although the usual convention is to view the relatively positive one of the two potentials as corresponding to the presence of a signal. One or both of the two potentials may be negative with respect to ground.

The particular potentials to be used are determined from engineering considerations as they apply to the electronic switching components and circuits selected for use in any given application. In most present-day transistor switching circuits the potential difference between the two signal levels may be as little as 0.5 volt or as much as 5 volts. Potential swings at the lower end of this range are encountered with the so-called ECL circuits (where ECL is derived from "emitter-coupled-logic"—other common terms being "current-mode" or "current-switching" transistor circuits). However, because a nonzero potential change is required to cause a change in the current flow, the presence or absence of a signal on an interconnecting wire is still indicated by the potential of that wire. Potential swings approaching 5 volts are commonly employed with the popular transistor-transistor-logic (TTL) and diode-transistor-logic (DTL) circuits.

A few exceptions to the situation described above exist. For one, a signal or its absence may be repesented by the presence or absence of a voltage pulse at a given time. Several early electronic digital computers were built around switching circuits that functioned in accordance with this so-called "dynamic" mode of operation, but most engineers found this mode inferior in cost, speed, ease of system design, and ease of servicing. Therefore, this method of signal representation is now avoided in switching circuits. Some switching circuits, notably those based on magnetic cores, require pulse-type signals, but such circuits have never been widely used, and significant future use of them now seems unlikely.

Certain tunnel diode switching circuits and the parametron represent other examples where something other than potential is used to represent a signal. In these examples the presence or absence of a digital signal is represented by whether an a-c signal is in phase or out of phase with respect to a reference a-c signal of the same frequency. This difference in signal representation is not particularly consequential with respect to the applicability of Boolean algebra, but these circuits have the additional feature of operating on a majority or threshold basis. This latter feature has a profound influence on network design procedures, but the subject will not be carried further because tunnel diode circuits, parametrons, and analogous circuits have not been found to be competitive with transistor circuits that perform the more basic switching functions.

Although a few other exceptions could be cited, the two-potential method of representing the presence or absence of a signal can, in summary, be regarded as being almost universally applicable to electronic switching.

The Notation to Be Used for Representing the Presence or Absence of a Signal

In engineering applications the presence or absence of a signal is most commonly represented by the digits 1 and 0, respectively, and this is the notation that will be used in this book.

This notation does, however, have a disadvantage that is worth at least a brief discussion. Most systems employing electronic switching have computing as an important, and perhaps major, objective. This computing is, of course, performed on numbers, and the numbers employ the same 1 and 0 characters. Because many control signals and other signals in the system do not represent numbers, some confusion occasionally arises as to whether a given 1 or 0 represents a part of a number or whether it merely represents the presence or absence of a signal at some miscellaneous point in the system.

Alternative notations are "true" and "false," or their abbreviations T and F, for representing the presence or absence, respectively, of a signal. The origin of these notations is obviously from the fact that Boolean algebra was originally developed for the mathematical analysis of logic where the truth or falsehood of statements is basic. Except perhaps in a digital system specifically designed to simulate logic problems, the presence or absence of a signal bears no relationship to the English language meanings of "true" and "false," thus the acceptance of this notation by machine designers has been limited. The letters P and N to correspond to "present" and "not present" or to "positive" and "negative" are more pictorial, and some usage of this notation is found also.

On the other hand, the presence or absence of a signal does so often represent the binary digit 1 or 0, respectively, that the notation is a decided convenience in many important switching circuit problems.

The Notation for a Signal, Regarded as a Variable

In the course of operation of a digital system, the signal status at any given wire will change back and forth between 1 and 0 from time to time, and the universal practice is to regard the signal status or just "the signal" as a two-valued or "binary" variable which can be represented by a symbol. The symbols most commonly used for this purpose are the alphabetic characters, either in capital or lower-case form. As in conventional algebra, a single letter with subscripts, such as A_1, A_2, A_3, and so on, may be used in the designation of a number of signals, especially when the number is large. When the signals are to be visualized in groups or sets, a notation such as A_i may be used to represent the signals in one set where i ranges from 1 to n, the number of signals in that set. A second set might then be represented by B_j, where j ranges from 1 to m, the number of signals in the second set.

A variable A is 1 or 0 according to whether the signal on the corresponding wire A is 1 or 0 (present or not present), respectively, at a given time. A designer may not know the value of a signal at a given time, but this situation does not necessarily mean that a binary variable can be an "unknown," especially not an "unknown to be solved for," in the same sense that these terms are used in conventional algebra. The concept of an "unknown" is not generally encountered in Boolean algebra, at least not as the algebra is used with switching circuits.

The AND Function, Its Notation, and an AND Circuit

The AND function is defined as that function performed by a switching device that accepts signals from two or more input lines and generates an output signal of 1 only when all of the input signals are 1. The derivation of the name is easily appreciated by observing that, for a two-input AND device receiving input signals A and B and generating an output signal C, the value of C is 1 only when A "and" B are 1.

In Boolean notation the AND relationship among two or more signals is indicated by writing their corresponding symbols in succession, as ABC for example to indicate A "and" B "and" C. This convention is the same as used for multiplication in conventional algebra, and the term "product" is in fact often used with reference to the AND function. A person must, of course, be cognizant of whether binary variables are being related by Boolean notation or whether continuous variables are

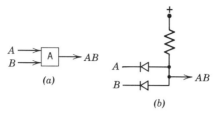

FIGURE 1-1. Symbol for an AND device and an AND circuit.

being related by conventional algebraic notation. (This notation for the AND function, as well as the notation for the OR function of the next section, seems to be the favorate of engineers, although several other notations are in use, especially by mathematicians.)

In the figures an AND device will be represented by a block labeled AND or by its abbreviation A, as in Figure 1-1a. The A for AND can easily be distinguished from any signal that might be designated A because the symbols for signals will not appear inclosed by a block. The input lines are distinguished from the output lines by arrowheads.

A diode circuit which performs the AND function is shown in Figure 1-1b, where 1's and 0's are represented by relatively positive and negative potentials, respectively. The supply potential connected to the upper terminal of the resistor is assumed to be more positive than either of the signal-representing potentials, and the output potential will be equal to (except for a relatively small voltage drop encountered across a conducting diode) to the more negative of the two input signal potentials. If one of the input signal potentials is more positive than the other, the corresponding diode will be in the high resistance condition, and the more positive input signal will therefore not affect the output potential. Thus, the output potential is relatively positive only when both of the input signals are relatively positive. Additional input lines can be added straightforwardly by adding a corresponding number of diodes.

The diode AND circuit, and especially arrays of diode AND and OR circuits (to be described in the next section), are presented in much more detail in Chapter 2 of Richards' book *Digital Computer Components and Circuits*, 1957, or in Chapter 2 of his more recent book *Electronic Digital Components and Circuits*, 1967. In fact, all descriptions of the physical devices needed to perform the various switching and storage functions will here be kept to a bare minimum needed to relate abstract subjects of switching, coding, and computing to the real world encountered by engineers. At all points where the physical devices become pertinent, a reference to these other books for details is implied.

The OR Function, Its Notation, and an OR Circuit

The OR function is defined as that function performed by a switching device that accepts signals from two or more input lines and generates an output signal of 0 when all of the input signals are 0 but generates an output signal of 1 when any one (or more) of the input signals is 1. Again, the derivation of the name may be appreciated by observing that, for a two-input OR device receiving input signals A and B and generating an output signal C, the value of C is 1 only when A "or" B (or both) is 1.

The Boolean notation for the OR relationship is to place "plus" signs between the corresponding symbols, as $A + B + C$ for example, to indicate A "or" B "or" C. Because of the similarity of this notation to that used for addition in conventional algebra, the term "sum" is often used with reference to the OR function.

In the figures an OR device will be represented by a block labeled OR or its abbreviation O, as in Figure 1-2a.

Figure 1-2b shows a diode OR circuit that is similar to an AND circuit except that the supply potential is more negative than either of the two signal-representing potentials, and the polarity of the diodes is reversed. By an action analogous to that described previously, the output potential is relatively positive only when one (or more) of the input potentials is relatively positive.

An "exclusive-OR" function is that function performed by a device which generates an output signal of 1 only when any one, but only one, of the input signals is 1. While seemingly basic from a logic standpoint, the exclusive-OR function is really a combination of AND, OR, and inversion functions (see next section), as becomes particularly apparent when attempting to devise an exclusive-OR circuit. Moreover, the need for exclusive-OR functions involving more than two input signals is encountered only rarely in practical applications whereas multi-

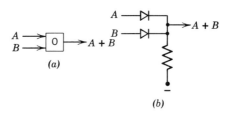

FIGURE 1-2. Symbol for an OR device and an OR circuit.

input OR functions are useful in virtually all applications. The two-input exclusive-OR function is a special case, both in the sense that a reasonably simple transistor circuit happens to be available for its generation and in the sense that a need for it is encountered in a number of important applications. The two-input exclusive-OR function and examples of its use will be presented in later chapters.

Inversion or the NOT Function, Its Notation, and an Inverter Circuit

Inversion is performed by a device that has only one input line and that causes a signal to be present or not on an output line when a signal is not present or present, respectively, on the input line. That is, a 0 is transformed to a 1, and a 1 is transformed to a 0. Whether the input signal is 1 or 0, the output signal is "not" the same as the input signal. If the input signal is A, the output signal is said to be NOT A.

In Boolean notation, inversion is represented by placing a bar over the symbol representing the signal being inverted, such as $\bar{A}$, for example. $\bar{A}$ may be recited as "A bar," "not A," or "the inverse of A," where all of these phrases are substantially synonomous.

A device which performs the inversion function is called an "inverter" (the terms "NOT device" or "NOT circuit" being used only rarely), and in the figures an inverter will be indicated by a block labeled INV or simply I as in Figure 1-3a. An inverter circuit employing an NPN transistor is shown in Figure 1-3b. If the input signal is sufficiently positive, the transistor will conduct so that the potential appearing on the output line will be relatively negative (although positive with respect

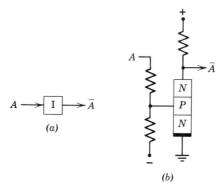

FIGURE 1-3. Symbol for an inverter and inverter circuit.

to ground). When the input signal is relatively negative, the transistor will not conduct, and the output potential will become relatively positive. The voltage divider is not essential, but it improves transistor response in the nonconducting condition when the input signal is supplied by the output of another transistor in a similar circuit.

Combinations of Switching Functions

Except when time is a parameter, any switching function can be represented by a combination of AND, OR, and NOT functions. Actually, as will be shown, either the AND or the OR function, but not both, can be eliminated from the set of basic functions required to represent any complex function. The procedure for finding the particular combination of functions required or appropriate for the representation of some given function is a problem that is fundamental to the whole subject of Boolean algebra as it pertains to switching circuits, and much attention will be given to this matter later. Here, the objective is limited to explaining the notation to be used when functions are to be combined.

In instances that would otherwise result in an ambiguity, an individual AND function or an individual OR function is specified by placing parentheses around the symbols representing the signals being combined. For example, $(AB)C$ means that A and B are combined in a two-input AND device with the output signal AB being combined with C in a separate two-input AND device. Thus, $(AB)C$ represents a different set of hardware (physical switching devices) than does ABC, which corresponds to a single three-input AND device, although functionally the final output signal would be the same in either case. Similarly, $A + (B + C)$, for example, would mean that signals B and C are applied as inputs to one two-input OR device, the output of which is applied as one input signal to a second two-input OR device which receives A as the signal on the other input line.

For more complex combinations, brackets and braces may be used to improve the clarity of the notation as in conventional algebra.

When the input to an inverter is derived from the output of another switching device, the fact is represented notationally by extending the bar over the entire Boolean expression for the signal that arrives at the inverter input. Thus, for example, $\overline{A(B + C)}$ would mean that B and C are applied as inputs to a two-input OR device, the output of which is combined with A in a two-input AND device, and the output of this device is in turn applied as the input to the inverter to generate a signal represented by $\overline{A(B + C)}$. For a slightly more complex example, Figure 1-4 shows in block diagram form the switching devices represented by the Boolean expression $\overline{(A\bar{B})C} + AD + E$. Any input variable can

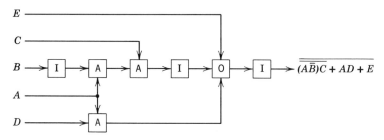

FIGURE 1-4. Example illustrating the relationship between a switching network and its Boolean notation.

appear in a Boolean expression any number of times, with A appearing twice in this example.

"Equality" in Boolean Notation

An indication that two Boolean expressions are "equal" to each other (and the conventional equal sign, =, is used for the purpose) means that the output signals from the two switching arrays being represented would be the same for all possible combinations of binary values of the input signals.

Sometimes an equality sign is used to portray two different physical networks that perform the same net switching function such as, for example, $(AB)C = ABC$, where each side of the expression is 1 only when all three of A, B, and C are 1.

At other times a Boolean equality means merely that a new symbol is being assigned to represent the output of a switching array. For example, $A + B = X$ would mean that a new symbol, X, is being assigned to represent the signal which appears on the output line of an OR device to which signals A and B are applied as inputs.

Some Basic Boolean Relationships

From the definitions of the AND, OR, and NOT functions, and from the notational conventions which were described, the following Boolean relationships follow directly.

$$00 = 0 \quad\quad 0 + 0 = 0$$
$$01 = 0 \quad\quad 0 + 1 = 1 \quad\quad \bar{0} = 1$$
$$10 = 0 \quad\quad 1 + 0 = 1 \quad\quad \bar{1} = 0$$
$$11 = 1 \quad\quad 1 + 1 = 1$$

When these relationships are added to the definitions and notational conventions, a set of basic relationships as listed in Table 1-1 may be derived directly. These relationships are extremely useful to a switching circuit designer, and the reader is advised to become closely familiar with them. The commutative relationships 1 and 2 are derived from the fact that the input connections to an AND or an OR device can be interchanged without affecting the output signal. Relationship 3 expresses the fact that, if a continuous 0 (the absence of a signal) is applied to one input of an AND device, the output must be 0 at all times. Relationships 4 through 11 follow from corresponding elementary arguments. Relationships 3 through 11 may alternatively be established by assigning the two possible values (0 and 1) to the variable A and by noting that both sides of each equality are in fact the same in each case.

In Table 1-1 associative relationships 12 and 13, distributive relationships 14 and 15, and relationships 16 and 17 (often called DeMorgan's laws or theorems) can be established either through a visualization of the operation of the switching devices they represent or by assigning values to the input variables in all possible combinations and noting that the same binary value for each side of each relationship is obtained in each case. For example, the left-hand side of relationship 14 means

TABLE 1-1. Some Basic Boolean Relationships

1.	$AB = BA$
2.	$A + B = B + A$
3.	$A0 = 0$
4.	$A1 = A$
5.	$AA = A$
6.	$A + 0 = A$
7.	$A + 1 = 1$
8.	$A + A = A$
9.	$A\bar{A} = 0$
10.	$A + \bar{A} = 1$
11.	$\bar{\bar{A}} = A$
12.	$(AB)C = ABC$
13.	$(A + B) + C = A + B + C$
14.	$A(B + C) = AB + AC$
15.	$A + BC = (A + B)(A + C)$
16.	$\overline{AB} = \bar{A} + \bar{B}$
17.	$\overline{A + B} = \bar{A}\bar{B}$

that an output signal is 1 when A is 1 "and" either B "or" C is 1, which is equivalent to the right-hand side which means an output of 1 when A "and" B are both 1 "or" when A "and" C are both 1.

Not all of the relationships in Table 1-1 are independent of each other. Some can be established through the use of others. Two examples of interdependence will be illustrated because the interdependence is interesting academically and because the examples provide good illustrations of Boolean algebra as used in designing switching circuits. Relationship 11, $\bar{\bar{A}} = A$, can be established as follows:

$\bar{\bar{A}} = \bar{\bar{A}}1 + 0$ \hspace{1em} (by using relationships 4 and 6)

$= \bar{\bar{A}}(A + \bar{A}) + A\bar{A}$ \hspace{1em} (by using relationships 10 and 9)

$= A\bar{\bar{A}} + \bar{\bar{A}}\bar{A} + A\bar{A}$ \hspace{1em} (by using relationships 14 and 1)

$= A(\bar{A} + \bar{\bar{A}}) + \bar{A}\bar{\bar{A}}$ \hspace{1em} (by using relationship 2 twice, then relationships 14 and 1)

$= A(X + \bar{X}) + X\bar{X}$ \hspace{1em} (by letting $X = \bar{A}$; that is, by assigning X to represent the signal that would be obtained from the output of an inverter to which A is applied as the input)

$= A1 + 0$ \hspace{1em} (by using relationships 10 and 9)

$= A$ \hspace{1em} (by using relationships 4 and 6)

The reader might suspect that relationship 11 could be established much more directly by letting $X = \bar{A}$ as a first step. However, a subtle problem is encountered in that both sides of this equality must be inverted to obtain $\bar{X} = A$ whereas, strictly speaking, the validity of this action has not yet been established.

Relationship 15, $A + BC = (A + B)(A + C)$, can be established as follows:

$A + BC = A(1 + C) + BC$ \hspace{1em} (by using relationships 7 and 2)

$= A + AC + BC$ \hspace{1em} (by using relationships 14 and 4)

$= A(1 + B) + AC + BC$ \hspace{1em} (by using relationships 7 and 2)

$= AA + AB + AC + BC$ \hspace{1em} (by using relationships 14, 4, and 5)

$= (A + B)A + (A + B)C$ \hspace{1em} (by using relationships 14 and 1)

$= (A + B)(A + C)$ \hspace{1em} (by using relationship 14 with the temporary substitution of $A + B = X$)

Some Other Boolean Relationships

Countless other relationships among binary variables can be devised and shown to be correct by using the 17 relationships of Table 1-1 in essentially the same manner as illustrated for the two examples in the previous section. Some of the simpler ones are listed below.

$$A + AB = A$$
$$A + \bar{A}B = A + B$$
$$A(A + B) = A$$
$$A(\bar{A} + B) = AB$$
$$AB + \bar{A}\bar{B} = (A + \bar{B})(\bar{A} + B) = \overline{A\bar{B} + \bar{A}B}$$
$$\overline{ABCD} = \bar{A} + \bar{B} + \bar{C} + \bar{D}$$
$$\overline{A + B + C + D} = \bar{A}\bar{B}\bar{C}\bar{D}$$
$$AB + CD = (A + C)(A + D)(B + C)(B + D)$$
$$AB + BC + AC = \overline{\bar{A}\bar{B} + \bar{B}\bar{C} + \bar{A}\bar{C}}$$
$$AB + \bar{A}C + BC = AB + \bar{A}C$$

When it is remembered that the expression on each side of each equation can represent a physical network, the importance of some of the above relationships becomes apparent. The left-hand side of the first equation, for example, corresponds to an AND circuit having A and B as inputs with the output signal AB connected as one input of an OR circuit to which A is supplied as the other input. In this example the final output signal will never depend on signal B. Therefore, the network can be eliminated except for a single direct connection from the A input line to the final output line. Boolean manipulation of the symbols can be used to establish the equality as follows: $A + AB = A1 + AB = A(1 + B) = A1 = A$.

Because $A = A1$, a variable which is common to two or more AND functions can be "factored out" by proceding through the same mental steps as in conventional algebra. Thus, the transition from $A + AB$ to $A(1 + B)$ may be viewed as a single step instead of the two steps shown.

Actually, the factoring process can be extended to more complex switching arrays with the same conventional algebraic rules being used. The extension results directly from the associative nature of the AND and OR functions as illustrated by the following example.

$$\begin{aligned} AB + AC + AD &= A(B + C) + AD \\ &= A[(B + C) + D] \\ &= A(B + C + D) \end{aligned}$$

To continue the analogy with conventional algebra, two or more AND-related variables may be factored simultaneously through using the commutative relationship as illustrated by this example:

$$ABC + ACE + ACDE = AC(B + E + DE)$$
$$= AC[B + E(1 + D)]$$
$$= AC(B + E)$$

Alternatively, in this example, ACE could have been factored out of the second and third AND functions in the original expression, and the final result would have been obtained even more directly as follows:

$$ABC + ACE + ACDE = ABC + ACE(1 + D)$$
$$= AC(B + E)$$

Strictly speaking, when the AND-related variables are factored simultaneously, a temporary substitution of variables must be made when using the relationships of Table 1-1 directly. That is, in this last example the temporary substitutions might be visualized as $ACE = X$ and $AC = Y$.

Note that in the example of the previous paragraph the physical switching array has been transformed from three AND devices, two with three inputs and one with four inputs, and a three-input OR device to an array consisting of a single three-input AND device and a two-input OR device.

The Use of a Binary Variable to Represent an Arbitrary Function of Other Binary Variables

The substitution of a new symbol to represent the output of an AND device or an OR device has already been illustrated. This idea can be extended to the point where each symbol is visualized as representing an arbitrary function of any number of variables. That is, A can be defined by $A = f(A_i)$ with f as the function of a set of variables A_i where i ranges from 1 to n, the number of variables in the set. Then B might be defined to mean $B = g(B_j)$, where g is a different function and the variables form a different set B_j, where j ranges from 1 to m, the number of variables in this set.

With this convention, either A or $f(A_i)$ would represent the signal which appears on the output line of a switching network determined in accordance with the AND, OR, and NOT functions that comprise f. Similarly, B or $g(B_j)$ would represent another output signal.

Now the source of the input signals was not specified for any of the relationships presented earlier in this chapter. In particular, these signals

14   BOOLEAN ALGEBRA

could have been generated as the outputs from other switching networks. Therefore, all of the relationships remain valid with this more generalized meaning of a binary variable or symbol. Just to cite one example, relationship 17 of Table 1-1 becomes

$$\overline{f(A_i) + g(B_j)} = \overline{f(A_i)}\,\overline{g(B_j)},$$

where $f(A_i)$ might be $A_1 + A_2$, and $g(B_j)$ might be B_1B_2 so that the relationship is in this instance

$$\overline{A_1 + A_2 + B_1 B_2} = \overline{A_1 + A_2}\,\overline{B_1 B_2}$$

Furthermore, in establishing this generalized meaning of a binary symbol, no reason is encountered why some or all of the variables in one set cannot be duplicated in another set. Thus, in the example just cited, signal A_1 might be called X, both A_2 and B_1 might represent the same signal called Y, and B_2 might be called Z. Then if the original switching array happened to be represented by the expression $X + Y\,\overline{YZ}$, this expression (and its corresponding switching array) could be transformed straightforwardly as follows: $X + Y\,\overline{YZ} = X + Y + \overline{YZ} = X + Y(1 + \overline{Z}) = X + Y$.

Truth Tables

A listing of the various input signal combinations together with the resultant output signal for each combination is called a "truth table," and the truth table for $X + Y\,\overline{YZ} = X + Y = f$ would be

X	Y	Z	f
0	0	0	1
0	0	1	1
0	1	0	0
0	1	1	0
1	0	0	0
1	0	1	0
1	1	0	0
1	1	1	0

as can be determined by computing f in accordance with either expression for each of the eight different combinations for three variables. In this example f happens to be independent of Z.

A truth table constitutes an alternative means for expressing a switching function. (Actually, only those combinations of input signals for which the output signal is 1 need be listed—000 and 001 in the above example.) The truth table form of notation is occasionally useful, particularly when a read-only storage unit is being employed as a substitute for a multi-output combinatorial switching function in the manner to be described in a later chapter. However, in this book the Boolean notation will be used almost exclusively because of its usefulness in finding advantageous arrays of switching modules to realize given switching functions.

The Absence of "Subtraction" and "Division" from Boolean Algebra

In spite of the similarity between Boolean "products" and "sums" to the corresponding features of conventional algebra, operations corresponding to subtraction and division are omitted from Boolean algebra. Consider the Boolean equality

$$AB + \bar{A}\bar{B} + \bar{A}C = AB + \bar{A}\bar{B} + BC.$$

The correctness of the equation can be established by forming the truth table for each of the two expressions and finding that the tables are identical. Although $AB + \bar{A}\bar{B}$ appears on both sides of this equation, these terms cannot be "subtracted out" because $\bar{A}C \neq BC$.

To illustrate why common factors cannot be "divided out" of both sides of a Boolean equation consider the previously listed equality $A(A + B) = A$. Contrary to conventional algebraic techniques, the expressions on the two sides cannot be divided by A, because $A + B$ is not necessarily equal to 1. For another example, consider

$$(A\bar{B} + B\bar{C} + C\bar{A})(\bar{A}\bar{B} + \bar{B}\bar{C} + \bar{C}\bar{A})$$
$$= (A + B + C)(\bar{A}\bar{B} + \bar{B}\bar{C} + \bar{C}\bar{A}),$$

where the common factor $\bar{A}\bar{B} + \bar{B}\bar{C} + \bar{C}\bar{A}$ cannot be "divided out" of the two sides of the equation because

$$A\bar{B} + B\bar{C} + C\bar{A} \neq A + B + C$$

It is instructive for the reader to show by means of the techniques already described that the equality is correct and that the inequality is also correct.

The Number of Functions of n Variables

In conventional algebra the number of possible different functions of just one variable is infinite. For example, the functions of x can be x, x^2, x^3, and so on to x^∞. However, in Boolean algebra, the number of different functions is finite and is reasonably small when the number of variables is four or less. Although the number of functions increases to astronomical figures very rapidly as the number of variables is increased beyond five, the number of functions is still finite.

Consider two binary variables, A and B. The corresponding signals can be present or not present (1 or 0) in just four combinations. That is, both A and B can be 0, A can be 1 when B is 0, A can be 0 when B is 1, and both A and B can be 1. That is, one and only one of the expressions $\bar{A}\bar{B}$, $A\bar{B}$, $\bar{A}B$, and AB can be 1 at any given time. A function of A and B results from specifying which of these four expressions must be 1 for the final output signal to be 1. None, any one, any two, any three, or all four of the expressions may be chosen, although if none are chosen the trivial result of a 0 output signal at all times is obtained, and if all four are chosen, the final output signal is 1 at all times. Table 1-2 sets forth the two-variable functions in detail.

In many cases the straightforwardly obtained expression of a function can be simplified by Boolean manipulation of the variables, and the simplified form of each expression is likewise shown in Table 1-2. In several instances one or both of the variables does not appear at all in the simplified form, and in these instances the function is not really of two variables but is either a constant (a "function of no variables") or is a function of only one variable.

TABLE 1-2. The 16 Boolean Functions of Two or Fewer Variables

$\bar{A}\bar{B}$	$A\bar{B}$	$\bar{A}B$	AB	
0	0	0	0	0
1	0	0	0	$\bar{A}\bar{B}$
0	1	0	0	$A\bar{B}$
0	0	1	0	$\bar{A}B$
0	0	0	1	AB
1	1	0	0	$\bar{A}\bar{B} + A\bar{B} = \bar{B}$
1	0	1	0	$\bar{A}\bar{B} + \bar{A}B = \bar{A}$
1	0	0	1	$\bar{A}\bar{B} + AB$
0	1	1	0	$A\bar{B} + \bar{A}B$
0	1	0	1	$A\bar{B} + AB = A$
0	0	1	1	$\bar{A}B + AB = B$
1	1	1	0	$\bar{A}\bar{B} + A\bar{B} + \bar{A}B = \bar{A} + \bar{B}$
1	1	0	1	$\bar{A}\bar{B} + A\bar{B} + AB = A + \bar{B}$
1	0	1	1	$\bar{A}\bar{B} + \bar{A}B + AB = \bar{A} + B$
0	1	1	1	$A\bar{B} + \bar{A}B + AB = A + B$
1	1	1	1	$\bar{A}\bar{B} + A\bar{B} + \bar{A}B + AB = 1$

THE NUMBER OF FUNCTIONS OF n VARIABLES 17

Furthermore, most electronic switching devices are of such nature that all input lines are designed to handle signals of the same nature (with regard to voltage levels, etc.) so that the input connections can be permuted in any desired manner. Therefore, the switching array needed to perform the function $A\bar{B}$, for example, can be identical to the array needed to perform $\bar{A}B$. A simple interchange of connections to the input terminals is all that is required for a single switching array (a single AND device with an inverter in series with one input terminal) to be capable of performing either function. When functions of fewer variables are eliminated and when due consideration is given to possible permutations of input connections, only eight different functions of two variables remain: $\bar{A}\bar{B}$, $A\bar{B}$, AB, $AB + \bar{A}\bar{B}$, $A\bar{B} + \bar{A}B$, $\bar{A} + \bar{B}$, $A + \bar{B}$, and $A + B$.

In many (but definitely not all) electronic switching applications, the signals are supplied by symmetrical circuits which provide two output signals where one signal is the inverse of the other. Thus, for a signal such as A, two wires are present with one carrying signal A and the other carrying signal $\bar{A}$. With each input signal and its inverse available on separate wires, the number of distinctly different switching functions is further reduced in that many different Boolean expressions become equivalent when inversions as well as permutations of the variables are permitted. In the two-variable case, the number of distinctly different switching functions is then only three: AB, $AB + \bar{A}\bar{B}$, and $A + B$. To obtain the function $A\bar{B} + \bar{A}B$ for example, the switching array for developing $AB + \bar{A}\bar{B}$ is selected, but the connections for the B and $\bar{B}$ input terminals are interchanged (or, alternatively, the symbols designating the B and $\bar{B}$ signals are interchanged).

In the general case, the total number of different combinations of n variables (or signals) is 2^n, where n is the number of variables. To generate a given function, the combinations can be selected in any of $(2)^{2^n}$ ways to produce a corresponding number of different functions. However, as already illustrated in some detail for the two-variable case, some of these functions represent output signals that do not depend on the values of certain input variables and are therefore really functions of less than n input variables. Thus, $(2)^{2^n}$ is the number of functions of n or fewer variables. In particular, when n is zero $(2)^{2^n} = 2$ to indicate that the number of "functions of no variables" is two, and these functions are the constants 0 and 1. If $n = 1$, then $(2)^{2^n} = 4$, and the functions of one or fewer variables are 0, 1, A, and $\bar{A}$. For $n = 2$, the number of functions of two or fewer variables is $(2)^{2^n} = 16$, and the functions were listed in Table 1-2. The number of functions of three or fewer variables is 256. For four or fewer variables the number of functions is 65,536. For five variables, $(2)^{2^n}$ jumps to over four billion, and for six variables it is over 10^{19}. Note that for an increase of one in n, the total number of functions is squared.

Determining the number of distinctly different switching functions that result after permuting or both permuting and inverting the input variables is a surprisingly difficult problem. For permuting only, there are two functions for exactly one variable: A and $\bar{A}$. For two variables, there are the eight functions which have already been listed. For three variables the number of different functions is 68, and these are listed in Table 1-3 in simplified form. In this table, the functions are grouped in accordance with the number of different combinations of input signals they represent. For example, $AB + AC$ is in the 3-group because this function is 1 whenever any of the three combinations $AB\bar{C}$, $A\bar{B}C$, or ABC is 1. That is, $AB + AC = AB\bar{C} + A\bar{B}C + ABC$. Although the function $AB + \bar{A}C$ appears similar, it is in the 4-group because $AB + \bar{A}C = ABC + AB\bar{C} + \bar{A}BC + \bar{A}\bar{B}C$. The significance of these relation-

TABLE 1-3. The 58 Functions of Three Variables

1 term:
ABC
$AB\bar{C}$
$A\bar{B}\bar{C}$
$\bar{A}\bar{B}\bar{C}$

2 terms:
$ABC + A\bar{B}\bar{C}$
$ABC + \bar{A}\bar{B}\bar{C}$
$AB\bar{C} + A\bar{B}C$
$AB\bar{C} + \bar{A}\bar{B}C$
$AB\bar{C} + \bar{A}\bar{B}\bar{C}$
$A\bar{B}\bar{C} + \bar{A}B\bar{C}$

3 terms:
$AB + AC$
$AB + A\bar{C}$
$A\bar{B} + A\bar{C}$
$A\bar{B} + \bar{B}\bar{C}$
$\bar{A}\bar{B} + \bar{A}\bar{C}$
$AB + \bar{A}\bar{B}C$
$AB + \bar{A}\bar{B}\bar{C}$
$A\bar{B} + \bar{A}BC$
$A\bar{B} + \bar{A}B\bar{C}$
$\bar{A}\bar{B} + ABC$
$\bar{A}\bar{B} + AB\bar{C}$
$ABC + A\bar{B}\bar{C} + \bar{A}BC$
$AB\bar{C} + A\bar{B}C + \bar{A}BC$
$AB\bar{C} + A\bar{B}C + \bar{A}\bar{B}\bar{C}$
$ABC + \bar{A}\bar{B}C + \bar{A}B\bar{C}$
$A\bar{B}\bar{C} + \bar{A}BC + \bar{A}B\bar{C}$

4 terms:
$AB + \bar{A}C$
$AB + \bar{A}\bar{C}$
$A\bar{B} + BC$
$A\bar{B} + \bar{A}C$
$AB + AC + \bar{A}\bar{B}\bar{C}$
$AB + A\bar{C} + \bar{A}\bar{B}C$
$AB + A\bar{C} + \bar{A}\bar{B}\bar{C}$
$A\bar{B} + A\bar{C} + \bar{A}\bar{B}C$
$A\bar{B} + A\bar{C} + \bar{A}BC$
$A\bar{B} + A\bar{C} + \bar{A}B\bar{C}$
$A\bar{B} + \bar{B}\bar{C} + \bar{A}BC$
$A\bar{B} + \bar{B}\bar{C} + \bar{A}B\bar{C}$
$\bar{A}\bar{B} + \bar{A}\bar{C} + AB\bar{C}$
$\bar{A}\bar{B} + \bar{A}\bar{C} + AB\bar{C}$
$\bar{A}\bar{B} + \bar{B}\bar{C} + ABC$
$ABC + A\bar{B}\bar{C} + \bar{A}B\bar{C} + \bar{A}\bar{B}C$
$AB\bar{C} + A\bar{B}C + \bar{A}BC + \bar{A}\bar{B}\bar{C}$

5 terms:
$A + BC$
$A + \bar{B}C$
$A + \bar{B}\bar{C}$
$\bar{A} + BC$
$\bar{A} + \bar{B}C$
$\bar{A} + \bar{B}\bar{C}$
$AB + \bar{A}\bar{B} + \bar{A}C = AB + \bar{A}\bar{B} + BC$
$AB + \bar{A}\bar{B} + A\bar{C} = AB + \bar{A}\bar{B} + \bar{B}C$
$A\bar{B} + \bar{A}B + AC$
$A\bar{B} + \bar{A}B + A\bar{C}$
$A\bar{B} + \bar{A}B + \bar{A}C$
$A\bar{B} + \bar{A}B + \bar{A}\bar{C}$
$AB + AC + BC + \bar{A}\bar{B}\bar{C}$
$AB + A\bar{C} + B\bar{C} + \bar{A}\bar{B}C$
$A\bar{B} + A\bar{C} + \bar{B}\bar{C} + \bar{A}BC$
$\bar{A}\bar{B} + \bar{A}\bar{C} + \bar{B}\bar{C} + ABC$

6 terms:
$A + BC + \bar{B}\bar{C}$
$A + B\bar{C} + \bar{B}C$
$\bar{A} + BC + \bar{B}\bar{C}$
$\bar{A} + B\bar{C} + \bar{B}C$
$AB + \bar{A}C + \bar{B}\bar{C}$
$A\bar{B} + \bar{A}C + BC$

7 terms:
$A + B + C$
$A + B + \bar{C}$
$A + \bar{B} + \bar{C}$
$\bar{A} + \bar{B} + \bar{C}$

ships will be more apparent after studying the next chapter on simplifying AND-OR networks. (The number of different functions of three or fewer variables is $2 + 2 + 8 + 68 = 80$.)

Note that when the functions are expressed in simplified form as in Table 1-3, two of the functions have alternative expressions. In each case, the two expressions are equal as can be shown by Boolean manipulation, and yet one expression cannot be obtained from the other by a permutation of the variables. For functions of more than three variables, many more such equivalencies are encountered. This situation means that when identifying a function (as for example when selecting a switching network from a catalog of networks) some less simple representation of each function would probably be preferable.

So far as is known, for the general case of n variables, no one has ever worked out a practicably usable formula for determining the number of functions that are different under permutations of input variables. In fact, so far as is known, no one has ever determined by any means the actual number of such functions when n is greater than three. The number is probably a matter of a few thousand for $n = 4$.

When inversions as well as permutations of the input variables are permitted, the number of distinctly different functions of one, two, three, and four variables is 1, 3, 16, and 380, respectively. (For four or fewer variables the number is 402, which is derived by adding the numbers just listed and by including the two binary constants 1 and 0 which are "functions of no variables.") The 16 such functions of three variables are listed in Table 1-4. D. Slepian [*Canadian Journal of Mathematics*, 5 (1953), pp. 185–193] reports that the number of such functions of five variables is 1,228,158 and is over 400 trillion for six variables. The number is not known to have been determined for n greater than six.

Catalogs of Switching Networks

A major conclusion to be drawn from the very large number of different functions that are possible with any significant number of variables (input signals) is that the preparation of a catalog of switching networks becomes totally impractical when the number of variables is greater than four—or five in an extreme case. A catalog of four-input vacuum tube switching networks was published in the early days of computers, and a few catalogs of three-input networks employing other components have been published. However, even when the number of input signals is only three, the catalogs tend to be regarded has having limited usefulness because of specialized circumstances accompanying the specific components and circuits implied.

TABLE 1-4. The 16 Three-Variable Functions
That Are Distinct Under Inversions As Well
As Permutations of the Variables

1 term: ABC

---

2 terms: $ABC + A\bar{B}\bar{C}$
$ABC + \bar{A}\bar{B}\bar{C}$

---

3 terms: $AB + AC$
$AB + \bar{A}\bar{B}C$
$ABC + A\bar{B}\bar{C} + \bar{A}\bar{B}C$

---

4 terms: $AB + \bar{A}C$
$AB + AC + \bar{A}\bar{B}\bar{C}$
$AB + A\bar{C} + \bar{A}\bar{B}\bar{C}$
$ABC + A\bar{B}\bar{C} + \bar{A}B\bar{C} + \bar{A}\bar{B}C$

---

5 terms: $A + BC$
$AB + \bar{A}\bar{B} + \bar{A}C = AB + \bar{A}\bar{B} + BC$
$AB + AC + BC + \bar{A}\bar{B}\bar{C}$

---

6 terms: $A + BC + \bar{B}\bar{C}$
$AB + \bar{A}C + \bar{B}\bar{C}$

---

7 terms: $A + B + C$

In particular, for each function, the network that is selected for inclusion in the catalog is the one that is "best" by some criterion, but the criterion for determining "best" is not the same in all applications. Examples of such criteria include the following: minimizing the number of switching devices or units, minimizing the total number of input lines to the assemblage of units (as this may minimize the number of subunits), minimizing the number of switching "levels" or devices through which an input signal must pass to reach the output line, minimizing the number of cross-overs required for the interconnections among the devices, and minimizing the total area required (as determined by engineering considerations pertaining to the switching devices under consideration) for the switching devices and their interconnections.

Moreover, any given catalog of switching networks must be prepared in the light of the constraints imposed by the switching devices under consideration. For example, if the devices are limited to the AND-INVERT or OR-INVERT (NAND or NOR, respectively) combinations of functions as described in a later chapter, the resulting network for any given function may bear no apparent relationship to a corresponding network comprised of different switching modules, of which there are many. Even with a single type of module, the AND-INVERT for example, the resulting switching network will be dependent on other constraints imposed, notably the fan-out and fan-in (where fan-out is the maximum number of other devices the output signal from a given device is capable of actuating and where fan-in is the number of input signals which can be combined in an AND relationship within one such device). In an extreme case, the fan-out may be one and the fan-in may be only two. At the other extreme, the fan-out and fan-in may be essentially unlimited. For each different set of specifications on fan-out and fan-in limitations, a different switching network may result for any given cirterion of "best."

In summary, apart from the very large number of different switching functions to be considered, the preparation of a catalog of switching networks becomes impractical in the light of (a) the many different criteria for "best," (b) the many different functional characteristics of electronic switching devices, and (c) the different design constraints that might be encountered with a switching device of a given type.

Computer-Aided Boolean Algebra

Because catalogs of switching networks tend to be impractical, the need is for a method for deriving switching networks from the Boolean expressions of the functions to be performed. Preferably the method should be simple enough for the networks to be derived quickly by manual means. For the case of two-level single-output diode circuits having no more than about eight input signals, manual methods have been devised that are reasonably suitable. However, for any significant departure from this "basic" diode switching problem, the required number of algebraic manipulations becomes so great that the steps must be programmed on a computer if the networks are to be derived in a reasonable time with reasonable assurance of a correct result.

Literally hundreds of technical papers have been published on this general subject. In some instances the authors merely claim that their proposals are programmable on a computer in a useful way and in other instances the work of actual programming has been carried to various stages of completion. In at least a few instances, the resulting programs have actually been established as design tools for engineers working with specific sets of circuit modules. However, computer programs for digital network design are still not available for most of the wide range of miscellaneous switching problems encountered in practice. Perhaps this situation will be quite different at some future date, but because of the great multiplicity of problem parameters (see the last paragraph of the previous section) any one program would have limited applicability. A more stable and standardized (that is, a less rapidly advancing) state of technology with regard to switching devices would appear to be necessary for the more widespread development of computer programs for network design.

Incidentally, engineering costs as well as technical considerations affect the choice of a network to perform a given switching function. For example, at the present state of the art the cost of one diode is roughly the same as the cost of one minute of an engineer's time. Thus, in a one-of-a-kind system virtually no effort at all may be justified in searching for a switching network that may contain fewer components than some more or less obvious network. For longer production runs, a correspondingly larger amount of engineering time can be justified in the search for component savings. In any case, minimizing the amount of engineering time can be just as important as minimizing the number of components. Therefore, automated design procedures can be important for relatively simple problems as well as being essential for the more complex problems.

In the network design procedures, as described in the immediately following chapters, manual techniques tend to be emphasized because they form the basis for understanding the problems. Also, designers constantly encounter problems which, with reasonable proficiency on the part of the designer, can be solved by manual methods virtually as rapidly as the data can be prepared for computer entry. However, wherever the techniques are expressible in terms of specific routines, they may be reduced to computer programs.

Exercises

1. Draw the switching networks in block diagram form as represented by the following Boolean expressions (AND, OR, and INVERTER blocks).

(a) $\overline{(A + BC)D + \overline{A + BC}}$

(b) $\overline{\bar{A}B\bar{C}D} + ABD + \overline{\bar{B} + \bar{D}}$

2. Prove relationship 17 of Table 1-1, $\overline{(A + B)} = \bar{A}\bar{B}$, by using other relationships in the same table.

3. By Boolean manipulations convert the expression in Exercise 1(a) to a different but functionally equivalent expression that represents an AND-to-OR network, that is, a network consisting of one or more AND devices the outputs of which are fed to the inputs of a single OR device which generates the final output signal. (If a given AND device requires only one input signal, that device may, of course, be replaced by a closed circuit.)

4. Similarly convert the expression in Exercise 1(b) but continue the Boolean manipulation to show that a single two-input OR device is sufficient to generate the function.

5. By Boolean manipulations, prove the correctness of all relationships listed on the top half of page 12.

6. Show by Boolean manipulations that the following equalities are correct:
(a) $A\bar{C} + \bar{A}\bar{B} + \bar{B}\bar{C} = A\bar{C} + \bar{A}\bar{B}$
(b) $AB + \bar{A}\bar{B} + \bar{A}C = AB + \bar{A}\bar{B} + BC$
(c) $AB + \bar{A}\bar{B}D + \bar{A}CD = AB + \bar{A}\bar{B}D + BCD$
(d) $A\bar{B} + B\bar{C} + C\bar{A} = \bar{A}B + \bar{B}C + \bar{C}A$
(e) $(A\bar{B} + B\bar{C} + C\bar{A})(\bar{A}B + \bar{B}C + \bar{C}A) = (A + B + C)(\bar{A}\bar{B} + \bar{B}\bar{C} + \bar{C}\bar{A})$

7. Show that the following inequality is correct by finding at least one combination of binary values for the input variables for which the two sides of the expression are not the same.

$$A\bar{B} + B\bar{C} + C\bar{A} \neq A + B + C$$

8. Determine the seven combinations of binary values of A, B, and C for which the expression $A + B + C = 1$ is correct.

2
AND-OR NETWORKS

In this chapter attention is directed to networks containing only AND and OR switching devices. The assumption is made that, whenever required, each input signal is available in inverted form on a separate wire.

The explanations and illustrations will be given primarily in terms of all-diode switching circuits because of their simplicity. Actually, all-diode switching circuits, especially in the multilevel forms, are of limited practical applications because of certain electrical (not logical) considerations. However, the principles illustrated by these circuits are of fundamental importance in the design of the more widely used diode-transistor or all-transistor switching circuits involving inversions, and an understanding of the all-diode circuits is virtually an essential prerequisite to an understanding of the more complex electronic switching circuits.

The Relationship between a Function and the Amount of Equipment Required to Perform That Function

Consider, for example, the following switching functions:

1. $AB + CD$ 6 diodes
2. $AB + C$ 4 diodes
3. $A\bar{B} + CD + \bar{A}E + \bar{B}\bar{C}$ 12 diodes
4. $ABCDE + FGHK$ 11 diodes

Because each AND or OR circuit input signal requires one diode, a total of six diodes is required for function 1 with two diodes being needed in each of two AND circuits and another two being needed for the OR circuit that combines AB and CD. The number of diodes required for each of the other functions is determined analogously. Function 2

TWO-LEVEL AND-TO-OR OR SUM-OF-PRODUCTS FORM OF A FUNCTION   25

differs from function 1 only in the absence of input signal D, but the number of diodes required is two less because one of the AND circuits is not needed—although in the interest of uniformity in engineering design a fifth diode in a one-input AND circuit may be included for signal C.

Another point to be observed is that some input signal lines may be called upon to actuate more switching devices than others. For example, in function 3 the line carrying signal $\bar{B}$ must actuate two AND devices, and the electrical loading on this input line may imply a need for components not indicated by the elementary count of diodes. In the general case the details of the loading effect depend upon the specific switching function under consideration and upon any "don't care" conditions that may exist, as will be explained later.

An even more important point with regard to the amount of equipment required to perform various switching functions is illustrated by comparing functions 3 and 4 in the above listing. Although function 3 contains eight appearances of variables and function 4 contains nine, function 3 requires one more diode for its realization. (Incidentally, this comparative relationship does not hold with certain other types of components, notably the series-parallel connections of contacts as might be on electromagnetic "relays." With series-parallel contacts, the number of contacts is always equal to the number of appearances of the variables in the Boolean expression—that is, eight contacts, some of which would be of the normally closed form, for function 3 and nine contacts for function 4.) Nevertheless, as will become more apparent later in this chapter, minimizing the number of diodes required to achieve a given switching function is a matter of minimizing the number of appearances of variables in the expression for that function. In other words, in realistic examples a given Boolean expression (such as function 3 requiring 12 diodes in the above listing) would never be equivalent to an expression (such as function 4) which has more appearances of variables but which requires fewer diodes.

The Two-Level AND-to-OR or "Sum-of-Products" Form of a Function

The number of "levels" in a switching network is commonly defined as the maximum number of switching devices through which an input signal must pass to reach the output line. Ordinarily, the networks are such that each signal encounters AND devices and OR devices alternately in its path from input to output.

As was illustrated in Chapter 1, any specified switching function may be generated by an appropriate set of AND devices, the outputs of which are transmitted to a single OR device. The resulting AND-to-OR network is said to have two levels of switching, and the corresponding Boolean expression is said to be in "sum-of-products" form. All four of the example functions listed near the beginning of this chapter were presented in this form. Apart from the relationship between the form of the Boolean expression and the interconnections among switching devices, the AND-to-OR or "sum-of-products" version of a switching function is a sort of "fundamental" or "natural" way in which to express a function, because it clearly specifies those combinations of input signals for which the output signal is to be 1. Each such combination is called a "term," where this word is used in a matter analogous to the usage of "term" in conventional algebraic notation.

In the following discussion the assumption will be made that any Boolean expression not initially in AND-to-OR form as encountered in the design of a digital system is first transformed to AND-to-OR form by algebraic manipulations as described in Chapter 1.

Testing for Superfluous Terms and Variables

In spite of the apparently fundamental nature of an AND-to-OR version of a switching function, an expression in this form may contain superfluous variables in the terms—or some of the terms themselves may be superfluous. A seemingly obvious first step in minimizing the number of diodes or other switching components in a network would be to detect the superfluous variables and terms and eliminate them.

An elementary example of a superfluous variable is in the Boolean expression $A + \bar{A}B$, where $\bar{A}$ in the second term is superfluous because $A + \bar{A}B = A + B$, as has already been shown.

For a slightly more complex example, consider the function $AB + A\bar{B}C$. The appearance of $\bar{B}$ in the second term is superfluous as can be shown as follows: $AB + A\bar{B}C = A(B + \bar{B}C) = A(B + \bar{B})(B + C) = A(B + C) = AB + AC$.

An elementary example of a superfluous term is $A + AB$ where the second term AB is superfluous because $A + AB = A$. Another example of a superfluous term is $AB + \bar{B}\bar{C} + A\bar{C} = AB + \bar{B}\bar{C}$. The term $A\bar{C}$ in the first form of the function is superfluous since it can be shown, by using the techniques described in the previous chapter, that the equality is correct. Note that, although $A\bar{C}$ is superfluous, neither of the variables A or $\bar{C}$, considered alone, is superfluous.

Superfluous terms and variables can be detected with relative ease after an appreciable amount of practice in Boolean manipulation, but a well-organized procedure for positive identification of them is preferable.

In the case of variables, a testing procedure can be used as follows. The variables are tested one at a time, and when a superfluous variable is found it is eliminated before testing any of the remaining variables. The test of a given variable in an AND-to-OR expression involves noting the particular combination of other variables which would cause the corresponding term to have a binary value of 1. Then with the other variables having this combination of values, an observation is made as to whether or not the final output signal is dependent on the given variable.

For example, in the expression $AB + A\bar{B}C$, the variable $\bar{B}$ in the second term can contribute to an output signal of 1 only when A and C are both 1. In this case the expression becomes $1B + 1\bar{B}1 = B + \bar{B} = 1$. That the resulting expression equals 1 is evidence that the tested variable $\bar{B}$ is superfluous. The reasoning behind this test is that without the appearance of $\bar{B}$ in the second term the value of the expression as a whole would be unchanged, namely 1, in the one instance (A and C both equal to 1) when $\bar{B}$ affects the value of the second term.

If a given variable or its inverse appears in two or more different terms, a separate test is performed for each appearance as though the variables were different.

An analogous test can be performed on entire terms. As with variables, the terms must be tested one at a time, with superfluous terms being eliminated as they are encountered. To test a term note the particular combination of binary signals which would cause the value of that term to be 1. Temporarily eliminate the term and assign this combination of binary signals to the same variables as they appear in the remaining terms. If the remainder of the expression then becomes equal to 1, the term being tested is superfluous. In other words, a given term is superfluous in an AND-to-OR network if at least one other term is 1 in every instance that the given term is 1.

For example, in the expression $AB + \bar{B}\bar{C} + A\bar{C}$ the term $A\bar{C}$ is tested by noting that it contributes to a final output signal of 1 only when A is 1 and C is 0. The remaining part of the expression then becomes $1B + \bar{B}\bar{0} = 1$, thereby indicating that term $A\bar{C}$ is superfluous.

After testing all variables and terms the process should, in general, be repreated as many times as necessary to test the terms and variables in all possible sequences that might result in a greater reduction in the number of components required. As an example consider the following function.

$$AB + \bar{B}C\bar{D} + \bar{A}\bar{C}\bar{D} + B\bar{C}\bar{D} + AC\bar{D} + \bar{A}\bar{B}D$$
$$\;\;1\quad\;\;\;\;2\quad\;\;\;\;\;3\quad\;\;\;\;\;4\quad\;\;\;\;5\quad\;\;\;\;6$$

In this expression of the function, none of the variables is superfluous but any of the terms except AB (term 1) will be found to be superfluous when tested first. For example, term 2 is tested by setting B to 0, C to 1, and D to 0, in which case the remainder of the expression becomes $A1 + \bar{A}01 + 001 + \bar{A}11 + A11 = 1$. However, if term 2 is eliminated, terms 5 and 6 are both found to be necessary as is one or the other (depending on which of the two is tested after the other) of terms 3 and 4. That is, either terms 2 and 3 or terms 2 and 4 will be found to be superfluous. A total of 15 diodes will be required. On the other hand, if terms 4, 5, and 6 are tested first (in any sequence with regard to these three), all three of them will be found to be superfluous, and the resulting minimized expression becomes $AB + \bar{B}C\bar{D} + \bar{A}\bar{C}\bar{D}$, requiring only 11 diodes.

The testing procedure fails when, in the example of the previous paragraph, the original switching function happens to be initially presented with either term 2 or term 3 eliminated. The testing procedure will indicate that one of the remaining terms is superfluous, but the true minimal form of the function will not be found. A more elaborate minimizing procedure is needed, as described in the next section.

Minimizing the AND-to-OR Version of a Switching Network

As explained in the previous section, the elimination of superfluous variables and terms in a given AND-to-OR expression of a switching function does not necessarily lead to the particular expression which represents the minimum number of components. A procedure for finding the minimum is given.

The expression is first expanded to its elemental form where "elemental" here means that the expression specifies in detail all input variable combinations that cause an output signal of 1. An elemental term is one in which all of the input variables appear, and all terms in the elemental form of a switching function are elemental terms. (The elemental form is commonly called the "canonical" form, but the relationship, if any, between the usual English language meaning of "canon" or "canonical" to a switching function is obscure.) The expansion to the elemental form can proceed straightforwardly by "multiplying" each term by factors of the form $X_i + \bar{X}_i = 1$, where the X_i represent the missing variables in any given term. Thus, for example, in a four-variable function, a term such as AB would be "multiplied" by $C + \bar{C}$ and $D + \bar{D}$ to produce

$$AB = AB(C + \bar{C})(D + \bar{D}) = ABCD + ABC\bar{D} + AB\bar{C}D + AB\bar{C}\bar{D}.$$

When all of the terms in an originally encountered AND-to-OR expres-

MINIMIZING THE AND-TO-OR VERSION OF A SWITCHING NETWORK   29

sion are thus expanded, many duplicate elemental terms may be encountered. The duplicates are simply discarded, because $X + X = X$.

Alternatively, the elemental terms can be derived from the given terms by adding the missing variables to each given term in all possible combinations. Thus, in the previous example, the two missing variables C and D can appear in four combinations CD, $C\bar{D}$, $\bar{C}D$, and $\bar{C}\bar{D}$, and these combinations when added to AB produce the four elemental terms directly. Duplicates are discarded as before.

The next step in the minimization process is nearly the reverse of the process used for determining the elemental terms. Each elemental term is compared with each other elemental term. For each pair of terms that are the same except for the inversion of one variable, a new term containing one less variable is derived. For example, terms $A\bar{B}CD$ and $A\bar{B}C\bar{D}$ in an expression can be combined to produce $AC\bar{D}$, because $A\bar{B}CD + A\bar{B}C\bar{D} = AC\bar{D}(B + \bar{B}) = AC\bar{D}$. The two terms entering into this process are not eliminated until each of them has been compared with all other elemental terms in an effort to find other terms containing one less variable. If a given elemental term cannot be thus combined with any other elemental term, it is not eliminated at all. Each resulting term containing one less variable is then compared with all other such terms to find pairs that are the same except for the inversion of one variable. For each such pair that is found, a new term containing two less variables than contained in the elemental terms is derived in like manner. This process is continued analogously until no new terms containing fewer variables can be found. The terms remaining after this process has been completed are called basic terms (or, commonly, "prime implicants"), although some of the basic terms may be superfluous.

The problem now is to find the particular set of superfluous basic terms which, when eliminated, will result in the expression representing the minimum number of components. Because all basic terms appear in the expression as derived above (whereas some of them may have been missing in the expression originally given), the testing procedure described in the previous section could be used. However, that procedure tends to be clumsy and requires an excessive amount of "enumerating" all possibilities. Much, although not all, of the enumerating can be by-passed by preparing a chart of elemental and basic terms as illustrated in Table 2-1 for the function $AB + A\bar{C}D + B\bar{C}\bar{D} + \bar{A}\bar{B}\bar{C}$. By either of the procedures already explained, eight elemental terms are found, and these are listed across the top of the table. The six basic terms, likewise found by a procedure already explained, are listed in a column in the left-hand part of the table. For each basic term, x's are placed in the columns corresponding to the elemental terms included in that basic term. The x's in a given column can be visualized as indicating those basic

TABLE 2-1. Table of Terms for the Function $AB + \bar{A}\bar{C}\bar{D} + \bar{B}\bar{C}D$

	$ABCD$	$ABC\bar{D}$	$AB\bar{C}D$	$AB\bar{C}\bar{D}$	$A\bar{B}\bar{C}D$	$\bar{A}BC\bar{D}$	$\bar{A}\bar{B}\bar{C}D$	$\bar{A}\bar{B}\bar{C}\bar{D}$
AB	ⓧ	ⓧ	ⓧ	ⓧ				
$A\bar{C}D$			ⓧ		x			
$\bar{A}\bar{B}\bar{C}$							ⓧ	x
$\bar{A}\bar{C}\bar{D}$						x		ⓧ
$B\bar{C}\bar{D}$				ⓧ		x		
$\bar{B}\bar{C}D$					x		ⓧ	

terms that are equal to 1 when the input signal combination is such that the corresponding elemental term is equal to 1.

Instead of finding the superfluous terms directly, the technique employed is to find those basic terms that are necessary or otherwise wanted in the final minimum-component expression. Any term not thus selected must, of course, be a superfluous term. The development of a suitable selection procedure has been one of the classic problems in the field of switching theory. In general, the objective has been to find a procedure which

(a) leads directly to the minimum-component versions of the function without any enumeration of alternative selections and which
(b) arrives at the result in a reasonable number of steps.

For many switching functions the objective can be realized quite well—in fact, for many functions the desired expression is more or less obvious from a brief inspection of the table.

However, for the general case, no procedure is known which entirely eliminates enumeration. In fact, a substantial amount of enumeration may be required for some functions, particularly if all alternative minimum-component expressions are to be found. A need to find all alternative minimum-component expressions is not just academic but is encountered in practical applications where other objectives are present, such as equalizing the loading on the input lines or such as minimizing the number of cross-overs in a wiring layout. As for part b of the objective, the number of required steps is also strongly dependent on whether all alternative minimum-component expressions must be found or whether the finding of just one is sufficient. Unfortunately, when the number of input variables is appreciable, say more than eight, the finding of even one minimum-component expression can be terribly tedious.

The term selection procedure to be described first is particularly suitable for problems of a size that can be handled manually. (Conceivably, this procedure might fail to find the minimum-component expressions for highly specialized functions of many variables, but no such function is known.) The basic terms which are first selected are those which are "essential," where an essential basic term is one which includes at least one elemental term not included in any other basic term. In a table, a column containing only one x is an indication that the corresponding basic term is essential. These essential basic terms and all elementary terms included in these basic terms may then be deleted from the table. Of the remaining basic terms, the one which includes the most elemental terms is selected next. This basic term and its included elemental terms are then likewise deleted. This procedure is continued until all elemental terms have been accounted for.

At some steps in the procedure, the designer may have a choice of two or more elemental terms in selecting the one that includes the most remaining elemental terms. In such cases, any one of the indicated terms may be selected and the problem is continued as before, but the entire procedure is repeated for each possible selection. Many sets of terms generated by this "branching" can be eliminated because they will be duplicates of each other. The minimum-component expressions will be those containing the fewest appearances of the variables or, as it also happens to be the case, the fewest basic terms.

In a few rare instances, an expression as derived above will contain a superfluous term (conceivably more than one superfluous term in problems with a large number of input variables), but such a term will be apparent from the table and can be eliminated.

The above procedure is applied to the example of Table 2-1 in the following manner. Basic term AB is selected because it is essential as indicated by the fact that it is the only one which includes elemental terms $ABCD$ and $ABC\bar{D}$. These two elemental terms and all other elemental terms included in AB are deleted from further consideration by circling the x's in the corresponding columns. Of the remaining x's, two are found in each of three different rows, namely, the rows corresponding to basic terms $\bar{A}\bar{B}\bar{C}$, $\bar{A}\bar{C}\bar{D}$, and $\bar{B}\bar{C}D$. Of these, term $\bar{A}\bar{B}\bar{C}$ is arbitrarily selected, and the included elemental terms, $\bar{A}\bar{B}\bar{C}D$ and $\bar{A}\bar{B}\bar{C}\bar{D}$, are then deleted by drawing dotted circles around the x's in the corresponding columns. Now, of the remaining x's, each corresponds to one of the four remaining basic terms $A\bar{C}D$, $\bar{A}\bar{C}\bar{D}$, $B\bar{C}\bar{D}$, and $\bar{B}\bar{C}D$. Any one of the four may be selected, and after deleting the elemental term, the designer still has a choice of one of two basic terms. After eliminating duplicates, the basic-term sets obtained by this process are indicated by the following array:

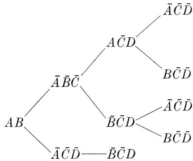

The one minimum-component version happens, in this example, to be obtained when $\bar{A}\bar{C}\bar{D}$ is selected on the second step, in which case the next selection must be $\bar{B}\bar{C}D$.

Actually, in the example shown very little experience is needed to determine by "inspection" of Table 2-1 that $AB + \bar{A}\bar{C}\bar{D} + \bar{B}\bar{C}D$ is the minimum-component expression of the given function. In fact, for the vast majority of functions encountered in digital system design, little more than inspection of the table is needed. On the other hand, many functions do exist (especially when the number of input variables is more than about six) which require a meticulous and tedious chart analysis to find the minimum-component expressions.

Another Example

As another example of minimizing the number of components in an AND-to-OR array consider the function

$$ACD + A\bar{C}\bar{D} + \bar{A}C\bar{D} + \bar{A}\bar{C}D + BC\bar{D} + B\bar{C}D + \bar{B}CD + \bar{B}\bar{C}\bar{D},$$

which, as given here, has no superfluous variables or terms. This function has 12 elemental terms and ten basic terms as shown in Table 2-2. None of the basic terms is essential, and the first term to be selected will be either AB or $\bar{A}\bar{B}$ because each of these terms includes four elemental terms while each of the other basic terms includes only two. The second term to be selected will be the other of AB and $\bar{A}\bar{B}$ for similar reasons. Then the only elemental terms not included will be $A\bar{B}\bar{C}\bar{D}$, $A\bar{B}CD$,

Table 2-2. Term Table for $AB + \bar{A}C\bar{D} + \bar{A}CD + \bar{B}CD + \bar{B}C\bar{D}$

	$ABC\bar{D}$	$AB\bar{C}D$	$A\bar{B}\bar{C}\bar{D}$	$A\bar{B}\bar{C}D$	$ABCD$	$A\bar{B}CD$	$\bar{A}BC\bar{D}$	$\bar{A}\bar{B}CD$	$\bar{A}\bar{B}C\bar{D}$	$\bar{A}\bar{B}\bar{C}D$	$\bar{A}\bar{B}\bar{C}D$	$\bar{A}\bar{B}\bar{C}\bar{D}$	
AB	x	x											a
$\bar{A}\bar{B}$				x					x		x	x	b
ACD					x	x							c
$A\bar{C}\bar{D}$			x										d
$\bar{A}C\bar{D}$							x		x				e
$\bar{A}\bar{C}D$				x			x						f
$BC\bar{D}$						x				x			g
$B\bar{C}D$								x		x			h
$\bar{B}CD$								x			x		i
$\bar{B}\bar{C}\bar{D}$			x									x	j
	1	2	3	4	5	6	7	8	9	10	11	12	

33

$\bar{A}BC\bar{D}$, and $\bar{A}B\bar{C}D$. Each of the remaining basic terms includes only one of these elemental terms, so that any one of the remaining basic terms may be selected next. If, for example, $\bar{B}\bar{C}\bar{D}$ is selected to include $A\bar{B}\bar{C}\bar{D}$, then $A\bar{C}\bar{D}$ may be deleted. After completing this process, the following two basic-term sets will have been derived:

$$AB + \bar{A}\bar{B} + \bar{A}C\bar{D} + \bar{A}\bar{C}D + \bar{B}CD + \bar{B}\bar{C}\bar{D}$$
$$AB + \bar{A}\bar{B} + ACD + A\bar{C}\bar{D} + B\bar{C}D + BC\bar{D}$$

However, neither of these expressions is of minimum-component form, and the previously described last step of the process must be utilized. As can be shown by testing, or as is apparent from the table, $\bar{A}\bar{B}$ is superfluous in the first expression and AB is superfluous in the second. After eliminating these terms, the resulting minimum-component expressions, each requiring 19 diodes, are as follows:

$$AB + \bar{A}C\bar{D} + \bar{A}\bar{C}D + \bar{B}CD + \bar{B}\bar{C}\bar{D}$$
$$\bar{A}\bar{B} + ACD + A\bar{C}\bar{D} + B\bar{C}D + BC\bar{D}$$

This particular example might be construed as more of an illustration of the need for a more foolproof minimization procedure than as an illustration of a procedure that is practical. However, except for functions derived from this one by permuting or inverting (or both) the variables, it is the only known four-variable function for which the indicated procedure does not yield the minimum-component expressions without a final testing for superfluous terms. More such functions exist for five or more variables, but they are very few in comparison with the total number of functions and, as mentioned, are not among the ones likely to be encountered in the design of a digital system.

A Method for Finding All Possible Basic-Term Sets Containing No Superfluous Terms

A straightforward but sometimes tedious method of finding all basic-term sets containing no superfluous terms will be explained by using the example of Table 2-2. First, the observation is made that no terms are essential. (If any terms were essential, these terms would be selected, and the corresponding rows and columns would be deleted.) For simplicity in notation, each basic term is designated by a lower-case letter, as shown in the right-hand part of the table. Also, the elemental terms are given numeric designations as shown at the bottom of the table.

From Table 2-2 the observation is made that term a or term d must be selected to include elemental term 1. Then, to include term 2, either term a or term h must be selected. To include both terms 1 and 2, a "or" d must be selected "and" at the same time a "or" h must be selected. This requirement can be represented in Boolean notation as $(a + d)(a + h) = a + dh$. Here, the OR and AND functions are more in the nature of true logic functions than switching functions. Although d "and" h can be selected to include elemental terms 1 and 2, the corresponding basic terms would be combined in the OR switching relationship $A\bar{C}\bar{D} + B\bar{C}D$.

For all 12 elemental terms, the basic terms would then be selected in accordance with the following expression:

$(a + d)(a + h)(j + d)(a + g)(a + c)(e + c)(e + g)(b + i)(b + e)$
$(f + h)(b + f)(b + j)$
$= (a + dh)(j + d)(a + g)(a + c)(e + c)(e + g)(b + i)(b + e)(f + h)$
$(b + f)(b + j)$
$= (aj + ad + dh)(a + g)(a + c)(e + c)(e + g)(b + i)(b + e)(f + h)$
$(b + f)(b + j)$
$= (aj + ad + dgh)(a + c)(e + c)(e + g)(b + i)(b + e)(f + h)$
$(b + f)(b + j)$
.
.
.

$=, abcdef + abcdeh + abcdfg + abcefj + abcehj + abcfgj + abcghj$
$+ abdefi + abdehi + abdfgi + abdghi + abehij + abfgij + abghij$
$+ aefij + bcdgh + cdefghij$

Of the 17 terms in this logic expression, two of them, $aefij$ and $bcdgh$, each contain only five variables, and these two terms correspond to the minimum-component basic-term sets derived in the previous section.

A Simplification in the Minimization Procedure—Applicable When Finding Any One Minimum-Component Expression Is Sufficient

In problems where the finding of just one minimum-component expression of a given switching function is sufficient (even though two or more such expressions may exist), certain of the terms can in some instances be deleted to produce a great simplification in the minimization procedure. The resulting procedure consists of the following three steps:

1. Select all essential basic terms (if any exist), and delete from the table all elemental terms included in the essential terms.
2. Determine the basic terms in which each elemental term is included. If the basic-term set for a given elemental term is the same as for

another elemental term except for the addition of one or more basic terms, delete the given elemental term. (Any basic term becomes deleted automatically if all elemental terms included in that basic term are deleted as a result of steps 1 and 2.)

3. If the set of elemental terms included in a given basic term is the same as the set for some other basic term except for the addition of one or more elemental terms in the set for the other term, delete the given basic term. The other basic term should not, however, be one which involves fewer variables, because that would correspond to fewer components.

These three steps are applied sequentially and repeatedly until no further selections or eliminations can be made, at which point the remaining x's in the table are handled as in previous sections.

As an example of this process, consider the function specified by the elemental terms across the top of Table 2-3a. This function has the basic terms listed in the left-hand part of the table, and as before the elemental terms included in each basic terms are indicated by x's.

In this example, step 1 is applied as follows. Basic term $\bar{A}CE$ is essential because it is the only one which includes elemental term $\bar{A}BC\bar{D}E$. Basic term $\bar{B}C\bar{D}$ is essential because it is the only one which includes elemental term $A\bar{B}C\bar{D}E$. Each of these terms is selected as indicated by an asterisk (*). All elemental terms included in these two basic terms are deleted as indicated by drawing circles around the x's in the corresponding columns.

A search is then made for elemental terms which meet the specifications of step 2. Three such terms are found in Table 2-3a: $\bar{A}B\bar{C}DE$, $AB\bar{C}\bar{D}\bar{E}$, and $AB\bar{C}DE$. These terms are deleted as indicated by the dotted circles around the x's in the corresponding columns.

Then a search is made for basic terms that may be eliminated in accordance with step 3. $\bar{B}\bar{D}\bar{E}$ may be thus eliminated, because all elemental terms included in it are likewise included in $\bar{B}\bar{C}\bar{E}$ (after giving due regard to the circled x's). Similarly, $\bar{A}\bar{B}E$, $\bar{A}DE$, and BDE may be deleted because other basic terms are present which include the same elemental terms in each instance. Each deleted basic term is indicated by a check mark (✓), and the corresponding x's are deleted by inclosing them in rectangles.

To facilitate the illustration of how the steps are applied a second time, the table is redrawn as Table 2-3b with the above-indicated deletions. The various terms are indicated only by their lower-case letter and numerical designations as indicated at the right-hand and bottom of the table. Basic term p is now essential because it is the only one that includes elemental terms 20 and 21. In step 2, elemental terms

TABLE 2-3. The Term Table as Used to Find One Minimum-Component Expression (Not All Such Expressions)

(a)

	$\bar{A}\bar{B}\bar{C}\bar{D}\bar{E}$	$\bar{A}\bar{B}\bar{C}\bar{D}E$	$\bar{A}\bar{B}\bar{C}D\bar{E}$	$\bar{A}\bar{B}\bar{C}DE$	$\bar{A}\bar{B}C\bar{D}\bar{E}$	$\bar{A}\bar{B}C\bar{D}E$	$\bar{A}\bar{B}CD\bar{E}$	$\bar{A}B\bar{C}\bar{D}\bar{E}$	$\bar{A}B\bar{C}\bar{D}E$	$\bar{A}B\bar{C}D\bar{E}$	$\bar{A}BCD\bar{E}$	$A\bar{B}\bar{C}\bar{D}\bar{E}$	$A\bar{B}\bar{C}\bar{D}E$	$A\bar{B}\bar{C}D\bar{E}$	$A\bar{B}CD\bar{E}$	$AB\bar{C}\bar{D}\bar{E}$	$AB\bar{C}\bar{D}E$	$AB\bar{C}D\bar{E}$	$ABC\bar{D}\bar{E}$	$ABC\bar{D}E$	$ABCDE$	
$AB\bar{E}$																⦗x⦘	x		x	x		a
$A\bar{C}\bar{E}$											x	x		⦗x⦘	x							b
$\bar{C}D\bar{E}$		x			x					x			x									c
√$\bar{B}\bar{D}\bar{E}$	⟦x⟧			ⓧ					⟦x⟧		ⓧ											d
$\bar{B}\bar{C}\bar{E}$	x		x					x	x													e
$AD\bar{E}$								x		ⓧ		⦗x⦘			x							f
√$\bar{A}B\bar{E}$		⟦x⟧		⟦x⟧		ⓧ	ⓧ															g
√$\bar{A}DE$			⟦x⟧		ⓧ	⦗x⦘	ⓧ															h
*$\bar{A}CE$				ⓧ	ⓧ		ⓧ	ⓧ														i
√BDE						⦗x⦘	ⓧ									⦗x⦘					⟦x⟧	j
$\bar{A}\bar{B}\bar{D}$	x	x		ⓧ	ⓧ																	k
*$\bar{B}C\bar{D}$			ⓧ	ⓧ							ⓧ	ⓧ										l
$\bar{A}\bar{B}\bar{C}$	x	x	x	x																		m
$\bar{A}\bar{C}D$		x	x					x	⦗x⦘													n
$B\bar{C}D$								x	⦗x⦘							x	⦗x⦘					o
ABD																x	⦗x⦘		x	x		p
	1	2	3	4	5	6	7	8	9	10	11	12	13	14	15	16	17	18	19	20	21	

(b)

	1	2	3	4	8	12	13	17	19	20	21	
√								ⓧ	⟦x⟧	ⓧ		a
√						⟦x⟧	⟦x⟧	ⓧ				b
			⦗x⦘		x			x	ⓧ			c
	⦗x⦘		⦗x⦘		x	x						e
									x		x	f
√	⦗x⦘	⟦x⟧										k
	⦗x⦘	x	⦗x⦘	x								m
			⦗x⦘	x	x							n
√							⟦x⟧			ⓧ		o
*									ⓧ	ⓧ	ⓧ	p

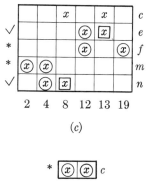

(c)

	2	4	8	12	13	19	
			x		x		c
√				ⓧ	⟦x⟧		e
*					ⓧ	ⓧ	f
*	ⓧ	ⓧ					m
√		ⓧ	⟦x⟧				n

(d)

	8	13	
*	ⓧ	ⓧ	c

1 and 3 are deleted because of their relationship to terms 2 and 4, respectively. In step 3, basic terms a, b, k, and o are deleted.

The table is redrawn again as Table 2-3c after making the indicated deletions. Basic terms f and m are now essential and are indicated by asterisks. Step 2 does not apply, but on step 3 terms e and n are deleted.

Again, the table is redrawn as Table 2-3d. Although the table is now trivial, step 1 applied, and term c is essential and is selected.

The "sum" (OR function) of the selected terms is one minimum-component expression as follows.

$$\bar{A}CE + \bar{B}C\bar{D} + ABD + A\bar{D}\bar{E} + \bar{A}\bar{B}\bar{C} + \bar{C}D\bar{E}$$

The above example was carefully selected to illustrate the procedure. For other functions the procedure does not necessarily produce a complete solution in the manner illustrated. For some functions, no simplifications at all result. For other functions a partial simplification results, but then the remaining x's "lock up" in a pattern which must be handled in some other way, such as by using the techniques described in previous sections.

Finding All Basic Terms without Finding the Elemental Terms

The procedures described so far can be laborious with some problems, and a person might wonder if a procedure exists for finding all basic terms without expanding the expression to its elemental form.

Such a procedure does exist, and it is based on the relationship

$$AB + \bar{A}C = AB + \bar{A}C + BC$$

which was included in Chapter 1. Each of the three terms on the right-hand expression of the function is a basic term as can be established by the methods described previously. Ordinarily, the problem would be to find the left-hand expression because it represents fewer components, but here the objective is to find all basic terms of a given switching function.

As the relationship will be used, a more general and illustrative form of it is

$$Xf(Y_i) + \bar{X}f(Z_j) = Xf(Y_i) + \bar{X}f(Z_j) + f(Y_i)f(Z_j),$$

where the function f is simply the AND function in this application. Any given variable can appear in both of the sets Y_i and Z_j. If a variable appears in the same form (with respect to being inverted or not inverted) in each set, that variable need appear only once in the term $f(Y_i)f(Z_j)$, because $AA = A$. However, if any variable appears in both sets but in

FINDING ALL BASIC TERMS WITHOUT FINDING THE ELEMENTAL TERMS   39

opposite form, the entire term $f(Y_i)f(Z_j) = 0$, because $A\bar{A} = 0$. For example,

$$ABCD + A\bar{B}D\bar{F} = ABCD + A\bar{B}D\bar{F} + (ACD)(AD\bar{F})$$
$$= ABCD + A\bar{B}D\bar{F} + ACD\bar{F}$$

where B is used as X in the general relationship. As can be shown by previous methods, the three terms in the derived expression are all of the basic terms in the function. For another example, $A\bar{B}CD + AB\bar{C}$ does not yield any new terms regardless of whether B or C (the only possibilities) is used as X in the general relationship. If B is X, then $C\bar{C}$ will cause the new term to be 0. If C is X, then $B\bar{B}$ will cause the new term to be 0. In other words, the relationship is applicable only to term pairs which have one and only one variable that appears inverted in one term and not inverted in the other term. For expressions that have more than two terms, the terms are paired in all possible combinations in the search for new basic terms.

In general, to find all of the basic terms, each new term must also be compared with each other term for the purpose of finding still more new terms. At each stage of the process any term is eliminated that is either (a) the duplicate of another term or (b) of the form AB in $A + AB = A$ or, in the more general notation, of the form $f(Y_i)f(Z_j)$ in $f(Y_i) + f(Y_i)f(Z_j) = f(Y_i)$. This process is continued until no retainable new terms can be found, at which point all basic terms will have been found.

As an example consider the function $\bar{A}B + ABC$. The variable A in the second term is superfluous, although the designer need not be aware of the fact. When the relationship given above is applied, a new term, BC, is found to yield an equivalent expression $\bar{A}B + ABC + BC$, where term ABC should be eliminated because the term pair $ABC + BC$ is in category (b) of the previous paragraph. The two terms of the resulting expression $\bar{A}B + BC$ represent all of the basic terms because they do not yield any other new terms, and no terms meet the requirements for elimination.

For a more complex example consider the function set forth in Table 2-2.

$$AB + \bar{A}C\bar{D} + \bar{A}\bar{C}D + \bar{B}CD + \bar{B}\bar{C}\bar{D}$$
  1   2   3   4   5

$$= AB + \bar{A}C\bar{D} + \bar{A}\bar{C}D + \bar{B}CD + \bar{B}\bar{C}\bar{D}$$
   1   2   3   4   5

$$+ BC\bar{D} + B\bar{C}D + ACD + A\bar{C}\bar{D} + \bar{A}\bar{B}C + \bar{A}\bar{B}\bar{D} + \bar{A}BD + \bar{A}B\bar{C}$$
  1-2   1-3   1-4   1-5   2-4   2-5   3-4   3-5

40 AND-OR NETWORKS

The terms are numbered for convenience, and the new terms are derived by pairing the terms indicated. Now when all of the terms are paired in all possible ways, more new terms will be found. Some of these can be eliminated because they will be duplicates. For example, when term 2 is paired with term 3-4, the new term is $\bar{A}\bar{B}C$, which is the same as term 2-4. Others of the new terms can be eliminated because of relationship (b) given above. An example here is the pairing of term 1-2 with term 1-4 to produce ABC, which would be eliminated because it is superfluous in view of AB, term 1. For other pairs, relationship (b) is encountered in a different way. When term 2-4 is paired with term 3-5, the new term is $\bar{A}\bar{B}$, whereupon both $\bar{A}\bar{B}C$ and $\bar{A}\bar{B}\bar{C}$ (terms 2-4 and 3-5 themselves) should be eliminated. Further, $\bar{A}\bar{B}\bar{D}$ and $\bar{A}\bar{B}D$ (terms 2-5 and 3-4) may then be eliminated directly because of similar comparisons with $\bar{A}\bar{B}$.

The resulting expression for the originally given function is

$$AB + \bar{A}C\bar{D} + \bar{A}\bar{C}D + \bar{B}CD + \bar{B}\bar{C}\bar{D} + BC\bar{D} + B\bar{C}D + ACD \\ + A\bar{C}\bar{D} + \bar{A}\bar{B}$$

and it contains all of the basic terms as listed in Table 2-2. When the last new term, $\bar{A}\bar{B}$, is paired with other terms, the only further new terms to be found are ones that are duplicates of terms already present.

After finding all of the basic terms, the designer still has the problem of selecting that set of terms which represents the function and which corresponds to the minimum number of components. Determining the elemental terms and forming a table is unattractive because avoiding the elemental terms was a primary reason for using this procedure to find all basic terms. Superfluous terms can be found by testing in accordance with the procedure described earlier in this chapter. As before, superfluous terms should be eliminated in all possible sequences. In the above example, terms AB and $\bar{A}\bar{B}$ will each be found to be superfluous even when tested after the other is eliminated. However, if both AB and $\bar{A}\bar{B}$ are eliminated, none of the remaining eight terms will then be superfluous, and the minimum-component version of the function will not have been found.

A disadvantage of this approach to switching circuit design is that it seems to offer no effective means for handling the "don't care" conditions to be described later.

The Number of Basic Terms Relative to the Number of Elemental Terms

For the large majority of switching functions encountered in practice the number of basic terms is less, usually much less, than the number of elemental terms. This relationship prevailed for all of the functions so far presented as examples. However, this relationship is not an inherent property of switching functions, and a better understanding of the nature of the overall problem is obtained through a realization that the number of basic terms can actually be greater than the number of elemental terms. One four-variable example is as follows where the first expression indicates all elemental terms, of which there are 11, and the second expression indicates all basic terms, of which there are 13.

$$ABC\bar{D} + AB\bar{C}D + AB\bar{C}\bar{D} + A\bar{B}CD + A\bar{B}\bar{C}D + A\bar{B}C\bar{D} + \bar{A}BCD$$
$$+ \bar{A}B\bar{C}D + \bar{A}BC\bar{D} + \bar{A}\bar{B}CD + \bar{A}\bar{B}\bar{C}\bar{D}$$
$$= AB\bar{C} + A\bar{B}C + \bar{A}BC + AB\bar{D} + A\bar{B}D + \bar{A}BD + AC\bar{D} + A\bar{C}D$$
$$+ \bar{A}CD + BC\bar{D} + B\bar{C}D + \bar{B}CD + \bar{A}\bar{B}\bar{C}\bar{D}$$

The last elemental term, $\bar{A}\bar{B}\bar{C}\bar{D}$, does not combine with any other elemental term, so it remains in the expression as a basic term. If the function were altered by deleting this term, the example would have 10 elemental terms and 12 basic terms.

For functions of more than four variables, the number of basic terms can be much greater than the number of elemental terms. In the following five-variable function the 22 elemental terms can be "combined" to form 32 basic terms.

$$ABC\bar{D}E + AB\bar{C}DE + AB\bar{C}\bar{D}E + A\bar{B}CDE + A\bar{B}\bar{C}DE + A\bar{B}C\bar{D}E$$
$$+ \bar{A}BCDE + \bar{A}BC\bar{D}E + \bar{A}B\bar{C}DE + \bar{A}\bar{B}CDE + \bar{A}\bar{B}\bar{C}\bar{D}E$$
$$+ ABCD\bar{E} + AB\bar{C}\bar{D}\bar{E} + A\bar{B}C\bar{D}\bar{E} + A\bar{B}\bar{C}D\bar{E} + A\bar{B}\bar{C}\bar{D}\bar{E}$$
$$+ \bar{A}B\bar{C}D\bar{E} + \bar{A}BC\bar{D}\bar{E} + \bar{A}\bar{B}\bar{C}D\bar{E} + \bar{A}\bar{B}CD\bar{E} + \bar{A}\bar{B}C\bar{D}\bar{E}$$
$$+ \bar{A}\bar{B}C\bar{D}\bar{E}$$

Finding all basic terms and then finding the minimum-component expression of the function is a bit tedious but instructive.

When n, the number of variables, is greater than five, the total number of basic terms can even be greater than 2^n, the total number of elemental terms possible (not just greater than the total number of elemental terms present in a given function). However, so far as is known, the maximum possible number of basic terms has not been determined for functions of more than five variables.

Separation of Variables

For a given number of input variables, the finding of the most advantageous form of a switching function becomes simpler whenever an AND-to-OR expression can be found wherein some of the terms contain only variables which do not appear in any of the other terms. In such instances, the problem divides into two or more separate problems, each of which is less laborious to handle because of the fewer variables involved.

As an example, consider the five-variable expression

$$A\bar{B}E + AC\bar{D} + BC + D + \bar{E}$$

The designer may note that the first and fifth terms may be considered together as $A\bar{B}E + \bar{E}$ which is a simple three-variable problem in which the appearance of E in the first term can easily be shown to be superfluous by any of the methods previously described. Similarly, the second and fourth terms may be considered together as $AC\bar{D} + D$ in which $\bar{D}$ is superfluous. The resulting expression,

$$A\bar{B} + AC + BC + D + \bar{E},$$

can be handled in two parts with variables A, B, and C appearing in one part and variables D and E appearing in the other part. In this example, term AC is found to be superfluous but no other simplifications result.

For another example, the function

$$A\bar{B} + BC + ACE + DE + \bar{D}\bar{E}\bar{F}$$

may appear to require a full six-variable expansion in the search for the simplest form, but the term ACE is superfluous, and after this term is removed the problem reduces to two elementary three-variable problems. That ACE is superfluous may be determined by testing. Alternatively, if the designer is adept at Boolean manipulation, he may notice that when the first two terms, $A\bar{B}$ and BC, are paired, a new term AC is generated in the manner described previously. When AC is paired with ACE the result is $AC + ACE = AC(1 + E) = AC$. Because AC was superfluous, ACE is likewise superfluous and may be eliminated.

Mechanizing the Simplification Process

Mechanizing the simplification process is largely a matter of programming a general purpose computer to proceed through the various term-manipulating steps. A full appreciation of the problem and its solutions requires some knowledge of computer organization and programming (which is the topic of Chapter 9). On the other hand, much of the present topic is simply a matter of finding an organized arithmetic procedure for handling the terms, and these procedures can be understood without a knowledge of computers or programming. The details of the mechanization would depend strongly on the simplification procedure which has been chosen and upon the physical design of the computer at hand, but the objective here is limited to a presentation of some basic considerations.

First of all, in a computer the variables would not be indicated by letter designations. Instead, a number or "word" in the computer would be used to represent a term. Each digit in a word would correspond to one variable in the term. To represent a variable, three different digit values are needed, one to represent the appearance of that variable in conventional form, a second to represent the appearance of that variable in inverted form, and a third to indicate that the variable does not appear at all in the term. In a binary computer the digits 1 and 0 may represent the variable and its inverse, respectively. The absence of a variable presents a problem. On paper the absence is indicated by a hyphen so that in a five-variable function, for example, a term such as $A\bar{B}E$ would appear as 10 - - 1. In the computer, two groups of digits (words) would be needed. One word contains 1's and 0's to represent the variables which are present, and the second word is used as a "mask" to indicate which of the digits have meaning. Thus, in the example cited, the mask word would be 11001. The term-representing word could have any combination of 1's and 0's in the third and fourth digit positions, but through the action of appropriate instructions in the computer program the effect of these digits would be nullified by the 0's in the corresponding positions of the mask word.

Alternatively, in a binary computer the digits may be mentally grouped by two's. Then each binary digit pair may represent four different conditions, 00, 01, 10, and 11, and these may be thought of as 0, 1, 2, and 3, respectively. Any three of these digits may be assigned with any permutation to represent the presence in conventional form, the presence in inverted form, and the absence of a variable, respectively.

In the discussion to follow, 2 will be used to represent a variable in conventional form, 1 for the inverted form, and 0 for the absence of a variable. Then, for example, $A\bar{B}E$ will appear as 21002.

When expanding a given term to generate elemental terms, the 0's are replaced, by means of a computer program, by 1's and 2's in all possible combinations to yield, in the above example, the terms 21112, 21122, 21212, and 21222. Although no digit will be greater than 2, the digit groups (words) may be visualized as decimal numbers suitable for entering into the arithmetic operations of addition and subtraction in a decimal computer.

A convenience in both the manual and mechanized manipulation of the elemental terms is to assign a decimal number to each term, where the number is derived by adding the corresponding weights of the variables that appear in conventional form and adding nothing for the variables in inverted form. Thus, the term $A\bar{B}C\bar{D}E$ at the top of the column is designated as term 21 which is derived by adding 16, 4, and 1. This term also has the representation 21212 as indicated in the table.

Table 2-4 illustrates the procedure to be used for finding all of the basic terms when the elemental terms are as listed in part *a* of the table—these terms having been given in the initial expression of the function or having been obtained from the initial expression in the manner indicated above. The three columns in part *a* list the decimal representations, the letter designations of the variables (signals), and the numeric representations of the variables for each term. The first step in finding the basic terms is to compare each elemental term with every other elemental term in a search for term pairs that are the same except for the inversion of one variable. Such pairs can be combined with that variable eliminated in the manner described previously. This first step

TABLE 2-4. A Programmable Scheme for Finding Basic Terms

		(a)				(b)				(c)	
	21	$A\bar{B}C\bar{D}E$	21212		17-21	21012					
3	19	$A\bar{B}\bar{C}DE$	21122	2	17-19	21102					
	14	$\bar{A}BCD\bar{E}$	12221		5-21	01212					
	7	$\bar{A}\bar{B}CDE$	11222		5-7	11202					
	17	$A\bar{B}\bar{C}\bar{D}E$	21112		3-19	01122					
2	9	$\bar{A}B\bar{C}\bar{D}E$	12112		3-7	11022					
	5	$\bar{A}\bar{B}C\bar{D}E$	11212	1	16-17	21110			1-5-17-21	01012
	3	$\bar{A}\bar{B}\bar{C}DE$	11122		1-17	01112	1	1-3-17-19	01102
1	16	$A\bar{B}\bar{C}\bar{D}\bar{E}$	21111		1-9	10112		1-3-5-7	11002
	1	$\bar{A}\bar{B}\bar{C}\bar{D}E$	11112		1-5	11012					
					1-3	11102					

can be simplified by noting that (when expressed in the numeric designations) the number of 2's in one term of a pair must be exactly one greater than the number of 2's in the other term. Therefore, in Table 2-4a the terms have been sorted in accordance with the number of 2's, as indicated. Presumably this sorting is accomplished by a part of the computer program. Within each group the terms are listed in numeric sequence. The same sequence results from considering either the decimal representation or the numeric-variable representation of the terms. The lower numbered terms are toward the bottom of the table to aid in the visualization of the subtraction process to be described.

After sorting the terms as shown, term 1 in the group containing one 2 is compared with each of the terms in the group containing two 2's. The comparison is accomplished by subtracting the numeric-variable representation of the lower numbered term from the corresponding representation of the other term. Thus, for example, terms 3 and 1 are compared by subtracting 11112 from 11122 to yield 00010. That the difference is an integral power of ten (has exactly one 1) is an indication that the term pair can be combined. The new term is obtained by subtracting the difference from the lower numbered term to yield 11102 ($\bar{A}\bar{B}\bar{C}E$), and this term is listed as term 1-3 in part b of the table. The comparing of term 1 with terms 3, 9, and 17 proceeds analogously.

When term 16 is compared with each of the terms having two 2's, a simplification in the procedure may result by noting that term 16 cannot possibly be combined with a term having a lower numbered designation. This situation results from the easily proved general situation that, if a term P has fewer 2's than a term Q and if at the same time P is represented by a larger number than Q, then at least two variables must be different in P and Q. However, terms 16 and 17 can be combined in the manner described to produce new term 16–17 as listed in Table 2-4b.

Similarly, term 3 in the group having two 2's combines with term 7 in the group having three 2's to produce new term 3–7. However, when terms 3 and 14 are compared the result is $12221 - 11122 = 01099$ (if the subtraction is performed on a decimal computer), and the fact that the difference is not an integral power of ten is an indication that the two terms cannot be combined.

After completing the listing in Table 2-4b each term in this list is compared with each other term in the same list in the search for further pairs that will combine. Again, the relative numeric values of the terms can be observed to determine pairs that will not combine. As before, a requirement of a combinable pair is that the term having the one more 2 must also have the greater numeric value (when expressed in

2's, 1's, and 0's). Therefore, a numeric sort of the terms in Table 2-4b might be helpful, but the terms have been left in the sequence in which they were derived to indicate other relationships. For example, in spite of the fact that term 5-7 with two 2's has a larger numeric value (11202) than the numeric value (11012) of term 1-5 with one 2, these two terms cannot be combined because the terms consist of different variables. That is, when term 5 was combined with term 7 the eliminated variable was not the same as when term 5 was combined with term 1.

Also, when term 1-17 is paired with term 3-19 the result is 01122 − 01112 = 00010 to indicate that a combination is possible. However, when the new term is derived, 01112 − 00010 = 01102 ($\bar{B}\bar{C}E$), this term is found to be the duplicate of the term obtained by combining terms 1-3 and 17-19. A thorough discussion of the reasoning behind the relationships would be beyond present purposes, but the result is that for a useful combination to be obtained, all numbers representing the original terms used in the formation of one "intermediate" term of a pair must be larger than all numbers representing the original terms used in the formation of the other intermediate term. A pair meeting this requirement must still be checked by subtraction to determine whether or not the further combination is possible. For example, term pair 17-21 and 1-9 meets the requirement and their numeric representations are in the correct sequence, but the terms do not combined because the difference between their numeric representations is 21012 − 10112 = 10900, which is not an integral power of ten.

When all terms are compared as outlined above, three new terms as indicated in Table 2-4c are found. No further combinations are possible. Therefore these terms, 11002, 01102, and 01012, are basic terms, as are terms from parts a and b that did not enter into further combinations.

The programmar still has the problem of programming the computer to select the particular set of basic terms which correctly express the original function and which meets minimum-component requirements or other requirements of a given application. The same sort of term designations may be used in following the term selection procedures previously described.

"Don't Care" Conditions

In the design of digital systems the switching functions encountered are often of a nature whereby the value of the output signal is immaterial for certain combinations of input signals. This situation is encountered in two different ways. In some instances the designer may know that the output signal will not be utilized when the input signals appear in these certain combinations. Alternatively, and perhaps more commonly, the designer may know that certain combinations of input signals will never be applied to the switching array in question. In either case, when minimizing the number of components required, the designer has the option of assigning an output signal of 0 or 1 to each such input signal combination, expressed as an elemental term, in any pattern that results in the greatest component saving. From the standpoint of system functioning, the designer does not care what the output signal is for these input signal combinations, and therefore these combinations or terms are called "don't care" conditions.

Fortunately, the assignment of 1's and 0's to the "don't care" conditions in all possible combinations and the examination of each resulting switching circuit is not necessary in the search for the one requiring the fewest number of components. Instead, a minor elaboration in the procedure described previously for finding the elemental terms and basic terms is sufficient. The elemental terms for the "don't care" conditions are found by the same procedure as was described for the intended 1-output conditions. Then, the elemental terms for both the 1-output and the "don't care" conditions are considered collectively in finding the basic terms. However, when preparing the table to be used in the selection of those basic terms which result in the minimum-component circuit, only the elemental terms corresponding to the 1-output conditions need be included. The reason for including both sets of elemental terms when finding the basic terms is that more possible combinations are likely to be encountered, and the resulting basic terms then tend to be simpler (represent AND devices with fewer inputs) and tend to include more 1-output elemental terms so that fewer basic terms are needed to produce the intended output signal. The reason for eliminating the "don't care" terms from the table is obviously that the designer doesn't care what the output signal will be for the corresponding input signal combinations.

As an illustration, consider the function $AB\bar{D} + BCD + \bar{A}CD$, which is expressed in minimum-component form. This function has five elemental terms which are listed across the top of Table 2-5. Assume that the input signal combinations represented by the expression $\bar{A}C\bar{D} + A\bar{C}D + A\bar{B}D + A\bar{B}\bar{C}$ are all "don't care" conditions. This expression corresponds to the six elemental terms $AB\bar{C}D$, $A\bar{B}\bar{C}D$, $A\bar{B}C\bar{D}$, $\bar{A}BCD$, $\bar{A}BC\bar{D}$, and $\bar{A}\bar{B}C\bar{D}$. When these elemental terms are grouped with the elemental terms of the original function and are combined in all possible ways, the six basic terms shown in the left-hand part of the table are obtained. After the appropriate x's are marked in the table, the pro-

TABLE 2-5. Term Table for the Function
$AB\bar{D} + BCD + \bar{A}CD$ Having
the "Don't-Care" Conditions
$\bar{A}C\bar{D} + A\bar{C}D + A\bar{B}D + A\bar{B}\bar{C}$

	$AB\bar{C}\bar{D}$	$ABC\bar{D}$	$ABCD$	$\bar{A}BCD$	$\bar{A}\bar{B}CD$
AB	x	x	x		
$A\bar{C}$	x				
AD			x		
BC		x	x	x	
CD			x	x	x
$\bar{A}C$				x	x

cedures described previously may be used to obtain the minimum-component versions of the function. Two such versions, $AB + CD$ and $AB + \bar{A}C$, are found. Note that in the first version the function appears to be dependent on D whereas in the second version the function appears to be independent of D. The cause for this apparent discrepancy is in the fact that the set of "don't care" conditions that correspond to an output of 1 is not the same for the two versions.

A Diagram Method for Representing and Simplifying Switching Functions

The diagrams to be described here are often called "Karnaugh maps" in recognition of a paper by M. Karnaugh (*AIEE Transactions on Communication and Electronics* November 1953). However, the basic concepts with only a relatively minor difference in notation were described earlier by E. W. Veitch (*Proceedings of the Association for Computing Machinery* May 1952, 127–133).

For two variables the diagram is little more than trivial, but it is illustrated in Figure 2-1a as an introduction. Each small square corresponds to an elemental term. The two top squares correspond to those terms in which $\bar{A}$ appears, and the two bottom squares correspond to those terms in which A appears. The left-hand and right-hand squares are analogously related to $\bar{B}$ and B, respectively. To represent a function, a 1 is placed in each square which corresponds to an elemental term of

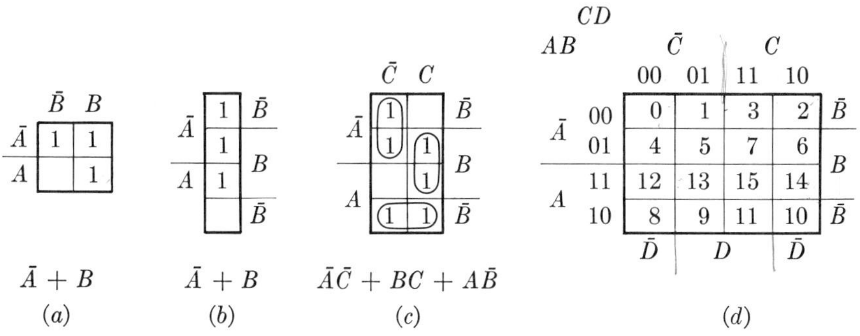

A DIAGRAM METHOD FOR SIMPLIFYING SWITCHING FUNCTIONS 49

	$\bar{B}$	B
$\bar{A}$	1	1
A		1

$\bar{A} + B$
(a)

$\bar{A}$	1	$\bar{B}$
	1	
A	1	B
		$\bar{B}$

$\bar{A} + B$
(b)

$\bar{A}\bar{C} + BC + A\bar{B}$
(c)

	CD	$\bar{C}$		C	
AB		00	01	11	10
$\bar{A}$	00	0	1	3	2	$\bar{B}$
	01	4	5	7	6	B
A	11	12	13	15	14
	10	8	9	11	10	$\bar{B}$
		$\bar{D}$	D		$\bar{D}$

(d)

FIGURE 2-1. Diagrams for representing two-variable, three-variable, and four-variable functions.

the given function. However, a major purpose of the diagram is to eliminate the need for writing out the elemental terms in detail. Thus, for example, the two-variable function $\bar{A} + B$ can be recorded in the diagram by marking 1's in the two upper and two right-hand squares directly. The upper-right square is affected by both $\bar{A}$ and B, but only a single 1 is written in this square. The writing of only a single 1 in a square coresponds to the elimination of duplicate elemental terms.

After writing 1's in the appropriate squares and thereby finding all elemental terms, the next step is to study the diagram in a search for basic terms, particularly the set of basic terms which will result in a circuit having the minimum number of components. This step is performed by pairing adjacent squares in all possible ways. In the example of Figure 2-1a the only possible pairing yields the original expression $\bar{A} + B$. To be consistent with the three-variable and four-variable diagrams to be described presently, the two-variable diagram may be drawn as in Figure 2-1b, where the top and bottom squares are "adjacent" by definition.

For three variables, a diagram as illustrated in Figure 2-1c is used. The 1's in the diagram correspond to the switching function $\bar{A}\bar{C} + BC + A\bar{B}$, where the looped pairs of 1's correspond to the terms in the function as given. In this example the 1's could be paired in a different way. That is, the 1's in the second row of squares could be paired to produce the term $\bar{A}B$, the 1's in the two bottom squares of the right-hand column produce the term AC, and the 1's in the top and bottom squares of the left-hand column produce the term $\bar{B}\bar{C}$. This alternative pairing includes all of the 1's to indicate that the expression as originally given is equal to $\bar{A}B + AC + \bar{B}\bar{C}$. Other features of the process will be explained subsequently with the four-variable examples.

The four-variable diagram with certain popular elaborations is illustrated in Figure 2-1d. As was explained for the representation of a term by means of a computer word, a 1 or a 0 may be used to indicate the presence of a variable or its inverse, respectively. For example, $\bar{A}BC\bar{D}$ would be represented by 0110. The two-digit numbers at the left of the diagram indicate the values of A and B for the corresponding rows, and the two-digit numbers at the top of the diagram indicate the values of C and D for the corresponding columns. Therefore, term 0110 would be indicated in the right-end square of the second row. Further, if the binary numbers are converted to decimal form, where the individual binary digits have the weights 8, 4, 2, and 1, respectively, the individual squares may be designated by the decimal numbers indicated in Figure 2-1d. The example of 0110 corresponds to square number 6.

In solving a four-variable problem, the function as given is first plotted on the diagram in the same manner as was illustrated for two and three variables. Then adjacent squares containing 1's are compared to indicate three-variable terms. Two squares are said to be "adjacent" if they have a common boundary (not just a common point), if they are at opposite ends of a row, or if they are the top and bottom squares of a column. (The diagram may be abstractly viewed as being rolled about a vertical axis to make the left and right boundaries in contact with each other and simultaneously rolled about a horizontal axis to make the top and bottom boundaries in contact.) Examples: if squares 4 and 5 both contain 1's, term $\bar{A}B\bar{C}$ is indicated; squares 3 and 7, term $\bar{A}CD$; squares 12 and 14, term $AB\bar{D}$; and squares 2 and 10, term $\bar{B}C\bar{D}$.

Two-variable terms are indicated in Figure 2-1d by adjacent pairs of squares containing 1's. Pairs are adjacent if they are either side-by-side or end-to-end but are not adjacent if they are in a "skewed" pattern. As in the previous paragraph, the left and right boundaries of the diagram can be viewed as being in contact with each other, as can the top and bottom boundaries. Therefore, some examples would be as follows: pairs 12-13 and 14-15, term AB; pairs 1-3 and 9-11, term $\bar{B}D$; pairs 8-12 and 10-14, term $A\bar{D}$; and pairs 0-2 and 8-10, term $\bar{B}\bar{D}$. (Each of these terms is indicated in an alternative way. For example, term $\bar{B}\bar{D}$ can alternatively be derived by combining pairs 0-8 and 2-10.) An example of pairs that do not combine is found in pairs 0-1 and 5-7. Although squares 1 and 5 are adjacent, the four squares together are in a skewed pattern, and no two-variable term results.

Single-variable terms in the function result when any two groups of four squares as derived in the previous paragraph are adjacent by essentially the same definition of "adjacent." For example, group 0-4-8-12 is adjacent to group 2-6-10-14 to indicate term $\bar{D}$ if all eight of these squares contain 1's.

Examples of Using the Diagram to Simplify Four-Variable Functions

After plotting the function on the diagram, the diagram is first studied in a search for all possible single-variable terms. Any 1-containing squares not represented by single-variable terms are then grouped with other 1-containing squares (which may have been used in the development of single-variable terms) in all possible ways to find all possible two-variable terms. Then any remaining 1-containing squares are analogously compared with other 1-containing squares in a search for three-variable terms. Any 1-containing square that cannot be paired with any other 1-containing squares represents a four-variable term. The terms thus derived are the basic terms.

In the general case, the selection of the basic term sets which correspond to the minimum-component versions of the function is fully as difficult with the diagram method as with any other method. In fact, the designer may wish to form a table of basic terms as before. However, with only four variables, the sets can be found easily through little more than a superficial inspection of the diagram.

Figure 2-2 shows four examples with minimum-component versions of the functions indicated in each case. In parts b, c, and d of the figure,

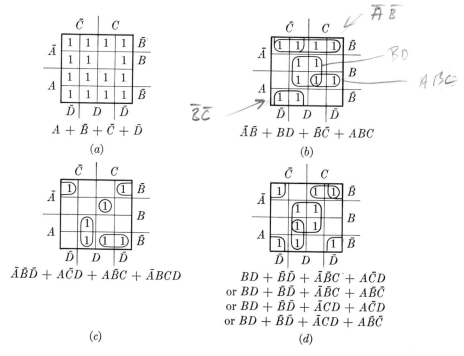

FIGURE 2-2. Examples of four-variable diagrams.

52 AND-OR NETWORKS

the basic terms utilized are indicated by looping the 1's in the corresponding squares. In d, the loops correspond to the first of the listed minimum-component versions, of which there are four. Note that in b term $\bar{C}D$ is also a basic term, but it is superfluous because all corresponding squares containing 1's are included in other terms. In c, terms $A\bar{B}D$ and $\bar{B}C\bar{D}$ are likewise superfluous basic terms. In this example, the original function might have appeared as $\bar{A}\bar{B}\bar{D} + \bar{B}C\bar{D} + A\bar{C}D + A\bar{B}D + \bar{A}BCD$ in which no terms are superfluous. After plotting the function, the pattern of 1's would lead directly to the new basic terms and to the simplified expression in the manner indicated by the loops in the diagram.

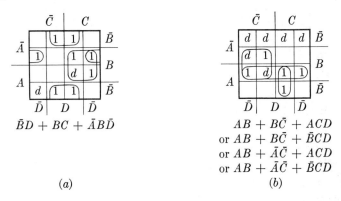

FIGURE 2-3. Diagrams of four-variable functions having "don't care" conditions.

Figure 2-3 shows two examples of the use of diagrams for problems that include "don't care" conditions. Each input signal combination for which the output signal is immaterial is indicated by the letter d in the corresponding square. In developing the basic terms, the d-squares are included, but when selecting the basic terms to form the minimum-component version of the function, only the 1-containing squares need be included. In the example in Figure 2-3a, one of the d-squares happens to be utilized in finding the simplest expression of the function, but the other d-square is not utilized. In Figure 2-3b, all squares in the top row contain d's, but the fact that no 1-containing square is included means that the corresponding basic term, $\bar{A}\bar{B}$, may be ignored.

Diagrams for More than Four Variables

Several different suggestions have been made for extending the diagrams to handle more than four variables. The complexity of the problems certainly increases rapidly with any increase in the number of variables, and as might be expected, opinions differ widely on the usefulness of the various types of diagrams for more than four variables. Probably the most commonly held opinion is that the usefulness of diagrams virtually stops at four variables. In this author's experience, most problems with up to eight variables can be handled quite effectively when the concepts of the four-variable diagram are extended in the manner illustrated in Figure 2-4—although the designer must be prepared to expend a reasonable amount of patience and diligence. In general, the diagram is straightforward in indicating the basic terms, but for certain carefully selected complex problems the determination of the appropriate set of basic terms can be awkward.

The eight-variable diagram in Figure 2-4 has 16 blocks, each of which contains a four-variable diagram. The blocks are shown numbered in the same pattern that was used previously for the numbering of the squares, but in this case the numbering pertains to the added variables E, F, G, and H. The interpretation of the diagram as a whole is essentially the same as before except that the concept of "adjacent" squares is extended to include squares in like-numbered positions in adjacent blocks. As with individual squares, two blocks are said to be adjacent if they are adjacent in the usual manner, if they are at the opposite ends of a row, or if they are at the top and bottom of a column of blocks.

An illustrative pattern of 1's is indicated in the figure, and the minimum-component version of the function is shown at the bottom of the figure. For an example, one term of the function is derived as follows. The 1's in the upper-right squares of blocks 4 and 5 can be paired to form the term $\bar{A}\bar{B}\bar{C}\bar{D}E\bar{F}G$. The 1's in the upper-right squares of blocks 6 and 7 form the term $\bar{A}B\bar{C}\bar{D}\bar{E}FG$. The two pairs are adjacent, and therefore the four 1's can be grouped to form the term $\bar{A}\bar{B}\bar{C}\bar{D}\bar{E}F$. It is suggested that the reader determine for himself the 1's which are included in each of the other listed terms. The function actually includes other basic terms, but they are superfluous. One such term, $\bar{A}B\bar{D}\bar{E}E\bar{G}\bar{H}$, is obtained by pairing the 1's in the upper-left and upper-right squares in block 4. Another superfluous term, $\bar{A}BCD\bar{F}\bar{G}$, is obtained by grouping the 1's in square position 7 of blocks 0, 1, 8, and 9.

The OR-to-AND Version of a Switching Network

All design procedures which have been described for AND-to-OR versions of switching networks have counterparts for OR-to-AND versions. Mathematically, the two network forms are exact analogies or "duals"

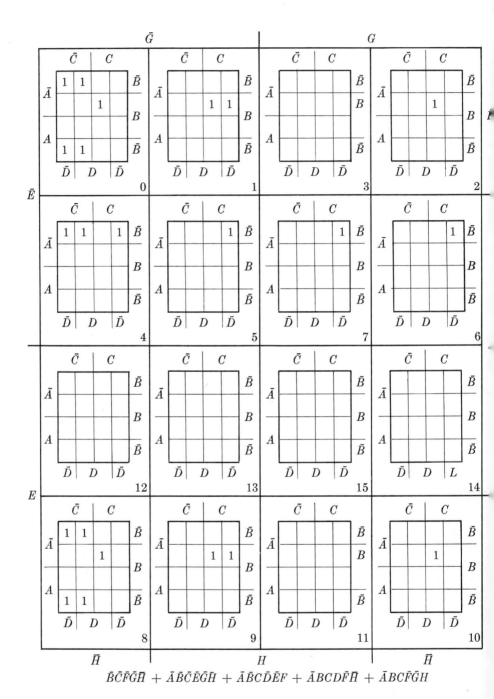

FIGURE 2-4. An eight-variable diagram.

of each other. However, in the vast majority of practical applications the switching functions are encountered in such a way that the AND-to-OR representations seem more "natural." That is, what is usually wanted is an output signal when one or more of certain specified combinations of AND-related input signals are present. The OR-to-AND version generally seems "unnatural" in that the function seldom appears to be such that an output signal is wanted when at least one signal in each of several sets of input signals is present. The OR-to-AND version seems particularly unnatural in those instances where some of the signals appear in two or more of the OR-related sets.

Nevertheless, each possible switching function, regardless of how it is encountered in practical applications, can be expressed in OR-to-AND form, and the component minimization techniques are essentially the same as for the AND-to-OR form. For a given randomly selected function, the minimum-component expression may be of the OR-to-AND form or it may be of the AND-to-OR form, and the probability is the same for finding the minimum-component expression to be of one form as the other. (The functions actually encountered in practice are probably most often such that the minimum number of components is realized with the AND-to-OR form, although this fact would be difficult to prove.)

Because of the "unnaturalness" of the OR-to-AND versions in most instances, the treatment of these versions will here be limited to showing how the minimum-component OR-to-AND version of a switching function can be obtained straightforwardly from an AND-to-OR version. The method is derived from the fact that a function is equal to "not" those terms for which the function is 0 (where a "term" is as defined previously, that is, an AND-related combination of input signals). The 0-terms are minimized by any of the procedures described for AND-to-OR expressions, and the selected minimum-component version is converted to OR-to-AND form by using the relationships $\overline{A + B + C + \cdots} = \bar{A}\bar{B}\bar{C} \cdots$ and $\overline{ABC \cdots} = \bar{A} + \bar{B} + \bar{C} + \cdots$.

For example, the OR-to-AND version of the function plotted in the diagram in Figure 2-5 may be determined by generating the terms that include all squares representing elemental terms for which the output is 0. Any square which represents a "don't care" condition, as indicated by a d, may or may not be included. Three two-variable terms, indicated by the loops shown in the figure, are sufficient to include all 0-containing squares. Because the function is "not" these terms, it may be represented as

$$\overline{A\bar{B} + B\bar{D} + \bar{C}D} = \overline{(A\bar{B})}\,\overline{(B\bar{D})}\,\overline{(\bar{C}D)} = (\bar{A} + B)(\bar{B} + D)(C + \bar{D})$$

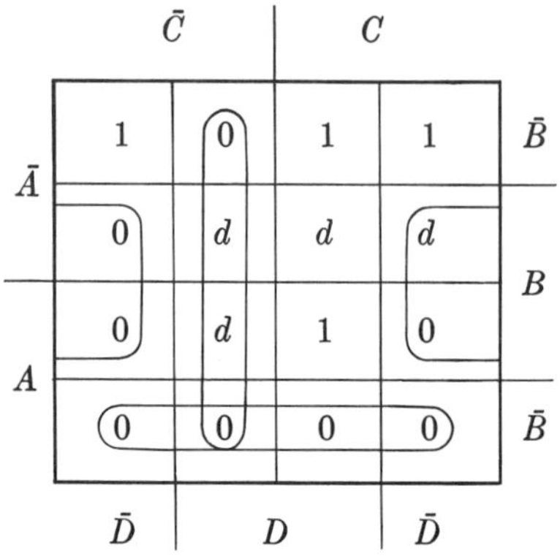

56 AND-OR NETWORKS

FIGURE 2-5. Diagram of a function that may be expressed as $(\bar{A} + B)(\bar{B} + D)(C + \bar{D})$.

where the last expression is the sought-for OR-to-AND version. When this expression is "multiplied out" the result will be a correct AND-to-OR expression of the function, but it will not be a minimum-component expression. It will contain a superfluous term and will not have utilized the appropriate "don't care" conditions. The minimum-component AND-to-OR expression happens to utilize all "don't care" conditions in this example and is $\bar{A}\bar{B}\bar{D} + \bar{A}C + BD$ requiring ten diodes compared with only nine diodes required for the minimum-component OR-to-AND version.

Three-Level OR-AND-OR and AND-OR-AND Versions of a Switching Function—Factoring

The number of required components can often be further reduced by using switching networks that have three or more levels instead of just the two levels considered so far. Unfortunately, no reasonably simple methods have been devised for finding the minimum-component versions of multilevel networks, although simple "factoring" is reasonably effective when the number of levels is limited to three. (The reader is reminded that engineering considerations often cause diode circuits having more than two levels to be undesirable—although with relay contacts no particular limitations are encountered in number of levels.) "Factoring" is similar to factoring in ordinary algebraic expressions, and it is the finding of common "multipliers" (AND-related variables) in the terms of an expression and then rearranging the variables in accordance with the relationship $AB + AC = A(B + C)$.

THREE-LEVEL VERSIONS—FACTORING

For example, a switching function expressed in Boolean notation as $ABC + ABD + ABE + FG$ requires 15 diodes in this AND-to-OR version, but it can be factored to yield the expression $AB(C + D + E) + FG$ which requires only 10 diodes in an OR-to-AND-to-OR sequence of devices.

Factoring does not always result in a reduction in the required number of components. An example is the expression $ABC + ADE + FG$ which represents a switching array having 11 diodes. Variable A can be factored from the first two terms to yield $A(BC + DE) + FG$, an expression corresponding to 12 diodes. Furthermore, this particular expression corresponds to four levels in an AND-OR-AND-OR sequence, not just three levels.

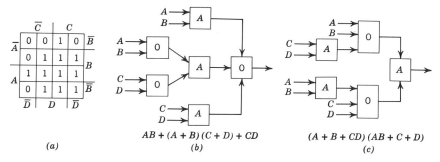

FIGURE 2-6. Three-level versions of the function $AB + AC + AD + BC + BC + CD$.

In a few instances, some of the terms in a two-level AND-to-OR expression can be grouped and converted to OR-to-AND form with the result being a three-level expression requiring fewer diodes. An example is the function $AB + AC + AD + BC + BD + CD$, which is shown plotted on a diagram in Figure 2-6a. A total of 18 diodes would be required. This is the minimum-component two-level AND-to-OR expression of the function. In this example the second, third, fourth, and fifth terms can be converted to $(A + B)(C + D)$ by the method described in the previous section. Alternatively, the terms can be factored as follows:

$$AC + AD + BC + BD = A(C + D) + B(C + D)$$
$$= (A + B)(C + D)$$

where $(C + D)$ is the common factor in the second step. The resulting OR-AND-OR expression for the function as a whole is $AB + (A + B)$

$(C + D) + CD$ and is shown in block diagram form in Figure 2-6b. Only 13 diodes are required.

The example of the previous paragraph happens to be realizable with still fewer diodes when a three-level AND-OR-AND expression is derived from the minimum-component two-level OR-to-AND expression. The OR-to-AND expression can be derived from the diagram in Figure 2-6a by considering the 0-containing squares. The steps are as follows:

$$\overline{A}\overline{B}\overline{C} + \overline{A}\overline{B}\overline{D} + \overline{A}\overline{C}\overline{D} + \overline{B}\overline{C}\overline{D}$$
$$= (A + B + C)(A + B + D)(A + C + D)(B + C + D)$$
$$= (A + B + CD)(AB + C + D)$$

The last expression cas be obtained by "multiplying out" the second and third and the third and fourth factors, respectively, in the second expression of the function. Alternative procedures could have been used, one of which employs a sort of Boolean "dual" of factoring and utilizes the relationship $(A + C)(B + C) = AB + C$. Regardless of the method used for deriving it, the resulting three-level AND-OR-AND expression of the given function requires only 12 diodes and is shown in block diagram form in Figure 2-6c.

In other examples, the expression must not be in its minimum-component two-level form when factoring to achieve a further reduction in the number of components required. An example of this situation is the function $ACE + \bar{A}BE + ADE + \bar{A}\bar{C}$, given here in one of its minimum-component AND-to-OR expressions. Fifteen diodes are required. No way is apparent for reducing the number of components by factoring this expression directly, but the designer may notice (such as by plotting the function on a diagram) that an alternative minimum-component expression is $ACE + BCE + ADE + \bar{A}\bar{C}$ to which a superfluous term BDE can be added to make the result factorable. The steps are as follows.

$$ACE + \bar{A}BE + ADE + \bar{A}\bar{C} = ACE + BCE + ADE + \bar{A}\bar{C}$$
$$= ACE + BCE + ADE + BDE + \bar{A}\bar{C}$$
$$= (AC + BC + AD + BD)E + \bar{A}\bar{C}$$
$$= [A(C + D) + B(C + D)]E + \bar{A}\bar{C}$$
$$= (A + B)(C + D)E + \bar{A}\bar{C}$$

Only 11 diodes are required for this OR-AND-OR version of the function.

Systemetizing the multilevel problem is largely a matter of proceeding through the indicated steps for all possible combinations of terms and factors. Although shortcuts can be devised, the procedures obviously can become laborious when the number of variables is greater than four and especially laborious when the number of levels is greater than three.

Multi-Output AND-to-OR Switching Networks

When two or more output signals are to be generated from the same set of input signals, component savings can often be realized by visualizing the array as a single multi-output network instead of a corresponding number of single-output networks as implied so far in this chapter. For example, consider the following expressions which are functions of input signals A, B, C, and D and which are in minimum-component AND-to-OR form:

$$f_1 = \bar{A}B + BD + C\bar{D}$$
$$f_2 = A + B\bar{C}D + \bar{B}C\bar{D}$$

The number of diodes required to generate the two indicated output signals is 18. However, f_1 may be expressed as $\bar{A}B + BC + B\bar{C}D + \bar{B}C\bar{D}$ (as would be apparent by plotting f_1 on a diagram), and the AND devices corresponding to terms $B\bar{C}D$ and $\bar{B}C\bar{D}$ may be used in the development of both output signals. The resulting network is shown in Figure 2-7. Only 17 diodes are required. Note that the expression used for f_1 is not only not in minimum-component form for f_1 alone, but also terms $B\bar{C}D$ and $\bar{B}C\bar{D}$ are not even basic terms for f_1.

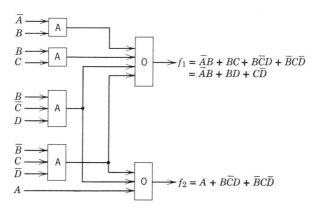

FIGURE 2-7. A two-output network requiring one less diode than the corresponding two single-output networks.

In general, the procedure to be used in finding the minimum-component AND-to-OR expressions for multi-output networks involves finding all basic terms that can be derived from elemental terms that are common to the output signals in all possible combinations. That is, in addition to the basic terms that are developed for the expressions of the output signals considered individually, the basic terms for each possible pair of output signals are determined, the basic terms for each possible set of three output signals are determined, and so on. The 1-producing input signal combinations that are common to two or more output signals may be derived by combining the output signals in an AND relationship. For the example of the previous paragraph the result is

$$f_1 f_2 = (\bar{A}B + B\bar{D} + CD)(A + B\bar{C}D + \bar{B}C\bar{D})$$
$$= \bar{A}B\bar{C}D + ABD + B\bar{C}D + AC\bar{D} + \bar{B}C\bar{D}$$
$$= ABC + ABD + B\bar{C}D + AC\bar{D} + \bar{B}C\bar{D}$$

where the second expression is obtained by "multiplying out" the factors in the first expression, and the last expression is obtained by converting to the complete set of basic terms (by any of the methods described earlier). Terms ABD and $AC\bar{D}$ are superfluous. The determination of the basic-term set that results in the minumum number of component requires a table that is an elaboration of the tables described earlier for single-output functions. However, in the multi-output problem the number of diodes or the "cost" of each basic term must be given more detailed consideration. This requirement and other features of multi-output tables will be illustrated by means of a slightly more complex example.

Assume that the network for which the number of components is to be minimized is a three-output network generating the following three functions:

$$f_1 = CD + ABC + ABD + \bar{A}\bar{B}\bar{C}\bar{D} \quad \text{(16 diodes)}$$
$$f_2 = AB + BD + \bar{A}\bar{B}\bar{C}\bar{D} \quad \text{(11 diodes)}$$
$$f_3 = A\bar{B} + AC + AD + \bar{B}C\bar{D} + \bar{B}C\bar{D} \quad \text{(17 diodes)}$$

Each function is expressed in its minimum-component form from the standpoint of single-output networks, and in this example no alternative minimum-component expressions are possible. The three functions f_1, f_2, and f_3 are shown plotted in the diagrams of Figure 2-8a, b, and c. Figure 2-8d is included as a reminder of the square numbering scheme described earlier, and elemental terms will be designated by their square numbers rather than by detailed signal combinations. The "product" function $f_1 f_2$ is shown plotted in Figure 2-8e. With diagrams, the "product" may be determined easily by noting those square positions which contain 1's in the diagrams for both f_1 and f_2—squares 0, 7, 13, 14, and 15 in this instance. Functions $f_2 f_3$ and $f_1 f_3$ are similarly plotted in Figures 2-8f and g, respectively, and $f_1 f_2 f_3$ is plotted in Figure 2-8h. In this example, $f_2 f_3$ happens to be equal to $f_1 f_2 f_3$. From this equality one might conclude that any basic term that is useful in generating output signals f_2 and f_3 would also be useful in generating output signal f_1, and that the $f_2 f_3$ diagram could therefore be eliminated from further consideration. However, this conclusion is not obvious and would be

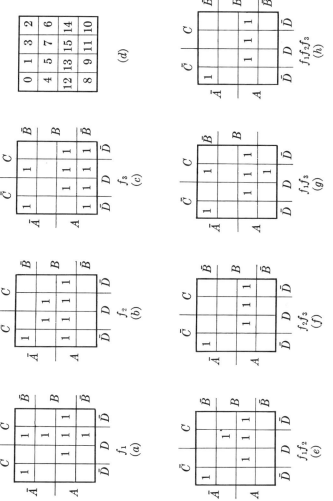

FIGURE 2-8. Diagrams for the three-output network for generating the functions $f_1 = CD + ABC + ABD + \bar{A}\bar{B}\bar{C}\bar{D}$; $f_2 = AB + BD + \bar{A}\bar{B}\bar{C}\bar{D}$; and $f_3 = A\bar{B} + AC + AD + \bar{B}CD + \bar{B}\bar{C}\bar{D}$.

incorrect in some examples because, as will be shown, the terms derived from the $f_1 f_2 f_3$ diagram are more "costly" than the corresponding $f_2 f_3$ terms.

The basic terms for the three originally given functions and for the four "product" functions are listed in the left-hand part of Table 2-6. The x's in the table are marked as for single-output functions except that each output signal has its own section of the table as indicated at the top. The basic terms for f_1 have x's only in the f_1 section of the table, the basic terms for $f_1 f_2$ have x's only in the f_1 and f_2 sections, and so on.

In Table 2-6 a number is listed at the right-hand end of each row corresponding to a basic term. This number is the number of diodes that would be required in the multi-output network if that particular term is utilized. For example, the first term CD, if used at all, would be used only in the generation of output signal f_1. It therefore requires three diodes, of which two are in an AND circuit that combines C and D and one is in the OR circuit that combines CD with other terms. The last term $\bar{A}\bar{B}\bar{C}\bar{D}$ requires seven diodes, of which four are for the AND circuit and three are in the OR circuits for output signals f_1, f_2, and f_1. Although $\bar{A}\bar{B}\bar{C}\bar{D}$ appears at several points in the left-hand part of the table, it is treated as a corresponding number of different terms, because the appearances apply to different output signals and have different "costs." Certain other terms similarly appear more than once for the same reasons.

In spite of the fact that the problem involves only four input signals and three output signals, the determination of all possible basic term sets that have no superfluous terms would be an extensive task. Therefore, a desirable thing to do is to find schemes whereby as many as possible of the x's may be eliminated from consideration. In the example of Table 2-6, various of the x's can be eliminated by several different arguments, and a different mark is used for each different argument. First, some of the terms may be essential as with single-output tables. A term is essential if it includes an elemental term which is not included in any other basic term. Elemental term 5 in f_2 is included only in basic term BD of f_2, and therefore BD is essential. Because BD will be in any finally selected set of basic terms, all elemental terms included in BD may be eliminated from further consideration. This fact is indicated by circling all x's in the BD row. Because all other elemental terms represented by these x's will be included (that is, terms 7, 13, and 15 in f_2), all x's in the corresponding columns may also be removed from further consideration, and these x's are likewise circled.

Further x's may be removed from consideration by observing that no given elemental term need be considered if it is included in all basic terms

TABLE 2-6. Term Table for the Three-Output Network of Figures 2-8 and 2-9

			f_1						f_2						f_3								
		0	3	7	11	13	14	15	0	5	7	12	13	14	15	0	3	8	9	10	11	13	14	15
f_1	CD		[x]	[x]	(x)		(x)																			a	3
	ABC					x	(x)																			b	4
	ABD				x		(x)																			c	4
	$\bar{A}\bar{B}\bar{C}\bar{D}$	(x)																								d	5
f_2	AB											(x)	(x)	(x)	(x)											3
	BD									(x)	(x)		(x)		(x)											3
	$\bar{A}\bar{B}\bar{C}\bar{D}$								(x)																e	5
f_3	$A\bar{B}$																	x	x	x	(x)				f	3
	AC																		x	(x)		x	(x)		g	3
	AD																		x	(x)	x			(x)	h	3
	$\bar{B}CD$																			[x]		(x)			i	4
	$\bar{B}C\bar{D}$																x		x						j	4
f_1f_2	ABC					[x]	(x)						(x)	(x)											k	5
	ABD				[x]		(x)					(x)		(x)											l	5
	BCD		[x]				(x)				(x)			(x)											m	5
	$\bar{A}\bar{B}\bar{C}\bar{D}$	x							x																n	6
f_2f_3	ABC												(x)	(x)									[x]	(x)	o	5
	ABD											(x)		(x)							[x]			(x)	p	5
	$\bar{A}\bar{B}\bar{C}\bar{D}$												(x)				(x)								q	6
f_1f_3	ABC					x	(x)																x	(x)	r	5
	ABD				x		(x)															x		(x)	s	5
	ACD			(x)			(x)													(x)				(x)	t	5
	$\bar{B}CD$		[x]	(x)														[x]			(x)			u	5
	$\bar{A}\bar{B}\bar{C}\bar{D}$	(x)											(x)												v	6
$f_1f_2f_3$	ABC					[x]	(x)						(x)	(x)									[x]	(x)	w	6
	ABD				[x]		(x)					(x)		(x)							[x]			(x)	x	6
	$\bar{A}\bar{B}\bar{C}\bar{D}$	x							x							x									y	7

63

in which some other one elemental term is included. The reason here is that, if basic terms are selected (as they must and will be) to include this one other term, the given term will certainly be included. The fact that the given elemental term may be included in other basic terms is of no consequence in selecting a minimum-component set of basic terms. An example of this situation is term 11 of f_1. This elemental term is included in CD of f_1 and in terms ACD and $\bar{B}CD$ of f_1f_3. However, term 3 of f_1 is included in CD of f_1 and in $\bar{B}CD$ of f_1f_3 and is not included in any other basic term. Therefore, either CD of f_1 of $\bar{B}CD$ of f_1f_3 must be selected to include term 3, and term 11 will then be included without giving any further consideration to it. The elimination of the x's in the term 11 column is indicated by dotted circles around them. Term 15 of f_1 is similarly eliminated because of its relationship to term 14 of f_1 (or to either of terms 7 or 13 of f_1). Terms 11 and 15 of f_3 are similarly eliminated because of analogous relationships to other terms. In each instance the x's in the corresponding columns are indicated by dotted circles.

Next, of the basic terms representing a given input signal combination, any one with relatively high cost may be eliminated if a lower cost term includes the same elemental terms. For examples, after eliminating x's as described in the previous two paragraphs, basic term ABD of $f_1f_2f_3$ includes exactly the same elemental terms as does basic term ABD of f_1f_3. Because the former costs six diodes and the latter only five diodes, the former is eliminated, and this elimination is indicated in Table 2-6 by drawing brackets around the otherwise unmarked x's in the corresponding row. Term ABC of $f_1f_2f_3$ is eliminated for similar reasons. Likewise, terms ABC and ABD of f_1f_2 can be eliminated when related to ABC and ABD, respectively, of f_1. The corresponding x's are also bracketed.

Terms $\bar{A}\bar{B}\bar{C}\bar{D}$ of f_1 and $\bar{A}\bar{B}\bar{C}\bar{D}$ of f_2 can be eliminated because there is a similar term, $\bar{A}\bar{B}\bar{C}\bar{D}$ of f_1f_2, which includes the same elemental terms but which costs less (six diodes) than the sum of the costs (10 diodes) of the two original terms. These eliminations are indicated in the table by parentheses around the corresponding x's. Terms $\bar{A}\bar{B}\bar{C}\bar{D}$ of f_2f_3 and $\bar{A}\bar{B}\bar{C}\bar{D}$ of f_1f_3 can also be eliminated, and the argument is essentially the same although slightly more complex. In this instance, the comparison is with $\bar{A}\bar{B}\bar{C}\bar{D}$ of $f_1f_2f_3$, which costs seven diodes. If, for example, $\bar{A}\bar{B}\bar{C}\bar{D}$ of f_2f_3 were selected, the only alternative method of including elemental term 0 of f_1 would be to select $\bar{A}\bar{B}\bar{C}\bar{D}$ of f_1f_2, and the comparative cost would be six plus six, or 12, diodes. Again, the eliminated x's are indicated by parentheses. (Note that $\bar{A}\bar{B}\bar{C}\bar{D}$ of f_1f_2 cannot be eliminated by this argument, because $\bar{B}\bar{C}\bar{D}$ of f_3 could be an accompanying term, and this term includes another elemental term which is as yet unaccounted for.)

Next, the designer may observe from Table 2-6 that elemental terms 3 and 7 of f_1 and 3 of f_3 are included in basic terms that do not include any other elemental terms (after eliminating those x's described in previous paragraphs). Therefore, the corresponding basic terms may be handled as a separate subproblem. The lower-case basic-term designations indicated in the right-hand part of the table will now be used. To include term 3 of f_1, either term a or term u must be selected; for 7 of f_1, either term a or term m; and for 3 of f_3, either term i or term u. The x's involved are indicated in the table by rectangles enclosing them. The selected terms may be recorded as follows.

$$(a + u)(a + m)(i + u) = ai + au + mu$$

The "cost" of ai (that is, a "and" i) is three plus four for a total of seven diodes. Because this cost is less than for either au or mu, terms a and i are selected.

The remaining unmarked x's are handled in the same manner, and the following expression is derived for the selection of the remaining basic terms.

$$(n + y)(c + s)(b + r)(n + y)(j + y)(f + j)(f + h)(f + g)(h + s)(g + r)$$

When the 10 factors (each of which happens to contain two term-representing variables in this example, but with the number not being limited to two in general) in this expression are "multiplied out" the number of terms to be encountered would appear to be $2^{10} = 1024$, but by choosing factors with like variables for the first multiplications, many simplifications can be achieved. For example, the three factors $(f + j)(f + h)(f + g)$ when "multiplied" in Boolean algebra yield $f + ghj$. When the "multiplication" of all 10 factors is complete, one of the resulting terms is found to be $fsry$. (Observe that the terms here are neither the elemental terms nor the basic terms of the original problem, but instead are AND-related combinations of basic terms.) The selection of the corresponding basic terms corresponds to a cost of $3 + 5 + 5 + 7 = 20$ diodes, which is less than for any of the other basic-term sets that are shown to produce the correct output signals.

When basic terms f, s, r, and y are added to terms a and i selected previously and to the essential terms selected near the beginning of the process, the resulting expressions for the three given functions become as follows.

$$f_1 = CD + ABC + ABD + \bar{A}\bar{B}\bar{C}\bar{D}$$
$$f_2 = AB + BD + \bar{A}\bar{B}\bar{C}\bar{D}$$
$$f_3 = A\bar{B} + ABC + ABD + \bar{B}CD + \bar{A}\bar{B}\bar{C}\bar{D}$$

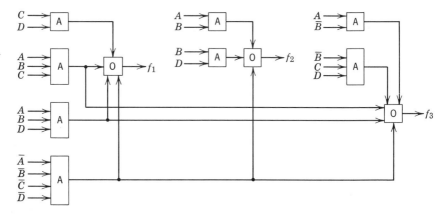

FIGURE 2-9. A three-output network corresponding to the functions illustrated in Figure 2-8 and Table 2-6.

The only change in the expressions is in the one for f_3, although the table indicates directly that the AND circuit for generating $\bar{A}\bar{B}\bar{C}\bar{D}$ is to be used for all three functions. The block diagram for the resulting three-output array is shown in Figure 2-9. The total number of diodes required is 33, a saving of 11 in comparison with the 44 required when the three functions are generated independently.

Remarks about the Multi-Output Problem

In spite of the many x's that were eliminated by one argument or another in the three-output example of the previous section, the final task of selecting the appropriate terms in accordance with the remaining x's is tedious. More complex arguments could probably be applied to eliminate additional x's and thereby simplify the final step, but the procedure is already so complex that the practical utility of the arguments has definite limitations.

Of course, the particular multi-output design procedure that was described is not the only procedure possible. In fact, several different procedures have been devised, but for the most part they differ from the one shown only in notational details that have no great effect on the basic character of the problem. The particular notation and procedure used in the previous section seemed to be the one that illustrates the character of the problem most lucidly.

An additional argument for eliminating x's in the table applies when the objective is limited to finding the minimum number of components and when the designer need not find all minimum-component expressions in the event that more than one exist. In this case, if any row (basic term) in the table includes all elemental terms included by some other row, that other basic term may be eliminated—provided that the other term is not of lower cost. In the example of Table 2-6, term s has an x in the column corresponding to elemental term 13 of f_1. Term c has an uneliminated x in this column and in no other column. Term c could therefore be eliminated except that it has a lower cost than term s. Thus, the argument does not happen to apply here in spite of the fact that term c is eventually eliminated for other reasons. The argument does, however, apply to the single-output problem of Table 2-1 earlier in the chapter. There, one or the other of terms $A\bar{C}D$ and $\bar{B}\bar{C}D$ and one or the other of terms $\bar{A}\bar{C}\bar{D}$ and $B\bar{C}\bar{D}$ may be eliminated, and one of the minimum-component expression will still be found. The argument also applies to the problem of Table 2-3 as previously explained in some detail.

In other multi-output problems the simplification in the procedure is dramatic when lifting the requirement of finding all possible minimum-component expressions. In fact, after thus eliminating some of the terms, the other arguments can sometimes be reapplied for still further simplifications. On the other hand, in applications where eliminating a few diodes is really important, the problem of equalizing loading on the driving circuits or of matching loads to the available driving signals is likely to be important also, in which case the designer would want to know all possible minimum-component expressions.

In the table, the x's were eliminated by marking them in some way. At the price of a substantial amount of additional labor, the table might be more easily interpreted at intermediate steps if it is redrawn at each step with the eliminated terms and x's omitted completely.

"Don't care" conditions are handled for multi-output networks in essentially the same way as for single-output networks, although a few elaborations in the procedure are required. If a given elemental term represents a "don't care" condition for all functions being combined in any given "product" function, this term is a "don't care" condition of the "product" function and is treated as before. If the elemental term is required in one or more functions but is a "don't care" term in one or more other functions forming a given "product" function, that elemental term is included (is not optional) when determining the basic terms of that particular "product" function. However, the elemental term (as represented by one or more columns in the table) is not then included in those sections of the table that correspond to the functions for which the term is a "don't care" condition.

Lastly, the most economical multi-output circuit is not necessarily realized with either a purely AND-to-OR or a purely OR-to-AND array.

Instead, many examples can be found where some combination of AND-to-OR and OR-to-AND switching produces the minimum number of components. For networks having more than two levels of switching, even more instances are encountered where component savings can be realized by departing from standardized alterations in the AND and OR functions. The particular array that is most economical further depends on the nature of the modules at hand, and in particular is affected by the availability of the inverted version of each signal, not only the input signals but also the signals at intermediate points throughout the array. This last feature is obtained with the so-called "current-mode" transistor switching circuits which have become popular in recent years. Some examples which illustrate the economies that can be obtained by employing both AND-to-OR and OR-to-AND switching in a single network are given in a paper by P. Weiner and T. F. Dwyer (*IEEE Transactions on Computers*, February 1968, pp. 184–186). Unfortunately, no organized scheme for finding the most economical array for a mixture of the two switching forms seems to have been devised, and of course such a scheme would be even more elusive when three or more switching levels are allowed.

Switching Networks (Matrices) for Transforming Binary-Coded Signals to One of N Signals

Switching networks for transforming a set of binary-coded signals to a single signal on one of N lines are specialized multi-output networks, but they are so commonly needed that particular attention will be given to them. If the binary-coded signals can appear in any possible combination, as will be assumed for the present, the number of output lines is $N = 2^n$, where n is the number of signals in the binary-coded set. Here, "binary-coded" is intended to mean virtually any set of signals where the individual signals may be visualized as having meanings or "weights" of 1, 2, 4, 8, and so on, respectively. The resulting binary number that is represented by a given combination of signals is (with respect to the networks to be described) seldom a number that is involved in the main information processing operations of a digital system but is more likely to be some auxiliary number such as the address of a location in the storage unit. The switching network is needed to transform the binary-coded representation of this number to a signal on one of N lines as needed for the actuation of the storage unit, for example. The inverse of each binary-coded signal will be assumed to be available on a separate input line.

A network for the purpose indicated is frequently called a "matrix" because, for some versions, the block diagrams or the physical layout (or both) of the components can appear in rows and columns vaguely analogous to the array of symbols in a mathematical matrix.

For only two binary-coded input signals A and B, the switching network for generating a signal on one of four lines is almost trivial. It consists of four two-input AND circuits which generate output signals $\bar{A}\bar{B}$, $\bar{A}B$, $A\bar{B}$, and AB. With diode AND circuits, eight diodes are required.

For three binary-coded input signals A, B, and C, two different forms of matrices are possible. One is a straightforward extension of the two-input matrix and consists of eight three-input AND devices which generate the eight output signals $\bar{A}\bar{B}\bar{C}$, $\bar{A}\bar{B}C$, $\bar{A}B\bar{C}$, $\bar{A}BC$, $A\bar{B}\bar{C}$, $A\bar{B}C$, $AB\bar{C}$, and ABC directly. The alternative form, called a "pyramid" (or sometimes a "tree") employs a two-input matrix for two of the three input signals, and each of the four signals generated by this matrix is combined with the third input signal and its inverse in eight two-input AND circuits. With either form of matrix, a total of 24 diodes is required.

For four binary-coded input signals A, B, C, and D, three significantly different forms of matrices are possible. The first form is an extension of the two-input matrix and the first form of the three-input matrix. It consists of 16 four-input AND circuits requiring a total of 64 diodes. The second form of four-input matrix is the pyramid and is illustrated in Figure 2-10a. With diode AND circuits, 56 diodes are required. In the various levels of the pyramid the input signals may be applied so that each level corresponds to a different input signal. Alternatively, in different parts of the pyramid the signals may be permuted in any of several different ways, one of which is as illustrated. Permuting the signals is desirable in some applications to distribute the load more evenly on the various input signal lines. The third form of four-input matrix is illustrated in Figure 2-10b. It consists of two two-input matrices, the outputs of which are combined in an array of 16 two-input AND circuits, and (as in the other forms) an output signal will be present on one of the 16 final output lines for each combination of input signals. The total number of diodes required is 48—which is 16 less than required for the straightforward matrix and eight less than required for the pyramid.

For more than four input signals, the straightforward matrix can be extended in an obvious way, and in general the total number of diodes is $n2^n$ for n input signals.

The extension of the pyramid to more than four variables is also fairly obvious, and in general the total number of diodes required is $2^3 + 2^4 + \cdots + 2^{n+1}$. Note that when $n = 2$, the pyramid reduces to the same circuit as the straightforward matrix.

70   AND-OR NETWORKS

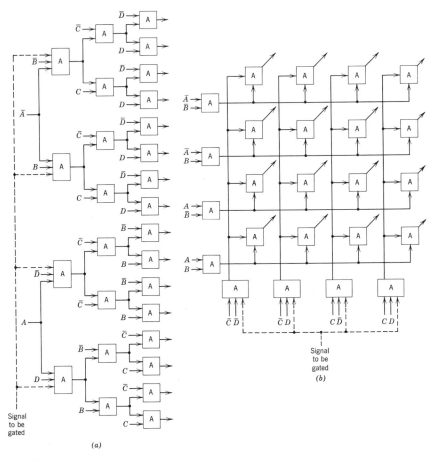

FIGURE 2-10. Matrices for transforming four-variable binary-coded signal sets to a single signal on one of 16 lines in *a* pyramid and *b* minimum-component forms.

The matrix concept of Figure 2-10*b* can likewise be extended to more than four input signals. For five input signals, three of the signals can be combined in a three-input matrix (either a straightforward matrix or a pyramid, either of which requires 24 diodes) and the other two input signals can be combined in a two-input matrix. The output signals thus generated may then be combined in an array of 32 two-input AND circuits to generate a signal on one of 32 final output lines. The total number of diodes required is 96.

For six input signals, the signals may be grouped in two sets of three, in which case two three-input matrices and 64 two-input AND circuits are required. The total number of diodes is 176. Alternatively, the input signals may be grouped in three sets of two, in which case three two-input matrices and 64 three-input AND circuits are required, and 216 diodes are needed. Therefore, this alternative grouping is clearly less desirable from the standpoint of the number of components required.

In general, for n input signals the matrix requiring the minimum number of components is derived by dividing the signals into two equal groups of $n/2$ signals if n is even, or by dividing the signals into two groups of $(n-1)/2$ and $(n+1)/2$ signals, respectively, if n is odd. Each group containing four or more signals is further subdivided in the same manner, and the subdividing is continued until each group contains either two or three signals. The signals for each group are then combined in a two-input or a three-input matrix, as the case may be, and the signals generated by these matrices are combined by appropriate arrays of two-input AND circuits.

TABLE 2-7. Numbers of Diodes Required in Various Matrices

	n						
	2 ($N=4$)	3 ($N=8$)	4 ($N=16$)	5 ($N=32$)	6 ($N=64$)	7 ($N=128$)	8 ($N=256$)
Straightforward	8	24	64	160	384	896	2048
Pyramid	8	24	56	120	248	504	1016
Minimum-component	8	24	48	96	176	328	608

The number of diodes required for the straightforward matrix, the pyramid matrix, and the minimum-component matrix for an n in the range between two and eight, inclusive, is summarized in Table 2-7. Of course, the reader should realize that elementary diode circuits of the type contemplated in determining the component counts are probably used only rarely. In the design of digital storage units especially, many specialized requirements are encountered and many "trick" circuits are available which can make the situation quite different from the idealized situation presented. Nevertheless, the three forms of matrices are good illustrations of the available overall approaches to the problem when the switching circuits to be used are not the simple diode circuits.

In many applications, the problem is a slight elaboration of the problem presented in that the purpose of the matrix is to "gate" a signal on a given input line to one of N lines under the control of n binary-coded signals. The obvious way to solve this problem would be to generate a signal on one of N lines as described and then use N "gate" circuits (where a "gate" is functionally the same as a two-input AND circuit) with the signal to be gated applied as one input signal to each "gate" circuit. However, many components can be saved by applying the signal to be gated to extra inputs in a certain few AND circuits as shown by the dotted connections in Figure 2-10a and b. A minor variation of the circuit in a is to employ two additional two-input AND circuits instead of third inputs to the AND circuits as shown. The two additional AND circuits would appear as a new level of switching in the pyramid, and the signal to be gated would be allowed to pass to one half or the other half of the pyramid under the control of A and $\bar{A}$. The total number of diodes is the same with either variation. A similar variation is possible with the matrix in Figure 2-10b when that matrix is used to gate an input signal to one of N output lines.

In problems where N is not an integral power of two, the fact that certain combinations of input signals will not appear can sometimes be used in the manner of "don't care" conditions to find networks that require fewer components than would be indicated by merely removing obviously unneeded components from the networks designed as described above.

Switching Networks for Transforming a Signal on One of N Lines to a Set of Binary-Coded Signals

The opposite problem, namely, that of transforming a single incoming signal on one of N lines to a set of binary-coded signals, is less commonly encountered in the design of switching systems, but a few applications for this function are encountered. Because finding the minimum number of required components is something less than obvious, the following brief presentation of this form of network is made.

A straightforward version of such a network can be derived by noting that, when N is an integral power of two, a given output signal should be 1 when an input signal is present on one or another of a certain $N/2$ input lines. The inverse of that output signal should be 1 when an input signal is present on any of the other $N/2$ input lines. Therefore, two $N/2$ input OR circuits are required to generate any given one of the n output signals and its inverse. With elementary diode OR circuits, the total number of diodes required is $2n(N/2) = nN = n2^n$, the same as the number of diodes required for the straightforward matrices described in the previous section.

An array requiring fewer components can be derived by dividing the n output signals into two equal groups with $n/2$ signals in each group if

TRANSFORMING TO A SET OF BINARY-CODED SIGNALS

n is even, or by dividing the signals into two groups of $(n-1)/2$ and $(n+1)/2$ signals, respectively, if n is odd. (However, contrary to the problem of the previous section, no subgrouping is performed.) Each input line is connected to an OR circuit in each of two sets of OR circuits. If n is even, each of these two sets consists of $2^{n/2}$ OR circuits each having $2^{n/2}$ inputs. If n is odd, one set consists of $2^{(n+1)/2}$ OR circuits each having $2^{(n+1)/2}$ inputs, and this set is used for developing $(n+1)/2$ output signals and their inverses. The other set consists of $2^{(n-1)/2}$ OR circuits each having $2^{(n-1)/2}$ inputs, and this other set is used for developing the remaining $(n-1)/2$ output signals and their inverses. In any case, the total number of input signals to the matrix as a whole is $2^n = N$. In either set of OR circuits, the pattern of connections is such that all input lines connected to a given OR circuit correspond to the same combination of output signals (with regard to those output signals being generated by the set of

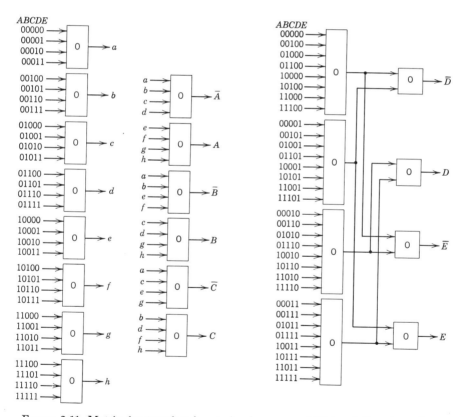

FIGURE 2-11. Matrix for transforming a signal on one of 32 input lines to a five-variable binary-coded set of signals.

which the given OR circuit is a part). Because each binary-coded output signal (or its inverse) is 1 for an input signal on any one of half of the input lines, each binary-coded output signal can be generated by combining in OR fashion the signals from one half of the OR circuits in the corresponding set.

The resulting array for 32 input lines and five binary-coded output signals, A, B, C, D, and E, and their inverses is shown in Figure 2-11. To help illustrate the pattern of connections, the combination of 0's and 1's to be generated by each input signal is shown at each input connection. For example, an input signal on the lines marked 11100 should cause output signals A, B, and C to each be 1 and output signals D and E to each be 0. Note that each input line is connected at two points in the array.

The total number of diodes required is $(2)2^n + n2^{n/2}$ when n is even, and is $(2)2^n + [(n+1)/2]2^{(n+1)/2} + [(n-1)/2]2^{(n-1)/2}$ when n is odd. These expressions work out to 8, 24, 48, 96, 176, 344, and 640 diodes as n ranges from two through eight, respectively. For an n up to six, the number of diodes happens to be the same as for the minimum-component matrices of the previous section, but for larger n the number of diodes required for the present function is greater.

Exercises

1. Show by test that one or the other (but not both) of the variables A and $\bar{B}$ is superfluous in the second term of $AB + A\bar{B}C + \bar{A}\bar{B}$.
2. Show by test that all of the first three terms or all of the last three terms (but no other three terms) are superfluous in $AB + \bar{A}C + \bar{B}\bar{C} + \bar{A}\bar{B} + A\bar{C} + BC$.
3. Find all of the elemental and basic terms in each of the following functions:
   (a) $AB + \bar{A}\bar{B} + \bar{A}BC$
   (b) $A\bar{B}\bar{C} + \bar{A}\bar{B}D + ABCD + \bar{A}B\bar{C}D + C\bar{D}$
   (c) $AC\bar{D} + ABC + \bar{A}BD + \bar{B}CD + \bar{C}D$
   (d) $ABE + A\bar{C}E + \bar{B}\bar{D}E + \bar{A}\bar{C}DE + \bar{A}C\bar{D}\bar{E} + BD\bar{E} + \bar{A}BC\bar{D}\bar{E}$
   (e) $ABE + A\bar{C}E + \bar{B}\bar{D}E + \bar{A}\bar{C}DE + \bar{A}C\bar{D}\bar{E} + BD\bar{E} + \bar{A}BC\bar{D}\bar{E}$
      $+ ABC\bar{D}\bar{E}$
4. Use the method explained on page 38 to find all basic terms in functions a and c of the previous exercise.
5. Use a table to find all minimum-component expressions for the five functions in exercise 3. (Note: For some functions, rule 2 on page 35 will be helpful in simplifying this exercise.)
6. Use the method illustrated in Table 2-3 to find one minimum-component expression for each of functions d and e in exercise 3.

7. Draw the diagrams corresponding to the five functions in exercise 3, and determine all basic terms from these diagrams.

8. Prepare an eight-variable diagram as in Figure 2-4, enter at least thirty 1's at random (in the total of 256 squares) and determine all basic terms of the function thus represented.

9. Use either tables or diagrams to find all minimum-component expressions of the following functions:
(a) $\bar{A}\bar{B} + \bar{A}C + \bar{B}C$, where $A\bar{C} + B\bar{C}$ represents "don't care" conditions.
(b) $A + BCD + \bar{B}C\bar{D}$, where $\bar{A}\bar{D} + \bar{A}B\bar{C}$ represents "don't care" conditions.
(c) $ABC\bar{D} + \bar{A}BCD + \bar{A}\bar{B}\bar{C}D$, where $BC + BD$ represents "don't care" conditions.

10. The function to be generated is $A\bar{B}\bar{C} + \bar{A}\bar{B}C + \bar{A}B\bar{C} + ABC$, but C will be 1 only when A and B are both 0 or both 1. Show the steps in finding a minimum-component equivalent expression. (Answer: $A\bar{B} + \bar{A}B + C$.)

11. Find all minimum-component OR-to-AND expressions of functions a, b, and c in exercise 3.

12. Find all minimum-component OR-to-AND expressions of the three functions in exercise 9.

13. Find a 13-diode AND-OR-AND expression of the function $ACE + \bar{A}BE + ADE + \bar{A}\bar{C}$. (Hint: First find the minimum-component OR-to-AND expressions.)

14. Find one minimum-component expression of the following two-output network.
$$f_1 = A\bar{C} + BD + C\bar{D},$$
$$f_2 = A\bar{D} + \bar{B}\bar{D} + \bar{A}BC.$$

15. Draw the block diagrams for the matrices corresponding to Figure 2-11 but with (a) 16 input lines and four binary-coded output signals and (b) 128 input lines and seven binary-coded output signals.

3
NETWORKS CONTAINING INVERTERS

In the previous chapter the inverse of each input signal was assumed to be available as needed. In applications where the inverted input signals are not available, inverters as described in Chapter 1 may be used to invert the input signals. However, this obvious design approach seldom produces the most economical network.

One approach to achieving economical networks might be to determine or assume some ratio between the cost of an inverter and the cost of an AND or OR device and then, for each given switching function, search for a network which utilizes AND devices, OR devices, and inverters in the numbers and proportion that minimize the cost. However, in practical electronic applications this approach tends to be totally unrealistic. The reason is that with diode AND and OR circuits the current "gain" (which might be better viewed as a "loss") is only a small fraction of unity in each level of switching with the result that a transistor current amplifier is needed after each level, or at best after each two levels, of diode switching. Then, because a transistor digital current amplifier circuit happens to be identical to an inverter circuit, the situation to be faced is that of having inverters distributed throughout the network for an extraneous reason. The problem is to utilize these extraneous inversions as advantageously as possible.

With certain transistor circuits the nature of the problem is modified by the fact that the AND and OR functions are not generated by separately identifiable diode circuits, but instead are generated through the manner in which the transistors are interconnected. In particular, two or more transistors may have their collectors connected to a common load resistor. This is a widely used scheme for combining inversion with the OR function. That is, with the input signals to the two transistors designated X and Y, respectively, the output signal will (with NPN transistors) be negative if either input signal is positive so that the function $\overline{X + Y}$ is generated. (Because $\overline{X + Y} = \bar{X}\bar{Y}$, the circuit may alternatively be viewed as two separate inversions followed by an AND function, but the OR-INVERT visualization is preferred—at least it corresponds with the OR-INVERT modules discussed later.) If PNP transistors are used, or if the opposite convention with respect to signal polarity is adopted (but not if both changes are made), the common collector load connection produces an AND-INVERT function.

In current-mode (ECL) transistor circuits, which are discussed near the end of this chapter, the design problem is further modified by the fact that each circuit contains, in effect, two inverters connected in tandem so that both the direct and the inverted version of each output signal is available with substantially equal facility. The OR function (or the AND function, as the case may be) is generated by a common collector connection substantially as described in the previous paragraph.

Minimizing the Number of Inverters in a Network

Before proceeding to the main topics of this chapter, one brief example will be given to illustrate the nature of what would appear to be the primary objective when designing switching networks containing inverters—namely, minimizing the number of inverters. Consider the following function.

$$\bar{A}\bar{B}C + \bar{A}\bar{B}DE + \bar{A}\bar{C}DE = \overline{A + BC}(C + DE)$$

If signals A, B, and C are not available in inverted form, three inverters are required to generate the function as expressed on the left-hand side of the equation. Only one inverter is required to generate the function in accordance with the right-hand expression. Moreover, four less diodes are required. Unfortunately, the resulting network is three-level from the standpoint of input signals D and E, and for most physical electronic realizations of the right-hand expression additional amplifiers would be needed anyway. Because of the need for added amplification, the right-hand expression of the function would not likely be used.

Apart from the disadvantages of the right-hand expression for use with electronic circuits, the finding of that expression (given the left-hand expression) is not easy. No practical generalized method for minimizing the number of inverters is known. The problem of minimizing cost for an arbitrarily assigned ratio of costs between inverters and AND and OR devices is even more remote from a practical solution.

One interesting fact in the matter of minimizing the number of inverters is that, for any multi-output switching function of n variables, the number of inverters need be no more than y, where y is the smallest integer for which the relationship $n < 2^y$ holds. For example, a network may have 63 input signals A_1, A_2, $\cdots$, A_{63}, and 63 output signals $\bar{A}_1$, $\bar{A}_2$, $\cdots$, $\bar{A}_{63}$. Although 63 inverters would appear to be required, one for each signal, the result can actually be achieved with only six inverters. Then with inverted versions of each input signal generated, any number of more complex functions may be generated without the need for additional inverters. Unfortunately, the number of AND and OR devices additionally required to generate the inverted input signals (with the number of inverters minimized) is so great as to be hopelessly impractical. Even for only three input signals, which can be inverted by means of only two inverters, nearly 20 AND and OR devices are also required. Because the required number of AND and OR devices is so great and because in virtually all realistically encountered devices the inversion function is integrally related to the other functions, the inverter minimization technique will not be described here. The technique can be found in a paper by S. B. Akers (*IEEE Transaction on Computers*, February 1968, pp. 134–135).

AND-INVERT and OR-INVERT (NAND and NOR) Switching Modules

In recent years much attention has been given to switching networks composed solely of modules that perform an AND function followed by an inversion or solely of modules that perform an OR function followed by an inversion. These modules have commonly, but far from universally, been called NAND and NOR circuits, respectively, where these names have been derived from the terms "negative-AND" and "negative-OR," respectively. Because the inversion definitely follows, rather than precedes, the AND or OR function, as the case may be, the modules will here be called AND-INVERT and OR-INVERT modules and will be abbreviated A-I and O-I, respectively. For modules that perform more complex combinations of functions, a correct indication of function sequence is particularly important to avoid confusion, and this notation is straightforwardly expandable to, for example, A-O-I.

The basic features of two A-I and O-I circuits are shown in Figure 3-1. These two circuits represent the most widely used form of switching, although many different elaborations are to be encountered in specific applications. Either circuit can be used to perform either the A-I or the O-I function, although with NPN transistors as shown and with the nearly universal convention of representing 1's and 0's by relatively positive and negative potentials, respectively, the circuit in Figure 3-1*a*

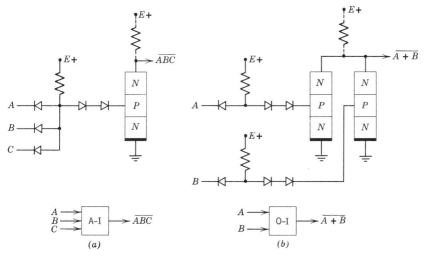

FIGURE 3-1. AND-INVERT and OR-INVERT circuits (where a relatively positive potential represents a 1).

performs the A-I function and the circuit in (b) performs the O-I function. In fact, this signal polarity convention and the use of NPN transistors (instead of PNP transistors) are now so extensive that the circuits in Figure 3-1a and b may almost be regarded as A-I and O-I modules, respectively, by definition.

The module in Figure 3-1a consists of a diode AND circuit driving a transistor inverter circuit, both as described in Chapter 1. A signal bias-shifting mechanism is needed between the two circuits, and the two diodes connected in series with the transistor base serve this purpose. If one or more of the input signals is at the relatively negative 0-representing potential (actually, slightly positive with respect to ground), the current from the left-hand E_+ supply terminal will be diverted to the corresponding input lines. The voltage across the diodes in series with the transistor base will not be great enough to cause them to be in the low resistance condition, and the transistor will be cut off. The collector potential then becomes highly positive as a result of its being connected through a resistor to the right-hand E_+ terminal. If all of the input signals are at the relatively positive 1-representing potential, the potential at the junction of diodes will be pulled sufficiently positive to cause current flow to the transistor base. The transistor will then conduct, and its collector potential will become relatively negative (actually, as before, slightly positive with respect to ground).

The connections to the resistor in the collector circuit are shown by dotted lines to indicate that the presence of this resistor is not necessary when the module output signal is used to actuate other modules of the same type, in which case the resistors in the diode AND circuits of the other modules serve as collector load resistors. With the direct collector load resistor omitted, the collector potential may not become strongly positive in the presence of a 1 output signal, but this result is not harmful. A cut-off of current to the diode AND circuits in the subsequent modules is sufficient.

The more or less obvious combination of a diode OR circuit (as described in Chapter 1—again with the convention of representing a binary 1 by a relatively positive potential) and an NPN inverter has proved to be unpopular for engineering reasons. Primarily, high speed operation requires that the resistor in the diode OR circuit be connected to a supply potential that is negative with respect to ground (the transistor emitter potential), and the resulting need for two supply potentials instead of just one is regarded as a serious disadvantage. Also, a collector load resistor is definitely needed in each module to pull the output potential positive when the output signal is 1, and the positive-pulling effect of the collector load resistor "works against" the negative-pulling effect of the resistors in the diode OR circuits in subsequent modules in a way which has the end result of seriously limiting the fan-out capability of a given module.

In a sense, the O-I module in Figure 3-1b is nothing more than two one-input A-I modules with a common collector connection. If input signals A and B are both binary 0's (relatively negative), the output signal is relatively positive as before. However, if either A or B is 1, at least one transistor will conduct, and the output potential will be pulled to its relatively negative 0-representing value. The module may be viewed as two separate inverters followed by a diode AND circuit where the transistor collectors simulate the diodes. That is, the module may be viewed as generating the signal $\bar{A}\bar{B}$. However, $\bar{A}\bar{B} = \overline{A + B}$, and the module is almost universally viewed in the latter way, that is, as performing an OR function followed by an inversion.

The diodes in the input circuits in Figure 3-1b are superfluous from a switching function standpoint and may therefore appear to introduce a serious disadvantage. However, with integrated circuit techniques, these diodes add very little to the cost, and the use of input circuits which are the same in both the A-I modules and the O-I modules allows the use of mixtures of both types of modules in a network.

Also, with the diode input circuits as in Figure 3-1b, an obvious elaboration is to add inputs to the one-input AND circuits to form multi-input AND circuits. The result is an A-O-I module which achieves two levels of AND-OR switching between inversions in an expeditious manner. A-O-I modules will be discussed in more detail later.

The popular TTL circuits are essentially the same as the circuits in Figure 3-1 except that specialized transistors are used to replace the diodes. For multi-input AND functions as in a, each such transistor has more than one emitter. Also, especially with integrated circuits, the inverter may consist of a multitransistor circuit instead of the single-transistor circuit shown, and in some designs the nature of the circuit is such that the connecting of final output terminals in common is not possible even though, internal to a given module, an O-I function is achieved in the manner of Figure 3-1b

The Universal Nature of A-I and O-I Modules

An attractive feature of A-I and O-I modules is that any given switching function can be generated with modules of only one type. This point is illustrated in Figure 3-2, part a of which shows how A-I modules may be interconnected to from the AND, OR, and inversion functions separately. Part b of the figure shows the same for O-I modules. In the figure, two-input modules are assumed. Dotted connections are indicated in those instances where only one input to a module is needed. With most circuits encountered in practice, this dotted connection may be eliminated. For example, with the DTL circuit of Figure 3-1a, an open connection to an input line would have the switching-function effect of applying a constant 1 to the corresponding input terminal. Although the potential of the floating input line may be relatively negative when

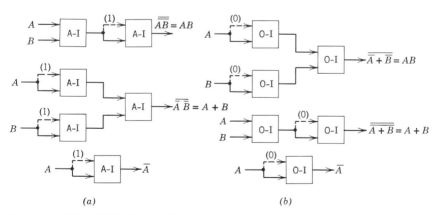

FIGURE 3-2. The AND, OR, and inversion functions as generated by A-I and O-I modules.

at least one other input potential is relatively negative, the output signal would not at this time be dependent on the potential of the open input line. When all other input potentials are relatively positive, the potential at the open input line is pulled positive by the action of current through the resistor.

With some types of circuits, an input line must not be left disconnected. An input signal must be connected as shown by the dotted lines in Figure 3-2 or else the input terminal in question must be connected to a steady-state supply potential of such value as to represent a constant 1 for an A-I module or a constant 0 for an O-I module.

The networks of Figure 3-2 can be straightforwardly extended to modules having three or more input lines. Also, even with modules having only two input lines, the networks can be straightforwardly extended to generate AND or OR functions for three or more input signals. With the circuits for forming separate AND, OR, and inversion functions known, the extension to functions that are complex combinations of these three basic functions is likewise straightforward, although finding the most economical or otherwise most desirable network for a given function is far from obvious.

Minimizing the Number of Components in Networks Composed of A-I or O-I Modules

A basic problem in the design of networks composed of A-I or O-I modules is minimizing the cost under the following conditions (simplifying assumptions):

1. Input signals available in both direct and inverted form.
2. No "don't care" conditions.
3. Modules with unlimited fan-in.
4. Modules with unlimited fan-out.
5. Maximum of three levels.
6. Cost a function of number of modules only (not number of inputs to the modules).
7. Single-output functions.
8. No mechanical restrictions (cross-overs, wire lengths, total area, etc.).

In some practical applications all of these simplifying assumptions are realistic, but in other applications none of them are. In the former case the cost minimization problem turns out to be essentially identical to the component-minimization problem for two-level AND-OR networks as described in some detail in the previous chapter. In the latter case the problem is so "messy" that virtually no design guidelines at all are available—even for functions of no more than three variables.

Before proceeding to the networks themselves, a few remarks will be made about simplifying assumptions 3 and 6 in the above list. A common practice is to mount the circuits in physical packages having standard characteristics. In particular, each package may have, say, 14 terminals. Two or more circuit modules may be mounted in a package, and the manufacturing techniques may be such that the network cost is determined directly by the number of packages—not the number of circuit modules. Moreover, to minimize overall cost, including the cost of stocking spare parts, the number of different types of packages may be strictly limited.

With the assumed 14 terminals on each package, two would be required for power supply functions and 12 would be available for input and output signals. The types of packages might then be limited to a type A containing two five-input A-I (or O-I) modules, a type B containing three three-input modules, and a type C containing four two-input modules. This practice not only limits the fan-in but it also requires that the designer find the form of expression which best fits the modules available. Even when each function is generated by means of a single-output network, the most appropriate network for a given function may not be the same at one point in a system as for the same function at a different point in the system. The reason is that the associated networks at the other point may result in a different set of "left-over" modules available for generating the given function. This situation, which is actually encountered in practice rather commonly, seems to eliminate all hope of ever finding an orderly and practical design procedure for the general case. Nevertheless, an understanding of the available design techniques for the simple problems is helpful when searching for appropriate networks where the simplifying assumptions are not applicable.

Basic Two-Level A-I and O-I Networks

Here the term "level" is used in the sense that the number of "levels" in a network is equal to the maximum number of circuit modules through which any one input signal must pass to reach the output terminal. The meaning is similar to that used with AND-OR networks except that here each level represents an A-I or an O-I function sequence, and the circuit modules in all levels are of the same form (in contrast to the alternating AND and OR circuits of the AND-OR networks).

In a two-level network of A-I modules the input signals to the modules in the first level may be designated A_1, A_2, and so on for one module and B_1, B_2, and so on for the second module, and this notation may be extended to include all modules in the first level. The signals generated by these modules would then be $\overline{A_1 A_2 \cdots}$, $\overline{B_1 B_2 \cdots}$, and so on. The second and last level in a two-level network contains only one module, and it generates the following signal.

$$\overline{(\overline{A_1 A_2 \cdots})(\overline{B_1 B_2 \cdots}) \cdots} = A_1 A_2 \cdots + B_1 B_2 \cdots + \cdots$$

In other words, a two-level network of A-I modules is equivalent to an elementary AND-to-OR network, where the first-level modules correspond to the AND devices and the second-level module corresponds to the OR device. In fact, the two-level A-I network may be visualized as an AND-to-OR network with an inverted signal convention (that is, a relatively negative potential for a binary 1) being used for the intermediate signals. The equivalence of a two-level A-I network and an AND-to-OR network is illustrated in Figure 3-3a.

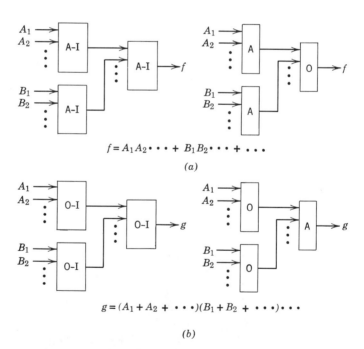

$f = A_1 A_2 \cdots + B_1 B_2 \cdots + \cdots$

(a)

$g = (A_1 + A_2 + \cdots)(B_1 + B_2 + \cdots) \cdots$

(b)

FIGURE 3-3. The relationship between two-level A-I and O-I networks and AND-OR networks.

In a two-level network composed of O-I modules the signals generated by the modules in the first level would be

$$\overline{A_1 + A_2 + \cdots}, \quad \overline{B_1 + B_2 + \cdots},$$

and so on, and the final output signal would be

$$\overline{(\overline{A_1 + A_2 + \cdots}) + (\overline{B_1 + B_2 + \cdots}) + \cdots}$$
$$= (A_1 + A_2 + \cdots)(B_1 + B_2 + \cdots) \cdots$$

as illustrated in Figure 3-3(b).

In the above networks, some of the signals in the sets A_i, B_j, and so on, may be duplicates of each other or inverted versions of each other. In any case, with all eight simplifying assumptions listed in the previous section, finding the minimum-cost network for a given function is a matter of finding the minimum number of basic terms as defined in Chapter 2, and the techniques described in that chapter are applicable.

A-I and O-I Networks for Applications Where the Inverted Forms of the Input Signals Are Not Available

As a first elaboration in the subject of A-I and O-I networks, consider the removal of simplifying assumption number 1. In applications where each input signal appears on only one line (not two) and where each signal appears in its direct (not inverted) form, a network containing only A-I modules or only O-I modules can be designed straightforwardly for any switching function by using the same two-level techniques except that modules are added to generate the needed inverted signals. The added modules are in simple inverter circuits, and they constitute a third level of switching. With n different input signals, as many as n added modules may be needed. (Actually, whether a given signal on a single input line is in direct or inverted form is merely a matter of the assignment of the symbol for that signal, but the expression for the function in question is affected accordingly.)

The network obtained in the elementary manner just described is not necessarily the minimum-cost network, however. Often, a single third-level module can be used to create the effect of inverting more than one input signal, although appropriate alterations may be needed in the second level of the network.

A thorough procedure has been worked out for finding the minimum number of A-I or O-I modules required in a network where no inverted input signals are available and where the number of levels is limited to two plus a third level for inverting. [See J. F. Gimpel, *IEEE Transactions on Electronic Computers*, February, 1967, pp. 18–38. For further discussion of the subject see K. K. Chakrabarti, A. K. Choudhury, and M. S. Basu, *IEEE Transactions on Computers*, June 1970, pp. 509–514]. However, the procedure is distressingly complex and tedious even when the number of input variables is only three. The practicality of the procedure is arguable for functions of only four or five variables and appears certainly impractical for six or more variables. Therefore, instead of describing this procedure, some patterns of input variables will be described for which a reduction in the number of third-level inverting modules can be realized. Finding these patterns admittedly requires some "cut-and-try" techniques, and the patterns do not necessarily lead to the exact minimum number of modules. On the other hand, the patterns do in fact lead to the minimum number for most (perhaps all) functions of a relatively few variables, and functions of many variables can be handled quite readily for at least a moderate saving in components.

The following presentation will be made for networks of A-I modules only. These networks correspond to the AND-to-OR networks, which as explained in the previous chapter, seem to be more "natural" and easily visualized. Duals of the patterns exist for networks of O-I modules, and corresponding O-I networks can be found just as readily from a mathematical standpoint, as will be discussed in more detail later.

One pattern is as follows, where the first form of the expression represents a conventional "sum of products" or AND-to-OR network, which can be realized in the manner of Figure 3-3a plus third-level inverters, and the last form of the expression indicates the input variables which appear in inverted form and which can be inverted in a single A-I module:

$$\bar{A}f(X_i) + \bar{B}f(X_i) + \bar{C}f(X_i) + \cdots \doteq (\bar{A} + \bar{B} + \bar{C} + \cdots)f(X_i)$$
$$= (\overline{ABC \cdots})f(X_i).$$

Here, $f(X_i)$ represents an AND function of any number of variables, any of which may appear in inverted form, but normally none of the X_i would be in the set represented by A, B, C, and so on. However, the equality is correct when $f(X_i)$ is any function of any variables. If $i = 0$, then $f(X_i)$ is assumed to be equal to 1.

As an example of the first pattern, consider the function

$$A\bar{B}\bar{C}D + A\bar{B}D\bar{E} + A\bar{B}D\bar{F} + G\bar{H}.$$

In this example $A\bar{B}D$ plays the role of $f(X_i)$ in the first three terms, and when this function is "factored" the result is

$$A\bar{B}D(\bar{C} + \bar{E} + \bar{F}) + GH = A\bar{B}D(\overline{CEF}) + G\bar{H}.$$

INVERTED INPUT SIGNALS NOT AVAILABLE   87

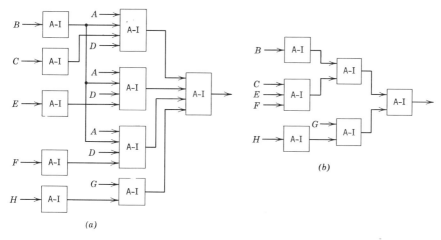

FIGURE 3-4. A-I networks for the function $A\bar{B}\bar{C}D + A\bar{B}D\bar{E} + A\bar{B}D\bar{F} + G\bar{H}$.

The networks formed with A-I modules in manners corresponding to the original and final expressions are shown in Figure 3-4a and b, respectively. A saving of four A-I modules has been achieved in this example.

A helpful technique for finding variables that fit the pattern is to write the expression as a column of terms and to use a 1 for each appearance of a variable in direct form, a 0 for each appearance in inverted form, and a dash (—) for each absence of a variable in a term. With this notation, the example of the previous paragraph becomes

A	B	C	D	E	F	G	H
1	0	0	1	—	—	—	—
1	0	—	1	0	—	—	—
1	0	—	1	—	0	—	—
—	—	—	—	—	—	1	0

Strips of paper may be used to cover (eliminate) individual columns and rows as desired in the search for patterns of variables. The last row, which corresponds to GH, may be eliminated to study the pattern of variables in the remaining terms. The columns corresponding to A, B, D, G, and H may then be eliminated because each variable appears in the same way (or does not appear at all) in each remaining term (row). The resulting pattern then is

C	E	F
0	—	—
—	0	—
—	—	0

where the appearance of one and only one 0 in each row and one and only one 0 in each column with dashes at all other positions reveals one of the sought patterns directly.

A second type of pattern which leads to a reduction in the required number of A-I modules is as follows:

$(\bar{A}BC \cdots N)f_1(X_i) + (A\bar{B}C \cdots N)f_2(X_i) + (AB\bar{C} \cdots N)f_3(X_i)$
$\quad + \cdots + (ABC \cdots \bar{N})f_n(X_i)$
$= (\bar{A} + \bar{B} + \bar{C} + \cdots + \bar{N})(BC \cdots N)f_1(X_i)$
$\quad + (\bar{A} + \bar{B} + \bar{C} + \cdots + \bar{N})(AC \cdots N)f_2(X_i)$
$\quad + (\bar{A} + \bar{B} + \bar{C} + \cdots + \bar{N})(AB \cdots N)f_3(X_i) + \cdots$
$\quad + (\bar{A} + \bar{B} + \bar{C} + \cdots + \bar{N})(ABC \cdots M)f_n(X_i)$
$= \overline{(ABC \cdots N)}[(BC \cdots N)f_1(X_i) + (AC \cdots N)f_2(X_i)$
$\quad + (AB \cdots N)f_3(X_i) + \cdots + (ABC \cdots M)f_n(X_i)]$

Here, each function $f_1, f_2, \cdots, f_n$ represents an AND function of the X_i variables, but the individual functions may differ from each other in that a given variable may appear in direct form in some functions, in inverted form in other functions, or not at all in still other functions. M is the next to the last variable in the set A through N.

An elementary example of this second pattern of variables is the expression $\bar{A}B + A\bar{B}$. No variables are in the set X_i so that f_1 and f_2 are each equal to 1. A brief study of the indicated pattern reveals that the given function may be expressed as $(\overline{AB})(B + A)$. The networks of A-I modules needed to generate the function in accordance with the two expressions are shown in Figure 3-5a and b, respectively.

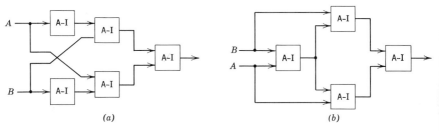

FIGURE 3-5. A-I networks for the function $A\overline{B} + \overline{A}B$.

The following function is a somewhat more complex and illustrative example of the second pattern type.

$B\bar{C}E + ABCD\bar{E}\bar{F} + A\bar{B}EF + \bar{B}C\bar{D}E$
$= B\bar{C}E + (BC\bar{E})AD\bar{F} + A\bar{B}EF + (\bar{B}CE)\bar{D}$
$= \overline{(BCE)}BE + \overline{(BCE)}ABCD\bar{F} + A\bar{B}EF + \overline{(BCE)}C\bar{D}E$

The first expression of the function indicates a need for five third-level inverters, one for each of $\bar{B}, \bar{C}, \bar{D}, \bar{E},$ and $\bar{F}$. The last expression indicates an A-I network requiring only four third-level inverting operations, one each for $\overline{BCE}, \bar{B}, \bar{D},$ and $\bar{F}$. Note that an inverter is needed for $\bar{B}$ in spite of its appearance in $\overline{BCE}$. This situation seems to be commonly encountered so that for randomly selected functions of many variables the attainable saving in modules is, on the average, substantially less than might be expected. This result is encountered with other component minimization techniques as well. Note further that the total number of inputs to the third-level modules has been increased from five to six in the course of eliminating one module, so that with some circuits or packaging conventions the "reduced" form of the network may actually be less desirable.

Finding the pattern in the above example may be aided by writing the original AND-to-OR expression in an array of 1's, 0's, and dashes as before.

	A	B	C	D	E	F			B	C	E
a	—	1	0	—	1	—		a	1	0	1
b	1	1	1	1	0	0		b	1	1	0
c	1	0	—	—	1	1		d	0	1	1
d	—	0	1	0	1	—					

Term c is eliminated (such as by covering the corresponding row with a strip of paper) by a "cut-and-try" process, although with a little practice, the inapplicability of c will be obvious to a designer. After the columns corresponding to variables A, D, and F are similarly eliminated, the resulting array is as on the right. The pattern is revealed by the presence of one and only one 0 in each column and row and the presence of 1's in all remaining positions.

For some functions, one or the other type of pattern will appear in two or more ways in the array of 1's, 0's, and dashes. In other instances, both types of patterns may appear, and the same set of input variables may be affected. An example of this last situation is the function

$$A\bar{B}\bar{C} + BC\bar{D} + A\bar{C}\bar{D} + \bar{B}CD$$

which may be represented as follows:

	A	B	C	D			B	D			B	D
a	1	0	0	—		a	0	—		b	1	0
b	—	1	1	0		c	—	0		d	0	1
c	1	—	0	0								
d	—	0	1	1								

the center array is obtained by eliminating terms b and d and columns A and C. The array on the right is obtained by eliminating terms a and c and the same columns. The resulting expression is

$$(\bar{B} + \bar{D})A\bar{C} + (B\bar{D})C + (\bar{B}D)C = (\overline{BD})A\bar{C} + (\overline{BD})BC + (\overline{BD})CD$$

The A-I networks corresponding to the original and final expressions are shown in Figure 3-6a and b, respectively.

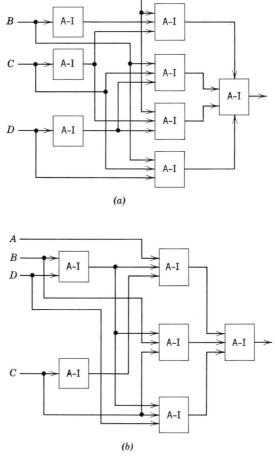

FIGURE 3-6. A-I networks for $\overline{AB}C + BC\bar{D} + A\bar{C}\bar{D} + \bar{B}CD$.

Additional Techniques for Eliminating Modules

In the previous section the assumption was implied that the given expression of each function was in minimum-component form from the standpoint of AND and OR devices. If the given expression is not in such form, it should be reduced to that form by the techniques described in Chapter 2. For those functions which have more than one minimum-component expression, each such expression should be examined in the search for ways to eliminate A-I or O-I modules in those instances where the inverted input signals are not available.

For example, in the last example in the previous section the given function, $A\bar{B}\bar{C} + BC\bar{D} + A\bar{C}\bar{D} + \bar{B}CD$, is in minimum-component form, but the function has two alternative minimum-component forms, $A\bar{B}\bar{C} + BC\bar{D} + AB\bar{D} + \bar{B}CD$ and $A\bar{B}D + BC\bar{D} + A\bar{C}\bar{D} + \bar{B}CD$, which should be similarly studied in an effort to eliminate modules. However, in this example, these alternative forms do not happen to lead to networks as economical as the one found previously.

A more seriously complicating factor is that all basic term sets that are equal to the given function should likewise be studied in the search for ways to find the minimum number of A-I modules required in a network that generates the given function.

Moreover, third-level inverting modules can sometimes be combined by adding ineffectual input signals to some modules so that the input signal sets to two or more such modules are the same.

Both of the above techniques are illustrated by the simple example $A\bar{B} + B\bar{C}$, which would require five A-I modules if generated straightforwardly as indicated in Figure 3-7(a). As may be observed readily by plotting it on a diagram, this function has three basic terms $A\bar{B}$, $B\bar{C}$, and $A\bar{C}$, so that it is equal to $A\bar{B} + B\bar{C} + A\bar{C}$, where $A\bar{C}$ is superfluous. The following array of 1's, 0's and dashes may be used to represent the function

A	B	C
1	0	—
—	1	0
1	—	0

If the second term (row) is temporarily eliminated, the first and third terms are seen to generate a pattern of the first type so that

$$A\bar{B} + B\bar{C} = A(\bar{B} + \bar{C}) + B(\bar{C}) = A(\overline{BC}) + B(\bar{C})$$

No reduction has yet been achieved in the number of modules required. However, input signal B may be added as an ineffectual input to the module which generates $\bar{C}$ in the second term. That is, the second term

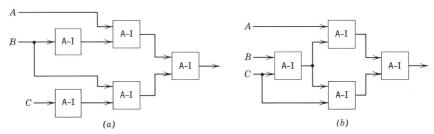

Figure 3-7. A-I network for $\overline{A\bar{B}} + \overline{B\tilde{C}}$.

then becomes $B(\bar{B} + \tilde{C}) = B(\overline{BC})$ which is equal to $B(\tilde{C})$ because $B\bar{B} = 0$. With this added input signal, the expression becomes

$$A\bar{B} + B\tilde{C} = A(\overline{BC}) + B(\overline{BC})$$

and a single A-I module can be used for generating BC. A saving of one module has been achieved as illustrated in Figure 3-7b.

The function $A\bar{B} + B\tilde{C} + B\bar{D}$ is a slightly more complex example which illustrates the techniques. This function has two other basic terms, $A\tilde{C}$ and $A\bar{D}$, both of which are superfluous. When these two terms are included the function may be represented as follows:

	A	B	C	D			B	C	D			C	D
a	1	0	—	—		a	0	—	—		b	0	—
b	—	1	0	—		d	—	0	—		c	—	0
c	—	1	—	0		e	—	—	0				
d	1	—	0	—									
e	1	—	—	0									

Terms a, d, and e generate a pattern of the first type on variables B, C, and D as shown in the center array. Terms b and c generate another pattern of the first type, but only on variables C and D as shown in the right-hand array. The function may be written as

$$A(\bar{B} + \tilde{C} + \bar{D}) + B(\tilde{C} + \bar{D}) = A(\bar{B} + \tilde{C} + \bar{D}) + B(\bar{B} + \tilde{C} + \bar{D})$$
$$= A(\overline{BCD}) + B(\overline{BCD})$$

By adding the ineffectual input signal B the number of A-I modules required has been reduced from seven to four in this example.

The above two examples can be generalized in the following manner:

$$f(A_i)\bar{B} + B\tilde{C} + B\bar{D} + B\bar{E} + \cdots = f(A_i)\bar{B} + B(\tilde{C} + \bar{D} + \bar{E} + \cdots)$$
$$= f(A_i)\bar{B} + B(\overline{CDE \cdots})$$
$$= f(A_i)(\overline{BCDE \cdots})$$
$$+ B(\overline{BCDE \cdots}),$$

where $f(A_i)$ is any function (but probably only an AND function in most instances) of any variables and where the last expression is obtainable directly from the relationship $A\bar{B} + B\bar{C} = A(\overline{BC}) + B(\overline{BC})$, with (A_i) playing the role of A and with $CDE \cdots$ playing the role of C.

For another example of using superfluous terms and ineffectual input signals for eliminating modules, consider the function

$$A\bar{B}\bar{D} + \bar{A}B\bar{D} + \bar{A}\bar{B}D + B\bar{C}\bar{D}$$

which is expressed in a minimum-component form from the standpoint of AND and OR devices. This function has another basic term, $A\bar{C}\bar{D}$, which when added to the given expression yields the following array:

	A	B	C	D			B	C			A	C
a	1	0	—	0		a	0	—		b	0	—
b	0	1	—	0		e	—	0		d	—	0
c	0	0	—	1								
d	—	1	0	0								
e	1	—	0	0								

Terms a and e generate a pattern of the first type on variables B and C, and terms b and d generate a similar pattern on A and C. The function may then be expressed as

$$A\bar{D}(\bar{B} + \bar{C}) + B\bar{D}(\bar{A} + \bar{C}) + \bar{A}\bar{B}D$$
$$= A(\bar{A} + \bar{D})(\bar{A} + \bar{B} + \bar{C}) + B(\bar{B} + \bar{D})(\bar{A} + \bar{B} + \bar{C})$$
$$+ (\bar{A} + \bar{D})(\bar{B} + \bar{D})D$$
$$= A(\overline{AD})(\overline{ABC}) + C(\overline{BD})(\overline{ABC}) + (\overline{AD})(\overline{BD})D$$

The ineffectual input signals are added as possible and as needed to match the sets of signals to be inverted. In this example, all possible such signals are needed. The function as originally expressed would require one first-level A-I module, four second-level modules, and four third-level modules, with the latter being required to generate $\bar{A}, \bar{B}, \bar{C}$, and $\bar{D}$. In the final expression the total number of modules has been reduced from nine to seven, where the seven include one first-level module, three second-level modules, and three third-level modules. In this example the third-level modules generate $\overline{AD}, \overline{ABC}$, and $\overline{BD}$.

Factoring

Factoring is another technique for finding ways of eliminating modules when the input signals are not available in inverse form. However, with A-I or O-I modules the nature of the technique is very different from that described in the previous chapter for AND and OR circuits. For

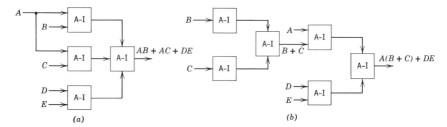

FIGURE 3-8. Example of the effect of "factoring" when using A-I modules and when input signals are not available in inverted form.

example, the function $AB + AC + DE$ may be realized with nine diodes in a two-level circuit, but by factoring the A from the first two terms to yield $A(B + C) + DE$, only eight diodes would be needed although the circuit becomes of the three-level OR-AND-OR type. With A-I modules, only four are required if the function is generated in accordance with the original expression, but six are required when A is factored. Moreover, the network becomes four-level as illustrated in Figure 3-8.

In general, to achieve a reduction in the number of modules, the function must be such that the portion of the expression remaining after factoring contains none or very few variables that appear in inverse form. This remaining expression should then be transposed to OR-to-AND form, and the result is examined to determine whether or not an improvement has in fact been attained. For example, the function

$$AB\bar{C} + AB\bar{D}\bar{E} + \bar{F}$$

requires six A-I modules if generated straightforwardly (note that no second-level or third-level modules are required for $\bar{F}$), but AB in the first and second terms may be factored to yield

$$AB(\bar{C} + \bar{D}\bar{E}) + \bar{F} = AB(\bar{C} + \bar{D})(\bar{C} + \bar{E}) + \bar{F} = AB(\overline{CD})(\overline{CE}) + \bar{F}$$

The last expression requires only four A-I modules as illustrated in Figure 3-9.

The transforming of $\bar{C} + \bar{D}\bar{E}$ to its OR-to-AND equivalent $(\bar{C} + \bar{D})(\bar{C} + \bar{E})$ in the above example is obvious to a person familiar with Boolean manipulation, although the more thorough procedure described in the previous chapter may be used. In a sense, the transformation of expressions of this complexity is the dual of the "factoring" that is analogous to the factoring in conventional algebra. Regardless of the viewpoint taken with respect to the process, the following example shows how

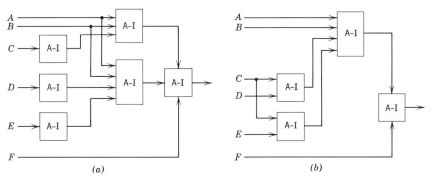

FIGURE 3-9. Module elimination by "factoring" $AB\bar{C} + AB\bar{D}\bar{E} + \bar{F}$.

a reduction in the number of modules can sometimes be achieved when most, but not all, of the variables appear in inverted form:

$$\bar{A}E + \bar{B}\bar{C}DE + \bar{D}F = E(\bar{A} + \bar{B}\bar{C}D) + \bar{D}F$$
$$= E(\bar{A} + \bar{B})(\bar{A} + \bar{C})(\bar{A} + D) + \bar{D}F$$
$$= E(\overline{AB})(\overline{AC})(\overline{A\bar{D}}) + \bar{D}F$$

The original and final expressions of the function require eight and seven A-I modules, respectively, as illustrated in Figure 3-10. A fourth-level module is used for D to make it appear in inverted form at the inputs to the third level. Note also that the output from this fourth-level module for D is applied as an input to a second-level module as well as to a third-level module. Making effective use of techniques such as this admittedly requires some experience and insight on the part of the designer.

"Don't Care" Conditions with Networks of A-I or O-I Modules

If inverted input signals are available, the handling of "don't care" conditions is accomplished in exactly the same manner with networks of A-I or O-I modules as with networks of AND and OR devices. However, if inverted input signals are not available, the need to consider all possible sets of basic terms in the search for patterns that will allow the elimination of modules causes the handling of "don't care" conditions to become a far more tedious task.

In most instances a reasonable expression to use as a starting point in the search would be that expression which is derived by utilizing the particular "don't care" conditions that result in the minimum-component AND-OR network (or all such expressions in instances where more than

96 NETWORKS CONTAINING INVERTERS

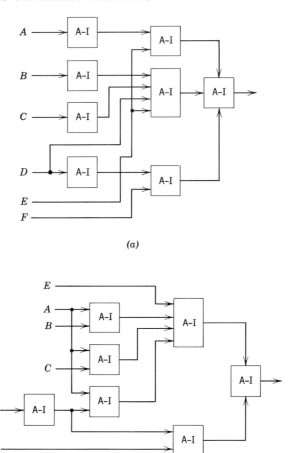

FIGURE 3-10. A-I networks for $\bar{A}E + \bar{B}\bar{C}DE + \bar{D}F$.

one exists). However, such an expression does not always lead to the minimum number of A-I or O-I modules. For example, consider the previously cited function $A\bar{B}\bar{D} + \bar{A}B\bar{D} + \bar{A}\bar{B}D + B\bar{C}\bar{D}$ (page 93) but with the added specification that $\bar{A}\bar{B}\bar{C}\bar{D}$ is a "don't care" condition. By utilizing this condition, input variable B in the last term becomes superfluous (as can be shown easily by plotting the function on a diagram), and one diode would be saved if the function were being generated by diode AND-OR circuits. However, when A-I modules are employed in a straightforward network, nine modules would be required whether or not the "don't care" condition is utilized. If the superfluous B in the last term is eliminated, the pattern of variables is disrupted so that no scheme is apparent for eliminating any of the nine—although, as already shown, the function can be generated by means of only seven modules. Discerning that the "don't care" condition should not be utilized in problems such as this requires a high degree of ingenuity or patience.

Networks of O-I Modules

Because of the "dual" relationship that exists between A-I and O-I modules as illustrated in Figure 3-2, essentially the same techniques may be used for finding suitable networks of O-I modules as were used for A-I modules. The major difference is that each function in question should be expressed in OR-to-AND form rather than AND-to-OR form. For minimizing the number of O-I modules, pattern-indicating equations analogous to those described for A-I modules can be derived—and deriving them would be a good exercise for the reader. However, present purposes will be served by illustrating the finding of an O-I network that generates the following function as an example:

$$AB + \bar{A}\bar{B} + A\bar{C}D + BCD$$
$$= (A + \bar{B} + C)(\bar{A} + B + \bar{C})(A + \bar{B} + D)(\bar{A} + B + \bar{D})$$

The function might have been originally expressed in OR-to-AND form as in the right-hand expression, or it might have appeared in AND-to-OR form as in the left-hand expression. In the latter case, the techniques described in the previous chapter would be needed to find the OR-to-AND expression. In this example, the OR-to-AND expressions include only one minimum-component version (although the AND-to-OR expressions include four such versions). If the variables in the OR-to-AND expression are designated by 1's, 0's, and dashes as before, the function may be represented as follows:

	A	B	C	D			A	B			A	B
a	1	0	1	—		a	1	0		c	1	0
b	0	1	0	—		b	0	1		d	0	1
c	1	0	—	1								
d	0	1	—	0								

Two patterns of the second type, both respect to variables A and B, are found as indicated by the center and right-hand arrays. The function may then be expressed as follows, where the expression is obtained in exactly the same manner as was used for A-I modules except that the AND and OR functions are interchanged.

$$\frac{(\overline{A+B}+A+C)(\overline{A+B}+B+\bar{C})}{(\overline{A+B}+A+D)(\overline{A+B}+B+\bar{D})}$$

Because the inversion of A and B is accomplished in a single module, only eight modules are required in comparison with the nine that would be required in a straightforward realization of the function.

However, by temporarily eliminating terms a and c from the array, a pattern of the first type is found as follows:

$$\begin{array}{ccc} & C & D \\ b & 0 & — \\ d & — & 0 \end{array}$$

This pattern leads directly to the expression

$$(A+\bar{B}+C)(A+\bar{B}+D)(\overline{C+D}+\bar{A}+B)$$

which can be implemented with only seven O-I modules as shown in Figure 3-11.

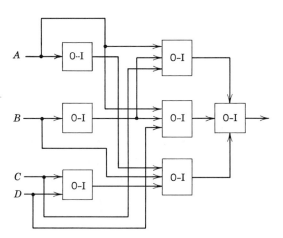

FIGURE 3-11. O-I network for $AB + \overline{AB} + A\overline{C}D + BCD$.

Networks of A-I or O-I Modules Having Limited Fan-In and Fan-Out

If the number of inputs to a module (the "fan-in") is limited or if the number of other modules a given module is capable of actuating (the "fan-out") is limited, the networks as derived previously in this chapter must be modified accordingly. Although a network that generates any given function can always be derived quite easily, finding the particular network which requires the minimum number of modules is an astonishingly difficult problem. Virtually no guidelines are known. Even the computer-assisted studies of all possible networks have been extremely limited in this aspect of the problem.

In view of the above situation, this presentation will be restricted to showing one straightforward method of finding networks that will perform any desired function for any reasonable limitation on fan-in and fan-out. The most stringent possible case is where the fan-in and fan-out are each limited to two. (A fan-in of one would imply the impossibility of generating the AND and OR functions at all, and a fan-out of one would imply the need for amplifiers in addition to the A-I or O-I modules.)

Figure 3-12a shows an array of A-I modules, each having a fan-in of two, which generates a three-input A-I function. The array is assembled simply by connecting two modules in tandem to one input of a third module. With the input signals designated A, B, and C as shown, the output signal is $\overline{ABC}$. The second input terminal to the second module is not used, although a signal X is indicated at this terminal, as shown by the dotted line. If such a signal were actually applied, the final output signal would be

$$A(\overline{\overline{BC}})X = \overline{ABC} + X,$$

which is something other than an A-I function with a fan-in of four. A fan-in of four can be obtained, however, by adding two more modules in tandem in any one of the input lines to which A, B, or C is connected. Most likely, line A would be selected, because the maximum number of levels would then not be any greater than the three levels required

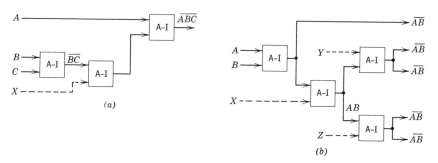

FIGURE 3-12. Adding modules to increase effective fan-in and fan-out.

for a fan-in of three, whereas if the modules were added in the B or C lines, the number of levels would become five. On the other hand, in a given application, signal A might always be the last one to change, in which case adding the modules in the B or C lines may be preferable because the speed of the circuit would then be dependent only on the one level through which A must pass, and the five levels through which certain other input signals must pass would not be detrimental to over-all system speed.

In general, the simulation of a module with a fan-in of n requires $2n - 3$ modules having an actual fan-in of two. If the modules have an actual fan-in of three, the effective fan-in is increased by two for each two modules that are added. With modules having an actual fan-in of four, the effective fan-in is increased by three for each two added modules, and so on.

The fan-out situation is slightly different as illustrated in Figure 3-12b for modules having a fan-out of two. Two modules in tandem can be added in one output line to create an effective fan-out of three, but an effective fan-out of five can then be obtained by the addition of only one more module as shown. Input signals X, Y, and Z would not be utilized if the objective is limited to increasing the fan-out from the first module (the one on the left), but they may be utilized in the generation of new functions. For example, if Y is utilized, the signal appearing at the two output lines from the corresponding module would be $\overline{ABY}$.

Adding one more module to the network of Figure 3-12b would not be useful in increasing fan-out, but adding two more (for a total of six) would allow the simulation of a module having a fan-out of six. Adding three modules in a pyramid arrangement in any one of the five output lines in the figure simulates an A-I module having a fan-out of eight. Further extensions are generated analogously. Of course, if the modules have an actual fan-out of three or greater, the adding of modules in the output lines produces a correspondingly greater increase in the effective fan-out.

The A-I network for any given function can be devised straightforwardly by first finding a network by the methods described earlier in this chapter and by then adding modules in series in the input, intermediate, and output lines as needed to simulate modules with the required fan-in and fan-out at each point in the network.

For an illustration, the network for generating $A\bar{B}\bar{C} + BC\bar{D} + A\bar{C}\bar{D} + \bar{B}CD$ as shown in Figure 3-6b is redrawn in Figure 3-13 for A-I modules having a fan-in of two and a fan-out of two. The resulting network is obviously unattractive, both from the standpoint of requiring three times as many modules and from the standpoint of requiring many

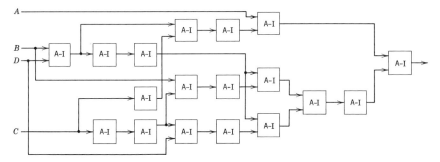

FIGURE 3-13. A-I network for $\overline{ABC} + \overline{BCD} + \overline{ACD} + \overline{BCD}$ where each module is limited to a fan-in of two and a fan-out of two—compare with Figure 3-6 b.

more levels of switching. However, no technique is obvious for finding an improved network, although the fact that the required fan-in or fan-out is only one for some modules is probably a hint that improved networks exist.

Of course, two is probably unrealistically low for the attainable fan-in and fan-out of most switching components and circuits, although some examples of magnetic core circuits, tunnel diode circuits, and one-transistor RTL circuits could be cited where the limitations were almost this stringent. Fortunately, most transistor-diode switching modules have reasonably high fan-in and fan-out ratios, and Figure 3-13 therefore serves primarily as an academic illustration.

Fan-in and fan-out problems are handled in essentially the same way with O-I modules.

Networks Having More Than Three Levels

Although a maximum of three levels of A-I or O-I switching is sufficient to generate any switching function, two examples involving more than three levels have already been presented: the network of Figure 3-10b for reducing the number of modules required and the network of Figure 3-13 for allowing the use of modules having limited fan-in and fan-out characteristics. The use of more than three levels is forced on the designer in situations of the latter type (limited fan-in and fan-out), but the designer might wonder about the magnitude of the economies attainable when considering networks of more than three levels for other reasons. No meaningful statistics on this point are known, but experience indicates that examples of the type represented by Figure 3-10b are rare in comparison with the total number of different switching functions that might be encountered. Instead, in the vast majority of instances, the network with the fewest modules seems to be obtained when the number of levels is exactly three (or two for functions for which no variables appear in inverted form, or one for the trivial case of an AND or OR function followed by an inversion).

Networks Where Cost is Realistically Measured

Although a 10-input A-I or O-I module might cost little or no more than a two-input module when all modules are separately manufactured and packaged, the modules are seldom manufactured and packaged separately. Instead, the physical package configuration is likely to be standardized so that a given package type may contain several two-input modules or just one or two 10-input modules, for example. Alternatively, the modules are sometimes designed with a standard number, say three, of input lines with provision for adding input lines by adding diodes, which may be packaged separately.

Widely differing variations in the cost situation are encountered when the components are manufactured on a large scale integration (LSI) basis where many, perhaps hundreds or more, modules are placed in a single package. Here, the total area consumed by all components, including the lines interconnecting the switching modules, may be the controlling cost factor.

In any of the above instances, cost is not simply proportional to the number of modules but is a function of both the number of modules and the total number of inputs to the modules and may also be a function of the interconnection pattern. Moreover, the interconnection pattern may be dependent on various physical characteristics of the components as well as on the required electrical connections to be made. In view of considerations such as these (and there are others), any detailed generalized analysis of cost minimization on a realistic basis is quite impractical, but for any given switching function the network with the fewest number of modules is probably the best network to use as a starting point in the search for the minimum-cost network even when cost is determined in some other way.

Multi-Output Networks of A-I or O-I Modules

With inverted input signals available and with no limits on fan-in or fan-out, the multi-output problem with networks of A-I or O-I modules is essentially the same as described in the previous chapter for networks of AND and OR circuits. If inverted input signals are not available, the multi-output problem is still the same if a separate inverter is provided for each input signal that needs to be inverted. However, minimizing the number of modules in a multi-output network when inverted input signals are not available would appear to be an almost hopelessly complex problem if any generalized solution is attempted, and no studies of this problem are known. (See pages 67 and 68 for one important aspect of the problem.)

A recently discovered fact about multi-output networks of A-I or O-I modules is that minimization of the number of modules is not necessarily achievable with loop-free networks, which are the topic of this chapter. In digital networks, feedback loops ordinarily serve the purpose of providing multiple stable states (digital storage) or sequential operation, as will be discussed in the next chapter. However, for at least a few multi-output combinatorial functions, feedback loops can be utilized to reduce the number of modules required. Generalized or practical methods have not yet been devised for employing loops to achieve minimization. In fact, at present the usefulness of loops appears limited to a relatively few highly specialized functions, but the discovery nevertheless has a profound effect on the theoretical aspects of minimization problem. For further information, reference is made to a paper by W. H. Kautz [*IEEE Transactions on Computers*, February 1970, pp. 162–164].

Networks of A-O-I Modules

If the collectors of two DTL modules containing NPN transistors are connected to a common load as indicated in Figure 3-14a, the resulting circuit will generate an A-I-A function as illustrated in Figure 3-14b. Again the convention of 1's and 0's being represented by relatively positive and negative potentials, respectively, is assumed. By using Boolean techniques as employed in connection with Figure 3-1, this A-I-A function is readily found to be equivalent to the A-O-I function illustrated in Figure 3-14c. The circuit module in *a* may therefore be visualized as being of A-O-I form. With other circuits, the A-O-I nature is even more readily apparent. For example, a single transistor may be preceded by two or more diode AND circuits as in Figure 3-14d where one of the bias-shifting diodes can be duplicated as necessary to form a sort of diode OR circuit.

With A-O-I modules, a single module having a suitably large fan-in at each AND circuit and at the OR circuit is capable of generating any switching function provided the inverse of each input signal is avail-

104 NETWORKS CONTAINING INVERTERS

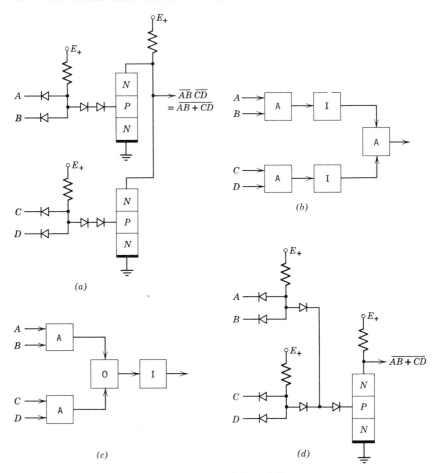

FIGURE 3-14. A-O-I modules.

able as needed. The appropriate connections may be found by determining those elemental terms for which the output signal is to be 0. The selection of the particular basic terms to be utilized is performed in exactly the same manner as with diode AND-to-OR circuits. "Don't care" conditions are handled essentially as before, except that a given "don't care" condition is included or not in accordance with whether it simplifies the expression of the inverse function, not the original function.

For example, if the given function is $AB\bar{C} + \bar{A}\bar{B}C + \bar{A}CD + \bar{B}CD$ with "don't care" conditions $ABCD$, $\bar{A}BC\bar{D}$, $\bar{A}B\bar{C}D$, and $\bar{A}\bar{B}\bar{C}\bar{D}$, all as plotted in Figure 3-15a, the desired expression is obtained by grouping

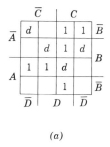
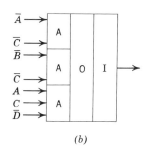

(a)   (b)

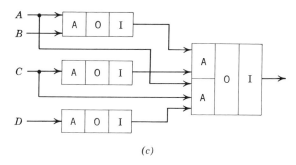

(c)

FIGURE 3-15. A typical switching function and A-O-I modules for realizing the function.

all 0-containing squares plus as many as needed of those marked d. The function may be expressed as $\overline{A\bar{C}} + \overline{B\bar{C}} + AC\bar{D}$, and the one-module network for generating this function is shown in Figure 3-15b.

If inverted input signals are not available, the straightforward realization of the above function would require four additional modules for inverting A, B, C, and D. The same techniques as described previously for networks of A-I modules are useful for finding ways to eliminate some of these added modules. In the example just cited, the first two terms form a pattern of the first type so that the function may be written as $\overline{(AB)}\bar{C} + AC\bar{D}$, and this expression may be realized with a total of four modules as illustrated in Figure 3-15c.

Networks of O-A-I Modules

O-A-I modules can be obtained by using PNP transistors instead of NPN transistors or by inverting the convention with respect to the relative polarity of signals representing 1's and 0's. (Changing both the

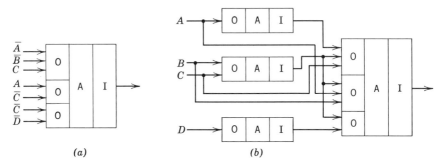

FIGURE 3-16. O-A-I networks for the function of Figure 3-15a.

transistor type and the polarity convention maintains the A-O-I character of the modules.)

Determining the proper O-A-I module connections is a matter of finding the OR-to-AND representation of the inverse function. Because, as discussed in the previous chapter, obtaining the OR-to-AND representation of a given function is accomplished through noting those terms for which the function is 0, the inverse is obtained by noting those terms for which the function is 1. Thus, for example, the function plotted in Figure 3-15a is $AB\bar{C} + \bar{A}C + CD$, where two of the four "don't care" terms are utilized. This expression equals

$$\overline{AB\bar{C} + \bar{A}C + CD} = (\bar{A} + \bar{B} + C)(A + \bar{C})(\bar{C} + \bar{D})$$

and the resulting "network" of one O-A-I module is shown in Figure 3-16a. Again, four additional inverting modules are required for a straightforward realization of the function if inverted input signals are not available. However, one module can be saved by using terms $\bar{A}\bar{B}C$ and BD instead of $\bar{A}C$ in the expression to produce (with different "don't care" terms being utilized)

$$\overline{AB\bar{C} + CD + \bar{A}\bar{B}C + BD}$$
$$= (\bar{A} + \bar{B} + C)(\bar{C} + \bar{D})(A + B + \bar{C})(\bar{B} + \bar{D})$$

When the notation described previously is used, the function may be represented as follows:

	A	B	C	D			B	C			B	C
a	0	0	1	—		a	0	1		b	—	0
b	—	—	0	0		c	1	0		d	0	—
c	1	1	0	—								
d	—	0	—	0								

The expression then becomes

$$(\overline{B + C} + \bar{A} + C)(\overline{B + C} + A + B)(\overline{B + C} + \bar{D})$$

and the corresponding network of O-A-I modules is shown in Figure 3-16b.

Another Aspect of A-O-I and O-A-I Modules

With some types of A-O-I modules the number of 1-signals simultaneously applied to the OR part of the module should be limited, and the limit is one in some cases. The circuit of Figure 3-14d is an example. The amount of current supplied to the base of the transistor is approximately proportional to the number of 1-signals generated at the outputs of the various diode AND circuits. Excessive current may damage the transistor or at least cause an unwanted delay in the change to the cut-off condition when the input signals change. Another example is found in certain current-mode switching circuits, which are the topic of the next section.

Limiting the number of 1-signals to the OR part is a matter of selecting the terms so that no elemental term is included more than once. The selected terms are not necessarily basic terms. Instead, the term selection procedure is largely a matter of: first, selecting any one-variable term that might be appropriate; second, selecting as many two-variable terms as possible whereby no elemental term is included in any two terms so far selected; third, similarly selecting as many three-variable terms as possible; and so on. The process is then repeated for each alternative choice that might have been made at any step.

For example, consider the function plotted in Figure 3-17a, b, and c where (because of the inversion in an A-O-I module) each 0-containing square must be included, but the inclusion of any d-containing square is optional. No one-variable terms are found, but this function has three two-variable terms, BD, $\bar{A}C$, and CD. However, because these terms overlap, only one can be selected. The selection of each with the corresponding subsequent selections is indicated in a, b, and c, respectively. The term set represented by a is undesirable because it corresponds to a relatively elaborate A-O-I module. The merits of the term set corresponding to b are substantially the same as the merits of the set corresponding to c. The A-O-I module for the c set is shown in Figure 3-17d. If inverted input signals are not available, the same techniques as described previously may be used in the search for ways of minimizing the number of inverting modules.

An analogous problem is encountered with O-A-I modules, but then

108 NETWORKS CONTAINING INVERTERS

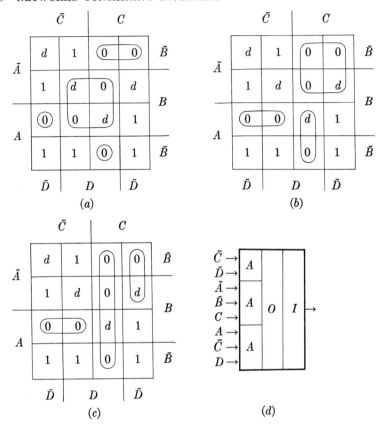

FIGURE 3-17. An illustration of the problem where only one 1-signal can be applied at a given time to the OR part of an A-O-I module.

the number of 0-signals to the AND part of the module is limited. Visualizing the problem may be somewhat more difficult, although essentially the same solution is applicable.

Modules Containing Transistors in Current-Mode Circuits

From a switching function standpoint, modules containing transistors in current-mode circuits are characterized primarily by the fact that, at very little additional cost, an output signal representing the direct (not inverted) OR or AND function can be provided. This output signal is in addition to the output signal which represents an A-I or an O-I function as already discussed at some length. With NPN transistors the function is O-I or A-I according to whether each 1 is represented by a relatively positive or a relatively negative potential, respectively. With PNP transistors the O-I and A-I roles are reversed for a given convention with regard to signal polarity. To reach the output terminal which corresponds to the direct OR or AND function, as the case may be, each input signal must pass through two transistors in sequence. The two transistors are needed to obtain the current-switching or "current-mode" effect. Although the circuit is not the same as two conventional inverters in tandem, the time required to generate the direct output signal is nevertheless dependent on the time required to change one transistor from the conducting to the cut-off condition and to change the other transistor from the cut-off to conducting condition, where these changes occur sequentially.

Certain obvious simplifications in network design result when both the O-I and OR (or both the A-I and AND) functions are obtainable from a single module. For one thing, if the signals to a given network are being supplied by other networks of current-mode modules, the inverted version of each input signal will be available, and the techniques described in earlier sections for minimizing the number of third-level inverters become unnecessary.

Also, reasonably obvious economies can be achieved with factorable functions such as $ABC + ABD + ABE + ABF$ which can be expressed as $AB(C + D + E + F)$, although an increase from two levels to three levels of switching is required when modules of only one type (as to NPN and PNP transistors) are available.

On the other hand, for most miscellaneous functions where two levels of switching are necessary and sufficient, the availability of the direct signals between the second and first levels is of no great value, and the design problem is essentially the same as described previously for two-level networks of O-I modules or A-I modules. About the only significant difference is that, for any given switching function, an alternative network exists. The alternative network requires the same amount of equipment, but it has the advantage that its functioning is more readily visualized. For example, consider the function $AB + C\bar{D}$ generated by current-mode modules as shown in Figure 3-18 for modules which generate the O-I and OR functions. The first-level module (the one on the right) is used to generate the OR function directly without inversion. The second-level modules generate the terms AB and $C\bar{D}$ straightforwardly, although the relationships $\overline{\bar{A} + \bar{B}} = AB$ and $\overline{\bar{C} + D} = C\bar{D}$ are utilized by inverting all input signals and employing the O-I functions of each second-level module.

110   NETWORKS CONTAINING INVERTERS

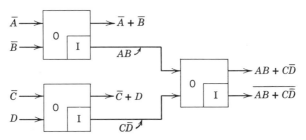

FIGURE 3-18. Network of current-mode modules.

Current-mode transistor circuits can also be expanded to create modules of A-O-I and O-A-I form, in which case the subject matter of the immediately preceding section applies. Such current-mode circuits, together with solutions to example switching function problems, are described in Richards' book *Electronic Digital Components and Circuits*, pp. 102–108.

Using Multiplexers to Generate Miscellaneous Switching Functions

A multiplexer is a device that transmits the signal from a selected one of a number of input lines to an output line. The selection is under the control of a set of binary control signals, and with n control signals, any one of up to 2^n input signals may be selected. Both analog and digital forms of multiplexers exist. A digital multiplexer is simply a single-output switching network, and the basic elements of a four-way multiplexer are illustrated in Figure 3-19a. The input signal on line M_0, M_1, M_2, or M_3 is transmitted to the output line in accordance with whether $\bar{X}\bar{Y}$, $\bar{X}Y$, $X\bar{Y}$, or XY, respectively is 1, where X and Y are the control signals. Figure 3-19b shows the symbol that will be used for the four-way multiplexer.

A four-way multiplexer is capable of generating any Boolean function of three binary variables in the following way. Two of the signals are connected to the X and Y control lines and for each of the four possible combinations of these two signals, the signal needed at the output line is connected to the corresponding multiplexed input line. In each instance, the signal needed at the output line may be either 1, 0, the third signal, or the inverse of the third signal.

The technique will be illustrated by using a four-way multiplexer to generate the function $\bar{A}\bar{B} + A\bar{B}\bar{C} + ABC$. (The appearance of A in the second term is superfluous in this expression, but this fact is substantially

USING MULTIPLEXERS TO GENERATE SWITCHING FUNCTIONS   111

irrelevant in this application.) Variables A and B happen to appear in each term of the expression and are therefore convenient signals to connect to the X and Y control lines, as shown in Figure 3-19c. By inspection of the expression, the output signal is seen to be 1 when $\bar{A}\bar{B}$ is 1. Therefore, a continuous 1 signal is connected to the M_0 input line, as shown in the figure. The combination $\bar{A}B$ does not appear in any of the terms, and because A and B (sometimes inverted) appear in all terms, the conclusion is that when $\bar{A}B$ is 1, the output signal should be 0. A continuous 0 signal is therefore connected to the M_1 input line. When $A\bar{B}$ is 1 the output signal should be 1 or 0 according to whether $\bar{C}$ is 1 or 0, respectively, because of the term $A\bar{B}\bar{C}$. Therefore, $\bar{C}$ is the correct signal to connect to the M_2 input line. Similarly, when AB is 1, the output signal is 1 or 0 according to whether C is 1 or 0, respectively, and C is therefore connected to the M_3 input line. The desired function is thus generated.

Note that because any one of four different signals, 0, 1, C, or $\bar{C}$, could have been connected to any one of the four multiplexed input lines, the total number of different signal combinations is $4^4 = 256$, which is

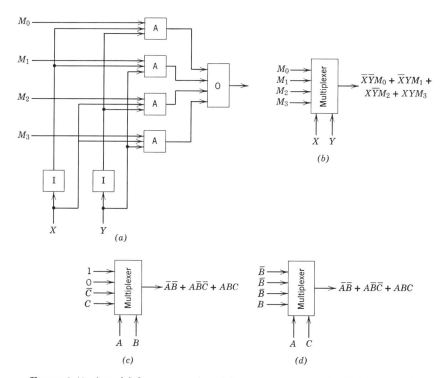

FIGURE 3-19. A multiplexer network and its use as a universal switching module.

exactly the number of different three-variable functions, as discussed in Chapter 1. In other words, each one of these different multiplexed signal combinations corresponds to a different three-variable switching function.

Any two of the three signals could have been selected for the control signals, but the character of the multiplexed signal pattern is strongly dependent on which two are chosen. Suppose in the above example of $\bar{A}\bar{B} + A\bar{B}\bar{C} + ABC$ that signals A and C are connected to the X and Y control lines as shown in Figure 3-19d. The determination of the correct signals to be connected to the multiplexed input lines is not quite so obvious but nevertheless may be made straightforwardly if the first term in the expression is "multiplied" by $\bar{C} + C = 1$ to produce $\bar{A}\bar{B}\bar{C} + \bar{A}\bar{B}C + A\bar{B}\bar{C} + ABC$. Signals A and C (sometimes inverted) now appear in each term, and the procedure described above may be used to determine that the correct signals to connect to the M_0, M_1, M_2, and M_3 input lines are $\bar{B}$, $\bar{B}$, $\bar{B}$, and B, respectively, as indicated in the figure.

By using the same technique an eight-way multiplexer having three control lines may be used to generate any Boolean function of four variables. A 16-way multiplexer having four control lines may be used to generate any five-variable function, and so on.

Incidentally, in the popular 5400/7400 series of integrated circuit modules, which are supplied by most manufacturers in the business, type number 152 (that is, 54152 or 74152 where the primary difference is in ambient temperature ratings) is an eight-way multiplexer. Type number 151 is the same but it also provides the inverted output signal on an added output line, and it further provides for a "strobe" input signal which, when 0, nullifies the effect of all other input signals. Type 150 is a 16-way multiplexer, also having a strobe input signal, but not an inverted output signal. Because of the ready availability of these multiplexer modules, economy of design is often achieved by abandoning the use of individual A-I or O-I modules even though the number of AND, OR, and inversion functions in the multiplexers may be much greater than the minimum otherwise required.

When multiplexers are used for miscellaneous switching functions in this way, they are sometimes called "universal" switching modules.

Multiple Levels of Multiplexers

When the number of variables in a function is greater than can be handled by the largest multiplexer modules available, the multiplexing action can be expanded by connecting several multiplexers in an array of two or more levels. The scheme is illustrated in Figure 3-20a for two levels of four-way multiplexers to provide a 16-way multiplexer.

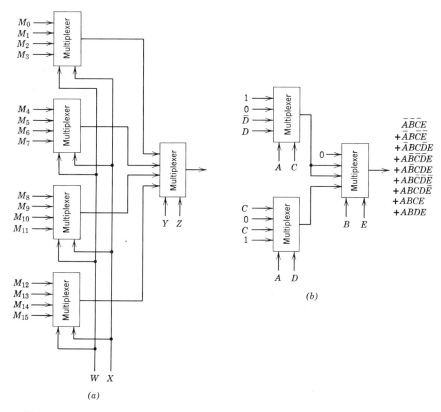

FIGURE 3-20. A switching network formed with multiple levels of multiplexers.

Extensions to larger modules and to more than two levels are straightforward. The four control signals W, X, Y, and Z select the signal on any one of the input lines M_0 through M_{15} for transmission to the output line. As with a single 16-way multiplexer, this network is capable of generating any Boolean function of five variables.

Any four of the five input signals may be connected with any permutation to the W, X, Y, and Z control lines, but with some functions a saving in multiplexers can be achieved by judicious selection of the permutation to be used. A thorough step-by-step procedure for selecting the best permutation for random functions has not been devised. However, for two-level networks the requirement is to determine the signals which, when used as the control signals for the first-level (right-hand) multiplexer, will maximize the number of instances where the multiplexed signals to that multiplexer are either 0, 1, or the same as some other multiplexed signal to that multiplexer. In each such instance the corresponding second-level multiplexer may be eliminated. With multiplexers that provide both direct and inverted output signals, the finding of two multiplexed signals that are the inverse of each other is equally useful—in fact, even without the inverse signals, a multiplexer can be replaced by a simple inverter in each such instance.

The minimization of the number of multiplexers will be illustrated by the example of the five-variable function.

$$\bar{A}\bar{B}\bar{C}E + \bar{A}B\bar{C}\bar{E} + \bar{A}BC\bar{D}E + A\bar{B}\bar{C}\bar{D}E + A\bar{B}CDE$$
$$+ AB\bar{C}\bar{D}\bar{E} + ABCD\bar{E} + ABCE + ABDE$$

This function is expressed in a form where B and E appear (sometimes inverted) in each term. If these two signals are used for the Y and Z control signals in the first-level multiplexer, the necessary signals to be multiplexed can then be determined by "factoring" the expression to yield the following.

$$\bar{B}E(\bar{A}\bar{C} + A\bar{C}\bar{D} + ACD) + B\bar{E}(\bar{A}\bar{C} + A\bar{C}\bar{D} + ACD)$$
$$+ BE(\bar{A}C\bar{D} + AC + AD)$$

Then, because $\bar{B}\bar{E}$ appears in none of the terms, the signal to the M_0 input of the first-level multiplexer is 0, and the corresponding second-level multiplexer is not needed. Because the factor for $\bar{B}E$ is the same as for $B\bar{E}$, the same input signal, $\bar{A}\bar{C} + A\bar{C}\bar{D} + ACD$, is used for both the M_1 and M_2 inputs to the first-level multiplexer, and this signal is generated by a single second-level multiplexer in the manner described previously for three-variable functions. The signal to the M_3 input of the first-level multiplexer corresponds to the case where BE is 1 and is $\bar{A}C\bar{D} + AC + AD$. This signal is formed in a similar way in another second-level multiplexer. (The fact that this last expression can be simplified to $C\bar{D} + AD$ is of no material consequence.) The resulting network of multiplexers is shown in Figure 3-20b.

Note that the W and X control signals for one second-level multiplexer are A and C whereas they are A and D for the other one. This permutation of variables is permissible but unimportant in two-level networks. For networks of three or more levels such permutations are helpful in minimizing the number of multiplexers, but the general minimization problem is thereby caused to be quite complex.

The use of multiplexers for switching is discussed with different examples in an article by J. L. Anderson (*Electronics*, October 27, 1969, pp. 100–105); see also S. S. Yau and C. K. Tang (*Proceedings of the Spring Joint Computer Conference*, April 1968, pp. 297–305).

Using a Storage (Memory) Unit to Replace a Combinatorial Switching Network

Although storage units and combinatorial switching networks are often viewed as being totally different from each other, they do in one sense have the same purpose. That is, for each given set of input signals, a predetermined set of output signals is generated. Because of this similarity, a storage unit can actually be used to replace a combinatorial switching network. In the past, the access time of economical storage units has been so great that such replacement was impractical, but with storage units formed with high-speed large scale integrated (LSI) techniques the access time and cost of storage units of significant capacity have become sufficiently low to allow the advantageous replacement of combinatorial switching networks by storage units in many instances.

One great advantage of using storage units for switching is that the "logical" design problem is totally eliminated. Instead, for each combination of input signals, a binary digit of 1 or 0 is stored according to whether an output signal is to be 1 or 0, respectively, for that combination. With n input signals (n Boolean variables) 2^n binary digits must be stored to represent a single-output function. For $n = 3$ for example, eight binary digits must be stored, and because each digit has two possible values (0 or 1), a total of $2^8 = 256$ different digit combinations is possible. This number is exactly equal to the number of different Boolean functions of three variables, as discussed in Chapter 1.

For multi-output functions of a given set of input signals, the storage unit is simply made to have a "word" length equal to the number of output signals.

The storage unit can be formed either with special components intended for storage, such as magnetic cores, or with switching devices in feedback networks as introduced in the next chapter. In either case the correspondence between a switching network and a storage unit is as illustrated in Figure 3-21. The block in a represents a five-input, three-output combinatorial network as discussed earlier in this chapter. The three output signals F_1, F_2, and F_3 represent any randomly selected functions of the five input signals A through E. The network may actually consist of three separate networks, or the networks may be intermingled in some way so as to reduce the number components required. In Figure 3-21b the input signals actuate an addressing matrix, the output signals from which are used to actuate appropriate binary storage cells in each of a number of storage "planes," one plane for each output signal. Most storage units additionally contain control signals for timing purposes, but these signals are not inherent in the storage function and are not found in some storage units.

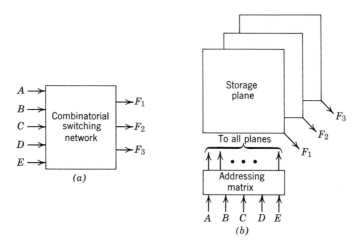

FIGURE 3-21. Illustrating the equivalence of a combinatorial switching network and a storage unit.

For fixed functions, a storage unit of the "read-only" type is satisfactory. That is, the stored digits are determined at the time of manufacture of the unit and are not subsequently changeable. Of course, with units for which the stored information is changeable as with the so-called "random access" or "read-write" units, the output switching functions can be altered as desired simply by "writing" appropriate new patterns of 1's and 0's in the storage unit—a capability totally absent in combinatorial networks.

Very high speed LSI storage units with capacities of 1024 words are available. At this capacity, 10 binary digits are required for addressing so that, in effect, functions of 10 variables can be generated. This capacity is not an upper limit, but as the number of variables increases the required capacity increases rapidly: 2048, 4096, 8192, and so on, words for functions of 11, 12, 13, and so on variables, respectively. "Don't care" conditions are not generally useful in reducing cost because, even though each "don't care" condition means that a binary storage cell may be eliminated, the primary cost generally arises from providing a storage array of a given basic size whether or not all binary cells are utilized.

Using an Associative Storage Unit to Replace a Combinatorial Switching Network

In an "associative" storage unit the input signals do not control an addressing matrix for the selection of an address. Instead, the input signals are compared with the word stored at each address. If a stored word in the unit is the same as the input word, a 1 signal is generated on a single output line. If no match is found, a 0 output signal is generated. (A generalized associative storage unit has many more features than indicated here. In fact, the name "associative" is derived from some of the other features, but this simplified version is all that is needed in this application.)

An associative storage unit can be used as a direct replacement for a combinatorial switching network in the following way. Each term in the Boolean expression for a given function is assigned an address in the associative storage unit. Each variable in the function is assigned a bit position. A variable is given the notation of 1 or 0 according to whether it appears in direct or inverted form, respectively, in a given term. In the most straightforward approach, the Boolean expression is expanded to its elemental terms so that each variable appears in each term. The number of bits of associative storage required is nt where n is the number of variables and t is the number of terms in the expression for the function.

For example, the Boolean function $\bar{A}\bar{B} + A\bar{B}\bar{C} + ABC$ would be expanded to $\bar{A}\bar{B}\bar{C} + \bar{A}\bar{B}C + A\bar{B}\bar{C} + ABC$, and the four three-bit words to be stored would be 000, 001, 100, and 111. Then, for any given combination of input signals, the output signal will be 1 or 0 according to whether the input combination does or does not, respectively, match any one of the stored words, as required for the realization of the Boolean function.

One minor elaboration in the associative storage unit is highly desirable. Namely, the storage cells should be designed so that any given cell can be caused to respond to either a 1 or a 0 input signal. Then, if a given term in the Boolean expression for the function does not contain all variables (that is, is not elemental), that term need not be expanded to its elemental terms but instead may be stored as a single term at a single address. In the above example, the term $\bar{A}\bar{B}$ could then be stored as 00- at a single address, where the hyphen means that the corresponding cell responds to either a 0 or a 1 input signal.

The above elaboration can be used to minimize the storage capacity requirements in much the same way that the number of components is minimized in a switching network, except that the objective is limited to minimizing the number of terms in the Boolean expression (the elimination of one or more variables from a term is of no help).

The primary reason for considering an associative storage unit instead of a conventional storage unit is that the required number of associative storage cells, nt, is for many functions significantly less than the required

118 NETWORKS CONTAINING INVERTERS

number of cells, 2^n, in a conventional storage unit (although, as is apparent from the material in Chapter 2, nt would be much greater than 2^n for many other functions). Also, the need for an addressing matrix is eliminated.

For functions having a particularly large number of terms for a given n, a reduction in the required number of associative storage positions can often be achieved by generating the inverse of the intended function and then inverting the output signal—in the manner discussed previously for switching networks.

Unfortunately, at the time of writing this book, associative storage units are in a relatively experimental stage of development, and neither the read-only nor the read-write versions are as available as are conventionally addressed storage units.

Exercises

1. With no inverted input signals available, find networks having the indicated number of A-I modules for the following functions.

(a) $\bar{A}\bar{C} + \bar{A}\bar{D} + \bar{B}\bar{C} + \bar{B}\bar{D}$ (four)
(b) $A\bar{B}\bar{C}D + A\bar{B}D\bar{E} + A\bar{B}D\bar{F}$ (six)
(c) $\bar{A}C + \bar{B}C + C\bar{E} + ACD\bar{E}$ (five) (Hint: Let $ADE = f(A_i)$ in the equation on page 92.)
(d) $A\bar{B}C + \bar{A}B\bar{D}E$ (five)
(e) $\bar{A}\bar{B}D + \bar{B}\bar{C}D + \bar{A}\bar{C}D$ (five) (Hint: Factor D and convert the remainder of the expression to OR-to-AND form.)

2. For the functions in Exercises 1(a), 1(b), and 1(d), derive corresponding networks of A-I modules that have a fan-in of two and a fan-out of two.

3. For all functions in Exercise 1, determine the corresponding one-module A-O-I network of simplest form. Assume inverted input signals are available.

4. Repeat Exercise 3 but with O-A-I modules.

5. With modules that generate both the O-I and the OR functions, determine the alternative network (see page 109) available for each function in Exercise 1. Assume inverted input signals are available. Minimize the number of modules required for each function where possible by factoring or by finding alternative expressions for the functions.

6. Repeat Exercise 5 but with modules that generate the A-I and AND functions.

7. Find the multiplexer connections for generating the function $\bar{A}\bar{B} + A\bar{B}\bar{C} + ABC$, with B and C used as the control signals.

8. Find a two-level network using only three four-input multiplexers for generating the function

$\bar{A}\bar{B}\bar{C}\bar{D}E + \bar{A}B\bar{C}\bar{D}\bar{E} + \bar{A}CD + \bar{A}BDE + \bar{A}\bar{B}D\bar{E} + A\bar{B}DE + A\bar{B}CD + ECD + AB\bar{C}D\bar{E}.$

(Hint: Use A and D as the control signals for the first-level multiplexer.) Assume that the inverted output of each multiplexer is available on a separate output line.

9. With a conventionally addressed storage unit, how many bits of storage capacity would be required to generate the function in the previous exercise?

10. With an associative storage unit, what is the minimum number of bits of storage capacity that would be required to generate the function in Exercise 8? Assume that the unit is such that any given cell can be caused to respond to a 0, to a 1, or to both.

4
FEEDBACK, DIGITAL STORAGE, AND TIME-DEPENDENT SIGNALS IN SWITCHING NETWORKS

In previous chapters, consideration was limited to networks to which steady-state signals are applied and from which one or more steady-state signals are obtained. In such networks, all signal paths lead from the input terminals to the output terminals with no loops or feedback paths of any kind. In nondigital (or so-called "linear") circuits, the "feeding back" of all or a portion of an output signal to an input terminal may be done for a variety of reasons according to the application. Examples include creating an oscillator, increasing the gain of an amplifier, and obtaining an amplifier that has a gain nearly independent of frequency. A multitude of other feedback purposes are found in other applications, particularly control applications. Of course, in each case the nature of the feedback signal and the details of the circuit are strongly dependent on the objective.

With digital networks, feedback can be used to cause a network to "lock up" in any of two or more different stable states, and the purpose is to store digital information. The digit thus stored need not be a true arithmetic digit. Instead, it may be merely a "bit" of digital information that is used to "remember" any one of such things as, for examples, whether an overflow has or has not occurred, whether a magnetic tape unit is being used for reading or writing, or whether or not some miscellaneous control signal has been received in a given recent interval of time.

Arrays of two-state storage networks can be used to store more conventional digital "numbers," and a common example is the use of four two-state networks to store a decimal digit in any one of several different codes to be described in a later chapter. Moreover, large arrays of two-state networks can be used to store many numbers, each of which is composed of several decimal or binary digits. "Scratchpad" storage units offering very high speed operation are often built in this way, although storage units other than those based on switching networks are generally found to be more economical for the storage of large amounts of digital information.

Further, in practical applications the over-all objective requires more than the mere combining and storing of steady-state signals. The input signals change from time to time (often at a very high frequency), and the output signals must be functions of the sequence of the applied signals as well as the combinations of the signals applied at any given time. The problem then is to devise a network—which may utilize storage mechanisms other than those based on switching network feedback—that will produce the desired relationship between the output and input signals. The resulting arrays of devices are commonly said to be sequential switching networks as opposed to the combinatorial (or combinational) switching networks of the previous chapters.

The theoretical aspects of sequential switching networks has been the subject of extensive study, and a substantial body of knowledge called "sequential circuit theory" has been assembled. Although much of this knowledge is interesting or even fascinating from a mathematical point of view, the simplifying assumptions that have been required to make progress in the theoretical studies have unfortunately been so restrictive that the results have tended to be remote from the real world as encountered by engineers. In the previous chapter the eight simplifying assumptions for combinatorial networks are seldom realistic. Yet, when only a few of these assumptions are removed, organized design procedures for attaining certain elementary objectives (such as minimizing the number of components) become virtually unattainable. Only a small amount of imagination is required to visualize that when feedback and time-dependent signals are added to the complexities, the development of straightforward design procedures becomes even more remote from realization.

For certain special problems, sequential circuit theory has been used to derive techniques for reducing the number of stable states a network must have, and a means for reducing the number of components is thereby suggested. However, these problems are not typical of the problems encountered in practice. Instead, the problems encountered in practice generally involve a hodgepodge list of details requiring, often, a hodgepodge network for effective solution. Nevertheless, a number of important basic principles pertaining to feedback, storage, and time-dependent signals have been worked out, and although they are on only the fringes of the realm called "sequential circuit theory," they are of great importance to the designer faced with the task of building a working machine. This chapter is devoted to these principles.

Incidentally, the terms "sequential circuit," "digital system," and "automaton" are synonymous in the sense that each means a machine which accepts digital input signals and which generates digital output signals that are functions of the input signals. However, the usage of these terms tends to be different (but with considerable overlap). "Sequential circuit" is usually selected when the physical components are of major interest and when the function performed by the components is limited to something such as a shift register, a counter, or a small segment of the control circuits of a larger set of components. "Digital system" is selected when a complete working machine such as a computer is implied. "Automation" is most often selected when the major interest is in the abstract mathematical properties of the machine, and the machine can range from the simplest "sequential circuits" to the most elaborate "digital systems" or it may have abstract properties unrealizable (at least not practically) by any physical machine at all.

An introduction to the purely theoretical aspects of sequential circuits is contained in Chapter 2 of Richards' book *Electronic Digital Systems*, 1966. The topic is one of continuing research interest, and more recent papers with particular emphasis on asynchronous operation are found in the *IEEE Transactions on Computers*, February, June, August, and December 1968, May, June, August, and December 1969, and March 1970.

A Two-State Network—the Set-Reset Flip-Flop

In a network consisting of only AND and OR devices the feeding back of an output signal to an input terminal ordinarily produces no useful results, although the particular results that are thus obtained depend on the nature of the devices as well as on the particular network connections under consideration. In a single diode OR circuit or a single diode AND circuit, for example, a feedback connection accomplishes nothing other than to short circuit one diode, and the effect of the other input signals is not altered. For more complex networks of diode AND and OR circuits the output potential may assume some intermediate value that depends on resistance values and other parameters. When a single inverter (which produces amplification as well as inversion) is in the feedback loop, the results are likewise generally not useful. The output signals, if affected at all by the feedback connection, will most likely settle at some intermediate potential that depends on other input signals, or the circuit may break into oscillation under certain combinations of stray capacitances and inductances.

For useful results, a feedback loop in a switching network must generally contain an even number of inverters. With an even number of inverters in the loop, the signal fed back will be of such polarity as to maintain the output signal generated (if the accompanying AND or OR devices are appropriately interconnected), and the resulting network will be capable of existing in any one of two or more stable states. In this way, digital information can be stored. Of course, for useful application the network must include provision for changing the network from one state to another as needed in the application at hand.

A network capable of existing in one of two stable states can be formed by simply connecting two inverters in a loop. If the input signal to one inverter is 0, the output signal from that inverter will be 1. This 1 applied as an input to the second inverter will cause the output there to be 0, and this 0 applied as an input to the first inverter will maintain this stable state indefinitely. The other stable state is maintained in the same way but with the 1's and 0's interchanged. With physical inverters another state of equilibrium exists when each signal is at some intermediate potential, but this state is unstable. Any slight disturbance of the potential at the input of either inverter will be amplified and will appear in inverted form at the input of the other inverter so that the network quickly assumes one of the stable states just described. By themselves, the two inverters do not include means for setting the network to one state or the other.

To allow the setting of the two-state circuit to one state or the other, an OR function is needed at the input of each inverter as shown in Figure 4-1. This network contains two output terminals and two input terminals. However, the two output signals from this network can be visualized as corresponding to a single binary variable F, because one signal is always the inverse of the other. That is, if one signal is designated F, the other signal must be $\bar{F}$. To set the network to the state represented by $F = 1$, a signal is temporarily applied to the input terminal at the lower-left part of the figure. This signal causes a 1 to be applied to the input of the left-hand inverter regardless of whether or not a signal already reaches that point from the output of the right-hand inverter. The output of the left-hand inverter becomes 0, the output of the right-hand inverter becomes 1, and the network remains in this stable state after the exterior input signal is removed. An input signal at the opposite exterior input terminal analogously causes the network to be set to the state represented by $F = 0$ (or $\bar{F} = 1$).

Signals applied to both input terminals simultaneously cause both

124   FEEDBACK, DIGITAL STORAGE, AND TIME-DEPENDENT SIGNALS

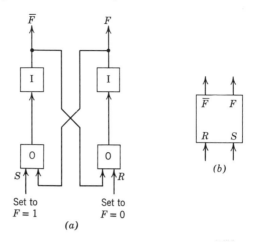

FIGURE 4-1. Two-state network—the set-reset (R-S) flip-flop.

output signals to be 0 throughout the duration of time that the input signals are applied. During this time the output signals are generally regarded as meaningless. If the input signals are terminated simultaneously, the network will assume one state or the other at random or in accordance with minor differences in the physical characteristics of the two halves of the network. Ordinarily this situation represents an improper use of the network and is avoided. Of course, if one input signal remains at least for a short period of time after the other has terminated, the network will assume the corresponding stable state. Input signals of this nature are permitted.

Any network including two inverters in a feedback loop is commonly called a "flip-flop," and when facilities are provided to set it to one state or the other in the manner described above, the flip-flop is said to be a "set-reset" flip-flop, sometimes abbreviated R-S flip-flop. The terms "set" and "reset" are derived from the fact that the flip-flop is often visualized as storing a binary digit or a binary signal. The flip-flop is "set" when the stored digit or signal is caused to 1 and is "reset" when caused to be 0.

The use of O-I modules to form the network of Figure 4-1 is obvious, but making a set-reset flip-flop with only two A-I modules is impossible. Four A-I modules are required.

Three-State Networks

Feedback connections can be used to form a three-state network straightforwardly as illustrated in Figure 4-2, where three four-input

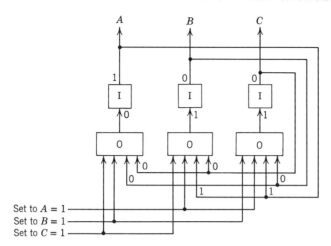

FIGURE 4-2. Three-state network composed of O-I modules.

O-I modules are used. With the three output signals designated A, B, and C, the relationships among the signals are $A = \overline{B + C}$, $B = \overline{A + C}$, and $C = \overline{A + B}$ in Boolean notation. These relationships have the effect of causing any one, but only one, of A, B, and C to be 1 at any given time with the other two being 0's.

The setting of the network to the state where A, B, or C, respectively, is 1 is accomplished by temporarily applying an input signal to the appropriate input line as indicated in the figure. The various 1's and 0's shown in the figure indicate the signals at the corresponding points when the network is in the state where $A = 1$. The operation of the circuit when changing from one state to another is a straightforward extension of the operation of the binary flip-flop.

A three-state network composed of three A-O-I modules is shown in Figure 4-3. The output signals from the inverters could be designated A, B, and C as before, but the character of the network seems to be indicated somewhat better if these signals are designated $\bar{A}$, $\bar{B}$, and $\bar{C}$, respectively. The 1's and 0's in the figure indicate the signals at the corresponding points for the state where $A = 1$ and $B = C = 0$. With the notation selected, signals representing A, B, and C are generated at the outputs of the AND parts of the respective modules. Setting the network to the desired state is accomplished, as before, by applying an input signal to an input line as indicated. In this instance a 1 is thereby applied to the input of the corresponding inverter, and the resulting 0 on the output line is applied to the AND devices in the other modules to cause 0's to be applied to the inputs of the other inverters. The signal from the AND part

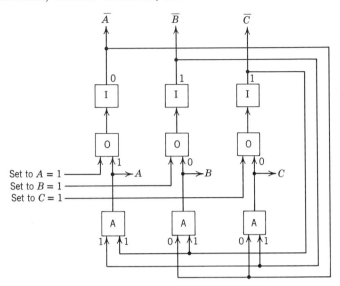

FIGURE 4-3. Three-state network composed of A-O-I modules.

of the original module is then 1 to maintain the network in a stable state after the input signal is removed.

For the network in either Figure 4-2 or 4-3, considerations pertaining to the simultaneous application of two or more input signals are analogous to those described for the two-state flip-flop.

Note that, although three inverters are used in each network, each closed-loop path includes only two inverters.

Four-State and Six-State Networks Containing Four Inverters

With four six-input O-I modules, the network of Figure 4-2 is expandable to one having four stable states. Alternatively, the network of Figure 4-3 is expandable to one having four stable states if four A-O-I modules are used. In this latter case, the only elaboration on the modules is that each AND part must have three instead of two inputs.

A different form of four-state network can be devised by using two two-state flip-flops as illustrated in Figure 4-4. The two flip-flops can be visualized as functioning independently so that the signals F_1 and F_2 are independent of each other. The network as a whole is in state A, B, C, or D according to whether F_1 and F_2 are 0-0, 0-1, 1-0, or 1-1, respectively. In some applications the two independent signals are

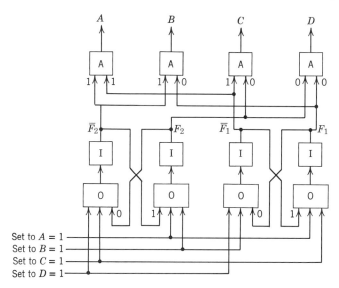

FIGURE 4-4. Four-state network based on two flip-flops.

appropriate for indicating the state of the circuit, but in other applications a separate output line for each state is required as shown. The AND devices across the top of the figure form a matrix as described in a previous chapter. The setting of the network to the desired state can be accomplished by connecting the input lines as shown in the figure, although in some applications the input signals may be in the form of two sets of two signals that operate the flip-flops independently.

With four inverters available for use in a feedback network, the 1's generated by these inverters can be in a 1-out-of-4 pattern as in the network form with O-I modules, they can be in a 3-out-of-4 pattern as with the network of A-O-I modules, or they can be in a 2-out-of-4 pattern. With four signals, six combinations of two 1's are possible as follows, where the individual signals are now designated X_1, X_2, X_3, and X_4.

	X_1	X_2	X_3	X_4
A	1	1	0	0
B	1	0	1	0
C	1	0	0	1
D	0	1	1	0
E	0	1	0	1
F	0	0	1	1

128 FEEDBACK, DIGITAL STORAGE, AND TIME-DEPENDENT SIGNALS

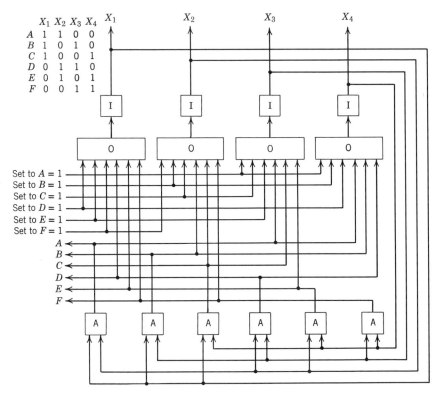

FIGURE 4-5. Six-state feedback network employing four inverters.

All six of these digit combinations can be utilized in forming a six-state network requiring only A-O-I modules as shown in Figure 4-5. Although the output from each AND device actuates two OR devices, this requirement can be eliminated if each AND device is duplicated, and then the modules would be exactly as described in the previous chapter. Again, the 1's and 0's in the figure indicate the signals at the corresponding points when the network is in state A, that is, when $X_1 = X_2 = 1$.

Feedback Networks Employing Five or More Inverters

If N inverters are available, the networks of Figures 4-2 and 4-3 can be extended straightforwardly to form N-state networks. The number of 1-signals generated at the outputs of the inverters are 1 and $N-1$, respectively.

With five inverters, the network of Figure 4-5 can be extended in either of two ways. In one way the 1's appear in a 2-out-of-5 pattern and in the other way they appear in a 3-out-of-5 pattern. In either case the total number of stable states possible is ten. The digit combinations for the X_1 to X_5 signals are as follows.

	X_1	X_2	X_3	X_4	X_5		X_1	X_2	X_3	X_4	X_5
A	1	1	0	0	0	A	1	1	1	0	0
B	1	0	1	0	0	B	1	1	0	1	0
C	1	0	0	1	0	C	1	1	0	0	1
D	1	0	0	0	1	D	1	0	1	1	0
E	0	1	1	0	0	E	1	0	1	0	1
F	0	1	0	1	0	F	1	0	0	1	1
G	0	1	0	0	1	G	0	1	1	1	0
H	0	0	1	1	0	H	0	1	1	0	1
I	0	0	1	0	1	I	0	1	0	1	1
J	0	0	0	1	1	J	0	0	1	1	1

With the 2-out-of-5 pattern, five 12-input OR devices and ten 2-input AND devices are required in addition to the five inverters. With the 3-out-of-5 pattern, the five OR devices each need only eight inputs, but the 10 AND devices each need three inputs.

With six inverters, the concepts of the previously described networks can be expanded and combined in various ways to produce an even greater number of variations. The maximum number of stable states attainable with six inverters is 20, and the required network is realized by expanding the network of Figure 4-5 where the X_1 to X_6 signals are 1's in a 3-out-of-6 pattern. Besides the six inverters, six 20-input OR devices and twenty 10-input AND devices are required.

If the concept is expanded to include M-out-of-N output patterns, where N is the number of inverters and M is not a constant but may range from 0 to N, the number of stable states for a given N can be increased still further. However, the network becomes impractical not only because of a very large increase in the number of AND and OR devices required but also because signal loops containing only AND and OR devices (no inverters) are encountered, whereas in available electronic circuits inverters are needed to maintain signal amplitude in a loop.

The usual design technique employed for storage networks is simply to group the inverters in pairs to form flip-flops. Input-output matrices of AND-OR devices are added as needed to actuate the flip-flops selectively and to generate signals indicating the state of the network as a whole. With n flip-flops each containing two inverters, the maximum

130 FEEDBACK, DIGITAL STORAGE, AND TIME-DEPENDENT SIGNALS

number of stable states then attainable is $2^n = 2^{(N/2)}$, where N is the total number of inverters. With flip-flops provided in separately packaged modules, this practice is doubtless the most appropriate in most instances even though $2^{(N/2)}$ is less (much less for large N) than the number of states attainable by the M-out-of-N approach, even for a constant M. On the other hand, with large-scale-integration (LSI) the separate identification of flip-flops is of no real consequence. Therefore, when the input-output matrices needed with the flip-flops are brought into consideration, the feedback networks described above become more competitive.

The Generalized Synchronous Sequential Circuit

The feedback networks described in the immediately preceding sections serve only to store digital information. For "processing" information as required in a complete digital system, an additional form of feedback is required as illustrated in Figure 4-6, where each heavy line represents a set of two or more wires. In this figure the storage network is usually a collection of flip-flops, which may be of the set-reset type, although in principle any of the multistate feedback networks may be used. The combinatorial network is as described in the previous two chapters, and it receives input signals from an input device as well as from the storage network. Similarly, the combinatorial network generates output signals, some of which are transmitted back to the storage network for subsequent storage and others of which are transmitted to the output device. For the present, the combinatorial network may be assumed to introduce no time delays. That is, any changes in the binary values of the signals generated by this network will be created instantly upon appropriate changes of one or more signals applied as inputs to this network.

The rectangle labeled "gates" in Figure 4-6 serves to prevent the ac-

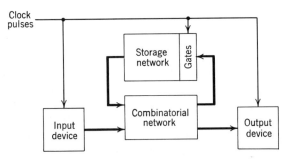

FIGURE 4-6. Generalized synchronous sequential circuit.

tuation of the storage network except when one of a more or less continuous series of "clock" pulses is applied. A network or system actuated by clock pulses is said to be "synchronous" as opposed to "asynchronous" networks or systems described later. The gates are of such nature that, when a given clock pulse is applied, the new signals arriving at the inputs of the gates will not be transmitted to the storage network at the time of this clock pulse. The need for this property of the gates is apparent from the fact that the newly stored digital information will create a new set of signals at the inputs of the combinatorial network, and a new set of output signals from this network will thereby be generated. Many different physical realizations of gates of the necessary properties have been invented, and some of these will be described later.

In Figure 4-6 the clock pulses are also applied to the input device and the output device. The input device may be visualized as a punched paper tape reader, and the clock pulse causes the paper tape to be advanced one position so that a new set of punched holes is sensed during the next actuation of the system as a whole. In the case of a magnetic tape input device, for example, the clock pulse may be visualized as causing an entire block of information to be read from the tape and stored in a buffer storage unit which is a part of the input device and which supplies the signals to the combinatorial network. In some systems the input device is actually operated in the manner indicated, although in advanced systems much more complex and versatile input control methods are employed, but the details of such methods are not of present concern. In the case of the output device, the clock pulse merely actuates the device to cause it to accept the signals arriving at the device at the time of the clock pulse. Although not shown in the figure, the output device must contain a "gate" function that serves a purpose similar to that of the gates at the inputs to the storage network.

The sequential circuit in Figure 4-6 is a generalized network in the sense that, with a storage network of sufficient capacity and with a suitably designed combinatorial network, the signals arriving at the output device can be made to be any physically specifiable function of the signals arriving from the input device. A classic theoretical problem pertaining to this generalized network is the finding of the functions to be performed by the storage and combinatorial networks when the output signals to be generated are specified for various sets of sequences of input signals. However, in practice this problem seldom arises except when the function relating the output signals to the input signals is relatively simple—as might be encountered when the sequential circuit is nothing more than a counter or a shift register. For a network or system of any complexity, such as a computer, the opposite situation tends to prevail. That is, the functions to be performed are specified, and the output signals are the unknowns. In fact, the designer never does determine the output signals—the system itself does that after it is built and put into operation, because that is the purpose of the system in the first place.

Regardless of whether the functions or the output signals are the unknowns in a given application, a complete computer or other system of considerable complexity can be based, both conceptually and physically, on the generalized sequential circuit block diagram of Figure 4-6. In fact, at least a few small binary computers have been designed and built in this way. The complete system can then be specified by means of a set of Boolean expressions which relate the output signals at a given clock pulse time to the input signals at this same time, where the input signals from the storage unit are some of the output signals generated at the time of the previous clock pulse. On the other hand, in computers or other significantly complex digital systems the storage network is generally replaced by a storage "unit" in which groups of digits are stored at "addresses" under the control of addressing signals sent to the unit. These addressing signals are in addition to the signals representing the information to be stored. Recovery of information is again by addresses, and the signals appearing on the output lines from the storage unit do not represent all of the stored information at any given time but only that information stored at the specified address.

Asynchronous Sequential Circuits

Conceptually, the use of clock pulses in Figure 4-6 is not an inherent requirement, and a system that functions without clock pulses is said to be "asynchronous." However, the design of asynchronous systems involves at least three important problems not encountered with synchronous systems.

For one thing, the physical devices in the combinatorial network do not really operate instantaneously. When even one input signal to this network is changed, the output signals may be temporarily incorrect with the detailed response being dependent on the details of the network and upon the relative operational speeds of the individual switching devices. The temporarily incorrect output signals may initiate unwanted action in the storage network, and the possibility of these signals is called a "hazard." Strictly speaking, hazards exist in synchronous systems as well as in asynchronous systems, but with synchronous systems the time interval between clock pulses is always made sufficiently great to prevent unwanted effects from temporarily incorrect signals at the outputs of the combinatorial network.

A second problem encountered with asynchronous systems is called "buzzing." For a given set of input signals to the combinatorial network, this network will generate a certain set of output signals. As indicated in Figure 4-6 (assume the clock pulses and gates to be omitted), some of these signals will be transmitted to the storage network and will set that network to a determinable stable state. The signals thereby generated at the output of the storage network will then be utilized as some of the input signals to the combinatorial network. Now these signals fed back from the storage network may not be the same as the original signals from that source, and therefore a new set of output signals will be generated. The storage network may then be set to a new stable state, and still another set of signals may be transmitted from the storage network to the combinatorial network. This process is called "buzzing," and it may continue for a finite number of steps or it may continue indefinitely in a cyclical pattern of some sort. The buzzing will cease if a state is reached whereby the signals generated by the storage network at a given step cause, through the action of the combinatorial network, a set of signals to be generated which, in turn, cause the storage network to remain in that same state. The details of the response obtained in any given instance depend on the initial state of the storage network and on the signals from the input device as well as on the design of the combinatorial and storage networks.

"Races" represent a third problem encountered with asynchronous systems. When the storage network is composed of flip-flops, these flip-flops may not consume exactly the same amount of time in changing state when the storage network as a whole is changed from one state to another. Alternatively, the individual signals for actuating the flip-flops may not be generated at precisely the same time. In either case, a "race" exists in determining the sequence in which the flip-flops change state. The character of the race depends on the initial and final states of storage network as a whole. When the network is changing state, it is in an unwanted state during that period of time when one or more, but not all, of the flip-flops have completed their change of state. When the network is in this unwanted state, the signals transmitted from the storage network to the combinatorial network will be incorrect, and incorrect signals may then be transmitted from the combinatorial network back to the storage network. The result will be unwanted and incorrect actuations of the flip-flops in the storage network.

Techniques have been devised for avoiding hazards, buzzing, and races or for obtaining wanted over-all response in spite of these conditions, but in general the techniques are impracticably complicated and at best applicable only to very elementary networks. Moreover, available design procedures are generally limited to the impracticably severe restriction of allowing only one input signal (from an input device in Figure 4-6) to change at a given time. Therefore, asynchronous systems based on fundamental considerations illustrated by Figure 4-6 (without the clock pulses and gates) are avoided in practice.

In spite of the above situation, asynchronously operating circuits of numerous miscellaneous forms are widely used, and in many cases the concepts of synchronous and asynchronous operation are intermingled in various ways that are difficult to describe in an organized manner. The simple complementing flip-flop described in the next section is an example, and more elaborate examples will be included later.

The Complementing Flip-Flop

Consider again the generalized synchronous sequential circuit of Figure 4-6. About the simplest imaginable version of such a circuit is obtained by assuming that the input signals are always 0 (input device eliminated entirely), that the storage network provides for only two stable states, and that the combinatorial network is such as to cause the storage network to change state each time a clock pulse is applied. The output device can also be eliminated, and only the signals appearing on the output lines need be considered. The resulting network with one possible means for performing the "gate" function is shown in Figure 4-7a. In this circuit the combinatorial network is reduced to two direct connections. Only one output signal, F, is generated. This signal and its inverse appear on the lines indicated.

When a clock pulse is applied, a signal is transmitted through one or the other of the AND devices and then through a "delay" device to one of the two input lines to the storage network. Each delay device may be assumed to be of such characteristics that the pulse appearing at its output is the same as the pulse (if any) applied to its input, but the time of appearance of the output pulse is later than the time of application of the input pulse. Thus, by the time a pulse arrives at an input to the storage network the clock pulse will have terminated. Even though the storage network may change state instantaneously, the new values of F and $\bar{F}$ will not result in any alteration in the signals passing through the gates.

A flip-flop which changes back and forth from one state to the other in response to each one of a sequence of pulses is called a "complementing" flip-flop—whether the pulses are "clock" pulses or whether the pulses

THE COMPLEMENTING FLIP-FLOP   135

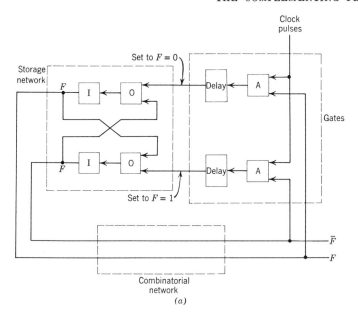

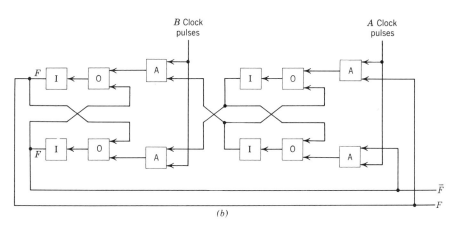

FIGURE 4-7. An elementary sequential circuit—a complementing flip-flop.

have some other significance. It is also called a T flip-flop, where T stands for "trigger" (because flip-flops were sometimes called "trigger circuits" in the early days of computers). A complementing flip-flop with separate resetting and setting input terminals is then called an R-S-T flip-flop. In the historical development of computers, the complementing flip-flop has often been viewed as having basic significance, although considering it as a particular and simple example of a synchronous sequential circuit now seems to place it in a more appropriate perspective with respect to switching networks generally.

Actually, separate delay devices as indicated in Figure 4-7a are virtually never used in physical realizations of complementing flip-flops. Instead, the delay function may be intermingled with other functions, such as in the basic flip-flop circuit consisting of two transistor inverters with speed-up capacitors and with capacitively coupled input signals. This type of flip-flop is described in almost any book on computer hardware—for example, see Richards' *Electronic Digital Components and Circuits*, pp. 133–137.

Another method of performing the gating function in a synchronous sequential circuit is to provide an additional set of flip-flops for temporarily storing the binary signals to be stored in the storage network. Two separate clock pulse lines are required, and the pulses appear alternately on these two lines. This method of gating is illustrated for the elementary complementing flip-flop in Figure 4-7b. An A clock pulse causes the right-hand flip-flop to be set to the state opposite to that of the left-hand flip-flop. A subsequent B clock pulse then causes the left-hand flip-flop to be set to the state of the right-hand flip-flop, so that the net effect is to invert the state of the left-hand flip-flop. A subsequent pair of A and B clock pulses produces a similar inversion of states and restores the network to its original condition. At the time of a B clock pulse, the new signals generated at the outputs of the left-hand flip-flop (which corresponds to the "storage network" of Figure 4-6) cannot affect the signals being entered into this same left-hand flip-flop because the A clock pulse is absent at this time. The absence of the A clock pulse prevents the passage of signals through the right-hand AND switches. An analogous situation prevails at the time of an A clock pulse.

The use of a second flip-flop as in Figure 4-7b would constitute an obvious cost disadvantage when the switching devices are composed of discrete components. However, with integrated circuits, the added switching devices may add virtually nothing to cost. In fact, with integrated circuits, the preferable approach to providing the gating function involves adding still more switching devices to the flip-flop as needed to eliminate the second clock pulse line. In effect, the second clock pulse is generated within each flip-flop module as illustrated in Figure 4-8.

The complementing flip-flop (still portrayed as a sequential circuit) in Figure 4-8a functions analogously to that of Figure 4-7b, but only a single clock pulse, here called an "actuating signal," is required. In the absence of an actuating signal, the signal at the output of inverter

THE COMPLEMENTING FLIP-FLOP 137

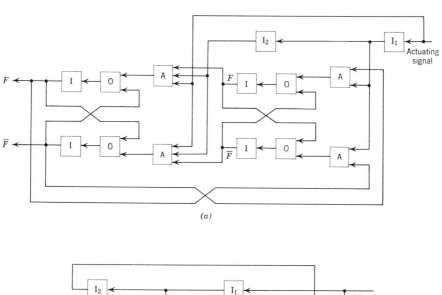

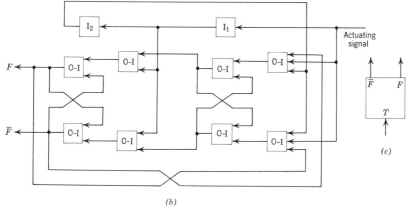

FIGURE 4-8. Asynchronous complementing flip-flop networks.

I_1 is 1, and the right-hand flip-flop is set to the state of the left-hand flip-flop. When the actuating signal is applied (becomes 1), the output from I_1 becomes 0 and prevents the passage of signals through the right-hand AND devices. The output from inverter I_2 then becomes 1 and causes the left-hand flip-flop to be set to the opposite state of equilibrium. When the actuating signal is removed (becomes 0), the first action is to prevent passage of signals through the left-hand AND devices, and then any changes in state of the right-hand flip-flop will not affect the left-hand flip-flop. The signal at the output of I_1 then returns to 1, and the right-hand flip-flop becomes set to the new state of the left-hand flip-flop. For proper operation of the network, inverters I_1 and I_2 must display a slight delay in the change in their respective output signals in response to a change in their input signals. Fortunately, transistors do display a slight delay in turn-on and turn-off characteristics, and because the only requirement is that the delay be greater than the delays in the associated AND switches (which may be diode circuits displaying extremely little delay), the design problem is not critical.

In Figure 4-8a the direct connection from the actuating signal input line to one input terminal in each of the left-hand AND switches can be eliminated if the designer can be assured that the delay through two inverters in tandem will be greater than the delay through a single inverter. In particular, when the actuating signal changes from 1 to 0 to cause the output from I_1 to change from 0 to 1, the resulting signal to one or the other of the two inputs to the right-hand flip-flop will first cause the output signal from the corresponding inverter in that flip-flop to become 0. This 0 signal will then cause the signal from the other inverter in this flip-flop to become 1, so that the two inverters operate in sequence. If the time required for this sequential operation of the two inverters in the flip-flop is greater than the time required for the I_2 output signal to change from 1 to 0, the left-hand AND switches will not pass signals during the short period of time in question, and the direct connection from the actuating signal input line may be omitted. Two-input AND switches are then sufficient.

The complementing flip-flop in Figure 4-8b is essentially the same as in a except that O-I modules are used, and the inverters I_1 and I_2 are connected in reverse sequence to compensate for the added inversions required by the O-I modules. An instructive exercise is for the reader to determine for himself the details of the signals that occur within the network when the actuating signal changes from 0 to 1 and from 1 back to 0.

Although the networks in Figure 4-8 have been developed from the concept of a synchronous sequential circuit, they may be equally well be viewed as asynchronous circuits where the actuating signal represents a single input signal from an input device.

In both networks of Figure 4-8, the actuating signal is assumed to change from one binary value to the other very quickly (when it changes at all). If the actuating signal potential should change very slowly or erratically from one binary value to the other, a new set of design problems is introduced. Specifically, for each device within the network, detailed consideration must be given to the input potential thresholds at which the output potentials begin to change. In some cases, the rate of change of output potential as a function of rate of change of input

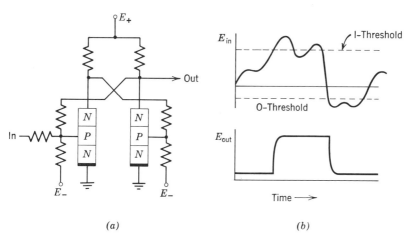

FIGURE 4-9. Flip-flop for transforming a slowly or erratically varying input signal to a fast-switching binary signal as needed for actuating a sequential circuit.

potential becomes important for some of the devices within the network. The design problem obviously then becomes very difficult. Usually, no attempt is made to solve problems of this type. Instead, the slowly changing input signal is suitably amplified and applied through a resistor to the base terminal of a transistor in a flip-flop. This extra flip-flop then acts as a threshold device that delivers a definite 0-signal or a definite 1-signal at all times except when changing from one state to the other, and these changes occur very rapidly. The output from this extra flip-flop is then used as the actuating signal to the complementing flip-flop or other sequential circuit. The flip-flop with resistor input is shown in Figure 4-9.

J-K Flip-Flops

Functionally, a *J-K* flip-flop is a complementing flip-flop having two added control signals, arbitrarily designated J and K, that serve to limit the conditions under which the flip-flop will change to the 1 and 0 states, respectively. Specifically, the flip-flop will switch from the 0 state to the 1 state only when J is 1 at the time of a clock or actuating pulse, although if the flip-flop is already in the 1 state (that is, "set"), the binary value of J will not affect the flip-flop. Analogously, the flip-flop will switch from the 1 state to the 0 state only when K is 1. In other words, when both J and K are 0, the clock or actuating pulses will have no effect on the flip-flop at all; when only one of J and K is

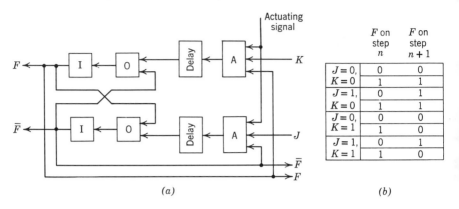

FIGURE 4-10. J-K flip-flop.

1, the flip-flop will be switched to the 1 or 0 state, respectively, if it is not already in the corresponding state; and when J and K are both 1, the flip-flop will act as a complementing flip-flop and will switch to the opposite state upon the application of each complementing pulse or actuating signal.

The switching functions in a synchronous J-K flip-flop are shown in Figure 4-10a, and they are the same as found in a complementing flip-flop except for added input lines to the AND devices. The states of the flip-flop for various combinations of signals are set forth in detail in the table in the b part of the figure.

An asynchronous J-K flip-flop is shown in Figure 4-11a. Note that if J and K are both 0, the signals at the outputs of inverters I_j and I_k will be 1 so that the output signals from the inverters in the right-hand flip-flop within the network (points X and Y) will both be 0. The AND devices in the left-hand flip-flop will then block the passage of the actuating signal. If both J and K are 1, the signals at the outputs of I_j and I_k will be 0, and the operation of the network will then be exactly the same as with the complementing flip-flop in Figure 4-8a. If $J = 1$, $K = 0$, and $F = 0$, an actuating signal will cause the flip-flop to change to the state where $F = 1$, because in this case the signal at point Y (but not point X) will be 1. If F was 1 initially, no flip-flop action will take place, because the signals at both points X and Y will be held at 0. Analogously, if $J = 0$ and $K = 1$, the flip-flop will be changed to the $F = 0$ state unless it was in this state initially.

An important property of J-K flip-flops is that the J and K signals may change almost simultaneously with the 0-to-1 change in the actuating ("clock") signal without affecting flip-flop action. That is, the signals

J-K FLIP-FLOPS 141

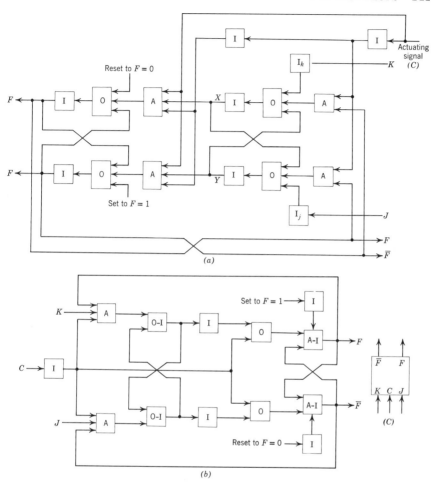

FIGURE 4-11. Asynchronous *J-K* flip-flops and symbol.

that are effective in determining flip-flop action are the values of *J* and *K* that are present immediately prior to the application of the clock signal. The *J* and *K* signals to a given flip-flop may therefore be derived from the outputs of other flip-flops that are actuated at the same time. The significance of this *J-K* flip-flop property will become more apparent in some of the examples of networks to be described in later sections.

A *J-K* flip-flop which is similar in most basic principles but which contains many differences in detail is shown in Figure 4-11*b*. With this variation, the double inverters in the actuating signal line are not needed, as can be appreciated by examining the step-by-step functioning of the

142   FEEDBACK, DIGITAL STORAGE, AND TIME-DEPENDENT SIGNALS

individual switching elements for various initial values for J, K, and F.

The flip-flops in Figure 4-11 are commonly called "master-slave" flip-flops, although each of the two R-S flip-flops within the J-K flip-flop exerts some control over the other, so the assignment of the roles of master and slave is not always obvious. Usually, the flip-flop on the right, as the flip-flops appear in this figure, is said to be the master, in which case the slave powers the output lines.

In Figure 4-11 additional input lines with signals designated "set" and "reset" are included as in an elementary R-S flip-flop. These lines, which are commonly included in physical J-K flip-flop modules, afford increased design flexibility as in, for example, a counter (such as to be described later) that is to be initially set to some predetermined value. The "set" and "reset" lines allow the initial setting directly, whereas awkward elaborations in the counter network would otherwise be required.

Pseudo-J-K Flip-Flops

Many widely used flip-flops are called J-K flip-flops although they have characteristics that are at variance, in one way or another, with the most widely accepted definition of J-K action. They are here called pseudo-J-K flip-flops. One example is that some flip-flops are actuated by the 1-to-0 change instead of the 0-to-1 change in the clock signal. This variation is minor, but it does affect the physical design of a network and it can be the source of considerable confusion.

Of more consequence, some pseudo-J-K flip-flops are internally structured so that changes in the J and K signals must normally be made only when the clock is 1 (for flip-flops actuated by 0-to-1 changes in the clock signal). J and K signal changes made when the clock signal is 0 will cause flip-flop actuations that are generally unwanted, although "trick" networks can sometimes be devised to make use of whatever action that does take place when the J and K signals are changed.

Other pseudo-J-K flip-flops contain two or more J input lines combined in an AND device and two or more J input lines combined in another AND device. No separate C input line is provided. By means of capacitors connected between the AND devices and the remainder of the flip-flop, as shown in Figure 4-12, the flip-flop is caused to be actuated by the 0-to-1 changes in the signals from the AND devices. Then, by connecting one J line and one K line together and using this as the C input line, a J-K flip-flop action of sorts is achieved in that, when the signal on this simulated C line is changed from 0 to 1, the flip-flop response depends on the other J and K signals in the pattern previously

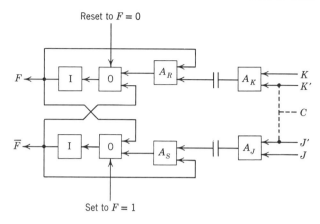

FIGURE 4-12. Pseudo-*J-K* flip-flop.

defined for *J-K* action. An important difference from conventional *J-K* action is encountered in that, if a *J* or *K* signal changes from 0 to 1 simultaneously with the 0-to-1 change in the *C* signal, the flip-flop response is though that *J* or *K* signal were 1 (not 0 as in conventional *J-K* action). A 1-to-0 change in a *J* or *K* signal simultaneously with a 0-to-1 change in the *C* signal may under certain circumstances cause an indeterminate flip-flop response. Therefore, the network employing the flip-flops must ordinarily be designed so as to avoid this situation. Also, when the *C* signal is 1, 0-to-1 changes in the other *J* and *K* signals can initiate flip-flop responses—a characteristic not in line with the usual *J-K* flip-flop definition. Again, these responses are generally unwanted, although "trick" networks can sometimes be designed to make use of the characteristics.

Once a designer becomes "mentally attuned" to any given pseudo-*J-K* flip-flop characteristics, he can probably design networks using them as well as when using *J-K* flip-flops that fit the more widely accepted definition of *J-K* action. In general, however, the commercial availability of flip-flops having differing logical characteristics, but all called *J-K* flip-flops, is a source of great confusion. Because the variants introduce no important new design principles, they will not be discussed further, and this section is included merely to alert the designer to the situation.

The *D* Flip-Flop

A type *D* flip-flop has an actuating ("clock") input signal line *C* as in a *J-K* flip-flop, but instead of the *J* and *K* input terminals, a single input terminal D is provided. The flip-flop output signal F becomes 1 or 0 according to whether D is 1 or 0, respectively, at the time of a 0-to-1 change in C.

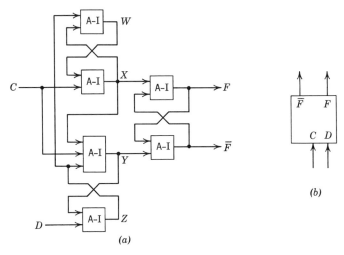

FIGURE 4-13. Asynchronous D flip-flop and symbol.

Various physical circuits have been devised to form a D flip-flop, but most integrated-circuit versions are based on the network of A-I modules illustrated in Figure 4-13a. When $C = 0$, the signals at points X and Y are both 1, and the storage part of the network (the two right-hand A-I modules) can exist in either stable state as required for the storage of a binary digit. If $D = 1$, the signal at point Z will be 0, and the signal at W will be 1. Now if C is changed from 0 to 1, the signal at X will become 0, but Y will remain at 1. Therefore, the flip-flop will be set to the state where $F = 1$ whether or not it was already in this state. Note that while $C = 1$, signal D can return to 0 without affecting the stored digit, because even though Z will become 1, the 0 at X will hold signals W and Y at 1.

If $D = 0$ when $C = 0$, signal Z will be 1 and W will be 0. Then at the time of the 0-to-1 change in C, the signal at Y will become 0, but the signal at X will remain 1. Therefore, the flip-flop will be set to the $F = 0$ state whether or not it was already in this state. If D is returned to 1 while C is retained at 1, the stored digit is not affected because the 0 now at Y will hold Z at 1.

The symbol for a D flip-flop is shown in Figure 4-13b.

Because of the characteristics of D flip-flops, the D signal to a given

D flip-flop may be derived from the output signals of other D flip-flops actuated by the same C signal.

Using R-S, T, R-S-T, J-K, and D Flip-Flops in a Sequential Circuit

The storage network in the generalized sequential switching circuit of Figure 4-6 may be composed of flip-flops of any of the five different types which have been described. However, the design of the combinatorial network depends strongly on the flip-flop type which is chosen.

With R-S flip-flops, the R signal for a given flip-flop must be 1 for each step on which that flip-flop is to be changed from the 1 state to the 0 state (that is, to be "reset"). The R signal must be 0 for each step on which the flip-flop is to be in the 1 state. If the flip-flop is initially in the 0 state and is to remain in the 0 state on the next step, a "don't care" condition is encountered in the R signal. The 1, 0, and "don't care" conditions for the S signal are analogous.

With T flip-flops, the T signal to a given flip-flop must be 1 for each step on which that flip-flop is to change state, either from 0 to 1 or from 1 to 0, and the T signal must be 0 on all other steps.

With R-S-T flip-flops, the designer has the choice of using any given flip-flop either as an R-S flip-flop or as a T flip-flop. Alternatively, the R, S, and T input lines may be utilized in a mixed manner as may be appropriate to achieve some objective, such as the minimization of the number of switching elements in the combinatorial network. With some R-S-T flip-flop circuits, indeterminate or at least unreliable action will occur if 1 signals are applied simultaneously to the T and R lines or to the T and S lines, and of course such simultaneous signals must then be avoided. More commonly, R-S-T flip-flop circuits are of such nature that the R and S signals (which must never be 1 simultaneously with each other) tend to override the T signal so that the simultaneous application of R and T signals (for a 1-to-0 flip-flop change) or of S and T signals (for a 0-to-1 change) is permitted. This feature introduces another type of "don't care" condition that can sometimes be utilized in minimizing the number of components in the networks that generate the flip-flop input signals. With some R-S-T flip-flop circuits the overriding action of the R and S signals is so great that on a step where the flip-flop is to remain in the same state, the T input signal may be 1 provided the appropriate one of the R and S signals is also 1. Opportunities for further reductions in the number of switching network components are thereby provided in some applications. However, because of the current preference for J-K or D flip-flops, this subject will not be pursued.

146   FEEDBACK, DIGITAL STORAGE, AND TIME-DEPENDENT SIGNALS

Whether $R\text{-}S$, T, or $R\text{-}S\text{-}T$ flip-flops are employed, gates as shown in Figure 4-6 are required. The gates are not required when $J\text{-}K$ or D flip-flops are used, because the gating function has been built into the flip-flop structures.

With $J\text{-}K$ flip-flops, the J signal to a given flip-flop must be 1 for each step on which that flip-flop is to be changed from the 0 state to the 1 state. The J signal must be 0 if the flip-flop is in the 0 state and is to remain in the 0 state. "Don't care" conditions on the J signal are encountered either when the flip-flop is in the 1 state and is to remain in the 1 state or when the flip-flop is to be changed from the 1 state to the 0 state. The 1, 0, and "don't care" conditions for the K signal are analogous.

With D flip-flops, the D signal to a given flip-flop must be 1 or 0 according to whether the flip-flop is to be in the 1 or 0 state, respectively, on the next step regardless of its state on a given state.

The above input signal requirements for the various flip-flop types are summarized in Figure 4-14, where each d represents a "don't care" condition.

Flip-flop type			Flip-flop state			
			$0 \rightarrow 0$	$0 \rightarrow 1$	$1 \rightarrow 1$	$1 \rightarrow 0$
	$R\text{-}S$	R	d	0	0	1
		S	0	1	d	0
	T	T	0	1	0	1
	$J\text{-}K$	J	0	1	d	d
		K	d	d	0	1
	D	D	0	1	1	0

FIGURE 4-14. Summary of input signal requirements for various flip-flop types.

The various "don't care" conditions described in the previous paragraphs are in addition to any "don't care" conditions that might be encountered as a result of the fact that certain combinations of input signals will never be present at the inputs of the combinatorial network of Figure 4-6.

An Example Showing the Use of $R\text{-}S$, T, $R\text{-}S\text{-}T$, $J\text{-}K$, and D Flip-Flops in a Sequential Circuit

Assume a sequential circuit as in Figure 4-6 but with no input and output devices. Assume further that the circuit is a six-step counter containing three flip-flops A, B, and C that are to cycle through the following state pattern.

$$\begin{array}{ccc} A & B & C \\ 0 & 0 & 0 \\ 0 & 1 & 1 \\ 1 & 0 & 0 \\ 1 & 1 & 0 \\ 1 & 0 & 1 \\ 1 & 1 & 1 \end{array}$$

The network may be assumed to be initially set to the 000 state by means that are not specified. The storage network is to step through the six states in the indicated sequence and is to return to state 000 after state 111. In the absence of physical malfunctioning of the network, it will never arrive at states 001 or 010, and the corresponding signal conditions, $\bar{A}\bar{B}C$ and $\bar{A}B\bar{C}$, respectively, represent "don't care" conditions at the inputs to the combinatorial network. The problem is to design the combinatorial networks that would be required for each of the five types of flip-flops that might be used in the storage network.

With R-S flip-flops, consider the generation of signal S_C, that is, the S signal for flip-flop C. Flip-flop C is to become 1 ("set") after steps 000 and 110. Therefore, $S_C = \bar{A}\bar{B}\bar{C} + AB\bar{C}$. Condition $A\bar{B}C$ is a "don't care" condition for S_C because flip-flop C is in the 1 state at step 101 and is to remain in the 1 state upon going to the next step, which happens to be 111. Then after giving consideration to the $\bar{A}\bar{B}C$ and $\bar{A}B\bar{C}$ "don't care" conditions that apply regardless of flip-flop type, signal S_C can be simplified to $S_C = \bar{A}\bar{C} + B\bar{C}$, where the simplification methods described in an earlier chapter may be used to achieve this result.

After all flip-flop input signals are similarly determined, the resulting sequential circuit with R-S flip-flops is then as shown in Figure 4-15a where the combinatorial network generates the following signals.

$$\begin{array}{lll} R_A = ABC & R_B = B & R_C = BC \\ S_A = \bar{A}B & S_B = \bar{B} & S_C = \bar{A}\bar{C} + B\bar{C} \end{array}$$

With T flip-flops, again consider the flip-flop in position C. This flip-flop is to change from 0 to 1 after steps 000 and 110 and is to change from 1 to 0 after steps 011 and 111. On each of these steps, T_C, the T signal to the C flip-flop, should be 1. That is, $T_C = \bar{A}\bar{B}\bar{C} + AB\bar{C} + \bar{A}BC + ABC$. With the "don't care" conditions of $\bar{A}\bar{B}C$ and $\bar{A}B\bar{C}$, this expression can be simplified to $T_C = \bar{A} + B$. T_A and T_B can be determined by the same

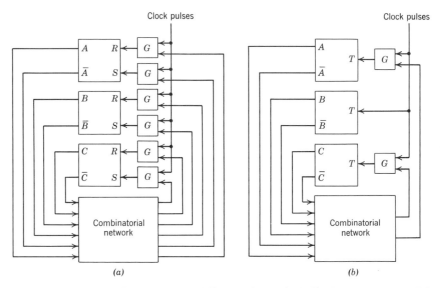

FIGURE 4-15. Using R-S (set-reset) and T (complementing) flip-flops in a sequential circuit.

procedure. The resulting sequential circuit is as in Figure 4-15b where the combinatorial network generates the following signals.

$$T_A = BC \qquad T_B = 1 \qquad T_C = \bar{A} + B$$

The 1 for T_B here means that a clock pulse is applied to flip-flop B on every step (not that signal to flip-flop B remains at a steady-state 1 potential).

Because the signals for T flip-flops are so simply generated in this example, no consequential advantage is gained by using R-S-T flip-flops. Some flexibility in design options would be realized if, for example, an R-S-T flip-flop were used in position C. Suitable signals would be $R_C = BC$, $S_C = B\bar{C}$, $T_C = \bar{A}$. Only AND devices (no OR devices) would then be needed in the combinatorial network.

With J-K flip-flops, again consider the flip-flop in position C. Signal J_C must be 1 for each step on which the flip-flop is to change from 0 to 1, that is, on steps 000 and 110 represented by $\bar{A}\bar{B}\bar{C}$ and $AB\bar{C}$. J_C must be 0 whenever flip-flop C is initially at 0 and is to remain at 0, that is, on step 100 represented by $A\bar{B}\bar{C}$. All other signal combinations to the combinatorial network are "don't care" conditions. Previously described techniques can then be used to determine that $J_C = \bar{A} + B$. The same

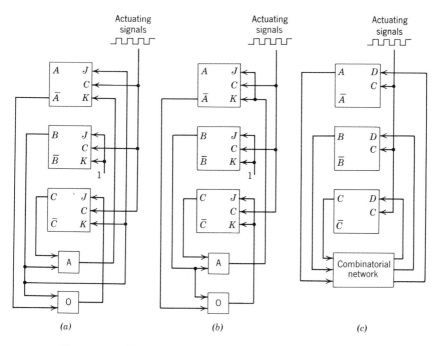

FIGURE 4-16. Using J-K and D flip-flops in a sequential circuit.

sort of analysis is used to determine that all of the J and K signals are as follows.

$$J_A = B \qquad J_B = 1 \qquad J_C = \bar{A} + B$$
$$K_A = BC \qquad K_B = 1 \qquad K_C = B$$

The resulting sequential circuit is shown in Figure 4-16a.

By using expressions for J_A and K_B that are not the simplest possible in this example of a six-step counter, the following relationships can be found.

$$J_A = K_A = BC$$
$$J_B = K_B = 1$$
$$J_C = K_C = \bar{A} + B$$

These signals are the same as were found for T flip-flops. In other words, the J-K flip-flops may be used as T flip-flops by connecting the respective J and K input terminals together, and in this particular example the number of AND and OR devices required happens to be the same as before. The resulting network in shown in Figure 4-16b.

With type D flip-flops, flip-flop C in the six-step counter previously set forth must be in state 1 after steps 000, 110, and 101. That is, $D_C = \bar{A}\bar{B}\bar{C} + AB\bar{C} + A\bar{B}C$. With "don't care" conditions of $\bar{A}\bar{B}C$ and $\bar{A}B\bar{C}$, the expression can be simplified to $D_C = \bar{A}\bar{C} + \bar{B}C + B\bar{C}$. The other D signals can be determined in like manner, and the three signals are then as follows.

$$D_A = A\bar{B} + A\bar{C} + \bar{A}C$$
$$D_B = \bar{B}$$
$$D_C = \bar{A}\bar{C} + \bar{B}C + B\bar{C}$$

The resulting sequential circuit is shown in Figure 4-16c where the combinatorial network generates the functions listed above.

Observe that, although the sequential circuits employing J-K and D flip-flops in Figure 4-16 were derived from the concepts of synchronous operation, the circuits are actually asynchronous with actuating signal C playing the role of an input signal from an input device.

In comparing J-K and D flip-flops the above example is reasonably typical in that the D flip-flops tend to require more complex switching functions. On the other hand, the need to transmit only one control signal (instead of two) to each flip-flop is a sufficient advantage to justify the selection of D flip-flops instead of J-K flip-flops in many applications.

Some Relationships Among R-S, T, R-S-T, J-K, and D Flip-Flops

Previous sections have shown how elaborations can be added to an R-S flip-flop to form a flip-flop of any of the other types.

A type T flip-flop (which by definition has only one input line) can be used in an R-S application by inserting in the T input line a switching network which generates the function $T = FR + \bar{F}S$, where F is the flip-flop output signal. The resultant R-S flip-flop could then be used, as above, to form other flip-flop types, but this practice would be wasteful of components and has no apparent advantages. Adding components to an R-S-T flip-flop to form a J-K or D flip-flop is similarly possible but generally impractical.

A J-K flip-flop can be used as a T flip-flop simply by connecting both the J and K signals to a constant potential that represents binary 1. The T input signal is connected to the C terminal as shown in Figure 4-17a. Then the result can be modified to an R-S flip-flop by adding a network that generates $C = FR + \bar{F}S$ or to an R-S-T flip-flop by expanding the network to $C = T + FR + \bar{F}S$. Figure 4-17b shows how a J-K flip-flop is connected to simulate a D flip-flop. The D input signal is connected to the J terminal and the $\bar{D}$ signal (generated by means of an inverter if not otherwise available) to the K terminal.

COUNTER CIRCUITS EMPLOYING J-K FLIP-FLOPS ONLY   151

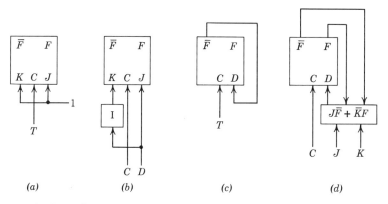

FIGURE 4-17. Examples of using a flip-flop of a given type to simulate a flip-flop of another type.

A D flip-flop can be used as a T flip-flop simply by connecting the $\bar{F}$ output terminal to the D input terminal with the T input signal connected to the C input line as in Figure 4-17c. The resultant T flip-flop can then be modified to an R-S or an R-S-T flip-flop as in the previous paragraph. A D flip-flop can be used in a J-K application by inserting in the D input line a network which generates the function $D = J\bar{F} + KF$ as in Figure 4-17d.

Counter Circuits Employing J-K Flip-Flops Only

The example used in the previous sections to illustrate the operation of flip-flops in a sequential circuit is actually a counter in that it proceeds cyclically through a set of stable states in response to a series of clock pulses or actuating signals. If the requirements are merely the stepping through a prescribed number of stable states, the designer may arbitrarily assign flip-flop states to the states of the circuit as a whole and thereby, in some instances, achieve a saving in switching devices in comparison with the number of such devices required when the individual flip-flop states are specified in the problem.

In particular, an intriguing problem of some practical importance is to devise counters requiring no switching devices other than the J-K flip-flops themselves. Figure 4-18 shows solutions to this problem for various numbers of steps in the cycling process. The detailed states of the individual flip-flops can be determined from the connections, although the states are listed for some of the counters. The 10-counter, for example, is formed by adding a flip-flop D to a circuit which is identical to the

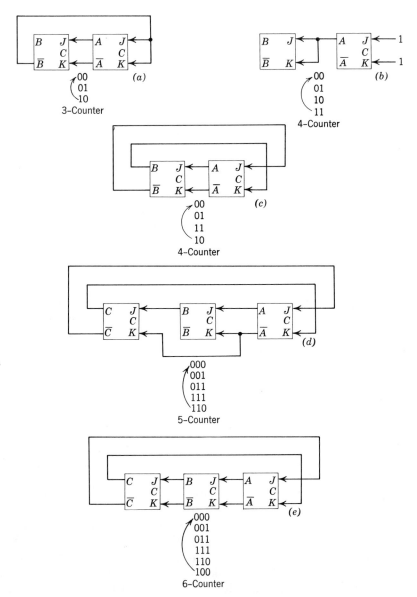

FIGURE 4-18. Counter circuits employing J-K flip-flops.

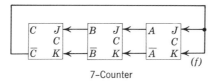

7-Counter

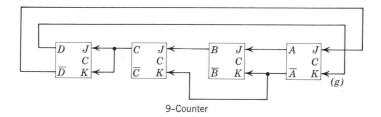

9-Counter

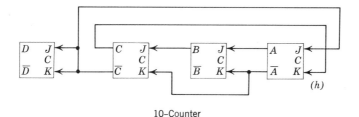

10-Counter

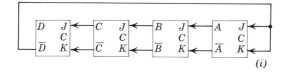

15-Counter

FIGURE 4-18. (CONTINUED)

5-counter. Flip-flop D receives both its J_D and K_D signals from the $\bar{C}$ output of the C flip-flop, and flip-flop D is thereby caused to change state after each step on which $C = 0$. A doubling of the number of states of the circuit as a whole is achieved because $C = 0$ an odd number of times in the cycling of the 5-counter. The C signal cannot be used in place of $\bar{C}$ because $C = 1$ an even number of times. (Alternatively, J_D and K_D could be derived from signals A or B, but not $\bar{A}$ or $\bar{B}$.)

By similarly adding a D flip-flop to the 6-counter or the 7-counter in Figure 4-18, a 12-counter or a 14-counter, respectively, may be formed.

However, an 8-counter formed with only three *J-K* flip-flops of the type assumed appears to be an impossibility. In particular, the adding of a third flip-flop in the manner just described to either of the 4-counters shown will not produce an 8-counter because A, $\bar{A}$, B, and $\bar{B}$ are each 1 on an even number of steps. An 8-counter with four flip-flops can nevertheless be formed by inserting an additional flip-flop in the shift-register loop of the 6-counter shown. Counters of the type under consideration but with 11 or 13 states, respectively, have not been found, although the impossibility of such counters is not apparent either.

J-K Flip-Flops with Multiple J and K Inputs

J-K flip-flop modules having more than one terminal for each of the *J* and *K* input functions are useful and are commonly manufactured. Internally, the several *J* input signals are combined in an AND function, as are the several *K* input signals. That is, the effective *J* input signal is 1 only when all of the individual *J* input signals are 1, and a similar remark applied for the *K* signals. In many *J-K* flip-flop designs the multiple *J* and *K* input terminals can be provided inexpensively merely by adding diodes to diode AND circuits which must already have been included to provide the *J-K* action. This situation is illustrated for the *J-K* flip-flop in Figure 4-10. The addition of AND-related *J* and *K* input terminals is not inherently so simple however, and the *J-K* flip-flop of Figure 4-11a happens to be an example where an inventor plus an input to an OR device is required for each added AND-related *J* or *K* input. (Combining the multiple *J* and *K* signals in other ways, such as by the OR function for example, would be quite feasible but is virtually never done in practice within the flip-flop modules.)

The purpose of the added inputs is to provide a more flexible module. In particular, many sequential circuits can thereby be designed with fewer accompanying AND and OR devices than when only one *J* and one *K* input is provided on each *J-K* flip-flop. This feature of multiple-input *J-K* flip-flop modules is illustrated in Figure 4-19 which shows a 16-counter requiring only four flip-flops with no added switching devices in *a* and a 10-counter with no added switching devices in *b*, and both counters count in a so-called "binary coded" manner. That is, a numeric representation of the counter state can be derived by assigning "weights" of 8, 4, 2, and 1, respectively, to the outputs of the four flip-flops and adding the weights of those flip-flops in the "set" state.

Ordinarily the modules are designed so that, in the absence of any connection to a particular *J* or *K* input terminal, the effect is to supply a continous 1 to that terminal. However, with some designs each unused

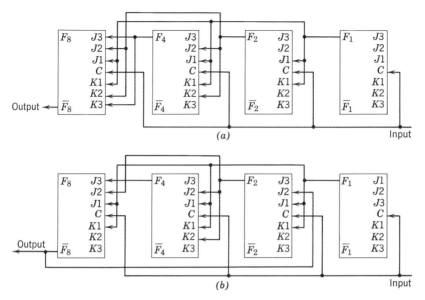

FIGURE 4-19. Binary-coded 16- and 10-counter.

input terminal should be connected to a suitable d-c potential source. In Figure 4-19 the former situation is assumed. In a, flip-flop F_1 performs a complementing action in response to each actuating signal on the input line. Flip-flop F_2 is complementing on steps when F_1 is 1. Flip-flop F_4 is complemented only on steps when both F_1 and F_2 are 1. Flip-flop F_8 is complemented only when all three of F_1, F_2, and F_4 are 1. The counter in b is similar except that it is modified to return to state 0000 after state 1001. The application of signal $\bar{F}_8$ to the $J2$ input on flip-flop F_2 allows flip-flop F_2 to function as before as long as $\bar{F}_8$ is 1, as it is from states 0000 through 0111 (decimal 0 through 7). On state 0111 the signals to inputs $J1$, $J2$, and $J3$ on flip-flop F_8 are all 1 (as is the signal to input $K1$ of this flip-flop) so that the next input signal to be counted causes F_8 to become 1 as desired for the next state 1000 (decimal 8). For stepping from 1000 to 1001 (decimal 8 to 9), only flip-flop F_1 is active. Then, although a 1-signal is applied to inputs $J1$ and $K1$ of flip-flop F_8, 0-signals are applied to $J2$ and $J3$ of this flip-flop. Therefore, the next input signal causes F_8 to become 0 as desired for the next state, 0000. The $\bar{F}_8$ signal (0 on step 1001) at the J_2 input of flip-flop F_2 holds F_2 at 0, as desired when stepping from 1001 to 0000.

When applying multi-input J-K flip-flops to more generalized problems as typified by Figure 4-6, the same techniques as before are used except that the expression for any given J or K signal is converted to OR-to-AND form. The signals represented by the individual factors in the expression are then applied to the respective individual J or K inputs, as the case may be. The AND function is performed internally in the J-K module. Of course, if the expression contains more factors than there are individual inputs available, appropriate additional switching elements must be provided.

Further complications are encountered with the modules of some manufacturers in that the J and K inputs are actually $\bar{J}$ and $\bar{K}$ inputs. This situation can be confusing. In particular, the designer must keep in mind as to whether, for a given module type, the $J1$, $J2$, etc., signals or the $\overline{J1}$, $\overline{J2}$, etc., signals are the ones which are internally combined in an AND relationship, but no new design principles are required.

Obtaining the Actuating Signals for Some Flip-flops from the Outputs of Other Flip-flops

Often a saving in components can be realized by utilizing some of the flip-flop output signals as the actuating pulses or signals for other flip-flops in the same sequential circuit. Figure 4-20 shows an elementary example. This circuit performs a counting function essentially the same as performed by the circuit in Figure 4-19a, but in the present circuit only complementing flip-flops are required (no J or K inputs) and only a single connection is needed from one flip-flop to the next in a chain

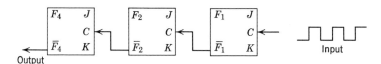

FIGURE 4-20. A network (a counter) in which the actuating signals for some flip-flops are obtained from the output signals of other flip-flops.

of flip-flops. The output from the right-hand flip-flop changes from 1 to 0 or from 0 to 1, as the case may be, whenever the input (actuating) signal changes from 0 to 1. No flip-flop response occurs when the actuating signal changes back to 0. Thus, for each two input pulses or signals, the output signal from the right-hand flip-flop changes back and forth between 0 and 1 once. An analogous action occurs in the other flip-flops so that the circuit as a whole proceeds through the stable states indicated in the listing in that figure.

With some flip-flop designs a minor problem with regard to resetting to 000 may occur. The flip-flops must be designed so that they may be reset when the respective actuating signals are 1's, because the $\bar{F}$ signals and not the F signals must be transmitted from one flip-flop to the next to obtain the state sequence indicated. In the case of the flip-flops shown in Figures 4-8a and 4-11, the "set" and "reset" inputs must be connected to the right-hand internal flip-flops. Alternatively, the circuit in Figure 4-20 may be initially set to 111 instead of 000 in applications where an initial setting of 111 is acceptable.

The counter in Figure 4-20 can be modified in many different ways to obtain counters that cycle through other than 2^n states, where n is an integer equal to the number of flip-flops. However, the more general problem of usefully utilizing some flip-flop output signals as the actuating signals for other flip-flops when the output signals of the network as a whole are some arbitrarily specified function of the input signals is, as yet, without an organized design procedure. Even the modifications of the 16-counter to obtain a 10-counter requires sufficient ingenuity to result in a patentable invention when such a modification is found. Six examples of 10-counters based on modifying a 16-counter of this category are shown in the previously mentioned book *Electronic Digital Components and Circuits,* Figures 9-8 and 9-11.

Controlling the Sequential Operation of Asynchronous Circuits

A complete digital system is seldom designed on the basis of a single elaborate sequential circuit (in spite of the theoretical possibility of doing so). Instead, the more usual and practical approach is to divide the system into subsystems, each of which functions more or less in the manner of a sequential circuit as already described. Then, in the operation of the system as a whole, a signal is applied to one subsystem (sometimes two or more subsystems simultaneously), and when the first subsystem completes a step of operation it generates a completion signal which initiates the operation of one or more other subsystems which in turn generate further completion signals for initiating the operation of still other subsystems. The transmission of any given completion signal from one subsystem to another (where it becomes an initiation signal) may be under the control of miscellaneous signals existing within the system as a whole, and conventional combinatorial switching networks may be used for this purpose.

In an asynchronous sequential circuit no particularly obvious means is available for signaling the completion of the change in the states of the flip-flops used in the storage network part of the circuit. The flip-flops which change state on some steps are not necessarily the same as the ones which change state on other steps. In fact, with some circuit designs and with some combinations of input signals, no state changes at all may occur in response to an actuating signal. Therefore, some additional means must be used to generate a completion signal. The particular form of network needed depends on the details of the system requirements, but the control network shown in Figure 4-21 has reasonably wide applicability in that it can be used with the generalized asynchronous sequential circuit described earlier. Moreover, the network responds properly to an initiating signal of any duration greater than a certain minimum duration, and the duration of the completion signal is independent of the duration of the initiating signal. This last property is needed for the proper sequential operation of many networks where the interconnecting completion-initiating signals may otherwise drift to longer or shorter durations.

In the quiescent condition of the control network, R-S flip-flop A in Figure 4-21 must be in the reset state because, if it were set, the 1 appearing on line A would pass through inverters I_2 and I_3 to the reset

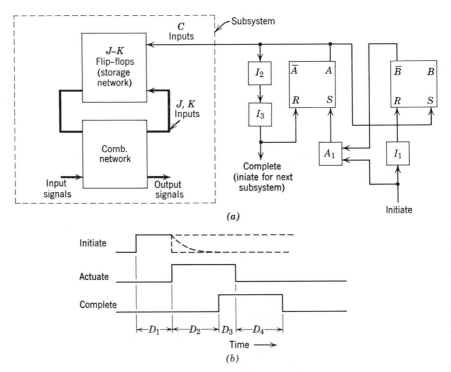

FIGURE 4-21. Control network for sequential operation of asynchronous circuits.

SEQUENTIAL OPERATION OF ASYNCHRONOUS CIRCUITS 159

terminal and reset it. Flip-flop B is also reset when the network is in the quiescent condition because the assumed 0 on the input line marked "initiate" appears as a 1 at the output of I_1, and this 1 resets B. In this control network, each inverter is assumed to display a slight delay. That is, when a signal to an inverter is changed from 0 to 1, or from 1 to 0, the resulting inverse change in signal at the inverter's output occurs slightly later in time. Although this delay may be extremely brief with very high speed circuits, it should be roughly the same for all inverters used, including the internal inverters in the flip-flops of the control network and in the flip-flops of the sequential circuit being controlled.

When an initiating signal is applied, the fact that $\bar{B}$ is 1 allows the initiating signal to pass through AND device A_1 to the set terminal of flip-flop B. As illustrated, after a time delay D_1 an actuating signal appears on the A line to actuate the sub-system, which is essentially in the form of a sequential circuit as described earlier. The duration of D_1 is the sum of a turn-on time of an inverter and the turn-off time of an inverter because (as will be recalled from R-S flip-flop action) the two inverters within a flip-flop must operate in succession to generate an output signal of the same polarity as the input signal. When A becomes 1, a "set" signal is applied to flip-flop B so that, after a time delay equal to the turn-off time of an internal inverter, $\bar{B}$ becomes 0 and AND device A_1 then blocks further passage of the initiating signal. Then, after a delay D_2 from the time when A became 1, this 1 reaches the reset terminal of flip-flop A and resets it. D_2 is likewise equal to the sum of a turn-off and a turn-on time, but in this case the times are a result of the delays in I_2 and I_3. After an additional delay D_3, signal A becomes 0 again, where D_3 is the turn-off time of an inverter in flip-flop A. Finally, after another delay D_4 the output of I_3 becomes 0 again. D_4 is also the sum of a turn-off and a turn-on time in I_2 and I_3, but the roles of these inverters are reversed with respect to their contribution to signal delay.

Provided only that the initiating signal was of sufficient duration to actuate flip-flop A, the response of the circuit did not in any way depend on the nature of this signal. However, to produce a second actuation of the sequential circuit, the initiating signal must be returned to 0 for a duration of time sufficient to cause flip-flop B to be reset again. The dotted lines for the waveform of the initiating signal in Figure 4-21b signify the lack of influence of this signal during the corresponding periods of time.

If the duration of the actuating signal is not sufficient with the network as shown in Figure 4-21a, it may be increased by adding an even number of inverters in series with I_2 and I_3.

For repeated actuations of the same subsystem, a second control network identical to the one shown may be added. The completion signal for each control network is used as the initiating signal for the other. However, at least one OR device and one AND device must ordinarily be included for switching purposes in the interconnecting completion-initiating lines. The OR device is needed to introduce a signal to get the network started. The AND device is needed for adding a control signal, the removal of which stops the operation. This added control signal may be derived from the J-K flip-flops assumed to exist within the subsystem. In a sense, the two control networks and the subsystem may then be viewed as a complete assembly with a single initiating signal (introduced through the OR device to get the operation started) and a single completion signal which is generated by taking the inverse of the signal which allows repeated actuations.

If certain errors are made in the design of the sub-system or if certain malfunctions in the components occur, the repeating operation described in the previous paragraph may fail to terminate. This result is typical of asynchronous circuit behavior, and in systems such as a computer a digitally operating time clock is needed to detect this type of error. An error-detecting switching network is used to sense whether or not expected results have been attained at predetermined times.

Sequentially Operating Asynchronous Shift Register

In many practical applications the sequential circuits are not of the so-called generalized form illustrated in Figure 4-6 but instead are any one of many miscellaneous forms which are usually much more complex in the manner in which feedback paths are found in the switching network. The miscellaneous sequential circuits are particularly useful in that part of a system which performs arithmetic operations where "carries" from one digit position to the next play an important role. The significance of many of these circuits cannot be fully appreciated until after a study of arithmetic operations as they are performed in a computer, but an elementary example of a sequential circuit in this category will be described to illustrate some of the basic considerations pertaining to sequential circuit interconnections.

The example is a simple shift register consisting of a set of binary storage elements (flip-flops) with means for "shifting" the stored binary digits (as represented by the states of the flip-flops) from each given storage element to the next storage element in succession in response to actuating signals. With J-K flip-flops the so-called generalized sequential circuit can be made to form a shift-register merely by connecting the output signals from each given flip-flop to the J and K input terminals, respectively, of the next flip-flop in the set. The actuating pulse is applied to the C input terminals of all flip-flops simultaneously. In fact, the 6-counter in Figure 4-18e is actually a shift register when the digit from the last flip-flop is entered in inverted form into the first flip-flop.

The sequentially operating shift register to be described here employs only R-S flip-flops and operates quite differently. Figure 4-22 shows the network for two stages of the register. Two R-S flip-flops are required for each stage. With integrated circuits, both flip-flops for a stage, together with the associated O-I modules and inverters can be fabricated in a single module that is little, if any, more complex than the circuit required for a single J-K flip-flop. In some circumstances the two R-S flip-flops may actually result in the simpler circuit that has various other advantages. In the quiescent condition when the actuating signal is 0 the outputs of inverters I'_1, I'_2, and so on, are 0. In this condition, the state of the right-hand flip-flop in each stage is set to the state of the left-hand flip-flop in that stage, and the left-hand flip-flop in any given stage is in effect isolated from the right-hand flip-flop in the next higher numbered stage (the one to the left).

When the actuating signal in Figure 4-22 is made 1 the output signals from the O-I circuits at the input of flip-flop F_1 both become 0 so that this flip-flop is in effect isolated from flip-flop F'_1. At essentially the same time (with the assumption that each inverter in the network introduces approximately the same amount of delay), the output of I_1 becomes 0. Because of the inversions in the O-I modules at the inputs to flip-flop F'_1, this flip-flop is set to the state of flip-flop F_2 in the second stage. Next, the output of I'_1 becomes 1 to cause an analogous action in the second stage of the shift register.

When the actuating signal is returned to 0, flip-flop F_1 is set to the new state of flip-flop F'_1 and then the 1 that appears at the output of I_1 again isolates flip-flop F'_1 from flip-flop F_2. The 0 that then appears at the output of I'_1 creates the same action in the second stage. Thus,

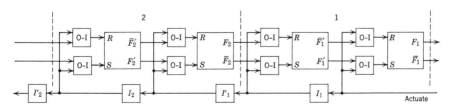

FIGURE 4-22. Sequentially operating asynchronous shift register.

after the actuating signal has been changed from 0 to 1 and then back to 0 the digits in each stage have been shifted to the next stage on the right. With the connections shown, the digit in the first stage has been shifted "out of" the register and is lost, but this digit may be shifted to the last stage by connecting the outputs of the first stage to the inputs of the last stage.

With respect to the major topic under consideration, an important feature of the shift register of Figure 4-22 is that the appearance of a signal at the left end of the chain of inverters along the bottom of the figure occurs significantly later than the application of the actuating signal. The total delay is, of course, the sum of the delays in the inverters. In some applications, the signal at the end of the inverter chain can be used as the completion signal to initiate the action of some other sequential circuit. However, with long chains the duration and waveform of the completion signal may be badly distorted in comparison with that of the actuating signal. Thus, apart from questionable actuation of the latter stages in the shift register, the completion signal may be unsuitable for transmission to other sequential circuits. Some means must be included to insure that the actuating signal is applied for a sufficient length of time and then removed in preparation for another shift operation, and the means is described in the next section.

In other kinds of sequential circuits, the completion signal may occur a substantial time after the initiating signal but for quite different reasons. For example, the "sequential circuit" might be an addressed storage unit that requires a considerable amount of time to gain access to the specified address and to transmit the stored information to a register for use. The generation of the competion signal would be accomplished quite differently than by merely passing a signal through a chain of inverters. The signal generation mechanism would depend on the characteristics of the storage unit, but the control of the unit could be the same as the control of the shift register just described.

Controlling a Sequential Circuit That Consumes a Substantial Time Interval

The characteristics of the network needed to control a sequential circuit that consumes a substantial period of time depend on the exact functional characteristics of the circuit. For purposes of illustration, the assumption will be made that the actuating signal must be applied continuously until a signal is generated at the "far" end of a chain of switching elements, at which time the actuating signal may be removed. A control network that meets this requirement appears to be widely applicable, although in some cases the actuating signal may be terminated sooner than the indicated time. In particular, the control circuit to be described will work with the shift register of the previous section even though the actuating signal may be applied longer than necessary. A possible purpose of terminating the actuating signal sooner would be that a second shifting operation could be initiated sooner. In fact, at a given time, two or more shifting operations may be "rippling" along, one after the other, in a sort of wave motion through very long shift registers.

In Figure 4-23a the block labeled "chain of switching elements" represents any sequential circuit, such as the previously described shift regis-

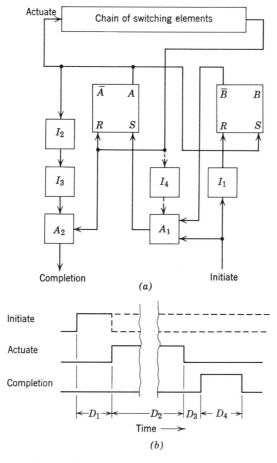

FIGURE 4-23. Modified control network—for asynchronous circuits requiring a substantial time interval.

ter, which generates an output signal after an input actuating signal has passed in sequence through a substantial number of delay-producing switching elements. The functioning of the network is essentially the same as before except in the resetting of flip-flop A and in the manner in which the completion signal is generated. Flip-flop A is reset by the signal returned from the chain of switching elements so that the actuating signal is applied continuously from the time that A is set to the time that the return signal is received. In some applications the returning signal itself may serve as a suitable completion signal, but in the network as shown, the completion signal is generated by combining the returning signal with a delayed version of the actuating signal. If the duration of the completion signal thus generated is not sufficient, it may be made longer by adding an even number of inverters in series with I_2 and I_3.

The waveforms of the pertinent signals as shown in Figure 4-23b are the same as in Figure 4-22b except that interval D_2 is longer and in the present instance is equal to the time required for a signal to pass through the chain of switching elements.

Inverter I_4, shown with dotted connections in Figure 4-23a, serves to prevent a second initiation of network operation during the time that the signal being returned from the chain of switching elements is still 1. The need for this inverter depends upon the detailed characteristics of the chain of switching elements and upon the nature of the source of the initiating signal.

Figure 4-24 illustrates a scheme for repeating the actuation of an asynchronous circuit by using the completion signal to generate a new initiating signal. The repeating is under the control of a "repeat" signal, which may be derived from an external source or may be derived from the sequential circuit itself.

The operation of the network in Figure 4-24 is the same as the operation of the network in Figure 4-23a except that flip-flop C, OR device O_1, inverter I_5, and AND devices A_3, A_4, and A_5 have been added. Flip-flop C may be assumed to be reset as a result of a previous operation of the control network. Alternatively, facilities not shown in the figure may be added to insure that C is initially reset. The original initiating signal is passed through O_1 to generate the first actuating signal as before. When the output of A_2 (formerly the completion signal) becomes 1, this signal is passed through A_3 or A_5 according to whether the repeat signal is 1 or 0, respectively. If the repeat signal is 0, no repetition is to take place, and the output of A_5 is thus the completion signal for the network as a whole. However, if the actuation of the sequential circuit is to be repeated, the repeat signal is 1 and the output of A_3 is used to set flip-flop C. When the output of I_4 becomes 1, this signal

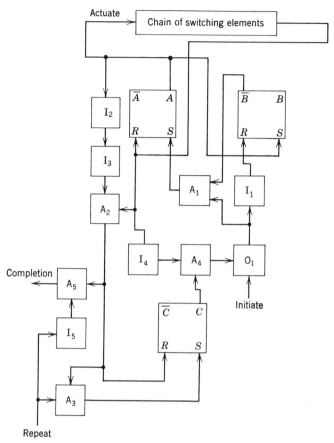

FIGURE 4-24. Control network for self-repeating the actuation of a sequential circuit.

is combined with signal C in A_4 to produce a 1-signal that is passed through O_1 to serve as the initiating signal for the second actuation of the sequential circuit. When A becomes 1, flip-flop C is reset, and the network is thereby prepared for a third sequence of operations in the event that the repeat signal continues to be 1. The process repeats as long as the repeat signal remains 1, but when the repeat signal is 0 a completion signal is generated at the output of A_5 at the time a signal appears from A_2.

Just to illustrate how the sequential circuit can control the repeating of itself, the chain of switching elements in Figure 4-24 may be a shift register as in Figure 4-22. The repeat signal may be $\bar{F}_1$ as derived from the flip-flop in the first stage of the shift register. Prior to the first initi-

166   FEEDBACK, DIGITAL STORAGE, AND TIME-DEPENDENT SIGNALS

ation of the network, any one (other than the first) of the shift-register flip-flops may be set to 1 with all others reset to 0. Then the operation will be repeated as long as $\bar{F}_1$ remains 1, but on the step when the 1 is shifted into the first flip-flop, $\bar{F}_1$ will become 0, and the repeating will be terminated. This particular set-up is probably of limited practical application, but it should be suggestive of the multitudinous ways in which sequential circuits can control themselves and each other.

Switching Networks Responsive to the Sequence of Applied Signals

In the design of complete digital systems the need arises for various sub-systems in which the parts function sequentially in all manner of different ways. Although all of these subsystems could conceivably be treated in the generalized manner discussed earlier in this chapter, such treatment is often impractical. Instead, the necessary networks can be derived more or less directly from the purpose of the individual subsystems. Although admittedly having intuitive characteristics, the design procedure to be described by means of examples in the following paragraphs leads reasonably directly to properly functioning networks.

For the first example, consider a network with only two input signals X and Y but which is to respond in accordance with the sequence in which these two signals are applied. In particular, assume that the output signal should be 1 when XY is 1 but only if, after the last point in time when X and Y are both 0, X is changed to 1 prior to (or substantially the same time as) the time when Y is changed to 1. A suitable switching network can be derived easily by noting that a flip-flop can be used to "remember" whether or not X became 1 first. This "remembering" is needed during the time that both X and Y are 1. An elementary R-S flip-flop can be used, and the "set" signal is obtained from the function $X\bar{Y}$. The flip-flop should be reset whenever X and Y are both 0, that is, the "reset" signal is $\bar{X}\bar{Y} = \overline{X + Y}$. The resulting network is shown in Figure 4-25a, where AND device A_2 allows the output to be 1 only when AXY is 1, where A is the signal from the flip-flop. The output signal that results for typical sequences in X and Y signal changes is illustrated in Figure 4-25b.

The network in Figure 4-25a can be simplified by grouping the individual switching functions within the R-S flip-flop with the switching functions external to the flip-flop. If the line that carries signal $\bar{A}$ is arbitrarily assigned the symbol G, an examination of the network reveals that A may be expressed as follows.

$$A = \overline{G + \overline{X + Y}} = \overline{G + \bar{X}\bar{Y}} = \bar{G}(\overline{\bar{X}\bar{Y}}) = \bar{G}(X + Y)$$

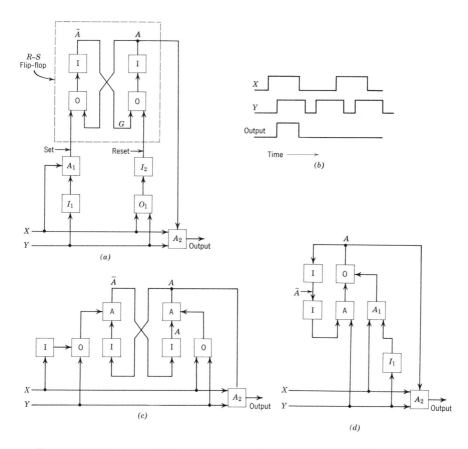

FIGURE 4-25. Network which is responsive to the sequence of applied signals.

By a similar examination, $\bar{A}$ as generated by the left-hand inverter in the flip-flop is found to be as follows.

$$\bar{A} = \overline{A + X\bar{Y}} = A\overline{(X\bar{Y})} = \bar{A}(\bar{X} + Y)$$

With A and $\bar{A}$ derived in accordance with these expressions the resulting network is as shown in Figure 4-25c. Note that signal A appears at two different network points, which on superficial examination would appear to carry signals that are not necessarily the same as each other. This apparent discrepancy is explained by the fact that, if the signals at the two points are not the same, a feedback action takes place within the network to make the signals the same. The final output signal in c represents the same function of X and Y as in a.

A further simplification of the network can be realized by expressing signal A as follows.

$$A = \overline{(\bar{X} + Y)\bar{A}}(X + Y) = (\overline{\bar{X} + Y} + A)(X + Y)$$
$$= (X\bar{Y} + A)(X + Y) = X\bar{Y} + AX + AY = X\bar{Y} + AY$$

The first form of the expression is determined by inspection of Figure 4-25c. The last form is determined by using the methods described in an earlier chapter to discover that term AX is superfluous. The resulting network is shown in Figure 4-25d. The fact that A is a function of itself does not mean that the output can be returned directly to an input. Instead, two inverters as shown are needed for amplification. The feedback path thus generated results in a network that has two stable states for certain combinations of X and Y, but the network is not a true flip-flop. Devices A_1 and I_1 in d serve essentially the same purpose as in a, but identically corresponding components are not found in c.

Any of the three networks in Figure 4-25 may be modified as necessary to suit the switching modules at hand, such as O-I, A-I, or O-A-I modules.

The example of Figure 4-25 is a good illustration of the usefulness of the flip-flop concept in devising sequential circuits, but it is also a good illustration of the fact that a flip-flop is not a basic switching module. A sequential switching circuit employs the same switching functions as found in a combinatorial network. Feedback paths are added, but these paths need not be in any standardized patterns as implied by the flip-flop concept.

Another Example Pertaining to Signal Sequence

As the next example, consider the problem illustrated in Figure 4-26a. In each specified time interval, short-duration signals X and Y may each appear once. If X appears first, Y should appear (possibly delayed slightly) on the output line, but if Y appears first, the output line should remain at 0. Between time intervals a separate reset signal is available to reset any flip-flops or multistate networks that might be used in the network as a whole. Flip-flop A and AND device A_1 in Figure 4-26b would constitute a sufficient solution for this problem except that an unwanted sliver signal may occur on the output line (output of A_1) if X and Y happen to appear at approximately the same time. The sliver problem is eliminated by using the output of A_1 to set a second flip-flop B. Although the pulse-like signal from A_1 may not be of desired

ANOTHER EXAMPLE PERTAINING TO SIGNAL SEQUENCE

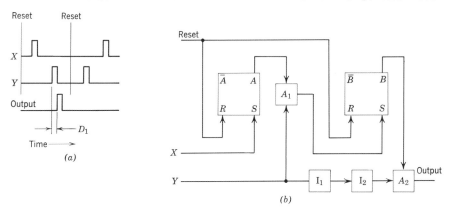

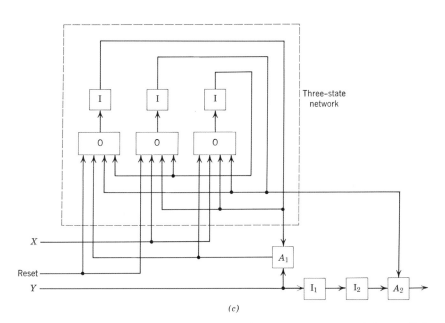

FIGURE 4-26. Another network which is responsive to the sequence of the applied signals.

amplitude and duration, this signal either will or will not set B. The output of B is then combined with Y in AND device A_2 to form the final output signal. Inverters I_1 and I_2 are needed to delay the passage of Y by an amount of time equal to the sum of a turn-off and a turn-on time of an inverter—D_1 in Figure 4-26a. This delay occurs in setting

B and is therefore needed in the path of Y to insure that a short-duration Y signal has not returned to 0 by the time that A_2 is made ready to pass the Y signal.

A study of the action of the network in Figure 4-26b reveals that B is never set at a time when A is reset. In other words, the two flip-flops when considered together exist in one of only three, not four, different stable states. Therefore, a three-state network requiring only three inverters (or only three O-I modules) can be substituted for the two flip-flops. The resulting network is shown in Figure 4-26c. The reset signal sets the three-state network to the state where signal C is 1, although C is not used except internally in the three-state network. X sets the three-state network to the state where signal A is 1. Then Y can set it to the state where B is 1. However, if the network is in state C at the time Y is 1, the action of A_1 prevents Y from reaching the three-state network. A_2 then prevents the passage of Y to the output terminal. In other respects the networks in b and c function in the same way.

A "One-Pulse" Network

Another example of a network which, in effect, responds to the sequence of applied signals is here called a "one-pulse" network. The function performed by this network is, in one variation or another, quite commonly required in the design of digital systems. The network has two input lines, one of which transmits a continuous series of "pulses" which for this example will be assumed to consist of a signal X that alternately changes back and forth between 1 and 0 with a more or less uniform time interval between transitions. The other input signal Y is 0 when the network is quiescent, but when Y is changed to 1, the network must allow one and only one X pulse to pass. The assumption is made that the output 1-signal should have a duration equal to the time interval that X is 1. This assumption implies that, if Y becomes 1 during an interval that X is 1, the output signal should not be responsive to X at this time but should remain at 0 until the next successive time interval that X becomes 1. Another assumption is that Y may remain 1 for only a very short period of time, and in particular may return to 0 prior to the time that X becomes 1.

A network that produces the indicated function is shown in Figure 4-27a. When Y is 0, flip-flop A will be reset as a result of the 1 that appears at the output of inverter I_1, and flip-flops B and C may be assumed to be initially reset as a result of previous operations of the type to be described. With C reset, input signal X cannot pass through

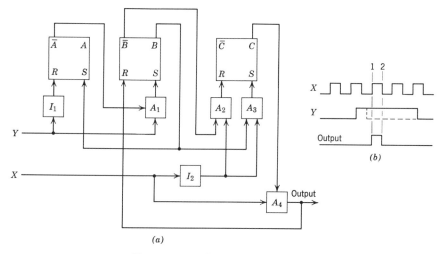

FIGURE 4-27. "One-pulse" network.

AND device A_4 to the output terminal. When Y becomes 1, this signal will pass through A_1 to set flip-flop B, because $\bar{A}$, the other signal to A_1, is 1 at this time. Signal B will then set flip-flop A, and Y can no longer pass through A_1. (Alternatively, the network could be designed so that both flip-flops A and B are set by the output of A_1.) When B becomes 1, A_3 is opened so that a 1 from the output of I_2 can pass to set flip-flop C. If X happens to be 0 at the time B becomes 1, the setting of C will occur immediately, but if X happens to be 1, the setting of C will occur later when the value of X changes from 1 to 0. In either case, an output signal from A_4 will not yet be generated. However, an output signal will be generated on the first change from 0 to 1 in X after the setting of C. This output signal is fed back to the reset terminal of flip-flop B so that, when X changes from 1 to 0 once more, the resulting 1 at the output of I_2 will pass through A_2 (which now allows the passages of signals because $\bar{B}$ is 1) and will cause flip-flop C to be reset. As long as Y remains 1, no further changes in flip-flop states can take place and no further X signals will pass to the output terminal.

When Y is returned to 0, the only action is the resetting of flip-flop A, and AND device A_4 continues to prevent the passage of X signals. However, the network is now ready for another actuation as will be initiated the next time that Y is made 1. If Y is made 1 for only a very short period of time, signals may be applied simultaneously to the set and reset terminals of flip-flop A from the time that Y is returned

172 FEEDBACK, DIGITAL STORAGE, AND TIME-DEPENDENT SIGNALS

to 0 to the time that B is reset, but no deleterious effects occur. One and only one full-duration X signal will pass.

The same techniques as illustrated in Figures 4-25 and 4-26 can be used to generate countless variations in the network showed in Figure 4-27. Still other variations can be found by substituting J-K flip-flops or pseudo-J-K flip-flops for the R-S flip-flops. Many of these variations may appear to or may even actually consume less equipment than the particular network shown. In particular, many versions requiring only two flip-flops can be found. However, with large scale integration (LSI) in circuit fabrication the number of individual switching elements tends to be more important than the number of flip-flops, so the more complex flip-flop modules may not create the components savings that are superficially indicated.

A Clock Network—Digital Cycles

Digital systems are commonly designed to operate in cycles where each cycle spans a predetermined interval of time. Within each cycle a sequence of signals or pulses designated T_1, T_2, and so on, are generated. These signals are used to actuate various components or subsystems at the corresponding points in time in each cycle. Of course, during any given cycle any individual component may or may not be called into play as controlled by switching networks to which all manner of internal controlling signals may be applied.

The use of cycles tends to be dictated by the nature of the digital storage unit employed in the system. In particular, the storage unit must first be supplied with an "address" number which designates the specific location in storage to be affected. After the appropriate signal lines have been brought to the proper potentials, the storage unit may be actuated for entering a word to be stored (writing) or it may be actuated for sensing the stored word (reading). In some storage units a reading operation must preceed a writing operation to insure that the specified storage location is first cleared of old information. In any event, both the reading and writing operations consume a time interval that is a function of the engineering design of the storage unit. Also, various miscellaneous functions incidental to the operation of the storage unit may require time periods interspersed with the major functions of supplying the address number, reading, and writing. The need for these miscellaneous functions arises from the physical characteristics of various storage units and also from the information processing to be done between the three basic storage functions just listed. In serial systems the individual clock pulses may correspond to digit positions within a word.

The switching network used to define the cycles and to generate the individual T_1, T_2, and so on, signals is called a "clock" (perhaps an unfortunate choice of a word, because digital versions of conventional clocks are also widely used in digital systems). Any one of the T_1 pulses can be used as a "clock" pulse as this term was defined earlier in this chapter, and the phrases "T_1 clock pulse," "T_2 clock pulse," and so on, are commonly used to distinguish one clock pulse from the others. Clock networks as found in actual systems vary tremendously from one example to the next because the detailed nature of the cycle is so strongly dependent on the various inventions and peculiarities that may have been incorporated in the system design. Even for a given set of specifications, the nature of the clock network would be strongly dependent on the nature of the switching modules available for use in the clock.

Nevertheless, a few of the more important considerations in clock network design can be illustrated with the example shown in Figure 4-28. A continuous series of pulses is generated by an oscillator of some sort. This oscillator may be a free-running multivibrator which generates square-wave signals directly, or for more accurate frequency control a crystal controlled oscillator may be used in which case the generated sine wave is amplified and limited (clipped) to produce a continuous square wave. In systems having a magnetic drum or a corresponding storage device, the source of pulses may be a "clock" track or "timing" track on the drum. A continuous series of 1's is recorded in this track, and the signals generated at the reading head on this track are amplified and clipped as appropriate. In this instance the frequency of the pulses is a function of the drum velocity, but this feature is often desirable for the actuation of other drum tracks. In the figure, the signal X represents the continuous sequence of pulses.

The individual clock pulses in a clock network are usually generated by a shift register of some sort, and in Figure 4-28a a three-stage shift register employing three J-K flip-flops T_1, T_2, and T_3 is indicated, and the corresponding clock pulse signals are similarly designated. A continuous sequence of X signals will cause the set condition to be shifted from one flip-flop to the next in a cyclic pattern, and the T_1, T_2, and T_3 signals will appear in sequence as desired. The problem in clock network design arises from the need to insure that, in the operation of a system, an integral number of cycles is generated. The random starting and stopping of the clock is seldom acceptable. In particular, especially when searching for malfunctioning parts in a system, an opera-

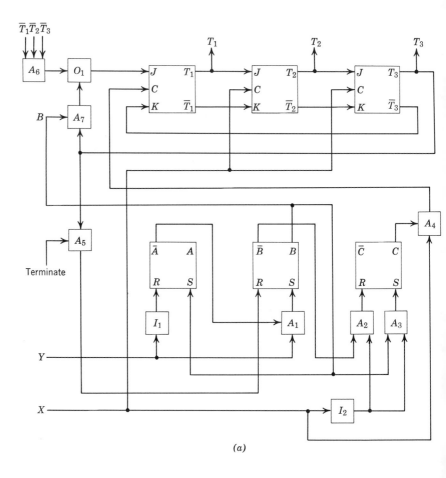

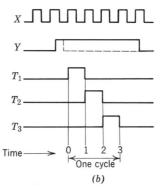

FIGURE 4-28. Clock network.

tor may wish to control the clock so that it proceeds through just one cycle in response to each manually supplied signal.

The network in Figure 4-28a represents one scheme by which the desired result can be obtained. The lower part of this figure is the same as the one-pulse network of the previous figure except for the manner in which flip-flop B is reset. In the operation of the clock network as a whole, all of the J-K flip-flops in the shift register may be assumed to have been reset by a previous operation of the clock—or separate resetting means, not shown, may be used. The starting of the clock is accomplished by changing input signal Y from 0 to 1. As before, the next full-duration X signal will then pass through A_4, and this signal is applied to the "clock" input terminal of J-K flip-flop T_1. Because of the action of AND device A_6, which senses the fact that all flip-flops in the shift register are reset, a 1-signal will be supplied to the J input of flip-flop T_1, and this flip-flop will be set. However, flip-flop B is not reset at this time, and another X signal is allowed to pass through A_4 to the shift register. This second X signal does not affect T_1 because both the J and K input signals to this flip-flop are now 0. However, flip-flops T_2 and T_3 are supplied at their "clock" inputs with a continuous series of X signals so that the 1 entered into T_1 will be shifted in the conventional manner.

When the 1 is shifted into flip-flop T_3 the subsequent action of the clock network will depend on whether the "terminate" signal is 0 or 1. If it is 1, a signal will be generated at the output of A_5 and flip-flop B will be reset. This action will terminate the passage of X signals through A_4. The next X signal applied to flip-flop T_3 will shift the 0 in T_2 to T_3, and all flip-flops in the shift register will thereby be left in the reset state, and no further action will take place. The resulting waveforms for this situation are as illustrated in Figure 4-28b. However, if the "terminate" signal is 0 during the time that T_3 is 1, flip-flop B will remain set, and X signals will continue to pass through A_4 to the clock input of flip-flop T_1. The outputs of flip-flops B and T_3 are combined in A_7 to produce a 1-signal that passes through OR device O_1 to the J input of flip-flop T_1. Therefore, T_1 becomes set again at the same time that T_3 is reset. In other words, the absence of a terminate signal allows the 1 from T_3 to be shifted to T_1, and another cycle of operation is allowed to proceed.

Whether only one cycle or many cycles of operation take place, signal Y may be returned to 0 at any time, as implied by the dotted line in Figure 4-28b, but a second initiation of the clock network will not occur unless the second 0-to-1 change in Y is made after the termination of the first one-cycle or multi-cycle sequence of operations.

176 FEEDBACK, DIGITAL STORAGE, AND TIME-DEPENDENT SIGNALS

In the operation of the entire system in which the clock network is only a part, conditions generated within the system may be such that operation should be stopped. The stopping can be accomplished by supplying an internally generated signal through an OR device to the "terminate" terminal in Figure 4-28a. If one or more different conditions can create stopping, the several control signals can be used to set corresponding flip-flops so that, after the system has stopped, the operator can quickly ascertain the cause of stopping by observing which flip-flop is in the set state. Countless other elaborations immediately suggest themselves.

For another example of a sequential circuit, the reader is referred to Figure 5-10 on page 301 of the book *Electronic Digital Components and Circuits* by Richards. This network is for writing the clock pulses on an initially blank clock track of a magnetic drum or disk—a problem that is considerably more difficult than might be expected—and the resulting network is a good illustration of the variety of sequential circuit problems encountered in practice.

Exercises

1. Draw a set-reset flip-flop network using only A-I modules (four).
2. Draw a 10-state network employing only five inverters. Include provision for setting the network to any one of its 10 states.
3. Find an eight-state synchronous network employing three complementing flip-flops which step through the states 000, 001, 100, 111, 110, and then back to 000 if initially set to any of these state combinations but which steps back and forth between 010 and 011 if initially set to either of these two state combinations. The remaining combination, 101, may be assumed to be a "don't care" condition.
4. In Exercise 3, add a single input D from an input device and redesign the network so that when $D = 1$ the network will step into any one of the state combinations representing the five-state loop and will stay in that loop as long as $D = 1$. When $D = 0$, the network must step into either of the states corresponding to the two-state loop and remain in that loop as long as $D = 0$.
5. Repeat Exercise 3 but with J-K flip-flips.
6. Repeat Exercise 4 but with J-K flip-flops.
7. In Figure 4-18h determine the detailed states of the individual flip-flops for each of the 10 states of the network as a whole. Do the same for the nine states of the network in g and the 15 states of the network in i.
8. In Figure 4-18g, modify the 7-counter by using B (instead of $\bar{B}$ as shown) for the K input to flip-flop C, and show that the resulting network is still a 7-counter.
9. Modify any of the networks in Figure 4-25 to maintain an output signal as long as Y remains 1 regardless of when X returns to 0.

10. In the "one-pulse" network of Figure 4-27, assume that if Y is made 1 it will remain 1 for several changes between 1 and 0 of the X signal. Utilize this assumption to derive a network which performs the same function but which requires only two R-S flip-flops. (Hint: With the flip-flops designated A and B, set A with a signal expressed by $\bar{X}Y\bar{B}$ and set B with A.)

11. In the "one-pulse" network of Figure 4-27, assume that when Y is 1 it remains 1 for only a short period of time in comparison with the duration of one X pulse. Utilize this assumption to derive a network which performs the same function but which requires only two R-S flip-flops. (Hint: The desired results can be attained with the principal step of eliminating flip-flop A in the network shown.)

12. Devise a network which will produce an output signal in response to three input signals X, Y, and Z such that, after all signals are 0, the output will become 1 when XYZ becomes 1 but only if the input signals become 1 in the sequence X, then Y, and then Z.

13. Modify the clock network of Figure 4-28 so that, in response to successive applications of signal Y, the shift register will step only one step at a time instead of proceeding through a complete cycle. After each step, all T_i should be 0. (Hint: One scheme is to add an appropriate number of flip-flops to "remember" which step was taken last. Another scheme, probably preferable, is to replace the shift register with a 3-counter that controls the transmission of signal C to one of three output lines.)

14. Assume that four flip-flops A_1, A_2, A_3, and A_4 are to step through the following state combinations in the indicated sequence upon applications of clock pulses: 0000, 1011, 1101, 0001, 0010, 1010, 0011, 1000, 0111. After the last state combination the sequence is to repeat. Find suitable formulations for the input signals to each of the four flip-flops when the flip-flops are of the (a) T type, (b) R-S type, (c) J-K type, and (d) D type.

5

DIGITAL CODING

The primary topic of this chapter is digital coding as it is used for error control in digital information processing and transmission systems—although the term "code" is widely used with other connotations as well. For example, any 10 binary digit combinations as used to represent the 10 decimal digits are commonly called a decimal "code," and such codes are discussed in Chapter 7 on decimal arithmetic. Cryptographic "codes" represent another example, and this aspect of coding is discussed in Chapter 8. Another type of "code" that deserves mention is the assignment of a different analog-type waveform to each of a finite number of different digital "messages" that might be transmitted. Although this last form of coding has little, if any, practical application, it is of profound importance in the theoretical study of digital information transmission systems.

Certain basic concepts of error-detecting and error-correcting codes can be traced to various manual computational techniques long known to mathematicians. In the 1940's an error-detecting code (a form of the so-called bi-quinary decimal code) was employed in some relay calculators built at the Bell Telephone Laboratories. The concept of adding a "parity check" digit to any group of other binary digits to obtain an error-detecting code appeared early in the development of computers, but the source of this concept is not known. The first known publication on error-correcting codes is a half-page item by M. J. E. Golay in the June 1949 issue of the *Proceedings of the IRE*. This note is unfortunately rather cryptically written, but it does seem to contain the kernel of the cyclic code concept, which has been the object of much study in recent years and which will be discussed later in this chapter. Golay found a 23-digit triple-error-correcting code having 12 information digits and nine check digits (binary). A paper which had more direct influence in launching the use of codes for error control was by R. W. Hamming in the April 1950 issue of the *Bell System Technical Journal*, and the particular codes described there are still commonly called Hamming codes, also to be discussed later in this chapter.

After the initial two papers cited in the previous paragraph, the subject of digital coding advanced very slowly at first, but in the late 1950's interest in the subject expanded rapidly. The subject became recognized, not only as having great practical importance, but as offering intense fascination and challenge from a theoretical standpoint. The general problem was to find codes which simultaneously tend to maximize the information content of the digits, minimize the probability of error in the final output of the system, and minimize the cost of implementing the codes.

Although advances are still being made, both in a practical and a theoretical sense, the basic concepts and limitations important to a digital systems designer now seem to be quite well understood. The most pressing present problem is in obtaining meaningful comparative evaluations of the many different codes that have been discovered.

The Distinction Between a Binary Digit and a Bit

Elsewhere in this book the terms "binary digit" and "bit" (or "bit of information") may be regarded as synonymous. However, in the realm of information theory of which coding theory is a part, the two terms are not synonymous, and an appreciation of the distinction between them is important.

A "binary digit" is still as the name implies, a symbol that can assume one of two forms or values, virtually always designated 0 and 1. "Binit" is sometimes used as an abbreviation for "binary digit" in those instances where "bit" would be incorrect.

A true "bit," on the other hand, is a measure of information. For present purposes a "bit" may be defined in the following way. A "message" is first defined as one of M states or conditions that can be stored or transmitted. Each such state may be assumed to occur with equal probability, $1/M$. The amount of "information" in the message is proportional to the logarithm of M, and in particular the number of bits in the message is defined as being $\log_2 M$.

That the quantity of information is proportional to the logarithm of M (not proportional to M) can be visualized by the following example. Consider the information written on pages where 50 different characters are used (such as the 26 letters of the alphabet, the 10 decimal digits, a space, a period, and an assortment of other special characters). If n characters may be written on each page, the total number of different messages that can be written on one page is $(50)^n$, but for two pages the number is $(50)^{2n}$. The ratio of the logarithms (to any base) of these two numbers is 2, which checks with the generally accepted fact that twice as much information can be written on two pages as on one page.

In the very special case where each message is composed of a single binary digit which has equal probability of being 0 or 1, the number of bits of information in a message is $\log_2 2 = 1$. Therefore, in this case, one binary digit carries one bit of information. When each message is composed of n binary digits where each individual digit is 0 or 1 with equal probability so that each possible combination of 0's and 1's is encountered with equal probability, the number of bits in a message is $\log_2 2^n = n$. Thus, the number of bits is still equal to the number of binary digits.

A binary digit may never carry more than one bit of information, but a binary digit may carry less than one bit. This point will be illustrated by means of an example. Assume that each message consists of four binary digits where the individual digits are really being used to represent one of the ten decimal digits in the well-known 8-4-2-1 code. That is, decimal digits 0, 1, 2, and so on to 9 are represented by the binary digit combinations 0000, 0001, 0010, and so on to 1001, respectively. Each of these combinations appears with equal probability, but the combinations 1010 through 1111 (decimal 10 to 15) appear with zero probability (that is, they never appear). In this example $M = 10$, and the number of bits in each message is $\log_2 10$. The number of bits per digit is $(\log_2 10)/4 \simeq 0.83$, so that, on the average, each binary digit carries only 0.83 bit of information in this example.

If the messages themselves occur with unequal probabilities (as when, say, decimal 0's each represented by 0000 are encountered much more frequently than any of the other nine decimal digits in the example of the previous paragraph), further refinements in the measurement of information content are required. However, for most practical purposes these refinements are not needed, because virtually all digital systems are designed to handle any message (i.e., any combination of digits) with equal facility. A rare exception to this statement is in the form of studies where attempts have been made to improve the performance of information transmission systems through making use of the fact that certain alphabetic characters (e.g., E) appear much more frequently in English language text than do certain other characters (e.g., J). The studies centered around the design of codes which employed

THE INFORMATION CONTENT OF A BINARY DIGIT   181

relatively few binary digits for each frequently encountered character and therefore required a relatively large number of binary digits for the infrequently encountered characters. For typical text material, the total number of binary digits required to represent a sequence of words can be reduced with such codes. A problem is introduced, however, in that means must be provided to distinguish, for example, two successive characters represented by 1101 and 011, respectively, from two successive characters represented by 110 and 1011, respectively, or from a single character represented by 1101011. This and other problems appear to be serious enough to discourage further attempts to make use of unequal message probabilities.

The Information Content of a Binary Digit Having a Nonzero Probability of Error

The previous discussion of the information content of a binary digit was based on the assumption that the digit was correctly represented, that is, that its probability of being in error was zero. However, in practical applications, especially in the application of digital information transmission, the probability of error is, in general, not zero. If a digit has a nonzero probability of being in error, it cannot carry as much information as it can when the observer of the digit knows the digit to be correct.

The expression relating information content to probability of error appears to have been first published by C. A. Shannon (*Bell System Technical Journal*, July 1948, pp. 379–423). The derivation of the expression is rather involved and will not be repeated here, but the relationship is

$$C = 1 + p \log_2 p + (1 - p) \log_2 (1 - p)$$

where C is the "content" in bits and p is the probability that the bit is in error under the assumption that p is the same for either a 0 or a 1. A plot of C as a function of p is given in Figure 5.1. Although no meaningful "word picture" seems to have been developed to explain why this particular mathematical formula should be correct, the reader may wish to observe that the formula at least produces the understandable result that C is 1 when p is 0. Also, the formula indicates that C is 0 when p is 0.5, which is understandable in view of the fact that a p of 0.5 implies that regardless of whether the correct digit is 0 or 1, it has an equal likelihood (probability) of incorrectly being the opposite value in which case it can carry no information. (If p were 1,

182 DIGITAL CODING

the implication would be that the observer could be certain that the opposite digit would be the correct one, and he would therefore be able to determine the information as certainly as when p is 0.)

When consideration is being given to a set of binary digits, each having the same probability of error p, the formula of the previous paragraph can be interpreted as yielding the information content in "bits per digit" for the set as a whole.

Much of the prior literature refers to the formula as an expression for the "capacity" of a "binary symmetric channel," where the word "symmetric" refers to the assumption that the probability of error is

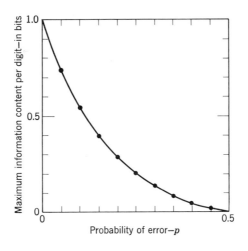

FIGURE 5-1. Information content of a binary digit as a function of probability of error.

the same for a 0 as for a 1. The word "channel" in this application should be interpreted broadly, and it could, for example, consist of a piece of paper on which the digits are written. "Capacity" here means the maximum information content, and this maximum occurs when any combination of binary digits is equally probable—as explained previously. (With only one digit, the capacity is reached when the digit has equal probability of being a 0 or a 1.) However, because "capacity" is also widely used to mean the maximum number of digits that can be transmitted over a true "channel" in a given unit of time and with arbitrarily low probability of error (all to be discussed in the next section) and because virtually all practical systems do provide for the handling of all digit combinations equally well, the C in the formula given in this section seems to be more properly termed the (maximum) information "content" of any set of binary digits having uniform probability of error p.

Note, in particular, that as p increases from 0, C drops away from 1 quite sharply. When p is only 0.1 for example, C has dropped to about 0.53. Thus, in a large set of digits, when only one tenth of the digits are in error, nearly one half (not just one tenth) of the information is lost. This seeming discrepancy results from the fact that the observer does not know which of the digits are in error. If he is to determine the original message (original set of 0's and 1's) precisely, he needs information about the positions of the erroneous digits, and this information is greater than the information that was carried by the erroneous digits themselves.

Apart from the desirability of preserving the information content of the digits, a small value for p is desirable for other reasons also, as will be more apparent later. These other reasons are generally so compelling in practical applications that the common practice is to design systems to have p no greater than perhaps 10^{-6} or less. In fact, at such small probabilities of error the loss of information content is not of itself serious at all, but serious problems can still arise in devising suitable codes for detecting and correcting those errors that do occur.

Channel Capacity

The formula, also attributed to C. A. Shannon, for expressing the capacity C of a channel in bits per second (not bits per digit as in the previous formula) is

$$C = W \log_2 (1 + P_S/P_N)$$

where W is the bandwidth of the channel in cycles per second and P_S and P_N are the signal and noise powers (not amplitudes), respectively. The noise is assumed to be Gaussian and white, where "Gaussian" means that at any given instant the probability of the noise being of a given amplitude is expressed by a certain mathematical relationship and where "white" means that the noise amplitude at any given instant is independent of the amplitude at any other instant, no matter how close in time to the given instant. Because a derivation of the channel capacity formula with an extensive discussion of it was presented in Chapter 5 on digital data transmission in Richards' book *Electronic Digital Systems*, 1966, only the following brief comments about the relationship of this formula to the present subject will be given.

The formula gives the maximum rate at which information (not binary digits) can be transmitted over a channel with an arbitrarily low probability that a given "message" will be received incorrectly. This probability is only elusively related to p, the probability of error of a binary digit, as defined in the previous section. This point will be illustrated as follows. The information to be transmitted can be visualized as being in the form of "messages" where each message is one of, say, 64 possible different messages. Each message might refer to one of 64 different typewriter symbols, one of 64 different analog-type waveforms, or one of 64 different cattle brands, to cite a few possibilities. However, if an attempt is being made to approach channel capacity, the information transmitted over a channel is most likely to appear as one of 64 analog-type waveforms regardless of its real meaning at the transmitting and receiving terminals of the channel. When any given waveform (message) is transmitted, the waveform will be disturbed by noise, and if the disturbance is great enough, the message might be incorrectly interpreted as being one of the 63 other possible waveforms (messages). Of course, the given waveform might be more similar to certain of the 63 alternatives than to others of the 63. Therefore, for each waveform that might have been transmitted, certain waveforms are more likely to be erroneously indicated than others. Nevertheless, in general, when an erroneous message is received, it could well be any of the 63 possibilities.

Regardless of the true significance of the 64 possible different messages, each message can be represented by a particular combination of six binary digits (because $2^6 = 64$), and the 64 combinations might be assigned to the 64 message possibilities in an arbitrary way. Thus, if the messages correspond to typewriter symbols (the alphabetic characters, the decimal digits, and an assortment of special characters), A might correspond to 110010, B might correspond to 000110, and so on. Moreover, the 64 binary digit combinations may be assigned in an arbitrary fashion to the 64 possible waveforms. Thus, if any message is received erroneously, the error might result in any number, from 1 to 6, of individual binary digits in error. Conceivably, the waveforms could be chosen and the binary digit combinations assigned in such a manner that most erroneously received messages would result in only 1 or 2 binary digits in error, but as a means for approaching channel capacity this possibility appears to be well beyond practical realization.

Apart from the considerations of the previous paragraph, the desired low probability of receiving a message in error cannot necessarily be achieved with messages of prespecified length, where "length" can be measured in bits of information content. In the example cited, the messages are each of six-bit length, and to achieve some low probability of error the implication is that the messages must be sent in groups of, say, 20 in each group. Each message then contains 120 bits of information, and of course the duration of time consumed in the transmission of each message is 20 times as great as before. Now, however, the number of different messages (different waveforms to be handled on the channel) is 2^{120}, an astronomical number that would introduce really serious practical problems. (The need for long message length to achieve low probability of error is not apparent from the present discussion, but it is apparent from the derivation of the channel capacity formula—see the reference cited.) Because message length is a design parameter that must be adjusted as needed in accordance with other system characteristics, the relationship between probability of message error and probability of digit error p is all the more elusive.

Another practical problem is that (except for a low-power "background" noise) the noise generally encountered is neither white nor Gaussian nor does it have a meaningful "average" power. Furthermore, the characteristics of most channels are such that, in addition to having noise signals added, the information signal becomes distorted as well, and the amount of the distortion can be a function of the intensity and nature of the noise. Thus, although white Gaussian noise can be shown to be the "worst" for a given average noise power, the wide fluctuations create still further difficulties in the selections of the waveforms needed to represent the various messages.

In summary, the channel capacity formula is of fundamental importance in setting an upper limit to the rate at which information can be transmitted, but attainable rates tend to be very much lower than this limit. The nature of the "coding" generally found to be practical is not the analog-type waveforms suggested by the formula derivation but is the transmission of individual signals, each of which represents a binary digit. The "coding" then becomes a matter of sending more digits than would be necessary to represent the information in the absence of noise. Vague parallels are still to be found between this digital coding and the analog coding, but rather than pursue this subject, attention will be directed to digital coding schemes whereby errors in individual binary digits can be detected and corrected.

Digit Error Probability as a Function of Transmission Channel Parameters

When information is transmitted in the form of a string of individual binary digits (in any code) with the digits being sensed individually at the receiver, the probability p that any given digit will be sensed in error is, of course, a function of the ratio of the signal power to the noise power, but it is also a function of the ratio of the bandwidth

186 DIGITAL CODING

of the channel to the rate (digits per second) at which the digits are transmitted. Further, the error probability is a function of the modulation method being used (the common methods being amplitude modulation, frequency modulation, phase-shift modulation, and various combinations and elaborations of these). For minimum probability of error the signals used for the representation of a 1 should be as different as possible from the signal used to represent a 0. This maximum signal difference is obtained when 1's and 0's are represented by full-amplitude pulses of opposite polarity, and this result is approximated with binary phase-shift modulation where the phase of a sine-wave-like signal is shifted 180° as appropriate to change back and forth between the representation of 0's and 1's.

For digit detection, the time of each pulse peak must be known at the receiver, and if this time is determined by sensing the peaking time of the previous pulse, the detection method is said to be incoherent. However, noise may shift the indicated peaking times so that this method does not yield the minimum probability of error. Minimum error probability is achieved with coherent detection, that is, any scheme whereby the pulse peaking times are indicated precisely through some other means—usually the transmission of a timing signal in addition to the information signal. With coherent binary phase-shift modulation, only white Gaussian noise, and no distortion from restricted bandwidth or other causes, the probability of digit error is

$$p = \tfrac{1}{2}\{1 - \text{erf}\,[(W/F)(P_S/P_N)]^{1/2}\}$$

where W, P_S, and P_N are as before, F is the signaling frequency in binary digits per second, and erf is called the "error function" defined as follows

$$\text{erf}\,A = \frac{2}{\sqrt{\pi}} \int_0^A e^{-u^2}\,du$$

For a much more extensive presentation of this formula for p, see the previously cited book *Electronic Digital Systems*, pp. 416–421.

For present purposes, the outstanding feature of the formula for p is that relatively small changes in F can produce dramatic changes in p. For example, if F is such that $(W/F)(P_S/P_N)$ is 8, p works out to about 0.00003, but if F is doubled so that $(W/F)(P_S/P_N)$ is 4, p is increased nearly two orders of magnitude to about 0.002.

Even though the noise in practical applications may not be white Gaussian and even though substantial distortion and other extraneous factors may be encountered, the above formula for p usually applies in at least a qualitative manner. Therefore, when evaluating the various error-detecting and error-correcting codes the system designer should keep in mind that the binary digits added to provide the detection and correction necessarily result in a greater F for any given information transmission rate. The resulting higher value of p may greatly outweigh the error detection and correction gains provided by the code. Alternatively, if in comparison with some trial design the designer is free to consider reducing the information transmission rate to achieve low probability of error, a reduction in F may prove preferable to changing to a code having increased error control properties.

Block Codes—Error Detection and Correction Capability as a Function of "Distance" Between Code Words

Most practical error-detecting and error-correcting codes are "block" codes. With these codes, information is transmitted or stored in "blocks" containing a fixed number of digits. Except in the trivial case where the code has no error-detecting or error-correcting properties, the number of bits of information in each block is less than the number of binary digits. Nonblock codes will be discussed briefly near the end of this chapter.

In a block code having n binary digits per block, the number of different digit combinations is 2^n, but only some of these combinations, called code words, are used to represent messages. If M is the number of different code words or messages, $2^n > M$, and generally $2^n \gg M$ for the code to have any significant error-detecting or error-correcting capability.

Incidentally, instead of "blocks," the sequence of digits to be transmitted could be said to be divided into "groups" where each group is essentially the same as a block. In fact, this terminology might be preferable except for the fact that certain block codes have been extensively studied with the aid of mathematical "group theory" where "group" is used to represent a very different concept. Although the term "group codes" will not be used in this text, this term has been widely used to mean a highly specific subclass of block codes—an unfortunately confusing situation.

The "distance" between two code words of a block code is defined as the number of binary digits that must be changed in one code word to make that word the same as the other code word. For example, with $n = 7$, the distance between 0011101 and 1110001 is 4 because these two words differ in four digit positions (the first, second, fourth, and fifth).

The "distance of a code" is defined as the minimum ("shortest") distance that exists between any two of the M code words in a block code.

If the distance of a code is only 1, the code is said to have no error-detecting or error-correcting capability at all. The reason is that, with such a code, at least one pair of code words exists for which, if one of them were transmitted, an error in a certain digit would result in the other code word. The observer at the receiver would have no information to indicate that an error had occurred. Actually, in any code for which $2^n > M$, an error that results in a digit combination that is not one of the M code words can be detected and can in some cases be corrected even though the distance of the code is only 1. However, the existence of even one possibility for an undetectable error in a single digit of a code word is generally regarded as sufficient for the assertion that the code has no error-detecting or error-correcting capability.

If the distance of a code is 2, the code is said to be single-error-detecting. That any single error can be detected is apparent from the fact that an error in any one digit of any code word will result in a digit combination that is not a code word. Many or nearly all instances of two or more erroneous digits in any one code word may also result in digit combinations that are not code words, but by definition of code of distance 2, at least one pair of code words exist for which a certain two erroneous digits will have the effect of changing one code word into another code word. Although the observer can detect that an error has occurred, he will not necessarily have been supplied with any information that indicates which digit is in error.

For codes of distance 3 or greater, an extension of the above considerations apply with regard to error detection capability. If d is the distance, the number of errors that can be detected is $d - 1$.

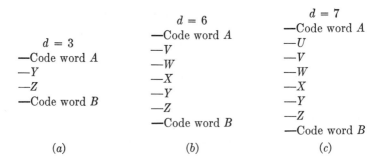

FIGURE 5-2. Diagram illustrating the "distance of a code."

BLOCK CODES—ERROR DETECTION AND CORRECTION CAPABILITY 189

In general, however, the observer will be unable to determine how many digits are in error. For example with $d = 3$, the situation is illustrated in Figure 5-2a where the heavy lines abstractly represent code words A and B, respectively. The light lines X and Y represent digit combinations intermediate between words A and B. Thus if the observer receives digit combination Y, he will not know whether A was transmitted with two errors or whether B was transmitted with one error (or some still other code word with a corresponding number of errors).

With regard to error correction capability, a code of distance 2 has none for reasons already explained. A code of distance 3 is said to be capable of correcting any single error or is said to be single-error-correcting. Thus, under the assumption that no more than one digit has been received in error, the reception of digit combination Z in Figure 5-2a would be an indication that code word B had been transmitted, because the change of only one digit would yield B. The change of some other digit to yield some other code word C could not be possible, because then the distance between B and C would be only 2, not 3 or more as specified. The coding scheme generally fails if more than one digit is erroneously received in any one code word. Thus, if B was transmitted, and Y was received as a result of two digits in error, the observer would incorrectly conclude that A had been transmitted with one error.

In general, the number of correctable errors is equal to $(d-2)/2$ when d is an even number and is $(d-1)/2$ when d is odd. This situation is illustrated in Figures 5-2b and c for the cases of d equal to 6 and 7, respectively. In b, the reception of digit combination W or Y will, under the assumption that no more than two digit errors have occurred, be indications that code word A or B, respectively, had been transmitted. If digit combination X is received, the observer will know that at least three digits had been received erroneously, but the received information will be insufficient to indicate which code word had been transmitted. Therefore, a code of distance 6 has triple-error-detection properties as well as double-error-correction properties. For more than three errors the code will fail in that, for at least some digit combinations, the observer will make erroneous conclusions about the code word which had been transmitted. In c, the reception of digit combination W or X will, under the assumption that no more than three errors have occurred, be indications that code word A or B, respectively, had been transmitted. As with codes of distance 6, a failure will generally occur when more than three digits are received erroneously.

Note that the detection-correction capability of a code is, in part, a function of the manner in which a code is used. Thus, for the code of distance 6 in Figure 5-2b for example, when this code is used for double-error-correction, the existence of no more than three errors can be detected, whereas when the correction capability is not being used, as many as five digit errors in one code word can be detected. Alternatively, the code may be used for the correction of single errors, in which case up to four errors in one code word can be detected.

Detection-Correction Capability for Nonrandom Errors—Error Bursts

The previous discussion was based on the assumption that the digit error probability p is a constant. In other words, the assumption was that the errors are randomly distributed among the transmitted digits. In many practical applications this assumption is not valid. The observation has often been made that errors appear to occur in "bursts." That is, in long sequences of digits, the errors appear to be bunched together in relatively short strings of consecutive digits with the transmission between these bursts being error-free or nearly so.

In some applications, physical phenomena can be found to explain the bunching of errors. One widely cited example is in high-density magnetic tape storage where a blemish or a dirt particle on the magnetic surface will cause erratic storage over several consecutive digit positions. Another widely cited example is in radio transmission where lightning will create electrical noise of sufficient intensity to disrupt communication for the time of several consecutive binary digit transmissions.

To handle bursts of errors, much attention has been given to the development of "burst-error-detecting" and "burst-error-correcting" codes. As is discussed in more detail later, such codes can indeed be devised. The capability of such a code may be specified, for example, as being capable of correcting bursts of up to three errors and detecting bursts of up to five errors. However, the detection-correction "mix" of a burst-error code appears to be an inherent property of each individual code and is not generally alterable by usage as in the case of codes designed for random-error codes. The concept of "distance" between code words apparently does not apply to burst-error codes, at least not in a useful way.

Unless specified otherwise, any given burst-error code is capable of detecting or correcting (as the case may be) only one burst of errors in any one code word. If so much as one error occurs outside that burst within the code word, the code may fail to detect the errors or may yield erroneous correction information (as the case may be). However, this situation is not all-inclusive, and codes can be designed to handle more than one burst in a code word (and the bursts need not be of equal maximum length) or to handle one burst plus a limited number of random errors in a code word. Thus, conceivably, a code might be designed and used as a burst-of-length-three-correcting, burst-of-length-five-detecting, double-random-error-correcting, quadruple-random-error-detecting code, for example. Although some codes which have been studied do, in fact, have their detection-correction capability divided between burst and random errors, only limited progress has been achieved in theoretical studies of such codes, and their practical application is limited.

Incidentally, a "burst" of errors is defined as a sequence of digits, the first and last of which are in error, where each intermediate digit may or may not be in error. This definition is more practical than a definition implying that all digits within a burst are erroneous, because even in the presence of severe noise, any given digit has a "50-50" chance of being received correctly.

A difficulty with either definition of "burst" will be illustrated as follows. In a long code word assume that errors occur in the first two digits and in the last three digits. Is this error pattern a burst of length five? Or is it two separate bursts, one of length two and one of length three? The preferable definition and corresponding code design seems to be to regard this situation as a single burst of length five even though the errors result from two different magnetic surface blemishes, two different lightning strokes, or what not. The reason for preferring this definition is that the disturbances can then come near the boundaries (in time) between consecutively transmitted code words, and provided only that no more than five errors occur in any one code word, the errors can be handled appropriately—with a properly designed code. The digits of each code word may be visualized as being placed in a circle, and the burst may occur in any set of consecutive digits as they appear in the circle. Some burst-error codes have this "end-around" feature, whereas others do not.

Unfortunately, even the assumption that errors appear in isolated bursts is remote from reality in most applications, including some of those cited for which a burst pattern might seem reasonable. At least, the bursts of errors, when they occur at all, are likely to appear in bursts. Codes for handling even a single "burst of bursts" of errors appear to be beyond attainment in a practical sense.

An often more realistic view of error patterns is simply to assume that the digit error probability p fluctuates greatly and erratically as a function of time. In some practical instances, p may be 10^{-10} or less over long periods of time but will sometimes be several orders of magnitude greater than this figure—and will be exactly 0.5 during those periods

192   DIGITAL CODING

of time when the channel is defective of completely blocked such as by a temporarily open connection. In such instances the preferable method of error control is to employ codes only for error detection. The correction is accomplished by retransmitting those blocks (code words) in which errors were detected.

Nonsystematic Block Codes

Block codes can be divided into two categories: nonsystematic and systematic. In a nonsystematic code the binary digit combinations that are used as code words can be selected more or less at random, and the information content of the message can be visualized as being distributed among all of the digits, although in general the information content of each digit is less than one bit. In systematic codes, the digits of each word are divided into two sets. The digits of one set carry all of the information, and in general each digit in this set carries one bit of information. The remaining digits are variously called "redundancy" digits, "parity-check" digits, or just "check" digits. These digits, which may be greater in number than the information-carrying digits of the first set, do not carry any additional information, but instead carry only information that can be deduced from the information-carrying digits (hence the name "redundancy"). The digits of the second set can be used to determine the correctness of the digits of the first set (hence the name "check"). Moreover, in all cases the checking is accomplished through a determination of the odd-even count of various sets of digits in the word as a whole (hence the name "parity"). Thus, the name "nonsystematic" is derived principally from a comparison with the obviously systematic procedures that can be used for coding and decoding with the "systematic" codes. Actually, "systematic" procedures can be used with the nonsystematic codes also, but these procedures tend to require much more elaborate physical facilities even though the concepts behind the procedures are, if anything, much simpler.

As an example of a nonsystematic code, Table 5-1 shows a 32-word code (that is, the number M of different messages is 32) having a length n of 12 binary digits. In this particular code, the distance between any two code words is at least 5. In other words, $d = 5$ as d was defined earlier. Although the fact that M is an integral power of 2 is a mere coincidence for this particular code, the fact is helpful in illustrating the considerable loss in information content that must be suffered to obtain error detection or correction capability. Because $2^5 = 32$, the in-

TABLE 5-1. A 32-Word Code Having $n = 12$ and $d = 5$

$n = 12$
000,000,000,000
000,001,111,111
000,110,111,010
000,111,000,101
001,010,110,101
001,011,001,010
001,100,010,011
001,101,101,100
010,010,010,110
010,011,101,001
010,100,100,111
010,101,011,000
011,000,001,101
011,001,110,010
100,010,101,100
100,011,010,011
100,100,011,101
100,101,100,010
101,000,101,011
101,001,010,100
101,110,000,110
101,111,111,001
110,000,110,001
110,001,001,110
110,110,001,011
110,111,110,100
111,010,011,000
111,011,100,111
111,100,111,110
111,101,000,001

$M = 32$

formation content of each word is only five bits. Therefore, to achieve the ability to detect up to four errors in any 12-digit word, the information rate must be reduced to $\frac{5}{12}$ the rate realizable under the assumption that no errors will occur.

The encoder required for a nonsystematic code requires, in general, a storage unit capable of storing all code words. Specifically, the unit must have M storage positions each of n-digit length. If the M different messages are originally expressed as binary numbers ranging from 0 to $M-1$, these numbers may be used directly as addresses for gaining access to the appropriate positions in the storage unit. Because digits may be stored in any pattern in the storage unit, the digits of each given code word need have no relationship at all to the digits of the corresponding message. Encoding is thereby accomplished with equal facility for any code. (A complexity is encountered when the messages are not initially represented by a uniform series of binary numbers. Switching matrices for translation are then required, and the practicality of these matrices depends on the magnitude of M and on the details of the translation required in any particular instance.)

The decoder required for nonsystematic codes must, in general, be a machine having most of the major features of a stored-program computer. For one thing, the decoder must contain a storage unit which, as in the encoder, stores all of the code words. As each digit combination corresponding to a message is received, this combination is compared with each stored code word, one at a time, until a code word is found that is identical with the received digit combination. The particular message that was sent can then be determined from the address at which the matching code word was found. The address number (appropriately translated, if necessary in the application) can then be used for the message representation.

If the received digit combination is found to be different from all stored code words, the decoder has detected the presence of at least one error in the received word. An appropriate signal can then be generated.

If the system is to be used for error correction, the decoder should be programmed so that, upon each comparison of the received digit combination with a stored word, a temporary record is made of the number of binary digits that are not the same. This temporary record may be destroyed when a comparison is found for which the number of different digits is less. After all comparisons are made, the address corresponding to the stored word for which the number of different digits is smallest is assumed to represent the intended message. (If two or more comparisons should happen to yield the same smallest number of different digits, the inference can be drawn that more errors have occurred than the code was designed to correct.)

In spite of the conceptual simplicity of this encoding and decoding method (which, incidentally, would work equally well with systematic codes), the required equipment is generally regarded as far too extensive for nearly all practical applications—although when computer facilities are in the system for other reasons anyway, this disadvantage may not be great. However, apart from the equipment requirements, the decoding process tends to be extremely time consuming for codes having a large M.

The finding of suitable nonsystematic codes for any given n and d tends to be a difficult task, and very little really useful information on this subject is known. A parameter of considerable interest would be the maximum value of M (that is, the maximum number of different messages or code words) that can be realized for each given n and d. This maximum number, here designated $M(n,d)$ has been the object of a considerable amount of research of a mathematical nature. A presentation of the details of this work would go far beyond present objectives, but the results, to the extent that they seem to be known, are summarized in Table 5-2. Each table entry consists of two numbers. The upper number gives an upper bound on $M(n,d)$ as determined from mathematical arguments, where several quite different arguments have been used for various relative values of n and d. The lower number gives $M(n,d)$ for actual codes that have been found, although in some instances the codes are not known to have been published.

For some combinations of n and d, $M(n,d)$ is greater for nonsystematic codes than for systematic codes. For other combinations of n and d, $M(n,d)$ is the same for the two types of codes. No instances are known where $M(n,d)$ is greater for systematic codes.

If the designer is restricted to an n and d for which a nonsystematic code produces the greater $M(n,d)$, the choice of a nonsystematic code will result in the possibility of a greater information transmission rate —if the error pattern is such that the number of errors in any one code word is never greater than the number for which the code was designed. However, an assumption of this type of error pattern is not necessarily valid. For random errors, instances can arise where more than this number of errors will occur in one code word even though p may be extremely small. Because the really important parameter is the probability of receiving a complete message incorrectly (after digit error correction), the preferable choice might be a systematic code with the same number of code words but having a smaller d. The distance between most pairs of code words might be sufficiently greater than d to result in a net reduction in message error rate.

In any event $M(n,d)$ is not known to be much greater for nonsystematic codes than for systematic codes for any combination of n and d. Because information content is proportional to the slowly increasing logarithm of the number of messages, the potential advantage, if any, of nonsystematic codes is extremely small. Therefore, practical interest in nonsystematic codes is minimal.

TABLE 5-2. The Maximum Number of Code Words $N(n, d)$ for Various n and d; Upper Numbers in the Table Entries are from Mathematical Arguments; Lower Numbers are from Codes Actually Found

d \ n	1	2	3	4	5	6	7	8	9	10	11	12	13	14	15	16	17	18	19	20
1	2 2	2^2 2^2	2^3 2^3	2^4 2^4	→etc.															
2		2 2	2^2 2^2	2^3 2^3	→etc.															
3			2 2	2 2	4 4	8 —	16 16	20 20	39 38	82 68	154 128	293 256	585 512	1024 1024	2048 2048	3616 2048	7090 4096			
4				2 2	2 2	4 4	8 8	16 16	20 —	32 —	112 —									
5					2 2	2 2	2 2	4 4	6 6	12 12	24 24	35 32	70 32	131 64	256 128	428 128	851 256			
6						2 2	2 2	2 2	4 4	6 —	12 —	24 —	28 —	38 —	72 64					
7							2 2	2 2	2 2	2 2	4 4	4 4	8 8	16 16	32 32	42 32	70 64	308 64		
8								2 2	2 2	2 2	2 2	4 4	4 4	8 —	16 —	32 —	36 —	45 —	70 64	
9									2 2	2 2	2 2	3 2	3 2	4 4	6 4	9 6	18 8	36 —	40 —	49 —

Elementary Parity Check Digit Error-Detecting Codes

In one of the most elementary and yet widely used codes for error detection a single parity check digit is appended to each block of information digits. This parity check digit indicates whether the total number of information digits is even or odd. Either convention may be used with regard to the assigning a 1 or a 0 to the indication for an even number of information digits, but if a 1 is used, the block including the check digit will always contain at least one 1, and this characteristic is sometimes desirable in applications where the digit combination of all 0's in a block cannot be readily distinguished from no information at all.

The code of the previous paragraph is single-error-detecting, because an error in any one digit will create a failure of the check digit to give a correct odd-even indication of the number of 1's in the block. An error in the check digit itself is detected in the same manner as an error in any of the information digits. Each block may contain any number of digits, but under the assumption that digit errors occur randomly with a probability p, the probability of receiving a message incorrectly increases as the number of digits per block increases. If p is sufficiently small and if N digits (including the check digit) are in each block, the probability of receiving an incorrect message is approximately pN. If p is not suitably small (that is, if pN is a substantial fraction of 1), the message error probability can be computed as follows. The probability of receiving any one digit correctly is $1 - p$. The probability of receiving all digits correctly is $(1 - p)^N$, and the probability of not receiving them all correctly, that is the probability of receiving the message incorrectly, is $1 - (1 - p)^N$.

However, an even more important parameter is the probability of receiving an undetected incorrect message. The probability of receiving the first digit incorrectly and in the same message receiving all remaining digits correctly is $p(1 - p)^{N-1}$, but because any digit may be in error the probability of receiving exactly one digit in error is $Np(1 - p)^{N-1}$. The probability of receiving two or more digits in error is determined by subtracting this probability from the probability of receiving the message incorrectly (any number of digit errors) and is therefore $1 - (1 - p)^N - Np(1 - p)^{N-1}$. If p is suitably small, this expression is an approximation to the probability of receiving an undetected incorrect message, because of the very small probability that three or more digits will be in error in any one message. Actually, the code will detect the fact that errors have occurred in three or any other odd number of digits. Also, if p is small, the expression just given is easily shown to be approximated by the much simpler expression $p^2N(N-1)$. Further, if N is large, the probability that the code will fail to detect an incorrectly received message is approximately by $(pN)^2$.

For example, if p is 0.01 and N is 10, the probability of receiving a message incorrectly is approximately 0.1 as determined by pN or is exactly 0.0956 (except for round-off error in the computation). The reason why the exact probability is less than 0.1 is that, even though an average of one out of each 100 digits is in error, occasionally two or more errors will occur in one 10-digit word or message so that the number of correctly received messages is slightly greater than would be the case if the errors were evenly distributed. The probability of receiving exactly one erroneous digit in a message works out to 0.0914 (except for round-off error), so the probability of receiving two or more erroneous digits in a message is 0.0042. This figure represents the probability of receiving an undetected incorrect message under the assumption (reasonably justified in this example) that the instances where three or more digit errors occur in one message are few enough to be neglected. In other words, on the average, of each approximately 250 ten-digit messages, one will contain errors that the code will fail to detect. Incidentally, in this example p is not sufficiently small for the approximation $p^2N(N-1)$ to be accurate.

Implementing the Single Parity-Check Code

If the digits of a word or message appear serially in time, the parity check digit may be determined by transmitting the information digits to the complementing input of a flip-flop which is initially set to a predetermined state, usually the 0 state. Each 1 digit is presumed to be represented by a signal that actuates the flip-flop, with the signals representing 0's having no effect. After all information digits have thus been transmitted to the flip-flop, the state of the flip-flop will indicate the odd-even count of the 1's in the message, and the output of the flip-flop may be used to generate the parity-check signal or digit. This process is called encoding.

Decoding is accomplished in essentially the same way except that the parity check digit is transmitted to the flip-flop along with the information digits. The parity check digit can be first, last, or at any point in the sequence of digits. After all digits of the word have been sent to the flip-flop, the output signal from the flip-flop signifies the odd-even count and is used as the signal for indicating whether or not an error has occurred. (More precisely, the indication is whether or not any odd number of errors has occurred.)

In addition to the complementing flip-flop, both the encoder and decoder usually require miscellaneous storage registers and control circuits as needed for the application, but the details of these other circuits tend to be too specialized for generalized treatment. In general, however, only relatively elementary timing and switching problems are encountered in the design of the necessary circuits.

If the digits appear in parallel, the switching networks illustrated in Figure 5-3 can be used for encoding and decoding. In this figure, each block labeled "Ex-OR" represents the "exclusive-OR" function, which is that function which generates an output signal when an input signal is present on one and only one of the two input lines. In Boolean notation, the output signal is $A\bar{B} + \bar{A}B = (A + B)\overline{AB}$, where A and B are the two input signals. In effect, an "exclusive-OR" function generates an odd-even count of two signals, and this count may be combined with another signal in another "exclusive-OR" function to generate an odd-even count for the three signals, and so on. Parts a and c of the figure indicate two different ways in which the blocks may be interconnected for encoding. The arrangement in a is probably the most straightforward, but the arrangement in c has the advantage that the number of blocks through which any signal must travel in sequence is fewer, and therefore

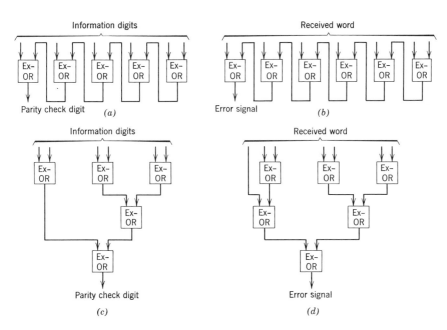

FIGURE 5-3. Encoders and decoders for single-error detection by means of a parity check digit.

200 DIGITAL CODING

a higher speed of encoding is provided. With either type of arrangement the number of "exclusive-OR" blocks required is $N - 1$, where N is the number of digits (including the parity check digit) per message.

The decoding arrangements in Figures 5-3b and d are essentially the same as the encoding arrangements in a and c, respectively, except that one switching block has been added in each instance to include the parity check digit in the odd-even count. Also, the output signal is the error signal with the decoder whereas it was the parity check digit with the encoder.

With many sets of switching module types, an exclusive-OR module is not available. However, this function may be generated by appropriate combinations of A-I, O-I, or other modules as explained in previous chapters. Although the encoder and decoder can still be realized in the general manner illustrated by Figure 5-3, various component-minimization techniques as previously described may then be necessary to achieve network economy.

Whether the arrangements in Figure 5-3 produce an odd count or an even count is in accordance with whether the number of digits per word is odd or even. If, for a given number of digits per word, the opposite type of signal is required, an inverter may be inserted an any point in the array to produce the desired result.

Note that if any one exclusively-OR function is generated erroneously in either the encoder or the decoder, the resulting error will affect the system as a whole in the same manner as though a single error had occurred in the storage or transmission unit being checked.

Employing Additional Parity Check Digits to Devise Single-error-correcting ($d = 3$) Codes—Hamming Codes

Although the parity check digits are redundant digits in the sense that they carry no information not carried in the information digits themselves, the parity check digits can be so organized that they carry useful information about the errors in a message. A single parity check digit can signal only a "yes-no" indication about the detection of errors, but two parity check digits considered collectively can carry four error-information messages, *oo, oe, eo,* and *ee*, where *e* and *o* mean error and no-error (ok), respectively. (The symbols 0 and 1 would be confusing here because the *e* and *o* signals, although binary, are the error signals generated by the decoder at the receiver and are not the parity check digits generated by the encoder at the transmitter.) One of these error-information messages is needed to signal the fact that no errors were detected, but the other three may be used to give information about the distribution or location of whatever errors that were detected. In particular, if each message as originally transmitted contains only three binary digits, the error-information messages can be used to indicate which digit is in error—under the assumption that no more than one error will occur in any one message.

The problem then is to devise a parity checking pattern which will yield the desired result. With so few digits per word, the problem is simple, however. Of each three-digit word, two digits y_1 and y_2 are check digits, so only one digit x_1 is left to carry the originally intended information. In this case, both y_1 and y_2 can be used to indicate the odd-even "count" of the single information digit x_1. The situation is illustrated in Table 5-3a where a check mark ($\checkmark$) signifies the information digits (only one in this case) which are included in the odd-even count of the parity check digits. Either convention may be used with regard to using a check digit value of 1 to represent an odd or an even number of 1's in the information digits. Of course, if a check digit of 1 is used to represent an even number of 1's in the checked information digits, the total number of 1's in the set in question is odd. At the receiver, if both check digits signal an error, the error must be in x_1, but if only one check digit signals an error, the error must be in the check digit itself. The four possible error messages and their meanings are listed in Table 5-3b.

Appending two check digits to each information digit to form three-digit messages is, as might be expected, utterly impractical in most appli-

TABLE 5-3. Using Two-Parity Check Digits to Form a Three-Digit Single-Error-Correcting Code

	x_1
y_1	$\checkmark$
y_2	$\checkmark$

y_1	y_2	Digit in Error
o	o	none
o	e	y_2
e	o	y_1
e	e	x_1

(a)   (b)

cations. On the other hand, if a large number of three-digit messages are interleaved in a manner to be described later, the resulting code has powerful burst-error-correcting capability. Also, the tripling of the number of signals to achieve high reliability is not always out of the question. Triple systems operating on a two-out-of-three majority basis are sometimes used where reliability is extremely important.

In general, with r parity check digits the number of different error messages obtainable is 2^r with one of these error messages required to indicate no errors, the other error messages can be used to identify a single erroneous digit when the original message contains no more than $2^r - 1$ digits including both information digits and check digits. The maximum number of information digits per message is then $2^r - 1 - r$.

The pattern by which the check digits are used to indicate the odd-even counts can be devised in various ways provided the following condition is fulfilled. No two digits should affect the odd-even counts of the same set of check digits. Also, in the interest of simplicity, each check digit should be determined by the odd-even count of selected information digits only (with other check digits not included). Then, if any one, but only one, check digit indicates an error, the error will be known to be in the check digit itself. A consequence of this requirement is that each information digit must affect the odd-even count of at least two check digits.

For the case where $r = 4$, for example, the maximum number of information digits is 11, so that each word or message contains a total of 15 digits. The check digits may be interspersed among the information digits in any way, but usually the check digits are visualized as being transmitted last. One possible pattern for generating the odd-even counts is illustrated in Table 5-4. Thus y_1, for example, indicates the odd-even

TABLE 5-4. Pattern of Odd-Even Counts for a Single-Error-Correcting Code Having Eleven Information Digits and Four Check Digits

	x_1	x_2	x_3	x_4	x_5	x_6	x_7	x_8	x_9	x_{10}	x_{11}	y_1	y_2	y_3	y_4
y_1	✓	✓	✓			✓	✓	✓		✓	✓	✓			
y_2	✓			✓	✓		✓	✓		✓	✓		✓		
y_3		✓		✓		✓	✓		✓	✓	✓			✓	
y_4			✓		✓	✓		✓	✓	✓	✓				✓

count of digits x_1, x_2, x_3, x_7, x_8, x_9, and x_{11}. The part of the table to the right of the dotted line is not necessary, but it illustrates the possible error messages more completely. If only y_1 indicates an error, the error must be in y_1 itself, because any other erroneous digit would cause a different set of odd-even count failures. If both y_1 and y_2 (but not y_3 or y_4) indicate odd-even count failures, the error must be in information digit x_1, because x_1 is the only digit which affects the corresponding odd-even counts.

Because the total number of digits per word is 15 in this example, the code is a (15,3) code when using the notation of Table 5-2. Each of the 11 information digits may be 0 or 1 arbitrarily, so the number of different messages or code words in $2^{11} = 2048$, the maximum possible for a code of this length and distance.

Incidentally, with any systematic code as defined earlier, $M(n,d) = 2^{n-r} = 2^k$, where k is the number of information digits.

Codes of this form (with or without further check digit to obtain $d = 4$, as described later) are called Hamming codes. Although Hamming codes are virtually always considered with error correction in mind, the error correction capability can be discarded in favor of increased error detection power, as is the case with any code of a given distance. By not attempting to correct errors, the $d = 3$ Hamming code can be used to detect any word which contains any one, any two, and many combinations of more than two errors.

The Equipment Needed to Implement the Hamming Codes

If the digits of a Hamming code are suitably sequenced, the code forms a cyclic code which can be implemented straightforwardly by means of shift registers in a manner to be described later for cyclic codes. However, the codes can also be implemented by combinatorial switching networks in a way that is at least conceptually simple, although the number of components required increases rapidly as increased word lengths come under consideration.

Figure 5-4 illustrates in block diagram form the combinatorial switching equipment needed for the decoder when using the Hamming code having seven digits per word of which four are information digits and three are parity check digits. The check marks in the lower-right part of the figure indicate the pattern by which the odd-even counts of the check digits were presumed generated in the encoder. In the decoder an odd-even counter circuit is needed for each check digit, and the network may be formed with exclusive-OR modules in the same manner as illustrated in Figure 5-3b or d. The three error-detecting signals thus

204 DIGITAL CODING

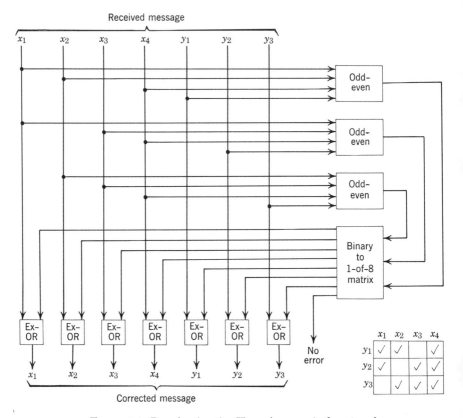

FIGURE 5-4. Decoder for the Hamming $n = 7$, $k = 4$ code.

generated are transmitted to a matrix network (such as described in an earlier chapter) which generates a signal on one of eight output lines. A signal on a certain one of the output lines indicates that no error was detected. A signal on any of the other output lines is used to correct the corresponding digit, and an exclusive-OR module is needed for each digit that might need to be corrected (inverted). In the figure, provision is shown for correcting any one of the seven digits in a word, although in some applications no need would arise for correcting the check digits y_1, y_2, and y_3.

Adding One Further Check Digit to Obtain Hamming Codes with $d = 4$

With a Hamming code of any length as described in the two previous sections, one further check digit can be added to provide double-error detection as well as single-error correction. This added check digit indicates the odd-even count of all other digits, both information digits and check digits, in the word. If any single error occurs in a word, the fact will be indicated by this final check digit, and the original check digits can be used to correct the error as before—except that if the error occurs in the final check digit itself, the correction will be accomplished by using the signal which otherwise indicated "no error."

If any two digits in a word are in error, a check failure will occur for at least one of the original check digits, but no check failure will occur for the final check digit. The resulting signals can be combined in an appropriate switching network to indicate the existence of two errors in one word.

By employing an extension of the argument used in the previous section, the resulting code can be shown to have $d = 4$. As with any code of this d, the error-correcting capability can be abandoned in favor of improved error-detection capability, in which case the code will indicate the presence of any one, any two, or any three (and many combinations of more than three) errors in a word.

Attempting to Extend the Concept of Hamming Codes to Codes with $d \geq 5$

Conceivably the concept of the Hamming codes could be extended in a manner whereby a more elaborate "error message" is used to indicate the locations of two or more errors in a single code word. However, when attempts are made to work out the details of such an extension, complexities are quickly encountered that render this approach to increased error detection-correction capability unattractive. In fact, so far as is known, the details have never been worded out completely, although certain characteristics that the code must have are reasonably apparent. For example, if the code is to be capable of correcting two errors and if eight parity check digits are provided, four of them could be used for indicating the location of one error in one of 16 digit positions, and the other four parity check digits could be used to indicate the location of a second error in one of the same 16 digit positions. Of the 16 digits per word, eight would remain available for carrying the originally intended information. The problem arises in finding a pattern of odd-even counts that yields the intended results.

Apparently the only known solution to the problem is to prepare a table of error messages containing, in the example cited, $2^8 = 256$ entries. Each combination of odd-even count failures corresponds to an entry in the table, and each entry contains an indication of which digits are in error. Unfortunately, even this solution requires a lengthy and tedious search for suitable odd-even count patterns for the eight parity check digits. Because of the obvious disadvantages involved in physically implementing the table, searches for suitable count patterns have been minimal. (Incidentally, the term "nonsystematic" has sometimes been used to described parity check codes requiring a table of this category, but this meaning of the term is at variance with the more commonly used meaning set forth earlier in this chapter.)

Instead of extending the concept of Hamming codes directly, the more attractive approach to increased error detection-correction capability is through multidimensional codes and cyclic codes, both to be described later.

Shortened Hamming Codes

If in a given application the desired number of digits per word is not $2^r - 1 - r$ for some integral value of r (i.e., is not 4, 11, 26, 57, 120, etc.), the Hamming code of length corresponding to the next larger integral value of r can be used except that an appropriate number of information digits are deleted from each block. Such a code, called a "shortened" Hamming code, has some of its error detection-correction capability wasted in the sense that each error message is capable of carrying more information than necessary (d is not increased, however), but of all possible systematic codes a shortened Hamming code still provides the greatest possible number of different code words for any given n and d, as defined earlier.

Multidimensional Codes

Consider a set of binary digits arranged in a rectangular array of rows and columns. (This arrangement need not be physical—a conceptual relationship of this form for any physically positioned digits is sufficient, but the number of digits should be $n_1 n_2$, where n_1 and n_2 are integers.) Assume that the digits in each row are coded in a code of distance d_r and that the digits in each column are coded in a code of distance d_c. The resulting possible digit combinations for the array as a whole represent a new code where each code word or message is of length $n_1 n_2$, and the question is to determine the distance of this new code.

Visualize the rectangular array and assume that the digits are in some pattern that represents a valid code word for the array as a whole. Then assume sufficient digits are changed in one row, say the top row, to transpose the code word of that row to some other word in the row code. The number of digits changed must be at least d_r. Unless $d_c = 1$, the new digit combination for the array as a whole will not be a valid code word for the code of the array. To arrive at another valid code word for the array as a whole, at least d_c digits must be changed in each column in which a row digit was changed. Therefore, the total number of digits changed when "traveling" from any code word to any other code word must be at least $d_r d_c$ so that the distance of this two-dimensional code is $d_r d_c$.

An extension of the above argument can be applied to codes having any number N of dimensions, and in general the distance of an N-dimensional code is $\prod_{i=1}^{N} d_i$, where the d_i are the distances of the codes used for the digits along the various dimensions of the array. Of course, if $N > 3$, the multidimensional nature of the code must be at least partially conceptual as opposed to physical. For example, if $N = 4$, the digits may be physically arranged in a set of three-dimensional blocks where the fourth dimension is realized by relating the digits in the corresponding positions of each block.

Multidimensional codes having large d are very easily devised. If the codes for the digits along any dimension are systematic (employ parity check digits), the encoder required for a multidimensional code is essentially an obvious extension of the encoder required for one-dimensional codes, although many variations are possible with regard to all-serial, all-parallel, and serial-parallel systems for encoding the various rows and columns of digits.

If error detection (not correction) is the only objective, the decoder needed for multidimensional codes is likewise an obvious extension of the decoder needed for one-dimensional codes.

However, if the multidimensional code is to be used for error correction, problems of considerable complexity arise in deducing the digits in error from the pattern of parity check failures. The problems are solvable, but the solutions generally imply a large number of components (high cost) for the decoder.

With either error detection or error correction, the multidimensional codes offer an ability to handle error bursts of substantial length when the digits are transmitted serially on either a row-by-row or a column-by-column basis. This burst-error capability is a sort of "bonus" feature, although an implication is that, for a given n and d, $M(n,d)$ as defined earlier is not reached with multidimensional codes.

The above points will be illustrated by means of two examples of two-dimensional codes in the following sections.

A Two-Dimensional Code with a Single Parity Check Digit for Each Row and Column

Figure 5-5 illustrates the manner in which any single error is located in a two-dimensional code with a parity check digit for each row and each column. The code is portrayed as having 20 information digits arranged in five rows, each having four digits. At the right-hand end of each row a parity check digit is appended. Also, a parity check digit is appended at the bottom of each column of information digits. The digit in the lower-right corner of the array can be used to form a parity check either on the right-hand column of parity check digits or on the bottom row of parity check digits, but the convention selected is that the check is on the bottom row. Thus, the dotted line is intended to indicate that the lower-right check digit is more closely associated with the bottom row than with the right-hand column.

Each x in Figure 5-5 represents an information digit, which may be

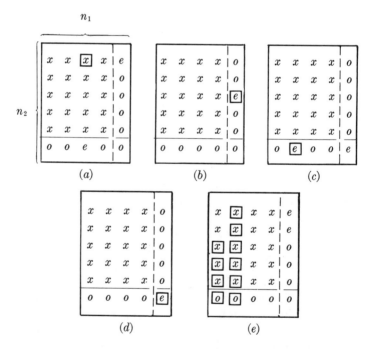

FIGURE 5-5. Parity check failures for various error locations in a two-dimensional code.

either 0 or 1. Each check digit is represented by o or e, either of which may likewise be 0 or 1, but o indicates that the parity count of the corresponding row or column is correct ("ok") and e indicates that an error is sensed by the check digit in question. A small rectangle enclosing a symbol signifies that the digit in question (whether it be a 0 or a 1) is in error.

In Figure 5-5a an error is assumed in one information digit, and the fact is signaled by a check failure in the corresponding row and column. In b the single error is assumed to be in one of the row check digits, in which case only the row check fails. In c the error is in a column check digit, and check failures occur for this column and for the check digit in the lower-right corner. If the single error is in the lower-right check digit itself, only this digit signals an error as in d. If two or more row checks fail (one of which may be the bottom row), if two or more column checks fail, or if only one column check fails where the lower-right check does not fail, the message is known to contain two or more errors—as can be established by considering various error patterns in detail.

Alternatively, the two-dimensional code of Figure 5-5 can be used for triple-error detection, because any combination of one, two, or three errors will produce at least one parity check failure. The distance of the code is 4, as expected from the products of the distances (each 2) of the individual row and column codes.

When the code is used for single-error correction with double-error detection, the desired results can be achieved by using switching networks to generate five signals A, B, C, D, and E with the following meanings.

A: one and only one row check failure, including the bottom row.

B: two or more row check failures, including the bottom row.

C: one and only one column check failure (does not include right-hand column).

D: two or more column check failures (does not include right-hand column).

E: corner check failure.

The digit to be corrected is then determined in accordance with the following Boolean relationships.

$AC\bar{E}$: error in information digit (at the intersection of the indicated row and column).

$A\bar{C}\bar{D}$: error in row check digit (the indicated row).

$\bar{A}\bar{B}CE$: error in column check digit (the indicated column).

$\bar{A}\bar{B}\bar{C}\bar{D}E$: error in the corner check digit.

$B + D + C\bar{E}$: two or more errors detected, but not located.

If error correction is accomplished solely by means of combinatorial switching networks in the manner suggested by Figure 5-5, an unattractively large number of components would be required for code words of the 30-digit length (20 information digits plus ten parity check digits) illustrated in Figure 5-5. For larger arrays, this disadvantage is even greater, because for a given increase in the number of digits a greater than proportionate increase in the number of switching components is required. On the other hand, with large scale integration (LSI) techniques, the code is reasonable. Also, serial operation can be substituted for portions of the all-parallel approach suggested, and great reductions can be attained in the number of required components at a price in speed of operation.

Additionally, the two-dimensional code has good burst-error detection capability as illustrated in Figure 5-5e. Assume that the digits are transmitted serially on a column-by-column basis. With n_2 digits per column, any number of sequential digits up to $2n_2 - 2$ can be in error and the conditions for the detection of two or more errors will be fulfilled. In the figure, the indication is that the first erroneous digit is the third digit in the first column. All succeeding digits in this column and all digits in the second column are subsequently received in error. In this particular instance, none of the column checks fail (because an even number of errors has occurred in each column), but two row checks fail. However, the burst-error capability of the code must be judged cautiously because, for example, a burst of only two errors combined with a strategically located single error in the same word will cause the code to make an erroneous "correction."

A Two-Dimensional Code Having $d = 16$

Figure 5-6 illustrates a two-dimensional code in which the digits along the rows and columns are in a Hamming $d = 4$ code as described previously. The distance of the code is then $(4)(4) = 16$. The information digits are at the intersections of rows and columns 1 through 4. The digits in rows 5, 6, and 7 in each of columns 1 through 4 correspond to check digits y_1, y_2, and y_3 in the pattern specified in the lower-right part of Figure 5-4, and the digits in row 8 of the same columns are check digits for the respective complete columns. The digits in columns 5 through 8 (all rows) similarly check the respective rows. As before,

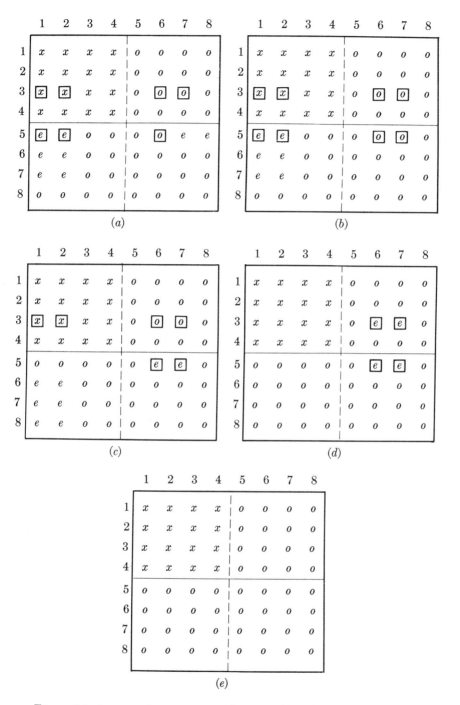

FIGURE 5-6. An example error pattern in a two-dimensional $d = 16$ code.

an x represents an information digit, and the symbols o and e indicate check digits in accordance with whether the parity count is correct ("ok") or in error, respectively. In each instance the actual value of any digit may be either 0 or 1 without affecting the characteristics of the code. A digit actually in error is enclosed by a small rectangle.

When $d = 16$, the code has seven-error-correction plus eight-error-detection capability. However, deducing the particular digits in error from the pattern of parity check failures by an extension of the techniques suggested for the two-dimensional code of the previous section appears hopelessly impractical. Instead, the erroneous digits can be located by working with individual row and column codes alternately and sequentially. Even this technique is rather complex. However, the technique is academically interesting and it may have at least some practical value. For this reason, certain of its characteristics will be explained by means of an example error pattern.

The pattern of seven errors illustrated in Figure 5-6a represents one of the more difficult patterns to correct. Assume that error correction is first attempted on the rows. Although the code as a whole is capable of locating up to seven errors, the row code is capable of locating only one and of indicating the presence of two. If more than two errors occur in any one row, the response of the error-detection-correction networks for that row will not be as desired, but at least some useful information will be generated. Specifically, for the pattern shown, the checks for row 3 will all be "ok," falsely indicating no errors in this row. In row 5, the check digits in columns 7 and 8 will indicate the presence of errors, but this combination of check digit failures will falsely indicate that digit 7 in this row is in error. After erroneously "correcting" this digit, a total of eight digits will actually be in error, and the pattern of check failures will be as indicated in Figure 5-6b. The decoder must "remember" the fact that a "correction" was made in row 5.

Next, error correction is attempted in columns 1 through 4 (not columns 5 through 8, because the check digits in the lower-right quarter of the array correspond to rows and not columns). In this instance, check digits in positions 5, 6, and 7 of columns 1 and 2 will signal errors. This combination of check failures signals the presence (correctly in this case) of two errors in each column. The locations of the errors are not directly indicated, but from the fact that a "correction" was made in row 5, the decoder may "assume" (correctly) that the correction was made erroneously, and that in fact the digits in row 5 of columns 1 and 2 are in error. The digits in these two positions are then corrected (inverted). Now each of columns 1 and 2 contain only one error, which can be located and corrected by the respective column code. The pattern of errors and check failures is presently as shown in Figure 5-6c. After correcting the digits in row 3 of columns 1 and 2, the array of erroneous digits and the pattern of check failures is as shown in Figure 5-6d. The decoder must "remember" that two digits were corrected in each of rows 3 and 5.

The next step is to use the row codes again. Two errors are signaled in each of rows 3 and 5 by the check failures in columns 6 and 7 of these rows. The row code cannot by itself indicate which digits are in errors, but from the fact that two digits in each row have already been corrected, the decoder can "deduce" that it must be the check digits themselves that are in error. After correcting these digits, all digits are correct and all check digits are "ok" as indicated in Figure 5-6e.

The "correcting," "remembering," "assuming," and "deducing" are all accomplished by means of switching networks in the general manner described earlier in this book. As can well be imagined, the details of the decoder design can assume any of countless widely differing variations.

Consider briefly one other pattern of seven errors. Assume that any seven digits in row 1 are in error. In this instance, the row code causes one digit to be altered, perhaps correctly and perhaps incorrectly. But then the column codes properly correct any errors in digits 1 through 4 of row 1. Then when employing the row code again, any check failures are known to signal errors in the respective check digits themselves, and all seven errors are thereby corrected.

It is instructive for the reader to determine the response of the code to other patterns of seven or fewer errors and to various patterns of eight errors. It is also instructive to find nine-error patterns for which the code fails. Because the code properly corrects most patterns of nine and even many more errors, the probability of receiving an incorrect message is extremely small for any reasonably small digit error probability p. Also, the code has very good burst-error-correction properties.

As before, all or part of the error-correction capability can be abandoned in favor of increased error-detection capability. With complete abandonment of error correction, up to 15 errors in any pattern can be detected. In fact, only highly specialized patterns of 16 or more errors will escape detection.

Of course, the fact that only 25% of the digits are information digits is an extremely detracting feature for the 64-digit (8-by-8) word length illustrated in Figure 5-6. With 256-digit (16-by-16) word length, for another example, the number of information digits would be $(11)(11) = 121$ for a substantial improvement in the proportion of information digits.

Interleaving for Burst-Error Detection-Correction

Consider a set of n-digit code words (messages) where the words are represented by $A_1A_2 \cdots A_n$, $B_1B_2 \cdots B_n$, $C_1C_2 \cdots C_n$, and so on. Each word may be represented in any code, but the assumption will be that $d > 1$ so that the code has at least single-error-detection capability. So far in this chapter, the implication has been that the digits of a given word are likewise transmitted one at a time (serially). In such instances, any burst of errors will appear in a string of successive digits, although when the physical processes that cause the error burst are realistically considered, the error burst must be defined broadly enough to include situations where some of the digits within the burst are, by happenstance, correct.

If, instead of transmitting the words one at a time, the words of the set are interleaved, the errors from the burst can be distributed over several words so that no one word contains very many errors. By this scheme the error detection-correction capability of the individual words is less likely to be exceeded, and the probability of receiving a message erroneously is reduced. The interleaving can be accomplished by visualizing the words to form a rectangular array of digits with each word occupying one row. Instead of transmitting the digits row-by-row, they are transmitted column-by-column so that for three words, for example, the digit sequence is $A_1B_1C_1A_2B_2C_2 \cdots A_nB_nC_n$.

If the number of words per set is m, the number of digits per block is nm. When $d = 2$ the resulting block code provides for the detection of any error burst up to m digits in length, where some of the digits in the burst may actually be correct. If any errors occur in addition to the m-digit burst, the code may fail, however. In fact, the code may fail in the presence of only a two-digit burst plus a single random error if the burst and the single error are strategically located with respect to each other. Corresponding remarks apply for codes of $d = 3$ when these codes are used for single-error correction.

In general for codes of distance d used only for error detection (not correction), the burst length that can be detected is equal to $m(d-1)$. If the burst is, in fact, shorter than this length the code will tolerate a determinable number of other errors in addition to the burst. Again, corresponding remarks apply to codes where all or part of the detection-correction capability is used for correction.

For any of the examples, if m is greater than the burst length, the code will be capable of handling more than one error burst. The burst-handling capability of the code then becomes a complex function of the relative timing of the bursts and of the details of the manner in which the code is used, but can be determined easily for any particular case.

When m is at least as great as the burst length to be encountered, the effect of interleaving is to cause the burst errors to appear as random errors in the individual code words. Then, if p is known, the probability of receiving a word in error can be computed or estimated in the manner indicated earlier in this chapter, and the fact that the errors are not random can be ignored.

Interleaving can be applied to any type of code. For example, the n_1n_2-digit code words formed by the two-dimensional array in Figure 5-5 are in themselves generated by a kind of interleaving regardless of whether the digits are transmitted row-by-row or column-by-column. Moreover, sets of such words can be further interleaved to produce mn_1n_2-digit block codes of still greater burst-error detection-correction capability. Interleaving can also be applied to the cyclic and nonblock codes to be described later even though some of those codes have moderately good burst-error properties by themselves.

Modulo 2 (Mod 2) Arithmetic

Before proceeding to the subject of cyclic codes a brief description of "modulo 2" arithmetic will be given, because that arithmetic is useful in explaining the codes. Here, "arithmetic" is used in the broad sense to mean a set of symbols together with a set of rules for manipulating those symbols. Although modulo 2 (often abbreviated "mod 2") arithmetic is not a numerical or computational system in the ordinary sense, it happens to be very similar to conventional binary arithmetic (to be described in a later chapter) but with the crucial exception that no "carry" or "borrow" digits are involved.

In modulo 2 arithmetic only two symbols are used. The symbols are usually 0 and 1 as in conventional binary arithmetic, but the meaning of the two symbols is only nebulously related to the meaning of conventional binary digits. Modulo 2 addition of two digits is defined to be essentially the same as the exclusive-OR function in that the sum (mod 2) of two digits is 1 when one but not both of the two digits being added (mod 2) is 1. That is, in detail: $0 + 0 = 0$; $0 + 1 = 1$; $1 + 0 = 1$; and $1 + 1 = 0$. Multiplication (mod 2) of two digits is the same as in conventional binary multiplication in that the product (mod 2) is 1 only when both of the digits are 1. That is: $(0)(0) = 1$; $(0)(1) = 0$; $(1)(0) = 0$; and $(1)(1) = 1$.

Digits may be grouped in sets where each set vaguely corresponds to a multidigit number in conventional arithmetic. However, the digit

216 DIGITAL CODING

sets do not have true numeric significance. Because (as will be explained shortly) the digits are often viewed as being the coefficients of a polynomial, the term "polynomial" will be used here to refer to a set of mod 2 digits. Two polynomials may be added (mod 2) by operating on the digits in corresponding positions. Thus, for example:

```
    0 0 1 1 0 1 0
+ 1 0 1 1 1 0 1
  ─────────────────
  1 0 0 0 1 1 1
```

If one polynomial has more digits than the other polynomial, 0's may be added to the "high-order" (left-hand) positions of the polynomial with the fewer digits to cause the total number of digits to be the same in both polynomials—as in conventional arithmetic.

Two polynomials may be multiplied (mod 2) in a manner that is essentially the same as binary multiplication except that no carries are considered when accumulating (mod 2) the partial products. For example:

```
                1 0 1 1 0 1
          ×         1 0 1 1
          ─────────────────
                1 0 1 1 0 1
              1 0 1 1 0 1
          ─────────────────
              1 1 1 0 1 1 1
            0 0 0 0 0 0
          ─────────────────
            0 1 1 1 0 1 1 1
        1 0 1 1 0 1
          ─────────────────
        1 0 0 0 1 1 1 1 1
```

As portrayed above, the individual partial products are handled one at a time, but they may be handled more or less simultaneously as illustrated below for the same example.

```
                1 0 1 1 0 1
          ×         1 0 1 1
          ─────────────────
                1 0 1 1 0 1
              1 0 1 1 0 1
        1 0 1 1 0 1
          ─────────────────
        1 0 0 0 1 1 1 1 1
```

Note that each product digit is actually a representation of the odd-even count of the 1's in the corresponding column of partial product digits. In other words, in mod 2 multiplication each product digit is an odd-even count of the 1's in certain positions of one polynomial where the positions are as specified by the 1's in the other polynomial. For the example cited, this relationship is illustrated in Table 5-5 where the x_i represent the digits of the multiplicand (101101) and the p_i represent the digits of the product. The multiplier (1011) is represented by the pattern of check marks that enter into the odd-even counts. The table indicates that p_5, for example, of the product is determined by the odd-even count of digits x_5, x_4, and x_2 of the multiplicand.

TABLE 5-5. Pattern of Odd-Even Counts in the Multiplication (Mod 2) of a Six-Digit Polynomial by 1011

	x_6	x_5	x_4	x_3	x_2	x_1
p_9	✓					
p_8		✓				
p_7	✓		✓			
p_6	✓	✓		✓		
p_5		✓	✓		✓	
p_4			✓	✓		✓
p_3				✓	✓	
p_2					✓	✓
p_1						✓

From the similarities between Tables 5-4 and 5-5, at least an intuitive relationship between modulo 2 arithmetic and error-detection-correction codes should now be apparent. However, an appreciation of a few further characteristics of the arithmetic are needed before it can become a useful tool in the study of the codes.

Addition (mod 2) and multiplication (mod 2) are both commutative. That is, $A + B = B + A$ and $AB = BA$, where A and B are polynomials. For addition, this property is more or less obvious, but for multiplication the property is not so obvious. A rigorous proof would go beyond present purposes, but the reader can quickly satisfy himself that $AB = BA$ by performing the indicated multiplications (mod 2) with some randomly selected polynomials. In the example that was cited, the following result is obtained by reversing the roles of the multiplier and multiplicand.

```
                1 0 1 1
       ×  1 0 1 1 0 1
                1 0 1 1
              1 0 1 1
            1 0 1 1
        1 0 1 1
        1 0 0 0 1 1 1 1 1
```

Addition (mod 2) and multiplication (mod 2) are both associative. That is, $(A + B) + C = A + (B + C)$ and $(AB)C = A(BC)$, where A, B, and C are polynomials. Again, rigorous proofs could be given, but the reader can be satisfied of the property by working out a few examples. The associative nature of addition (mod 2) has already been suggested in the two different manners, in which partial products were accumulated in a multiplication (mod 2).

The two polynomials entering into a multiplication (mod 2) need not have the same number of digits. In fact, in the example cited, one polynomial had six digits whereas the other had only four digits, where the number of digits is determined by the left-most digit position containing a 1. A characteristic of multiplication (mod 2) is that the number of digits in the product is always $n_1 + n_2 - 1$ (conventional notation), where n_1 and n_2 are the numbers of digits in the two polynomials. This result is in contrast with conventional arithmetic (either binary or decimal) where the number of digits in the product may be either $n_1 + n_2 - 1$ or $n_1 + n_2$ in accordance with the magnitudes of the numbers being multiplied.

Subtraction (mod 2) is characterized as follows. If 1 is subtracted from 1 the difference (mod 2) is 0, as might be expected. That is, $1 - 1 = 0$. But $1 + 1 = 0$. The surprising, but correct, conclusion is that in mod 2 arithmetic $-1 = +1$. Also, therefore, $0 - 1 = 1$. With these relationships among the symbols, the subtraction (mod 2) of one polynomial from the other produces exactly the same result as the addition (mod 2) of the two polynomials. To cite two examples:

```
    0 0 1 1 0 1 0            1 0 1 1 1 0 1
  - 1 0 1 1 1 0 1          - 0 0 1 1 0 1 0
    1 0 0 0 1 1 1            1 0 0 0 1 1 1
```

Division (mod 2) is the inverse of multiplication (mod 2) in analogy with the corresponding operations in conventional arithmetic, but because of the nature of subtraction (mod 2) the concept of relative magnitudes is quite different from the corresponding concepts in conventional arithmetic, and the details of division (mod 2) may seem unusually peculiar to the unitiated. For one thing, fractions are not generally defined in mod 2 arithmetic (at least not insofar as the arithmetic is used for coding), so the division process is always terminated when the number of digits in the remainder is less than the number of digits in the divisor. Of course, if the dividend has fewer digits than the divisor, the quotient is immediately known to be 0, and the dividend becomes the remainder. Also, because $-1 = +1$, a given quotient digit can be determined to be 1 whenever the highest order digit of the dividend or partial remainder is 1, regardless of the values of all other digits. These points are illustrated by the following example.

```
                              1 0 1 0 0 1   quotient
    divisor  1 0 1 1)1 0 0 1 1 0 1 1 1      dividend
                    -1 0 1 1
                     ─────────
                     0 1 0 1
                    -0 0 0 0
                     ─────────
                       1 0 1 0
                      -1 0 1 1
                       ─────────
                       0 0 1 1
                      -0 0 0 0
                       ─────────
                         0 1 1 1
                        -0 0 0 0
                         ─────────
                         1 1 1 1
                        -1 0 1 1
                         ─────────
                         1 0 0    final remainder
```

The first (left-hand) quotient digit is written over the first dividend digit to indicate that the division (mod 2) of the first four dividend digits (1001) by the divisor (1011) as 1 as determined by the first dividend digit. The remaining steps of the division operation proceed as indicated—although additions instead of subtractions could be used to obtain the intermediate remainders and the final remainder. The division can be checked by multiplying (mod 2) the quotient and the divisor and then adding (mod 2) the final remainder.

For another example, the division (mod 2) of 1001 by 1011 produces a quotient of 1 and a final remainder of 10 as can be checked analogously.

The polynomial nature of the digit sets (called "numbers" in conventional arithmetic) is derived by assuming that the 1's and 0's represent coefficients of powers of X, where X is any number other than 0 (conventional arithmetic 0), although it is sometimes convenient to visualize X as being equal to 2. When used in this way the symbols 1 and 0 have their conventional meanings of one and zero, or unity and nothing, respectively. However, because the value of X is indefinite, the positional significance of the symbols is not exactly the same as in conventional arithmetic—either decimal or binary. In a string of 1's and 0's the symbol at the right-hand end is the coefficient for X^0, and the powers increase by unity in each successive digit position to the left. Thus, for example:

$$11001011 = (1)X^7 + (1)X^6 + (0)X^5 + (0)X^4 + (1)X^3 + (0)X^2 + (1)X^1 + (1)X^0$$
$$= X^7 + X^6 + X^3 + X + 1$$

In other words, each 1 is represented by a power of X where the power corresponds to the position of the 1 in the digit set, and each 0 is represented by nothing.

Arithmetic operations, either conventional or mod 2, can be performed on the polynomials as expressed in the previous paragraph. The conventional arithmetic operations correspond to algebraic manipulations with which the reader is assumed to be familiar. The mod 2 operations are essentially the same except, as before, for the absence of carries and borrows. To present one example, the mod 2 division of $X^3 + 1$ by $X^2 + X$ proceeds as follows.

$$\begin{array}{r}X+1\\X^2+X\overline{)X^3+1}\\\underline{X^3+X^2}\\X^2\\\underline{X^2+X}\\X+1\end{array}$$

The remainders can be obtained either by subtractions (mod 2) or by additions (mod 2) because $-X^n = +X^n$, where n is any integer. That the quotient is $X + 1$ with a final remainder of $X + 1$ can be checked by showing that $(X + 1)(X^2 + X) + X + 1 = X^3 + 1$ in mod 2 arithmetic. The reader may wish to rework this same example by dividing (mod 2) 1001 by 110.

One of the advantages to be gained by using the algebra-like polynomial form is encountered for digit sets such as 1000000001, for example. A person has difficulty in determining the number of 0's, but when the notation is $X^9 + 1$, the position of the "high-order" 1 is instantly recognizable. Also, in some types of problems the position of the 1's are variable, in which case the notation can be X^n or X^{n-3}, for examples, where n is an integer specified in some way. However, in a physical system only the coefficient 1's and 0's are represented by signals.

An Introduction to Cyclic Codes

Conceivably, a useful scheme for encoding a set of binary digits would be to view the set as representing a conventional binary number and to multiply that number by some other binary number. With M different digit sets (messages), the multiplication would produce M new sets which might form a code having $d > 1$ and might have characteristics that would be useful for error detection-correction. For example, with three-digit sets, eight different messages are possible, and if these are multiplied by 11 (decimal 3) the following code words result.

Original Message	Code Word Obtained by Multiplying by 11
000	00000
001	00011
010	00110
011	01001
100	01100
101	01111
110	10010
111	10101

In this particular instance a five-digit code having a distance $d = 2$ does in fact result. Decoding can be accomplished by dividing the received word by 11. If the remainder is 0, no errors have been detected and the quotient represents the transmitted message. If the remainder is not 0 (is either 01 or 10), an error has been detected but it cannot be corrected. However, as was shown earlier, a single parity check digit is sufficient to produce a $d = 2$ code having only four digits per code word when the number of different messages is eight. Therefore, the coding scheme suggested here appears to be inferior. A person might wonder about the possibility of better codes with longer messages and with multipliers other than 11. This possibility seems not to have been investigated thoroughly, but no competitive codes of this nature are known.

However, if the multiplication for encoding and the division for decoding is performed with mod 2 arithmetic, a profoundly different situation results. Consider the same example of messages containing three information digits multiplied (mod 2) by 11.

Original Message	Code Word Obtained by Multiplying (mod 2) by 11
000	0000
001	0011
010	0110
011	0101
100	1100
101	1111
110	1010
111	1001

By examination the code is found to have $d = 2$, and each code word contains an even number of 1's. Error detection can be accomplished by means of an odd-even count essentially as was illustrated in Figure 5-3. Unfortunately, if no error is detected, the particular message that was transmitted is not readily apparent from the received code word. The message can, however, be determined by dividing (mod 2) the received message by 11. If this division produces a nonzero remainder, an error has been detected. Thus, the division accomplishes both decoding and error detection. Because the division is needed for decoding, the checking for errors by means of an odd-even count is superfluous.

With a multiplier of 11, the above coding scheme works for messages of any length, but in general for longer multipliers (to be discussed in more detail later), a certain maximum message length is associated with each different multiplier when generating codes with various error detection-correction properties.

The above code is nonsystematic. To obtain a systematic code, a minor variation involving a mathematical "trick" can be employed in the encoding procedure. To encode a message having k information digits, r digits (which will be parity check digits) are added to form code words having $k + r = n$ digits. The message, which may be viewed as a polynomial, is first multiplied (mod 2) by X^r. This multiplication has the effect of shifting the message digits r positions to the left, because X^r is equivalent to a digit set consisting of a 1 followed by r 0's. Note, in particular, that when $r = 1$, the equivalent digit set is 10 (not just 1). After this multiplication or shifting operation, the result is divided (mod 2) by the same constant which was used previously as a multiplier. This constant is now called a "generator" polynomial. In general, a nonzero remainder will be obtained, and the operations may be expressed as follows.

$$\frac{X^r \text{ (message polynomial)}}{\text{generator polynomial}} = \text{quotient} + \frac{\text{remainder}}{\text{generator polynomial}}$$

If, as will be assumed to be the case, the generator polynomial contains $r + 1$ digits (that is, if the highest-order nonzero term in the generator polynomial is X^r), an r-digit remainder will result. This remainder can be added to the shifted message digits to produce a code word that is an exact multiple of the generator polynomial. In other words:

$$\text{code word} = X^r \text{ (message polynomial)} + \text{remainder}$$

Error detection is accomplished by dividing (mod 2) the received code word by the same generator polynomial. If the remainder of this division is zero, no error is detected, and the intended message is found in the first k digits of the received word. In other words, decoding is accomplished merely by shifting the digits of the right r positions, although a division operation is required to check for errors, which are indicated by a nonzero remainder. For codes having $d > 2$, an error-correcting capability is present as before, but the correction procedure is a more complex matter to be discussed later.

Codes developed as described above are called cyclic codes, where the term "cyclic" seems to have come from two different sources. One source is in the fact that when a word of a cyclic code is shifted any number of positions "cyclically" (that is, where the digits shifted "out" at one end of the word are shifted "in" at the other end) another valid word of the code is formed. As an example, for the code presented in Table 5-7a and to be discussed later, consider code word 0100011. If the digits of this word are shifted one position to the right with the right-hand 1 being shifted to the left-hand end, the resulting word is 1010001, which is also listed as a code word in Table 5-7a. This cyclic property of the codes is a fortuitous result of the manner in which the codes are constructed. This property is of use in one particular decoding scheme (to be described later) but is more likely to be a nuisance in the face of synchronization errors (also to be described later).

The other source of the term "cyclic" is derived from the over-and-over ("cyclic") procedure that can be used to realize multiplication and division (mod 2 in this instance). Therefore, physically, the encoding and decoding can be accomplished by means of a relatively small amount of equipment, primarily shift registers, which perform a relatively simple operation repeatedly in contrast to a large amount of equipment performing a single complex operation. This cyclic property of the codes is highly attractive and is the major reason so much attention has been given to cyclic codes—although the rate at which information can be encoded and decoded is correspondingly reduced.

Consider again the example of eight messages 000 through 111 and assume that the coding is to be accomplished with the generator polynomial (previously called "multiplier") $X + 1$, that is, 11 in digital notation. In this case $r = 1$. To encode 010, for example, it is first multiplied (mod 2) by 10 to produce 0100. This set of digits is then divided (mod 2) by the generator polynomial 11 to produce a quotient of 011 and a remainder of 1. The 011 is not used, but the code word is formed by appending the remainder 1 to the original message 010 (or by adding 1 to 0100) to produce 0101. The list of all eight possible messages and their coded representations is as follows.

Original Message	Code Word Obtained as Described Above
0 0 0	0 0 0 ! 0
0 0 1	0 0 1 ! 1
0 1 0	0 1 0 ! 1
0 1 1	0 1 1 ! 0
1 0 0	1 0 0 ! 1
1 0 1	1 0 1 ! 0
1 1 0	1 1 0 ! 0
1 1 1	1 1 1 ! 1

The broken line is used to indicate the systematic nature of the code. The digits to the left are the original information digits, and the digits to the right are the parity check digits—although the scheme for obtaining these digits is very different from the odd-even counting networks discussed earlier. (The purpose of using this seemingly complicated scheme for accomplishing a simple result is probably not apparent in this elementary example. The purpose will become more apparent with more complex codes where a relatively straightforward shift-register technique can be used for encoding and decoding.)

Note that this coding scheme produces the same set of code words as when the message was merely multiplied (mod 2) by 11. However, a given code word does not necessarily correspond to the same message as before. Thus, for example, the code word 1111 corresponds to message 101 under simple multiplication (mod 2), but it corresponds to message 111 when the coding method described here is used.

The study of cyclic codes revolves principally around determining the code characteristics that result from generator polynomials that are more complex than the simple 11 used for illustration in this section. In general, for an r-digit generator polynomial, 2^{r-1} different generator polynomials might be considered (not 2^r because the "highest-order" digit must, by definition, be 1). For each r-digit generator polynomial, a code with $r-1$ parity check digits will result. Many of these codes will not have useful properties. Much is now known about many generator polynomials which do produce useful or potentially useful codes, but in the light of all possible generator polynomials that might be brought under consideration, much is still unknown.

A More General Description of Cyclic Codes Having $d = 2$

If a single error occurs in a given n-digit word, the difference (mod 2) between the correct word and the erroneous word will be 1 at some position in the digit set. In other words, if the erroneous word is subtracted (mod 2) from the correct word, the difference can be expressed as X^i in polynomial form, where i is an integer and $n > i \geq 0$. For this error to be detectable, this difference must not be an exact multiple of the generator polynomial (or "multiplier") as defined in the previous section. However, X^i cannot be an exact multiple of any generator polynomial having two or more terms (having two or more 1's when expressed in digital form). Therefore, any generator polynomial having two or more 1's will produce a code with a distance $d \geq 2$. Of course, if single-error detection is the only objective, generator polynomials other than 11 would probably be pointless, because generator polynomials having a greater number of digits would produce an unnecessarily large number of parity check digits (more than one). The purpose here is merely to introduce cyclic code considerations in a more generalized way.

Besides having $d = 2$, the elementary code produced by a single parity check digit is capable of detecting any odd number of errors. This code was shown to be the same as the code obtained with a generator polynomial $X + 1$ (that is, 11 in digit form). In the general case, a code developed by any generator polynomial having $X + 1$ as a factor (mod 2) will be capable of detecting any odd number of errors. That is, if the generator polynomial $P(X)$ can be expressed as $P(X) = (X + 1)P_1(X)$, where $P_1(X)$ is any other mod 2 polynomial, the resultant code will be the same as the code obtained with a generator polynomial of $P_1(X)$ but with an overall parity check digit added as a result of the influence of the $X + 1$ factor.

Among the polynomials containing $X + 1$ as a factor is the polynomial $X^a + 1$, where a is any integer greater than 0. Specifically,

$X^c + 1 = (X + 1)(X^{a-1} + X^{a-2} + \cdots + 1)$. Therefore, a code developed by any generator polynomial having $X^a + 1$ as a factor will likewise be capable of detecting any odd number of errors.

Cyclic Codes Having $d = 3$—Hamming Codes Expressed as Cyclic Codes

If a code word contains two digits in error, the difference (mod 2) between the erroneous word and the correct word can be expressed as $X^i + X^j$, where i and j are integers and where $n > i > j \geq 0$, with n being the number of digits in the code word. For detecting the presence of two errors by means of a cyclic code, $X^i + X^j$ must not be an exact multiple of the generator polynomial for any i and j. To determine the characteristics that the generator polynomial must have to meet this requirement, an observation is first made that the generator polynomial, in any case, should have at least two 1's (one of which is in the "lowest-order" position). In other words, if the generator polynomial contains only one 1, the effect would be merely a shifting of digits with no useful coding properties resulting. Next, the observation is made that $X^i + X^j$ can be factored (mod 2) to produce $(X^{i-j} + 1)X^j$. With the generator polynomial properties as specified, the factor X^j will certainly not be an exact multiple of the generator polynomial. Now because the maximum value of i is $n - 1$ and the minimum value of j is 0, the maximum value of the exponent $i - j$ is $n - 1$. Therefore, for a given value of n, the requirement on the generator polynomial is that it must not be a factor of $X^e + 1$, where e is any integer from 1 to $n - 1$.

Alternatively, if a given generator polynomial (and not a given n) is under consideration, the generator polynomial will produce a double-error-detecting $(d = 3)$ code if $n \leq e$, that is, if $(n - 1) < e$, where e is now defined as the smallest integer which causes $X^e + 1$ to be an exact multiple of the generator polynomial.

The relationship between each polynomial (whether or not it is used as a "generator") and its corresponding value of e as defined in the previous paragraph is a topic of mathematical study. For present purposes, it will only be stated that, for each value of e satisfying the relationship $e = 2^m - 1$, where m is any positive integer, at least one polynomial can be found for which $X^e + 1$ is an exact multiple of that polynomial and which has X^m as the "highest-order" term in that polynomial. (Actually, for large m, many such polynomials can be found.) Each such polynomial is said to be a "primitive" polynomial. Moreover, each primitive polynomial is "irreducible," which here means that it cannot be evenly divided (mod 2) by any other polynomial except itself and 1. Irreducible polynomials in mod 2 arithmetic correspond vaguely to prime numbers in conventional arithmetic. However, an irreducible polynomial is not necessarily primitive. For a given m, a polynomial which is irreducible but not primitive can be evenly divided (mod 2) into some polynomial of the form $X^{e'} + 1$, where $e' < 2^m - 1$.

Table 5-6 lists various irreducible polynomials. Because of the symmetry of mod 2 arithmetic, the digits of any given polynomial taken

TABLE 5-6. A Listing of Some Irreducible Polynomials. All Except Those Marked with an Asterisk (*) Are Primitive as Well as Irreducible

m	e	Polynomial		m	e	Polynomial	
2	3	111	7	7	127	10,010,001	121
3	7	1,011	13			10,000,011	203
		1,101	15			10,001,001	211
						10,001,111	217
4	15	10,011	23			10,011,101	235
		11,001	31			10,100,111	247
		11,111*	37			10,101,011	253
						10,111,001	271
5	31	100,101	45			10,111,111	277
		101,001	51			11,000,001	301
		101,111	57			11,011,011	313
		110,111	67			11,010,011	323
		111,011	73			11,011,101	325
		111,101	75			11,100,101	345
						11,101,111	357
6	63	1,000,011	103			11,110,001	361
		1,001,001*	111			11,110,111	367
		1,010,111*	127			11,111,101	375
		1,011,011	133	8	255	100,011,101	435
		1,100,001	141			101,110,001	561
		1,100,111	147				
		1,101,101	155	9	511	1,000,010,001	1021
		1,110,011	163			1,000,100,001	1041
		1,110,101*	165				
				10	1023	10,000,001,001	2011
						10,010,000,001	2201
				11	2047	100,000,000,101	4005
						101,000,000,001	5001
				12	4095	1,000,001,010,011	10123
						1,100,101,000,001	14501

in reverse sequence have the same properties as the given polynomial with respect to being irreducible or primitive. The table is believed to be complete for values of m from 2 through 7. For each of the values of m from 8 through 12, only one such polynomial and its reverse have been selected for listing. Except for the polynomials designated by an asterisk (*), all listed polynomials are primitive as well as irreducible. The role of these polynomials in coding will be seen in a later section. Curiously, for some m, all irreducible polynomials are also primitive, but not for the other m.

For example, as indicated in Table 5-6, for $m = 3$ and $e = 7$, $X^7 + 1$ is an exact multiple (mod 2) of either $X^3 + X + 1$ or $X^3 + X^2 + 1$. In other words, 1000001 is an exact multiple of either 1011 or 1101. Neither 1011 nor 1101 is reducible. Either 1011 or 1101 can be used as a generator polynomial to develop a $d = 3$ code provided that the total number of digits per code word is no greater than seven. Because the use of a four-digit generator polynomial will add three digits to the information digits, the maximum number of information digits in the original message is $7 - 3 = 4$.

The polynomial 10000001 (that is, $X^7 + 1$) is also evenly divisible (mod 2) by 10111 or by 11101, and these polynomials will likewise produce double-error-detecting $(d = 3)$ codes provided that the total number of digits per code word is not greater than seven, but the number of information digits per word is reduced by one because the number of parity check digits is increased by one in comparison with the generator polynomials of the previous paragraph. However, these particular generator polynomials actually produce codes having $d > 3$ as will be shown subsequently. For these polynomials $m = 4$, and neither polynomial is primitive. Also, neither is irreducible, because each is an exact multiple (mod 2) of 11.

The cyclic $d = 3$ codes obtained by using the primitive polynomials as generator polynomials are identical to the $d = 3$ Hamming codes described earlier provided that the digits of the Hamming codes are written in the appropriate sequence. A rigorous mathematical proof of the equivalence would go beyond present purposes, but the equivalence is easily demonstrated by means of an example as illustrated in Table 5-7 which shows the cyclic codes having $n = 7$ as obtained by the generator polynomials 1011 and 1101 in a and b, respectively. As a specific example in b the original message 0110 is multiplied (mod 2) by X^r (where $r = m$ with both being as defined earlier and both being equal to 3 in this example) to produce 0110000, which is divided by the generator polynomial 1101. The quotient 0100 is discarded, but the remainder 100 is added to 0110000 to produce a code word 0110100.

CYCLIC CODES HAVING $d = 4$ 229

TABLE 5-7. Cyclic $d = 3$ Codes Formed by Generator Polynomials
1011 and 1101, Respectively

Generator Polynomial 1011 $x_1\ x_2\ x_3\ x_4\ y_1\ y_2\ y_3$	Generator Polynomial 1101 $x_1\ x_2\ x_3\ x_4\ y_1\ y_2\ y_3$
0 0 0 0 0 0 0	0 0 0 0 0 0 0
0 0 0 1 0 1 1	0 0 0 1 1 0 1
0 0 1 0 1 1 0	0 0 1 0 1 1 1
0 0 1 1 1 0 1	0 0 1 1 0 1 0
0 1 0 0 1 1 1	0 1 0 0 0 1 1
0 1 0 1 1 0 0	0 1 0 1 1 1 0
0 1 1 0 0 0 1	0 1 1 0 1 0 0
0 1 1 1 0 1 0	0 1 1 1 0 0 1
1 0 0 0 1 0 1	1 0 0 0 1 1 0
1 0 0 1 1 1 0	1 0 0 1 0 1 1
1 0 1 0 0 1 1	1 0 1 0 0 0 1
1 0 1 1 0 0 0	1 0 1 1 1 0 0
1 1 0 0 0 1 1	1 1 0 0 1 0 1
1 1 0 1 0 0 1	1 1 0 1 0 0 0
1 1 1 0 1 0 0	1 1 1 0 0 1 0
1 1 1 1 1 1 1	1 1 1 1 1 1 1

	x_1	x_2	x_3	x_4	y_1	y_2	y_3		x_1	x_2	x_3	x_4	y_1	y_2	y_3
y_1	✓	✓	✓		✓			y_1	✓		✓	✓	✓		
y_2		✓	✓	✓		✓		y_2	✓	✓	✓			✓	
y_3	✓	✓		✓			✓	y_3	✓	✓	✓				✓

In the lower portion of a and b in Table 5-7 the pattern of check digits is shown with the same notation as used previously for the Hamming codes.

Cyclic Codes Having $d = 4$

Earlier in this chapter the appending of an over-all parity check digit to a Hamming $d = 3$ code was shown to produce a $d = 4$ code. In the immediately preceding sections, generator polynomial $X + 1$ was shown to have the effect of adding an overall parity check digit. Because a code, produced by a primitive polynomial is the same as a Hamming code,

230 DIGITAL CODING

the multiplication (mod 2) of any primitive polynomial by $X + 1$ produces a polynomial which generates a code having $d = 4$ provided the length of the code word is suitably limited. In particular, the length of the code word as generated by the primitive polynomial can be the same as before, but the $X + 1$ factor has the effect of adding one digit. Therefore, the total length n can be as great as $e + 1 = 2^m$, where m is the power of the "highest-order" term in the generator polynomial.

In the case of $m = 3$, for example (see Table 5-6), the primitive polynomial $X^3 + X^2 + 1$ can be multiplied (mod 2) by $X + 1$ to produce the generator polynomial $X^4 + X^2 + X + 1$ (that is, 10111) which in turn produces a $d = 4$ code having $n = 8$, $k = 4$, and $r = 4$, where k and r are, as before, the number of information digits and parity check digits, respectively. Alternatively, the primitive polynomial $X^3 + X + 1$ can be multiplied (mod 2) by $X + 1$ to produce a generatory polynomial $X^4 + X^3 + X^2 + 1$ (i.e., 11101), which in turn produces a code having the same characteristics.

Cyclic Codes Having Odd d Greater than 3—Multiple-Error-Correcting Cyclic Codes

Inasmuch as a primitive polynomial produces a single-error-correcting ($d = 3$) cyclic code provided the length of the code word is suitably restricted, a person might wonder if a second use of the same polynomial might produce a code capable of correcting a second error provided, perhaps, that suitable additional restrictions were placed on the length of the code word. Perhaps the primitive polynomial could be multiplied by itself or "squared" (mod 2) to produce a generator polynomial suitable for the formation of the double-error-correcting code. Without going into the reasons why, the statement will be simply made here that this scheme does not work. However, if the original primitive polynomial is multiplied (mod 2) by a certain different irreducible polynomial (which need not be primitive in every instance), a code capable of correcting a second error does indeed result. Furthermore, as the generator polynomial is successively multiplied (mod 2) by suitably selected irreducible polynomials, new generator polynomials are obtained which produce codes of correspondingly higher error detection-correction capability.

Consider, for example, the $d = 3$ cyclic code for which $n = 15$, $k = 11$, and which has the generator polynomial 10011, which is one of the primitive polynomials listed in Table 5-6. If this polynomial is multiplied (mod 2) by 11111, which is one of the irreducible but not primitive polynomials listed, the polynomial 111,010,001 is obtained, and this polynomial is, of course, neither irreducible nor primitive. However, it may be used as the generator polynomial for a double-error-correcting ($d = 5$) cyclic code. The length n of each code word must be no greater than 15 as before, and because a generator of this length produces eight instead of just four parity check digits, the number k of information digits per code word is reduced to seven.

From the standpoint of the information content of the added four parity check digits, these added digits may be visualized as indicating the location of a possible second erroneous digit, in analogy with the discussion earlier in this chapter. However, the parity check pattern is too complex for this elementary visualization to apply. The actual parity check pattern can be determined by computing the parity check digits for selected messages and their resulting code words as listed in Table 5-8a. Each message has one 1. If the message as represented by digits x_1 through x_7 is 0000001, for example this polynomial is shifted eight positions to the left and divided by the generator polynomial to produce a remainder of 11,010,001 which produces the parity check digits p_1 through p_8. From this result, x_7 is obviously included in the parity checks of p_1, p_2, p_4, and p_8, and check marks are correspondingly placed in the x_7 column of Table 5-8b.

If a code word is received with one or two (but no more than two) digits in error, the pattern of parity check failures will indicate the positions of the digit or digits in error. In particular, if only one parity check failure occurs, the conclusion is that the corresponding parity check digit itself is in error as in the single-error-correcting codes. If exactly two parity check failures occur, the conclusion is similarly that two parity check digits are in error. If more than two parity checks fail, the situation becomes slightly more complex. In instances where the failures are exactly as listed in one of the columns corresponding to information digits x_1 through x_7 in Table 5-8b, the conclusion to be drawn is that a single error has occurred and that this error is in the corresponding information digit.

For other patterns of parity check failures, an inspection of the table must be made to find, if possible, a combination of two errors which will produce that pattern. Of course, if two digits both affect a given parity check digit, errors in both of the two digits will not cause a check failure in the corresponding parity check digit.

Thus, for example, if digits x_1 and x_2 are in error, the parity counts for p_2 and p_3 will not fail. Instead, the counts which do fail will be those for p_1, p_4, p_5, and p_6, all as can be determined by inspection of Table 5-8b. Because no other single or double error will produce this pattern of parity check failures, this pattern signals the fact that

TABLE 5-8. Double-Error-Correcting ($d = 5$) Cyclic Code Developed by the Generator Polynomial 111,010,001: (a) Partial Listing of Code Words and (b) Pattern of Parity Check Counts

x_1	x_2	x_3	x_4	x_5	x_6	x_7	p_1	p_2	p_3	p_4	p_5	p_6	p_7	p_8
0	0	0	0	0	0	1	1	1	0	1	0	0	0	1
0	0	0	0	0	1	0	0	1	1	1	0	0	1	1
0	0	0	0	1	0	0	1	1	1	0	0	1	1	0
0	0	0	1	0	0	0	0	0	1	1	1	0	1	
0	0	1	0	0	0	0	0	1	1	1	0	1	0	
0	1	0	0	0	0	0	1	1	1	0	1	0	0	
1	0	0	0	0	0	0	1	1	1	0	1	0	0	0

(a)

	x_1	x_2	x_3	x_4	x_5	x_6	x_7	p_1	p_2	p_3	p_4	p_5	p_6	p_7	p_8
p_1	✓			✓		✓	✓								
p_2	✓	✓		✓	✓	✓			✓						
p_3	✓	✓	✓	✓	✓					✓					
p_4		✓	✓	✓		✓	✓				✓				
p_5	✓		✓	✓								✓			
p_6		✓		✓	✓								✓		
p_7		✓		✓	✓									✓	
p_8			✓		✓	✓									✓

(b)

digits x_1 and x_2 are in error. For another example, parity check failures for p_2, p_3, and p_5 would signal that digits x_1 and p_1 are in error, as can be determined from the table. In one instance, when digits x_4 and x_5 are in error, a total of seven parity check failures occur: for p_1, p_2, p_3, p_4, p_5, p_7, and p_8. However, for all other combinations of seven parity check failures, for most combinations of six failures, and for some combinations of fewer failures, the conclusion to be drawn is that at least three errors have occurred in the transmitted code word but that the positions of the erroneous digits cannot be determined.

Now if the generator polynomial 111,010,001 for the above double-error-correcting code is multiplied (mod 2) by 111, which is an irreducible polynomial that happens to be primitive, the product is 10,100, 110,111 which can be used as the generator polynomial for a triple-error-correcting ($d = 7$) code. As before, $n = 15$, but k is now reduced to five (because the number of parity check digits is increased from eight to ten) so that only 32 different code words (messages) are possible. The pattern of parity checks is even more complex than in Table 5-8b, and probably the easiest way to show that the resulting code is in fact triple-error-correcting is to compute all 32 code words and note that each word differs from each other word in at least seven digit positions.

Observe that when going from a double-error-correcting $n = 15$ code to a triple-error-correcting $n = 15$ code, only two (not four) parity check digits are added. Although four bits of information are required to specify the location of a single error, less than four bits (not digits) are required to specify the location of the second error because, once the location of the first error is specified, the number of possible locations for the second error is less than 15 (even after taking into consideration the possible absence of an error). Nevertheless, a double-error-correcting $n = 15$ code having less than eight parity check digits happens to be an impossibility. When going to a triple-error-correcting $n = 15$ code, the number of bits required to specify the location of the third error is still less, but in this case a code that utilizes this fact does happen to be possible.

For $n = 15$, no meaningful further extensions of the above technique are possible.

Table 5-9 lists generator polynomials that can be used to obtain various degrees of error correction for $n = 15$, $n = 31$, and $n = 63$. A feature to be noted in the listing is that in some instances where a given generator polynomial is multiplied (mod 2) by an irreducible polynomial to produce a generator polynomial offering greater error-correction capability, the number of correctable errors is increased by more than one. For example, for $n = 63$ and $k = 24$, the generator polynomial is g_7 and a seven-error-correcting code results. However, when g_7 is multiplied (mod 2) by 1,011,011 to produce g_8, a 10-error-correcting code results—with the number of parity check digits being increased by six and the number of information digits being decreased by six in each code word. For a listing of generator polynomials for $n = 127$, and $n = 255$ cyclic codes of this form reference is made to a paper by J. P. Stenbit in the *IEEE Transactions on Information Theory*, October 1964, pp. 309–391. This paper also lists the "multiplied out" versions of the generators listed in Table 5-9.

Incidentally, the name "Bose-Chaudhuri" (B-C) codes is derived from a 1960 paper by R. C. Bose and D. K. Ray-Chaudhuri. The codes are

TABLE 5-9. A Listing of Generator Polynomials for Cyclic Multiple-Error-Correcting Codes; in Each Instance the Subscript on g is Equal to $(d - 1)/2$, the Number of Correctable Errors

n	k	d	Generator Polynomial
15	11	3	$g_1 = 10{,}011$
	7	5	$g_2 = (11{,}111)g_1$
	5	7	$g_3 = (111)g_2$
31	26	3	$g_1 = 100{,}101$
	21	5	$g_2 = (111{,}101)g_1$
	16	7	$g_3 = (110{,}111)g_2$
	11	11	$g_5 = (101{,}111)g_3$
	6	15	$g_7 = (111{,}011)g_5$
63	57	3	$g_1 = 1{,}000{,}011$
	51	5	$g_2 = (1{,}010{,}111)g_1$
	45	7	$g_3 = (1{,}100{,}111)g_2$
	39	9	$g_4 = (1{,}001{,}001)g_3$
	36	11	$g_5 = (1{,}101)g_4$
	30	13	$g_6 = (1{,}101{,}101)g_5$
	24	15	$g_7 = (1{,}011{,}011)g_6$
	18	21	$g_{10} = (1{,}110{,}101)g_7$
	16	23	$g_{11} = (111)g_{10}$
	10	27	$g_{13} = (1{,}110{,}011)g_{11}$
	7	31	$g_{15} = (1{,}011)g_{13}$

also commonly called Bose-Chaudhuri-Hocquenghem (BCH) codes in recognition of a 1959 paper by A. Hocquenghem. Although Hocquenghem's paper is perhaps the more lucid of the two, it was written in the French language and appeared in a journal of relatively limited circulation. Also, his treatment of the subject was much less extensive than that of the other authors.

For a given n, generator polynomials other than those shown in Table 5-9 could probably be employed to obtain codes having somewhat different characteristics, but this aspect of the subject has apparently not yet been studied thoroughly.

A different aspect, which has been studied extensively but for which the results are not yet conclusive, is the true error-correction power of the codes formed in the manner of Table 5-9, especially for larger n. Some evidence exists that d is greater than anyone has so far proved for some of the codes obtained in this way. Alternatively, codes derived in some other, as yet unknown, way may have a greater d than these codes for certain combinations of n and k.

Cyclic Codes Having an Even d Greater Than 4

Insofar as is known, all of the codes developed as described in the previous section have a distance d which is an odd number. The multiplication (mod 2) of a generator polynomial by 11 to obtain the effect of adding a parity check digit on all digits of a code word has already been explained. Also, the use of a parity check to detect any odd number of errors has already been explained. Then, because any code having a given odd distances d_o is capable of detecting any number of errors equal to or less than $d_o - 1$, which is an even number, the multiplication (mod 2) by 11 of the generator polynomial for that code will produce a generator polynomial which will form a code having one more digit per code word but which will detect any number of errors equal to or less than $(d_o - 1) + 1 = d_o$. In effect, the distance of the new code thus developed is $d_o + 1 = d_e$. (The new code will also detect $d_o + 2$, $d_o + 4$, and so on, errors—but not necessarily $d_o + 1$, $d_o + 3$, etc., errors.)

Therefore, each code listed in Table 5-9 can be transformed to a code having an even d that is 1 greater than the odd d listed for that code. The transformation is made by multiplying (mod 2) the generator polynomial by 11. For any given code, the number of information digits k remains the same, but the number of parity check digits r and the total number of digits per code word n are each increased by 1.

Shortened Cyclic Codes

In many applications the system designer may need to use code words for which the number of digits per word is not exactly $2^m - 1$, where m is an integer (or exactly 2^m in the case of codes having an even d). For any n, cyclic codes having the desired error-detection-correction capability can be developed by assuming that sufficient 0's are added to the "high-order" end of the information digits to make a code word which has the desired characteristics and which does have a number of digits equal to $2^m - 1$ for some integral value of m. With the code words generated as described previously, these dummy digits will always appear as 0's in the code words, and therefore they will not affect any of the parity check counts, and they need not be transmitted. The resulting codes are called "shortened" cyclic codes.

236 DIGITAL CODING

Ordinarily the shortened cyclic codes may be expected to be less efficient than the regular cyclic codes in terms of the proportion of information digits to the total number of digits for a given distance d. However, this loss is not inherent, because the shortening may in some instances have the effect of increasing d. Moreover, d is not necessarily a true measure of ability to detect and correct errors. That is, for example, a code having $d = 6$ will detect any combination of five or fewer errors but may fail for nearly all combinations of more errors. On the other hand, a different code having $d = 5$ may fail on a few combinations of five errors but may detect nearly all combinations of six, seven, eight, and so on, errors. Thus, a specific $d = 5$ code may actually allow fewer erroneous messages to pass than a specific $d = 6$ code. For this reason and other subtleties, a realistic measure of the "efficiency" of shortened cyclic codes (and other codes too, for that matter) is still largely in the realm of the unknown.

Cyclic Codes for Burst-Error Detection

A burst of errors is here defined as a sequence of b digits of which the first and the last are erroneous. Each intermediate digit in the sequence may or may not be erroneous—presumably the intermediate digits signals were substantially disturbed by noise during the word transmission, but the noise may fortuitously cause some or all of the intermediate digits to be received correctly.

If the received word containing the error burst and the transmitted word are substracted (mod 2) one from the other, an "error polynomial" is obtained which may be represented by the notation $X^i E(X)$, where $E(X)$ is a polynomial that represents the pattern of errors and which has X^{b-1} as the highest nonzero term, and where i corresponds to the position of the error burst within the word. Because r, the number of parity check digits per word, is equal to the exponent of the highest-order term in the generator polynomial $G(X)$ for that code, the highest-power term of $E(X)$ will be of lower degree than the highest-power term of $G(X)$ provided that b is no greater than r. Therefore, if $b \leq r$, $E(X)$ cannot be an exact multiple (mod 2) of $G(X)$. Then, because the lowest-order term in $G(X)$ is always 1, the shifting of $E(X)$ as represented by $X^i E(X)$ cannot produce a polynomial which is an exact multiple of $G(X)$. That is, when the received word containing this error burst is divided (mod 2) by the generator polynomial, a nonzero remainder will always result.

In other words, any cyclic code is burst-error-detecting for any single burst in a word provided the length of the burst is no greater than the number of parity check digits per word.

Note, however, that if this burst-error-detecting capability of cyclic codes is utilized, the error-correcting capability is reduced. For example, consider the cyclic $d = 3$ code in Table 5-7a. Errors in digits x_4, y_1, and y_2 would constitute a burst-error of three digits. The parity checks corresponding to y_2 and y_3 would fail, and an error would thereby be detected—but this pattern of check failures must not be used as the signal to "correct" x_2 as would be the case when this code is employed as a single-error-correcting code. For cyclic codes having large d, the trade-off between burst-error-detecting capability and random-error-correcting capability is more dramatic.

Cyclic Codes for Burst-error Correction

Conceivably a cyclic code could be devised whereby b parity check digits would be used to indicate the error pattern in an error burst of length b and where additional parity check digits (approximately $\log_2 n$ in number, where n is the total number of digits per word) would be used to indicate the location of the burst within the word. Apparently no such code has been invented, but cyclic codes having generator polynomials of the form $G(X) = (X^c + 1)I(X)$, where $I(X)$ is an irreducible polynomial do have burst-error-correcting characteristics vaguely analogous to those suggested. Unfortunately, the manner in which these codes achieve burst-error correction is difficult to explain or understand, as the reader may appreciate by working out the parity check patterns for a few randomly selected generator polynomials of the form indicated. Even a mathematical proof that the codes do have burst-error-correcting capability requires more steps than is deemed appropriate to include here. Therefore, the properties of the codes will be merely stated without proof.

A cyclic code having a generator polynomial of the form $G(X) = (X^c + 1)I(X)$ generates a code that is capable of correcting an error burst of length b and is additionally (not just alternatively) capable of detecting a second burst of length b' in the same word provided that b and b' satisfy the relationships $b \leq m$, $b \leq b'$, and $b + b' \leq c + 1$, where m is the power of the highest-order term in $I(X)$. Note that if $c = 0$, the generator polynomial is simply the irreducible polynomial $I(X)$, and the relationships indicate that $b = 0$. Likewise, if $m = 0$, then $b = 0$. Moreover, because the number of parity check digits per word is $r = c + m$, the maximum length of the correctable burst is approximately $r/3$, and this maximum is achieved when $c \simeq 2m$.

The maximum length of the code word is $n = ce$, where e is the smallest integer which causes the polynomial $X^e + 1$ to be an exact multiple of $I(X)$. The maximum value for e is $2^m - 1$, and for many $I(X)$, e does in fact equal this maximum, but for other $I(X)$, $e < 2^m - 1$. The number of information digits per word is $k = n - r$, but the code can be shortened in the manner described previously. Shortening reduces k and n by an equal amount but leaves r unchanged.

Alternatively (not additionally), a cyclic code with a generator polynomial of the form $G(X) = (X^c + 1)I(X)$ can be used to detect a single error burst of length $c + m$, as established by the argument used in the previous section.

As an example, consider the generator polynomial

$$G(X) = (X^6 + 1)(X^4 + X^3 + 1) = X^{10} + X^9 + X^6 + X^4 + X^3 + 1$$

Here, $c = 6$, $m = 4$, and $e = 15$ (see Table 5-6 and associated text) so that code words having $n = 90$, $r = 10$, and $k = 80$ can be formed. Any single burst having $b \leq 4$ can be corrected, and a second burst having $b' \leq 7 - b$ can be detected. Alternatively, any single burst having $b \leq 10$ can be detected.

The codes will detect many single or double error bursts of lengths greater than those indicated. Further, the codes will detect most sets of three or more random errors or bursts of errors. In fact, even when the received digits represent "pure hash," the probability of error detection failure is only $1/(2^r)$—although this probability is the same as with any parity check code.

In many practical applications where bursts of errors are a problem, a reasonably high probability exists that, when any error bursts at all occur, many bursts will occur within a short interval of time. In these applications the burst-error-correction capability of this and other known burst-error-correcting codes is seldom great enough to be useful.

Codes of the category described in this section are commonly called Fire codes in recognition of the work of one P. Fire.

Implementing Cyclic Codes—Encoding

To encode a word in a cyclic code, the principal piece of equipment needed is a network which performs mod 2 division and preserves the remainder. The quotient developed by the division operation is discarded, but the remainder forms the parity check digits in the manner described previously. A general purpose binary computer with the carry propagation circuits disabled (by means of special built-in instructions, or otherwise) will serve the purpose, but most of the interest in cyclic codes is occasioned by the possibility of using relatively simple shift-register techniques for this mod 2 division.

Consider the following example of mod 2 division.

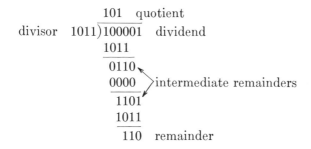

By studying this and other examples, the division process is seen to start by aligning the highest-order nonzero divisor digit with the highest-order dividend digit. The first intermediate remainder is determined by an exclusive-OR relationship between each pair of corresponding digits, and the process continues in a manner essentially analogous to that of conventional division. Each quotient digit is determined in accordance with whether the highest-order digit of the intermediate remainder (dividend for the first quotient digit) is 1 or 0, respectively. Note that, if the dividend contains k digits and the divisor contains $m + 1 = r + 1$ digits, where k, m, and r may have the same meanings as in previous sections, the quotient will contain $k - r$ digits, and the remainder will contain r digits, some or all of which may be 0's.

A block diagram for a mod 2 division shift-register network is shown in Figure 5-7a. All flip-flops of the shift register are assumed to contain 0's initially. The digits of the dividend are shifted, high order first, into the right-hand end of a shift-register having $m + 1$ stages interconnected through exclusive-OR modules. Because of the AND devices, only a shifting action takes place until the highest-order dividend digit reaches the left-hand stage of the shift register. The output from this left-hand stage then indicates the first quotient digit. If this digit is 0 as a result of the highest-order dividend digit being 0, the next step is similarly just another shifting operation with, in effect, 0's being subtracted (mod 2) from the dividend. For each quotient digit that is a 1, the AND devices are actuated to allow the divisor digits (shown in polynomial form in the figure) to pass to the exclusive-OR modules as required for subtraction (mod 2). Because the highest-orded divisor term X^m is always 1, no connection need be made to the X^m input line. (Certain complications, irrelevant to the present application, are encountered if X^m is allowed to be 0.) After all quotient digits have been determined, the remainder resides in the X^m to X stages of the shift register. The X_0 stage will contain a 0. The remainder may be extracted from the register by opening the connection at the point marked with a cross and by actuating the shift register r steps. The remainder then appears on the same output line as did the quotient.

When the divisor is always the same set of binary digits, as would be the case in coding, the exclusive-OR module and the AND device can be eliminated for each digit position for which the divisor digit is 0. Moreover, the AND device can be eliminated for all divisor digits that are 1's. The simplified network for the specific case of a divisor of 100101 is illustrated in Figure 5-7b.

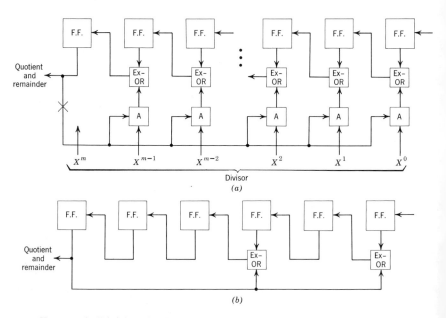

FIGURE 5-7. Division circuit (mod 2): a general and b for divisor of 100101.

Figure 5-8 illustrates the manner in which the division (mod 2) network is incorporated into an encoder for cyclic codes. The input consists of a message of k digits followed by r 0's. A control signal that is

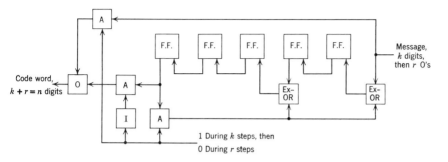

FIGURE 5-8. Encoding network for a cyclic code.

1 for k steps allows the input to pass directly to the output line. For the first r steps, the first r digits of the message also enter unaltered into the shift register of r (not $r + 1$) stages. The lowest-order (right-hand) stage of the previous division network has been eliminated, and the signal that would otherwise be taken from the eliminated stage is taken directly from the input line. The puropse of eliminating this stage is to allow the remainder (the parity check digits) to be developed in time to follow the transmission of the message digits without any intervening blank digits. After the first r steps, the next $k - r$ steps will be used in developing the quotient, which is discarded because the 0 appearing at the output of the inverter in Figure 5-8 will prevent the quotient from reaching the output of the encoder. After $r + (k - r) = k$ steps, the control signal is changed from 1 to 0 and is held at this value while the shift register is actuated for r additional steps during which the contents of the shift register (the parity check digits) are shifted to the output line. During these last r steps of the encoding operation, the control signal disables one input to each exclusive-OR module so that 0's are entered into the shift register in preparation for the encoding of a subsequent message.

Figure 5-8 illustrates the encoder for the specific generator polynomial of 100101, but the shift register within the encoder may be altered straightforwardly as needed for any other generator polynomial.

Implementing Cyclic Codes—Decoding for Error Detection

For error detection, the equipment needed to decode a message in a cyclic code is very similar to that needed for encoding. The r parity check digits are redetermined from the k information digits as received, and the redetermined parity check digits are compared with the parity check digits as received. Any discrepancy in the two sets of parity check digits signals at least one error in the $k + r = n$ code word digits as received.

The decoder is shown in Figure 5-9, again for the generator polynomial 100101 as an example. The flip-flop in the decoder is initially "reset" to 0. The received digits are entered into a mod 2 division network as in the encoder except that an AND device has been added at the input. This AND device is actuated by the control signal in such a way that the k received information digits enter the shift register, but

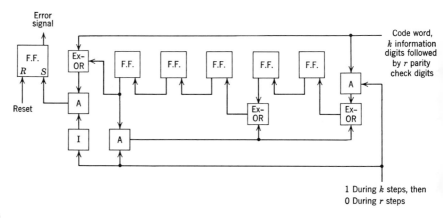

FIGURE 5-9. Error-detecting decoder for a cyclic code.

the r received parity check digits are prohibited from entering, as required for preparing the decoder for the subsequent message. After the first k steps of operation, the redetermined parity check digits are compared with the parity check digits as received in an exclusive-OR module in the top part of the figure. For any digit pair for which the digits are not the same, a signal is caused to be transmitted to the "set" input terminal of a flip-flop so that after n steps of operation, the output from the flip-flop indicates the presence or absence of a detected error.

The error patterns detectable by the decoder are a function of the generator polynomial being used and are not a function of encoder design, except that the details of the shift register must correspond to the particular generator polynomial being used, of course.

Implementing Cyclic Codes—Decoding for Single-Error Correction

In general, error correction is very much more difficult (more costly to implement) than error detection, although when the requirement is limited to single-error correction, reasonably economical decoding schemes are available. In any case, the problem is to relate the pattern of parity check failures to the locations of the errors.

As was shown previously, for any number r of parity check digits, a cyclic single-error-correcting code of length $n = 2^r - 1$ can be devised. With r parity check digits, 2^r different patterns of parity check failures are possible. One pattern (all 0's, where 0 represents no failure and 1 represents a failure) indicates that no errors are detected. The other $2^r - 1$ patterns may be used to indicate which of the $2^r - 1$ digits in the code word is in error. The parity check failure pattern (sometimes called a "syndrome") may be treated as a number to be entered into a counter as the initial setting of that counter. The counter may then be stepped "up" or "down" until it reaches the zero state, and if the message to be corrected is simultaneously stepped along a shift register, the erroneous digit will appear on the same step on which the counter reaches zero—provided the counter steps through its various stable states in the appropriate sequence.

Fortunately, the same sort of shift register that is used for determining the parity check digits when encoding can be used as a suitable counter when decoding for error correction. The reader can verify that such a shift register will function as a counter of $2^r - 1$ states by manually dividing (mod 2) $1000\cdots$ by any of the primitive polynomials listed in Table 5-6. Moreover, by using the same primitive polynomial in the decoder as was used in the encoder, the sequence of stable states in the shift-register counter corresponds exactly to the sequence of parity check failure patterns corresponding to successive digit positions in the code word.

The error-correcting decoder based on the above facts is shown in Figure 5-10, again with the generator polynomial 100101 as an example. For the first k steps, the decoder functions in essentially the same manner as in the error-detecting decoder described in the previous section. For the next r steps, the functioning is again similar except that the output of the exclusive-OR module is not entered into an error-detecting flip-flop but is instead fed back through the OR device at the right-hand part of the figure to the shift-register input. The digits entered into the shift register during these r steps represent the parity check failure pattern (not the parity check digits themselves). Also, during the first $k + r = n$

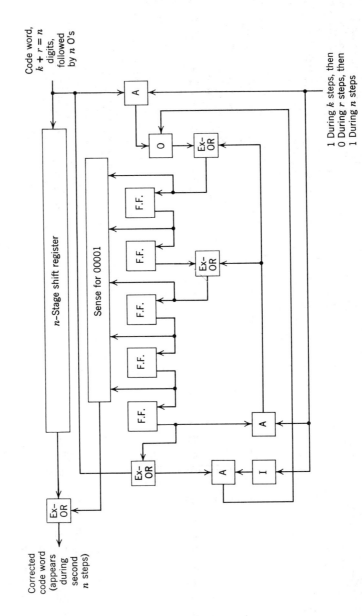

FIGURE 5-10. A single-error-correcting decoder employing a cyclic code.

IMPLEMENTING CYCLIC CODES—DECODING   245

steps the input message, which may contain an error, is entered into an n-stage shift register so that after n steps the message is again available (at the output of this register) for any correction that may be necessary.

For the next n steps, the decoding shift register acts as a counter by repeatedly dividing the parity check failure pattern by the generator polynomial. When the counter reaches the state 00001, as represented by the digits in the respective five stages of the decoding shift register, the next digit to emerge from the n-stage shift register is the erroneous digit. The block labeled "sense for 00001" (which is essentially a five-input AND device) generates an output signal which, when applied as one input to the exclusive-OR module in the upper-left part of the figure, has the effect of inverting (correcting) the erroneous digit. The inputs to the "sense for 00001" device are taken from the inputs instead of the outputs of the shift-register stages so that, on any given step, the digit currently emerging from the n-stage shift register can be corrected. Otherwise, an $(n + 1)$-stage shift register would be needed, and various complications would be encountered in the decoding of successive words.

For example, with the generator polynomial 100101, $n = 31$, $k = 26$, and $r = 5$. Information digit x_1 (with the notation employed earlier in this chapter) will be checked by parity check digits p_1 and p_4 (as can be determined by manually working out the parity check pattern). Thus, if x_1 is in error, the parity check failure pattern to be placed in the shift-register counter will be 10010 with an implied 0 in the missing low-order stage. Then when 100101 is subtracted (mod 2) in the division or counting process, the shift-register counter will contain 00001 on the very next step as the signal to correct digit x_1. The correction will actually take place a step sooner because of the manner in which the "sense for 00001" device is connected, as explained previously. Other digits, including the parity check digits themselves are corrected analogously, although the parity check digits should be transmitted in the sequence $p_5p_4p_3p_2p_1$ instead of the $p_1p_2p_3p_4p_5$ sequence indicated previously.

If the received code word contains no errors, the "sense for 00001" device will not generate an output signal because the decoding shift register will contain all 0's for each counting step, and the word as received will appear on the "corrected message" output line.

For decoding a string of successive messages, the simplest scheme is to employ two complete decoders of the type illustrated in Figure 5-10. The two decoders are needed because, while one decoder is correcting a message, 0's must be applied to the input line. Added control elements are used to transmit successive code words to the two decoders alternatively, and the outputs of the two decoders are then combined in an OR device for transmission of the corrected message string on a single output line.

Except for obviously needed alterations in shift-register details, the error-correcting decoder of Figure 5-10 will correct single errors when the generator polynomial is of the form $G(X) = (X^c + 1)I(X)$ as discussed previously for burst-error-correcting codes. However, the correcting capability of this decoder is limited to single errors, that is, a "burst" of $b = 1$. The error-detecting feature of the decoder in Figure 5-9 can be added to provide detection of all additional error patterns within the detection capability of the particular code being used. If an error pattern which is detectable but not correctable (not a single error) is encountered, the digit combination of a 1 in the lowest-order shift register stage and 0's in all other stages ($\cdots 0001$) will not occur during the counting process, but it will occur if a correctable error is present. Therefore, the output from the "sense for $\cdots 0001$" device can be used to reset the error detection flip-flop and thereby nullify the error detection signal when a correctable error is encountered.

Implementing Cyclic Codes—Decoding for Burst-Error Correction

The correction of a burst of errors, where $b \geq 1$, can be accomplished with the decoder of Figure 5-10 provided a significant elaboration is incorporated. The "sense for $\cdots 0001$" device is applied not to all r stages of the shift register but to only the $r - b + 1$ highest-order stages, and the n-stage shift register is lengthened to have $n + b - 1$ stages ($n + b$ stages if the "sense for $\cdots 0001$" device is connected to the outputs instead of the inputs of the mod 2 dividing shift register). The characteristics of the $(X^c + 1)I(X)$ codes and the decoder happen to be such that when the digit combination $\cdots 0001$ is sensed during the counting process, the digits in the b lowest-order stages of the shift register represent the burst-error pattern.

The decoder is modified to cause the "sense for $\cdots 0001$" signal to terminate the dividing action, although the shifting action is allowed to continue. The decoder is further modified to cause the digits then in the b lowest orders of the shift register to be transmitted to the exclusive-OR module in the upper-left part of Figure 5-10 for correcting the erroneous digits in the message.

An additional flip-flop is needed to "remember" whether or not a "sense for $\cdots 0001$" signal has been received. The output from the flip-flop can then be used to gate the appropriate digits from the shift register to the exclusive-OR module. Alternatively, the output from the flip-flop can be used with an AND device to, in effect, break the connection from the bth stage to the next higher order stage in the shift register, and the exclusive-OR module can be actuated directly by the "sense for $\cdots 0001$" device as in Figure 5-10, because the $r - b + 1$ highest-order stages will then contain $\cdots 0001$ for each step on which an erroneous digit appears at the output of the shift register having $n + b - 1$ stages.

Error patterns that are detectable but not correctable can be handled as described in the previous section.

Implementing Cyclic Codes—Decoding for Multiple Random-Error Correction

For applications where the number of information digits k per code word is relatively small (say, $k \leq 7$, for which the number of different messages is no more than $2^7 = 127$) the most practicable method for correcting multiple random errors is probably to compare the received word with all 2^k words that may have been transmitted. With a total of n digits per word, n exclusive-OR operations are required for each comparison, although this function may be performed in parallel with n exclusive-OR modules. Means are provided for counting the number of mismatched digit pairs in each comparison. If this number is zero for any of the 2^k comparisons, the word entering into that comparison is assumed to be the error-free word which was transmitted. If, for any given comparison, the number of mismatched digit pairs is within the error-correcting capability of the code, the digits at the positions where mismatches are encountered are assumed to be in error and are corrected. If all of the 2^k comparisons result in more mismatches than the code is capable of correcting, an uncorrectable error pattern has been detected, and means are included to signal this fact. As always a given code need not be used to its full error-correcting capability, and the excess may be used for added error-detection capability. Also, as always, if more errors occur than the code is capable of handling, the decoder may fail to detect any errors or may erroneously "correct" the word.

The decoding scheme described in the previous paragraph will work for any code—even nonsystematic codes. In principle, any kind of a storage unit may be used to present all possible code words, one at a time, to the comparator.

A hardware simplification results with cyclic codes where a cyclic shift of any code word produces another code word (although, in general, not all code words will be obtained by cyclicaly shifting a given code word—see Table 5-7). A cyclically connected shift register is loaded with a digit pattern that represents one code word, and more code words are generated by stepping the shift register. The shift register is reloaded with new digit patterns and the operation is repeated as necessary to present all different code words to the comparator.

With some cyclic codes only one shift register is needed, with certain code words being obtained from the 1-outputs of the shift register stages and other words being obtained from the 0-outputs of the stages. The two words represented by all 0's and by all 1's may be added. An example is the code in Table 5-7a. A seven-stage shift register initially containing 0001101 will, when cyclically shifted, generate seven code words in conventional fashion and will generate another seven when the signals are taken from the 0-outputs instead of the 1-outputs. The last two of the sixteen code words in this particular code are 0000000 and 1111111. Although this code is only single-error-correcting (not multiple-error-correcting), the error-correcting procedure is illustrated in that the received word may be compared with all 16 words of the code generated as suggested here. Any single mismatch represents an error that can be corrected.

Alternatively, only eight comparisons need be made. The comparisons are with 0000000 and the seven words generated at the 1-outputs of the shift register. If a word is correctly received but is one of the remaining eight code words, one of the comparisons will result in mismatches in all seven digit positions. If one digit is in error, the number of mismatched pair will be six. Thus, when making the comparisons, a mismatch in any one digit position or mismatches in all but one of the digit positions will indicate a single correctable error. In the latter case, the correction is made by inverting the digit where a match is encountered, of course.

An error-correcting procedure which is more directly related to the nature of parity check codes is to determine the pattern of parity check failures, because this pattern is a direct indication of the positions of the erroneous digits. Although, as was shown, the relationship between the parity check failure pattern and error pattern can be made quite simple for single-error-correcting codes, the relationship is unfortunately complex and difficult to understand for codes capable of correcting two or more random errors—an important disadvantage in itself when equipment servicing must be considered. Of more consequence, the equipment needed to implement the relationship rapidly becomes extensive (expensive) as the number of correctable errors per word increases above one. In fact, the complexity and cost disadvantages are strikingly apparent for decoders capable of correcting only two or three random errors per message. For these reasons, and because the two-dimensional $d = 16$ codes described earlier in this chapter can be used to correct up to seven random errors per message in a manner which, as complex as it is, is relatively easy to understand, the equipment needed for computing the relationship between the parity check failure pattern and the error locations in cyclic codes will not be described here.

Instead of computing the error locations from the parity check failure pattern, an alternative technique is to use a random-access storage unit to store the precomputed relationships. With r parity check digits per word, a storage unit having 2^r addresses is needed. In the most straightforward version, the number of digits per stored word must be k, the number of information digits per message (or n if errors in the parity check digits themselves must be corrected), although $k + 1$ digits are used for reasons to be explained later. Of course, when r is relatively large, say more than about ten, as would be the case for multiple-error-correcting codes, the necessary capacity of the storage unit is unattractively large. Specifically, for example, a triple-error-correcting cyclic code having 63 digits per word would require that 18 of them be parity check digits (see Table 5-9), so a storage unit having a capacity of $2^{18} = 262,144$ addresses would be required. On the other hand, the storage unit may be of any high speed permanent or "read-only" type, and the system will locate all errors in the time of one cycle of storage unit operation. Therefore, a single storage unit could be time-shared among many incoming transmission lines on which the same code is used.

The general scheme for decoding with correction of multiple random errors is illustrated in Figure 5-11 for cyclic codes. The k information digits of the incoming message are divided (mod 2) in the same manner as in an encoder. The parity check digits thus generated are then compared with the parity check digits received on the incoming transmission line. The resulting pattern of parity check failures forms an r-digit binary number (the "syndrome") which is used as an address number to locate the particular word in storage which indicates which the k information digits, if any, are in error. With a stored 0 and 1 indicating no error and error, respectively, the word obtained from storage is added (mod 2) to the k digits as received to obtain the corrected information digits.

With nearly all codes, many parity check failure patterns ("syndromes") will signal the detection of more errors that the code is capable of correcting. Therefore, an additional or $k + 1$ digit position is used in each stored word. A 0 or a 1 is used in this position to indicate whether or not, respectively, a correctable error pattern is present. For

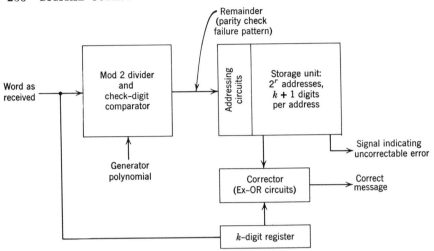

FIGURE 5-11. An error-correcting decoder.

each stored word containing a 1 in this $k + 1$ position, all other digits of this word are irrelevant and may all be 0's. As the decoder is used, a 1 appearing in this position of the storage unit output signals the detection of errors which are beyond the capability of code to correct (or which are not to be corrected—when some of the code's error-correcting capability is being used for increased error-detecting capability).

Because of the general nature of the above described decoder, the full error correcting capability of any cyclic code can be utilized. For example, a code capable of correcting any three random errors may actually be capable of correcting many combinations of four, five, and more errors. For each such correctable combination, an appropriate set of 1's is stored in the storage unit address corresponding to the associated parity check failure pattern. Of course, the necessary precomputing of the patterns of 1's at each address is a lengthy task that would need to be performed on a computer for most codes. In fact, the finding of all correctable error patterns is not necessarily practicable even with the aid of a computer. Nevertheless, for those correctable patterns which can be found, the actual correction in the decoder takes place with equal simplicity and rapidity without added equipment regardless of the number of correctable errors in a word.

As an alternative to the possibility suggested in the previous paragraph, the required capacity of the decoder storage unit can be decreased by storing numbers that indicate the error positions. For example, in the triple-error-correcting code having $n = 63$, three six-digit numbers at each address can be used to indicate the locations of three errors. For each number, the digits 000000 indicate no error. Three six-digit matrices are needed to translate the numbers to signals on one, two, or three (according to the number of errors present) of 63 lines when correction is in parallel as implied by Figure 5-11, but decoders that are more serial in nature can employ counters in any of a variety of ways to translate the stored numbers to error-correcting signals. Thus, in this example, the number of digits per stored word can be decreased from 63 to $(6)(3) = 18$ except for the $k+1$ digit which is added in either case.

Synchronization Errors—General Discussion

In the operation of physical equipment in which the digits are transmitted serially, an obvious need arises to provide means to determine the positions of the various digits within each word. For the most part, this requirement is a matter of identifying the first digit within each word. With most transmission methods, each digit, whether it is a 0 or a 1, is represented by a detectable signal of some sort (in contrast to representing 1's by signals and 0's by the absence of signals) so that all digits after the first digit in a message can be identified merely by counting the digits received. If the assumption could be made that no errors will occur in the counting process, the principal problem would reduce to providing means for identifying the first digit in a continuous sequence of messages, because the first digit in each subsequent message could then be identified by a counting process from this first digit. The number of digits in each message would need to be known, of course.

Unfortunately, for one reason or another errors do seem to occur in the counting process, so that in virtually all practical applications some means other than counting from the first digit in the first message must be provided to identify the first digit in each message.

In many applications a practical approach to this problem is to provide a signal different from that used for the 0's and 1's that carry the intended information. One example is a signal on a separate transmission line. This signal is timed to correspond to the time of the first digit in each message. Such a scheme is commonly used in the form of a timing track on a magnetic tape, disk, or drum, for instance. Another example is the use of a signal of larger amplitude for the first digit in each message, or alternatively a large-amplitude non-information pulse of some sort may be transmitted just prior to the first digit in each message. Although the large-amplitude timing signal is probably impractical with magnetic tapes, disks, and drums, it is quite practical when the transmission path is a more elementary electromagnetic medium such as a coaxial cable or mere space as in a radio system. A third example of the use of a different kind of signal to control timing is simply to send no signal at all for a short duration of time between each pair of successive messages. This scheme is feasible with any of the non-return-to-zero (NRZ) digit-representing methods wherein each digit is represented by at least one polarity change of the signal. Such methods (see *Electronic Digital Components and Circuits* by Richards, pp. 285–287, for detailed descriptions) are variously called phase-shifting or frequency-doubling and are applicable to magnetic tapes, disks, and drums as well as to other transmission media.

However, in a sort of "ideal" sense, the capability of the transmission path should be limited to the transmission of nothing but a continuous series of 0's and 1's, and the encoder and decoder should be designed so that the timing or identification of the digits, as well as the information content of the digits, can be determined solely from the digit sequence. In some applications the transmission path has indeed been constructed with this limitation, although the net merit in comparison with the schemes mentioned in the previous paragraph is difficult to establish. In any case, the use of coding to control synchronization errors as well as information errors opens a topic of considerable fascination from a mathematical standpoint, and it has been studied extensively.

One clear and important advantage of using coding to control synchronization errors is that the physical equipment can be in the form of the same AND, OR, and storage devices used for other digital applications—no specialized hardware is needed. Moreover, when the decoding is accomplished by means of a stored-program computer, the error control facilities can be nothing more than a program for the purpose.

If a synchronization error occurs in a message through the deletion of a digit or through the addition of an extraneous digit, all subsequently received digits of that message will be shifted one position (to the "right" or "left," as the case may be). If these subsequent digits happen to be all 0's or all 1's, none of them will be in error, but if the sequence happens to be alternating 0's and 1's (that is, 010101 · · ·), all of them will be in error. With random digits, the error pattern will be more or less random. If the error is near the end of the message, the number of digit errors may be within the information-error handling capability of the code. If the synchronization error is near the beginning of the message, the error may be practicably corrected by shifting all digits to the right or left on a trial basis to see if a digit sequence results whereby a correctable number of errors occur—except that with cyclic codes the fact that a cyclically shifted message produces a digit sequence that represents a different message so that in this instance a synchronization error would not necessarily be detectable. However, the synchronization error may occur near the middle of a message in which case the number of digits that appear to be in error will, in general, be far too great for practical correction by any technique.

On the other hand, even though the message containing a deleted or added digit may not be correctable, the digits of the subsequent message are all merely shifted to the right or to the left. The detection and correction of this condition by means of suitable codes is both possible and practical under certain conditions. In many applications the regaining of synchronization after an occasional loss or gain of a digit (or "burst" of digits) is adequate for useful information transmission, and the codes for synchronization control serve this purpose. The messages actually containing the deleted or added digits must, in general, be retransmitted unless the person or machine at the receiver can tolerate an occasional loss of a word.

The Use of a Prefix or "Comma" for Synchronization Control

A simple and practical scheme for synchronization control is to insert a special sequence of 0's and 1's between each pair of successive messages. Usually, in practice, this sequence of digits is visualized as being a prefix added to the beginning of each message. The decoder "looks for" this sequence of digits in the continuous stream of incoming 0's and 1's. When the sequence is detected, the next subsequent digit is known to be the first digit of a message. This digit sequence is also commonly called a "comma," especially by persons interested in the mathematical aspects of codes, in which case the visualization is that the specialized sequence follows each message to indicate a dividing point between a given message and the next one.

A desirable feature of the prefix (or comma) is that the digit sequence be one which is never encountered in any message or which is at least encountered only rarely. The degree to which this feature can be attained is a function of the code being used for information-error control. Of course, the feature cannot be attained at all when the messages contain no redundancy (no error control) except by using an objectionably large number of digits in the prefix. At the other extreme, for codes that utilize relatively few of the 2^n possible digit combinations, where n is the number of digits per message, the positive identification of the prefix can be possible even for relatively short prefixes—under the assumption that no digit errors occur in the prefix itself. Unfortunately, this last-mentioned assumption is not valid in general, so the use of prefixes for synchronization-error control is considerably less straightforward than it might hopefully be.

The simplest prefix would be a single digit, say a 1. With this prefix, the decoder would not be designed to "look for" the prefix, because 1's would be too commonly encountered in the messages. Instead, the decoder would merely check to determine that the digit preceding each message is, in fact, a 1. If a synchronization error occurs, the decoder may not detect the fact at the beginning of the next message because of the "50-50" chance that the digit shifted into the prefix position is a 1. In fact, many messages may be transmitted before a 0 is by chance shifted into the prefix position. However, over a series of many messages a significant proportion of them will, with cyclic codes, contain apparent errors in the first or last (as the case may be) digit position. For any given code, the proportion will depend on the relative frequency of 0's and 1's in the first and last positions of the words of that code. With noncyclic codes, nearly all messages transmitted after the synchronization error may appear to contain large numbers of digit errors. In either case, the condition can be used to indicate the possibility of a synchronization error, and all messages can then be shifted on a trial basis to find a synchronization that produces no indicated errors in the prefix or in the information digits. Although a single-digit prefix can be helpful, it obviously falls considerably short of providing attractive synchronization-error detection and correction capability.

If the prefix is formed by a sequence of like digits, say three 1's (that is, 111), the probability of finding that particular sequence in a message is somewhat reduced, so the decoder can more meaningfully "look for" the prefix in obtaining proper synchronization, but the situation is not really improved when the first one or two digits of a message are 1, because in this instance the decoder will not be able to distinguish which sequence of 1's actually constitutes the prefix. Also, an error in synchronization will produce essentially the same effect and problems as described in the previous paragraph.

The most positive indication of message synchronization is obtained when the prefix consists of a sequence of like digits followed by a single digit of opposite value such as, for example, three 1's followed by a 0 (that is 1110, where the digits are visualized as moving from right to left with the left-hand digit being transmitted first). With such a prefix, a shift in either direction as would be caused by the deletion or addition of a digit in the previous message will cause a detectable shift in the prefix regardless of the digits that happen to comprise the message immediately preceding and following the prefix. The problem is to determine the approprite number of 1's to be used in the prefix.

If the prefix contains only one 1 followed by a 0 (i.e., if it is 10), any single-digit synchronization error can be detected positively by observing that either a 0 will occur in the first prefix digit position in the case of a deletion or a 1 will occur in the second prefix digit position in the case of an addition. However, 10 is still such a commonly encountered sequence in the body of a message that "looking for" this prefix is still not meaningful. Also, the correction of a synchronization error, especially when the shift may be more than one digit position, involves essentially the same uncertainties as with a single-digit prefix.

If the prefix contains two 1's followed by a 0 (i.e., if it is 110), any shift of no more than one position in either direction can not only be positively detected, it can be positively corrected. For example, if the digits in the expected prefix positions are observed to be $x11$, where x represents either a 0 or a 1, an extraneous digit is known to have been added to the previous message, and all subsequent digits should be advanced one position. (The next digit to be sensed will be a 0—if it is not, a kind of error other than a single-digit synchronization error has occurred.) If the digits in the expected prefix positions are observed to be $10x$, a digit is known to have been deleted from the previous message, and all subsequent digits should be retarded one position. Of course, the correction procedure may fail if the synchronization error is by more than one digit or if other kinds of errors have occurred.

Alternatively, the prefix 110 can be used to detect a shift of one or two digit positions as would be caused either by the addition or the deletion of one or two digits in a previous message. In the case of an addition, the digits observed at the expected prefix positions will be either $x11$ or $xx1$ in accordance with whether one or two digits have been added. With the three prefix digits designated p_1, p_2, and p_3, respectively, the fact that p_3 is 1 can be used as an indication of a synchronization error. In the case of a deletion, the observed digits will be either $10x$ or $0xx$ in accordance with whether one or two digits have been deleted. The fact that p_1 or p_2 is 0 can be used as the synchronization error indication. Thus, in Boolean notation $E = \bar{p}_1 + \bar{p}_2 + p_3$, where E is the synchronization error signal.

Note, however, that when the prefix 110 is used to detect two-digit synchronization errors, it cannot be used to correct one-digit synchronization errors. For example, in the case of two added digits, the last three digits of the previous message might be 110, in which case the digit sequence appearing at the expected prefix positions would be 101 (preceded by a 1 as the apparent last digit of the previous message) and would erroneously indicate the deletion of a digit. Similarly, in the case of two deleted digits, the first digits of the subsequent message might be 110 in which case the observed digits would be 011 (followed by a 0 as the apparent first digit of the subsequent message) and would erroneously indicate the addition of a digit.

Codes can be designed in which the sequence 110 never appears in the body of a message. With such codes the prefix 110 is adequate for establishing or reestablishing synchronization regardless of the number of digits that may have been erroneously added in or deleted from a previous message. The decoder "looks for" the prefix and assumes that the first digit after the 0 is the first digit of a message. Although such codes have not been studied extensively, they almost certainly would be non-systematic in virtually every instance and their information-error detection-correction properties are not expected to be attractive. Because two successive 1's could not be permitted in a message, the prefix in this instance could be shortened to two 1's (with no 0), and the message would start immediately after the second 1.

If the prefix is 1110, any extraneous addition or deletion of one, two, or three digits in a previous message can be detected positively. Alternatively, any one-digit synchronization error can be corrected with any two-digit error detected. In either case the techniques are essentially as described previously. However, the prefix 1110 is not long enough for the correction of two-digit synchronization errors. For example, with two erroneously added digits, the digits appearing at the prefix positions would be $xx11$ (followed by 10 as the apparent first two digits of the subsequent message), but in the event the last digits of the previous message are 1110, the decoder would not be able to distinguish the sequence from the $10xx$ sequence that would be obtained with a two-digit deletion where the first digits of the subsequent message are 1110.

In general, with d' digits in the prefix, the maximum number of added or deleted digits that can be detected is $d' - 1$. For synchronization-error correction, the correctable number of digits is $(d' - 1)/2$ when d' is odd and is $(d' - 2)/2$ when d' is even—all in close analogy to the relationship between the distance d and the number of detectable and correctable information errors. Moreover, in continuation of the analogy, any portion of the synchronization-error correction capability can be traded for detection capability so that, for example, with a prefix of suitable length up to three positions of shift can be corrected with up to four additional positions of shift being detected.

The above results are attainable when the information digits are random, that is, when all 2^n digit sequences represent valid messages, with n being the number of digits per message. As already suggested, the synchronization-error detection-correction capability of the prefix can be increased when constraints are placed on the digit sequences, particularly if certain digit sequences near the beginning and end of each message are forbidden by the code in use. Of course, if $p > n$, where p is the number of prefix digits, the digit sequence represented by the prefix cannot be the same as any message, and the decoder can be made capable of restoring synchronization after the addition or deletion of any number of digits regardless of the code in use, but ordinarily the implied halving (or worse) of the information transmission rate would be considered an excessive price to pay for synchronization control.

In many applications a practical compromise approach to synchronization control is to employ a relatively short prefix (say $p = 8$ for n up to 60) and to design the decoder to "look for" this prefix. Occasionally, a digit sequence in a message will be erroneously sensed as a prefix, but the indicated error rate, especially for synchronization errors, in subsequent messages will then be high. The decoder can be designed to utilize this fact to make another search for the prefix. Alternatively, the new search can be commenced whenever an uncorrectable synchronization error is detected, and the decoder will eventually "lock in" with the correct synchronization. To maximize the number of correctly received messages, facilities must be included to store the received messages until synchronization is assured.

When a digit error occurs in a prefix digit, the usual immediate effect is an improper synchronization-error signal. The procedure suggested in the peceding paragraph can be used to handle this situation. Another technique for detecting and correcting such errors is to design the decoder to relate each prefix to the previous prefix and to the subsequent prefix. If both the previous and subsequent prefixes were sensed correctly, the "conclusion" the decoder draws is that the error was in the prefix itself.

For combinations of synchronization errors and digit errors (some of which may be in the prefix and some in the code word digits) further complications are encountered. The variations are too numerous to consider in detail here, but the limitations of the prefixes and the codes and the schemes for handling errors within those limitations can be determined by extensions of the above arguments.

Synchronization Scheme for Cyclic Codes

With cyclic codes, where a cyclic shift of the digits of any message produces a valid (but generally different and therefore erroneous) message of the code, the detection and correction of synchronization errors involves problems and possible solutions not encountered with noncyclic codes.

Consider a sequence of messages in some cyclic code. If a digit has

258 DIGITAL CODING

been erroneously added into or deleted from some prior message in the sequence, a given message will be shifted by one digit. However, the shifting will not be cyclic. Instead, a digit at one end of the message will have been shifted to the adjacent message and apparently lost, and a different digit from the adjacent message in the other "direction" will have been shifted into the adjacent message at the opposite end. If the digits are random and binary as is assumed to be the case here, the digit shifted in at one end has a "50-50" probability of being the same or the opposite with respect to the digit shifted out at the other end. Thus, the code may or may not be capable of detecting that an error of some sort has occurred. But if the digits do happen to be different so that the error is detected and if the code has at least single-error-correcting properties, the result will be the erroneous "correction" of the end digit. That is, although the resulting set of digits represents a valid message of the code, the message will not, in general (although it may in particular cases), be the message which was transmitted.

A possible way to solve the problem described in the above paragraph is to build a system which makes a statistical study of the apparent errors encountered. If the percentage of messages which contain errors is relatively high (roughly 50% over a long sequence of messages) and if most of the errors are found to be in the digit position at one end or the other of the respective messages, the decoder may "assume" that the digits themselves were not in error but that a synchronization error has occurred. The decoder may then shift all messages accordingly and recheck for errors. A dramatic decrease in the percentage of messages in error would be the indication that the true trouble was a synchronization error.

If the error had been a two-digit synchronization error, approximately 75% of a long sequence of messages would contain apparent digit errors which would appear in the two digits at one end of each message. For three-digit synchronization errors, approximately 87.5% of the messages would be in error, and so on.

For synchronization errors of increasing length, the required error-correcting capability of the code becomes impracticably great for a positive indication of the number of digits by which the words must be shifted to obtain correct synchronization. However, in any case where errors are encountered in most of the messages (with noncyclic as well as cyclic codes) the decoder may assume that the true problem is synchronization. The decoder then may shift all words one digit at a time with a recheck for errors after each shift. If a shift position is found which results in none or very few messages in error, that position may be assumed to correspond to proper synchronization, especially if (in the case of cyclic codes only) that position followed a position which produced errors largely limited to the digit position at one end of the messages.

Another Synchronization Scheme for Cyclic Codes

Another approach to synchronization control with cyclic codes is to alter each code word (message) so that it is not cyclic. This alteration can be accomplished quite simply by inverting one or more digits in each word or, which amounts to the same thing, by deliberately causing one or more digits in each word to be in error. The positions of the inverted ("erroneous") digits must, of course, be predetermined and known at the receiver.

Consider, for example, nine-digit messages where the digits are designated a_1 through a_9, and where the fifth digit, a_5, is inverted at the transmitter so that the received word is $a_1 a_2 a_3 a_4 \bar{a}_5 a_6 a_7 a_8 a_9$ with no errors in transmission, $a_2 a_3 a_4 \bar{a}_5 a_6 a_7 a_8 a_9 x$ with a one-digit synchronization error, or $a_3 a_4 \bar{a}_5 a_6 a_7 a_8 a_9 x\ x$ with a two-digit synchronization error. Here the digits may be visualized as moving from right to left so that the indicated synchronization errors are caused by deletions of digits in previous words, and x represents a digit shifted into a given word from the next succeeding word.

To detect that a word is correctly received, the code must be at least single-error-correcting ($d \geq 3$) to indicate that an apparent error in the a_5 position has occurred. Actually, the code must additionally have double-error-detecting properties ($d \geq 4$) to detect the presence of a one-digit synchronization error, because two apparent digit errors may occur in the presence of a synchronization error of one digit. In the example above, the two apparent errors will be in the a_4 and a_9 positions, with the digits being $\bar{a}_5$ and x, respectively. (Note that arbitrarily re-inverting or "correcting" the digit in the a_5 position at the receiver worsens the situation. The reason is that, in the event of a synchronization error, three instead of just two digits may then be in error.) To correct a one-digit synchronization error, the code must be at least double-error-correcting ($d \geq 5$).

When a code is used for synchronization-error control, at least some of its digit-error control capability must be sacrificed. In the above example, digit errors in positions a_4 and a_5 would result in the same received message as would the one-digit synchronization error (if x fortuitously has the same value as a_1). Thus, a code having $d = 5$ would not necessarily be capable of distinguishing these situations. A $d = 5$ code would, however, be capable of correcting any one digit error in a properly synchronized message. In other words, the code may be used to correct either a digit error or a synchronization error, but if a digit error occurs in a message that is out of synchronization, the code may fail.

The rules for indicating a digit error are as follows. No apparent errors at all would be the indication that digit a_5 is in fact in error. Two apparent errors, one in the a_5 position and one in any other position, would be the indication that the digit in the other position is actually in error. If the digit in the a_4 position is apparently in error, with or without an apparent error in the a_9 position, the decoder "assumes" that no actual digit error has occurred. Instead, the indication is that a digit has been erroneously deleted from a previous message. Similarly, an apparent error in the a_6 position, with or without an apparent error in the a_1 position, is the indication of an erroneous insertion of a digit in a previous message.

To be capable of correcting two-digit synchronization errors, the code must have at least triple-error-correcting capability ($d \geq 7$), because apparent digit errors may occur in as many as three positions in a word. This code will be capable of correcting a single digit error in a properly synchronized message, for which the rules of the previous paragraph apply. In general, for the correction of s-digit synchronization errors, the code must be capable of correcting $s + 1$ digit errors. Also, in the general case, the code will be capable of correcting a single digit error but only in a properly synchronized message (because any s-digit synchronization error is indistinguishable from an appropriately selected two-digit error).

If a synchronization-error-correcting code is additionally to be capable of either (a) correcting a digit error in an out-of-synchronization message or (b) correcting more than one digit error in a properly synchronized message, more than one digit per message must be inverted. The detailed capabilities and requirements for such codes and the rules for using them can be derived by reasonably straightforward extensions of the considerations already described. In general, as the number of inversions (forced "errors") per message is increased, the required distance d for a given amount of synchronization-error-correcting capability is increased. In other words, for a code of a given d, some or all of the digit-error-correcting capability can be traded for synchronization-error-correcting capability. In systems where the mere detection of some of the errors is sufficient the trade-off is four-way among (a) synchronization-error correction, (b) synchronization-error detection, (c) digit-error correction, and (d) digit-error detection.

When the synchronization error is so great that the inverted digit is "shifted out" of the given message, the above scheme of synchronization-error control may still function properly (with codes of sufficiently large d), although the correction will yield an adjacent message. In the nine-digit message used above as an example, an erroneous five-digit deletion in a prior message will result in the digit set

$$a_6 a_7 a_8 a_9 b_1 b_2 b_3 b_4 \bar{b}_5$$

where the b_i are the digits from the next succeeding message. To the decoder this digit set will appear the same as would occur as a result of an erroneous four-digit insertion, or

$$xxxx b_1 b_2 b_3 b_4 \bar{b}_5$$

In attempting to correct this error, the decoder would make a shift to the left instead of to the right, and one complete message will have been deleted. Such a result may be acceptable in some applications and unacceptable in others. However, to achieve this degree of synchronization-error-correcting capability with nine-digit messages, the code must have $d = 9$, for which the maximum number of different messages is only two (see Table 5-2). In other words, even though the decoder need not be capable of distinguishing which message in a series is being received, the development of a code which will, for any synchronization error, positively establish synchronization within the "space" of one message is possible but impractical in virtually all applications that can be imagined.

A Variation in the Cyclic-Code Synchronization Scheme

A variation in the above synchronization scheme is to invert the digits at each end of each message (instead of inverting a single digit near the middle). The received word will then be $\bar{a}_1 a_2 a_3 a_4 a_5 a_6 a_7 a_8 \bar{a}_9$ with no errors in transmission, $a_2 a_3 a_4 a_5 a_6 a_7 a_8 \bar{a}_9 x$ with a one-digit synchronization error, or $a_3 a_4 a_5 a_6 a_7 a_8 \bar{a}_9 x\ x$ with a two-digit synchronization error. The code must be at least double-error-correcting to indicate that no errors have occurred, which is an apparent disadvantage in comparison with the previous variation. However, when a one-digit synchronization error has occurred, this double-error-correcting capability is sufficient to provide one-digit synchronization-error correction (under the assumption that the code is not additionally being used for digit-error correction). In general, for the correction of s-digit synchronization errors, the code must be capable of correcting $s + 1$ digit errors, and when used for synchronization-error correction it can correct one digit error in a properly synchronized message. This capability is the same as for the previous variation, although the rules for deducing the correct message are modified.

The advantage of this variation is derived from the fact that, in the presence of a synchronization error, the apparent digit errors appear in a "burst" at one end of the message. The length of the burst may be as great as $s + 1$ digits for an s-digit synchronization error. Because, as explained earlier in this chapter, any cyclic code is capable of detecting a burst of r errors, where r is the number of parity check digits, any cyclic code can be used to detect any synchronization error of no more than $r - 1$ digits. Therefore, any cyclic code for which $r > (n+1)/2$, where n is the total number of digits per message, can be used for detecting any synchronization error, although if the messages are out of synchronization by more than $n/2$ digits for even n or $(n-1)/2$ digits for odd n, the entire message series will be advanced or retarded by one or more complete messages.

The above variation has a minor difficulty in that a message containing no errors will appear to contain a cyclic burst of two errors, one in the first digit and one in the last digit. This difficulty can be overcome by inverting these two digits and passing the resulting word through the decoder. A synchronization error will have been detected only if the decoder detects an error in both the altered and unaltered words.

The use of a cyclic code to detect synchronization errors by means of the scheme described in this section does not impair the code's ability to detect a burst of digit errors in the ordinary way. However, the code does not necessarily have the ability to distinguish between the two kinds of errors. If the detected errors are relatively rare in a series of messages, the decoder may "assume" that the errors are digit errors. If roughly half or more of the messages are found to contain errors, a synchronization error of one or more digits may be "assumed," and the decoder may then shift all messages in a search for a synchronization that produces a dramatic decrease in the percentage of messages in error.

When the code is used only for error detection (not correction) in the manner described here, a single digit error in a message out of synchronization may cause the code to fail. However, for those codes capable of detecting two bursts of errors (as some codes can do—see discussion earlier in this chapter) an appropriate number of digit errors can be tolerated in out-of-synchronization messages.

Synchronization Scheme for Shortened Cyclic Codes

Consider a cyclic code having distance $d \geq 7$ (capable of correcting at least three random errors) which has been shortened by five digits so that a message (code word) may be represented by

$$(0\ 0\ 0\ 0\ 0)\overbrace{k_k\ \cdots\ k_3 k_2 k_1 r_r\ \cdots\ r_3 r_2 r_1}^{n}$$

Only the k information digits and the r parity check digits are actually transmitted, and the number of digits per message is therefore n as indicated. As before, the digits may be visualized as moving from right to left so that the first and last digits transmitted are k_k and r_1, respectively. Because of the nature of cyclic codes, any cyclic shift of a code word is another valid code word. That is, for example, the following sequence of digits would represent a message in the unshortened code.

$$r_1 0\ 0\ 0\ 0\ 0\ k_k\ \cdots\ k_3 k_2 k_1 r_r\ \cdots\ r_3 r_2$$

In this instance the parity check digits would be taken as $k_1 r_r\ \cdots\ r_3 r_2$. However, a shortened cyclic code does not have this cyclic property so that, for example,

$$r_1 k_k\ \cdots\ k_3 k_2 k_1 r_r\ \cdots\ r_3 r_2$$

would not, in general, represent a valid message.

When transmitting a sequence of messages in a shortened cyclic code, the erroneous insertion of a digit in a previous message will cause the apparent digits of a given message to be

$$x\ k_k\ \cdots\ k_3 k_2 k_1 r_r\ \cdots\ r_3 r_2$$

where the x may be either 0 or 1 (is actually the r_1 digit from the previous message). The problem is to devise a scheme whereby this situation and the analogous situation resulting from an erroneous deletion of a digit in a previous message can be detected and corrected. In particular, the decoder must distinguish this one-digit synchronization error from an unknown number (which may be only one, two, or three in some instances) of digit errors which fortuitously result in the same digit set.

The desired result can be achieved by encoding each message on the assumption that the central one of the five untransmitted digits is a 1, so that the message becomes

$$(0\ 0\ 1\ 0\ 0)k_k\ \cdots\ k_3 k_2 k_1 r_r'\ \cdots\ r_3' r_2' r_1'$$

The assumed 1 is in the position which may be designated the k_{k+3} position. The parity check digits r_i' ($1 \leq i \leq r$) differ from r_i in those positions for which the k_{k+3} digit has an influence on the parity check count as established by the unshortened code. Physically, the encoder may function by operating in the conventional manner (a mod 2 division by the generator polynomial) with the unshortened code and with the five "high order" digits as assumed above. Alternatively, the encoder can perform the mode 2 division on only the k information digits and can subsequently add (mod 2) a digit set equal to the parity check failure pattern corresponding to the k_{k+3} position.

At the receiver the decoder "assumes" that all five untransmitted digits were 0's. Then, if only one digit appears to be in error and this digit is in the k_{k+3} position, the indication is that no errors at all, either in synchronization or in the digits themselves, have occurred.

If, on the other hand, an erroneous insertion of a digit occurred in a previous word, the word received would be

$$(x\ 0\ 0\ 1\ 0)x\ k_k \cdots k_3 k_2 k_1 r_r' \cdots r_3' r_2$$

where x is either 0 or 1 and the parentheses serve to remind that the indicated digits are not actually transmitted. The x in the k_{k+5} position may be visualized as being r_1' and the x in the position normally occupied by k_k is actually the r_1' digit of the previous message. Because

$$r_1' 0\ 0\ 1\ 0\ 0\ k_k \cdots k_3 k_2 k_1 r_r' \cdots r_3' r_2'$$

is a valid code word in the unshortened code, the received word will produce and error indication for the position normally occupied by k_{k+2} but not for the position normally occupied by k_{k+3}. This result is a positive indication of a one-digit shift to the right. Errors may or may not be signalled in the positions containing x's in accordance with the binary values the digits in these positions happen to have. Nevertheless, the possibility of apparent errors in these positions must be considered, and therefore the code must be at least triple-error-correcting.

The synchronization error, once detected, is corrected by shifting all digits to the left one position. If the digit r_1' cannot be recovered from the next subsequent message (to which it will have been erroneously shifted by the synchronization error), this digit can be recovered by the code itself by means of a second error-correcting operation on the properly synchronized message.

The response of the code to an erroneous deletion of a digit in a previous message is analogous. The x's in this case appear in the k_{k+1} and the r_1 positions of the message.

This synchronization control scheme for shortened cyclic codes has a number of variations. The variation that was presented was selected because it seems to be the easiest to explain and understand. However, essentially the same results can be achieved through a shortening of only four digits if the four untransmitted digits are assumed to be 1001.

That is, the number of information digits k can be one greater than previously implied. In this case the transmitted word is

$$(1\ 0\ 0\ 1)k_k \cdots k_3 k_2 k_1 r_r'' \cdots r_3'' r_2'' r_1''$$

where each r_i'' differs from its corresponding r_i if it is in a position for which one but not both of the digits in the k_{k+4} and k_{k+1} positions influence the parity check count. With this scheme, the absence of any actual errors will produce apparent errors in the untransmitted k_{k+4} and k_{k+1} digits.

If an erroneous digit insertion occurred in a previous message, the received digits will be

$$(x\ 1\ 0\ 0)x\ k_k \cdots k_3 k_2 k_1 r_r'' \cdots r_3'' r_2''$$

When decoding, this synchronization error is positively identified by the apparent error in the k_{k+3} position. The assumed 1 in the k_{k+1} position can be visualized as having been shifted into what is normally the k_k position. The digit actually appearing in this position will be the r_1'' digit of the previous message, but because this digit might create an apparent error anyway, the situation is substantially the same as in the previously described variation.

By analogy, an apparent error in the k_{k+2} position plus possible apparent errors in the positions normally occupied by the k_{k+1} and r_1'' digits of a given message would be the indication that a digit had been erroneously deleted from a previous message.

Alternatively (not additionally), the four-digit shortening scheme can be used to detect two-digit synchronization errors, but as can be easily shown, the ability to correct one-digit synchronization errors is sacrificed and the code must be capable of correcting up to five random digit errors. An apparent error in either the k_{k+3} or the k_{k+2} position is the signal for a synchronization error of one or two digits.

In general, for the correction of s-digit synchronization errors, the number of digits by which the code must be shortened is $2s + 2$, and the code must be capable of correcting $2s + 1$ random errors. Conceivably, this latter requirement could be relaxed somewhat through the development of special codes particularly well adapted to the control of digit errors in those positions of the word in which apparent errors are caused by synchronization errors.

When used for one-digit synchronization-error correction, the price in selecting the four-digit shortening variation in comparison with five-digit shortening is a reduction in the ability to correct random digit errors when no synchronization errors occur. Specifically, in the case of the assumed triple-error-correcting codes for example, only one random digit error can be corrected when no synchronization error is sensed with the four-digit shortening whereas with five-digit shortening two random digit errors can be corrected in the absence of synchronization errors.

The following variation achieves full random-error-correcting capability in the absence of synchronization errors, and it is adaptable to either the five or four-digit shortening technique described above (or to their extensions for multiple-synchronization-error control). The code and the encoding are the same as before, but at the receiver the decoder first adds (mod 2) to the received word the same parity check failure pattern used at the transmitter to develop the coded word. If no errors in transmission occur, the word thus generated is the same as the original encoding with all 0's in the untransmitted digit positions. Therefore, in the absence of synchronization errors, any digit errors which do occur will occur in the n transmitted digits and can be detected and corrected accordingly. Like everything else in coding, this capability comes at a price, and the price here is the requirement for a code which (in the instance of single-digit synchronization-error correction) is capable of correcting four instead of just three random errors.

Other variations include "lop-sided" patterns of assumed 1's in the untransmitted digits of the shortened code. In particular, only the first or only the last of these digits may be made 1 to favor erroneously inserted or erroneously deleted digits, as the case may be, in accordance with whichever kind of error is considered more probable in some physical system. The details of the interplays between detection-correction capabilities of erroneously inserted, erroneously deleted digits, digit errors, and the distance d and other properties of a code are far too numerous to set forth here, but by following the arguments already presented, the details can be worked out for any particular case as needed.

Snake-in-the-Box Codes

When each code word represents the amplitude of a signal derived from some analog-type source another desideratum of codes is encountered. Because each code word must correspond to at least a small range in amplitude, a problem arises in translating the amplitude to its digital representation when the amplitude is near the boundary of the ranges represented by two different code words. When the analog-to-digital conversion is made by a device which generates the various digits independently of each other, some digits of the resulting code word may correspond to one range while other digits correspond to the other range. Thus, the desideratum is that, for any two adjacent analog ranges, the corresponding two code words should be the same except for the digit in one position.

The so-called Gray codes or reflected binary codes have the desired property and are widely used in analog-to-digital converters—see Richards' book *Electronic Digital Components and Circuits*, 1967, pp. 490, 500–504, and 510–511. However, these codes have $d = 1$ and have no error-detecting or error-correcting capability. Further, although the code will not produce a "wild" representation of amplitude when only the "boundary" digit is in error, an error (as might be caused by a malfunctioning converter or by an error in transmission) in any other digit of the code word would create the unwanted "wild" representation.

Error control can be incorporated if the code words are selected so that two or more digits must be changed when proceeding from one analog range to the next—provided that proceeding from one code word to the next (by changing one digit at a time) no other valid code words are encountered. The number of digits that must be changed corresponds to the distance as before. Codes of this form are called "snake-in-the-box" codes because in certain geometrical representations of them each code word corresponds to a point in n-dimensional space and when a line is drawn connecting successive words in the amplitude-representing sequence, a vague illusion of a snake in a box is obtained.

Unfortunately, when attempting to apply snake-in-the-box codes having $d > 1$ to analog-to-digital converters, certain difficulties are encountered that render the codes less attractive than might be hoped. In particular, for $d = 2$ an outright malfunction is difficult to distinguish from the situation where one but not both digits have been changed when proceeding from one analog range to the next. For a further description of the codes, reference is made to a paper by R. C. Singleton in the *IEEE Transactions on Electronic Computers*, August 1966, pp. 596–602.

Nonblock Codes

A quite different form of coding for error control is obtained by interspersing parity check digits among the information digits in a uniform (or at least predetermined) pattern wherein the odd-even count for any given parity check digit is with respect to digits a specified number of positions "away" from (usually limited to previously transmitted digits, and not digits to be transmitted subsequently) the given parity check digit. Thus, the odd-even counts are in a sort of continuously overlapping pattern, and the digits are not divided into "blocks." For any given parity check digit, the digits in the odd-even count may include other parity check digits as well as information digits. When starting the transmission of a sequence of digits, all previously transmitted digits are assumed to have been 0's.

The codes as described in the previous paragraph have been variously called "convolutional" codes, "recurrent" codes, or "tree" codes. Although some authors may draw subtle distinctions among these three terms, they are essentially synonymous. The characteristics of the codes are not really close to the conventional meanings of "convolve" and "recur," respectively. The term "tree" is derived from the fact that a portrayal of all possible digit sequences in the code resembles a tree (analogous to a "tree" switching matrix). That is, for the first binary digit only two possibilities exist (the "trunk" being absent) but as second and successive digits, some of which may be parity check digits, are considered, the number of possibilities increase in a manner analogous to the branching of tree limbs.

Although nonblock codes have generated much interest, the practical advantages of interspersing the parity check digits in this nonblock fashion are at best obscure. The most commonly claimed advantage is simplicity and resulting economy of encoding and decoding. Encoding can indeed be accomplished quite simply and economically but not really more so than with the cyclic codes already described. Comparisons for decoding are not easily made, because the decoding procedures required for nonblock codes are so different from those for block codes.

One technique for decoding nonblock codes is called "sequential decoding." As the term implies, the information digits in the received sequence are ascertained one at a time, and in the absence of errors the output from the decoder "keeps up" with the received information on a digit-by-digit basis. When an error is detected, the decoder chooses the valid subsequent digit sequence (or tree "branch") which differs from the received sequence by the fewest number of digits, that is, which has the minimum "distance" from the received sequence. If the distance between the received sequence and the closest valid sequence becomes greater than some allowable maximum, the decoder must halt its output temporarily and return to previous branching points in a search for a "path" through the tree which yields a distance less than the allowable maximum.

If no such path is found, an uncorrectable error pattern has been detected. Of course, in the presence of excessive errors the decoder may fail to detect any errors at all or may make erroneous "corrections"—the same as with block codes.

A characteristic of sequential decoding is that when an erroneous "correction" has been made or when an uncorrectable error pattern is detected, unwanted effects of this situation may continue over a substantial sequence of subsequently received digits. The decoder will, however, restore itself to correct operation after a sufficiently long subsequent sequence of error-free or nearly error-free digits have been received.

Although certain basic concepts of block codes, such as the relationship between distance and error detecting and correcting capability, are vaguely applicable to nonblock codes, the interplay of the various factors are far more complex with nonblock codes. To quote one person who has worked in the field: "Most propositions concerning [nonblock codes] have been extremely difficult to develop, even with access to an electronic computer." Even the basic relationship between the parity check pattern and the density of detectable and/or correctable errors becomes a complex subject. Other parameters that are important with nonblock codes and which are difficult to ascertain are the frequency of having to search through previous branching points, the number of computations needed for this search, the number of previously received digits that must be stored to facilitate the search, and in the event of a code failure the number of digits (hopefully but not necessarily completely error-free) over which the decoder must operate to restore itself to proper operation.

The problem of synchronization errors is particularly "sticky" with nonblock codes. Although the decoder will eventually restore itself after the erroneous insertion or deletion of one or more digits, in nearly all practical applications the intended information is not in the form of an abstract continuous sequence of digits but is, in fact, in "blocks" or "messages." Thus, even after correct decoding from a nonblock standpoint, the correct synchronization of the digit sequence must still be found. Certain previously described techniques, such as the use of prefixes, are applicable for the detection and correction of synchronization errors but only at the expense of considerable complications arising from the nature of nonblock coding.

Because of the above-mentioned difficulties in analyzing the characteristics and capabilities and because their net advantages, if any, are elusive, nonblock codes will not be described in detail here. For further information, the reader is referred to the *IEEE Transactions on Information Theory*, various issues of which contain papers treating miscellaneous aspects of the subject.

Using a General Purpose Computer as a Decoder

Much attention has been given to the possibility of designing codes so that the construction of encoders and decoders is relatively simple and economical. However, although an encoder can be so constructed, a decoder tends to be an unfortunately complicated and expensive piece of equipment for all but the relatively elementary error-detecting codes of limited capability. For any significant degree of sophistication in the matter of error correction, particularly synchronization-error correction, the designer may wish to consider using a general purpose computer for the purpose, in which case the decoding is not a matter of digital design at all but is a matter of writing a program in the same sense that solving any other mathematical problem is a matter of writing a program. The computer need not, of course, be used only for decoding the information received on a single input line but may be time-shared with many input lines or with accounting, process control, and other problems totally unrelated to decoding.

Exercises

1. Besides sextuple-error detection or, alternatively, triple-error correction, what error-detection-correction properties can be realized with a code of distance $d = 7$ as illustrated in Figure 5-2c.

2. Explain why the approximation $p^2 N(N-1)$ is not accurate in the example on page 198. Use the third term of the binomial expansion $(1-x)^n = 1 - nx + (\tfrac{1}{2})n(n-1)x^2 - \cdots$ to obtain a more accurate approximation for the probability of receiving a message containing an undetected error when a single parity check digit is being used for each message.

3. By "cut-and-try" techniques find a nonsystematic code having four code words where the number of digits $n = 5$ and the distance $d = 3$.

4. Extend the decoder of Figure 5-3 as necessary to provide for double-error detection under the assumption that an added parity check digit y_4 is available to indicate the odd-even count of all digits in a word.

5. For the $n = 15$, $d = 3$ Hamming code, write down the 16 code words when the parity check digits are determined as set forth in the lower-right part of Figure 5-3. Use the convention that a parity check digit of 0 indicates an even number of 1's in the digits checked.

6. In the two-dimensional code of Figure 5-5, find at least one pattern of three errors which will cause the code to fail when used for single-error correction.

7. Consider a three-dimensional code composed of a set of n_3 planes of digits where $n_3 - 1$ of the planes are in the form of two-dimensional codes as in Figure 5-5 and where the last plane contains nothing but parity check digits where each such parity check digit indicates the odd-even count of the digits in the corresponding position of all planes. How many digit errors per message is this code capable of correcting?

8. Multiply (mod 2) 1110110 by 101101 and check the results by dividing (mod 2) the result by each of the factors.

9. Divide (mod 2) $X^{10} + X^5 + X + 1$ by $X^4 + X^3 + X$ and check the result by multiplying (mod 2) the result by the divisor and adding (mod 2) the final remainder.

10. For the example on page 217, find the pattern of parity check odd-even counts when the factor 101101 is regarded as the multiplier of a four-digit multiplicand.

11. Show that generator polynomials 10001, 10101, 11011, and 11111 in digital notation do not result in useful codes. (Hint: Do this through finding the smallest e for which $X^e + 1$ is an exact multiple in each instance.) (Note that all other five-digit polynomials for which the first and last digits are 1's are discussed in the text.)

12. Determine the length of the prefix needed to correct two-digit synchronization errors and to detect three digit synchronization errors. Determine the switching functions (in terms of p_1, p_2, etc., as defined in the text) for developing the necessary error signals: one digit deleted, two digits deleted, one digit inserted, two digits inserted, and three digits deleted or inserted.

13. Show that a prefix containing at least two 1's and at least two 0's is inferior to a prefix containing only one 0 with the remainder all 1's. (Prefixes having at least four digits are assumed.)

14. Show that in a prefix containing one 0 with the remainder all 1's the 0 must be either the first or the last digit to achieve maximum synchronization-error control capability.

15. With respect to the synchronization control scheme which was described for shortened cyclic codes, explain why a shortening by three digits is insufficient for the positive correction of one-digit synchronization errors.

16. In the synchronization control scheme for cyclic codes as described on pages 259–261, determine the requirements on the distance d and the associated error control capabilities when two adjacent digits near the middle of each message are inverted. Repeat this exercise but with the two inverted digits not adjacent.

6
IMPLEMENTING BINARY ARITHMETIC

Virtually all early digital systems were constructed primarily to perform computations and were accordingly called "computers." Computers are still prominent in the realm of digital technology although in an increasing proportion of digital system applications the computing capability of the system is little more than incidental to the major purpose which is more broadly called "digital information processing." A few examples of such applications are cryptographic analysis, information retrieval, telephone switching, abstract mathematical studies, process control, traffic control, and many aspects of business accounting such as record sorting. Nevertheless, even in the applications where computing is relatively incidental, the system as a whole is likely to be designed with essentially the same form of arithmetic and control unit (plus appropriate storage) as used in a computer. Moreover, although the familiar operations of addition, subtraction, multiplication, and division may not be objectives in many digital information processing applications, the need for these operations is nearly always strongly felt when working out the details of the problem. Accordingly, the techniques available to perform arithmetic operations are likely to be of great concern to the machnie designer regardless of the application at hand.

As extensively implied in earlier chapters, the physical devices found to be most suitable for digital systems employ two-valued (binary) signals. Thus, the first consideration in implementing arithmetic in a digital system is to determine a scheme whereby quantities can be represented by sets of binary signals. A great amount of thought and study has been given to this question with the net result that the most appropriate scheme is a number system called the binary system and is essentially the same as the familiar decimal system except that only two instead of ten different digits are used. The positional significance of each digit is related to an integral power of two instead of ten. In other words, the "radix" is two instead of ten.

Interestingly, the binary number system has been known to mathematicians for centuries, but because of its lack of application it was a thing of great novelty and considerable mystery to laymen and even to engineers when it became prominent with the advent of electronic computers. Now, however, because of the importance of computers, children in many schools are introduced to binary numbers at an early age.

Binary Numbers

In the binary number system the digits are 0 are 1 (although in principle any two different symbols could be used), and counting proceeds as follows.

     0 zero
     1 one
    10 two
    11 three
   100 four
   101 five
   110 six
   111 seven
  1000 eight
  1001 nine
  1010 ten
  1011 eleven
     .
     .
     .

The names listed for the various integral values are the familiar names used in the decimal system—no binary names having been developed.

In physical systems a fixed number n of digits are usually assigned to the representation of each quantity so that, if $n = 6$, for example, the representations would be 000000, 000001, 000010, and so on for zero, one, two, and so on.

The counting can be visualized as being accomplished by the addition of 1 to the right-hand digit when proceeding from one number to the next. If the digit in this position is initially 0, the addition of 1 merely changes that digit to 1. However, if the right-hand digit is initially 1, the addition of 1 changes it to 0 and creates a "carry" having a "weight" of 2 to be added to the digit in the next position. This carry, although it has a weight of 2, is really only a binary 1 which is to be added into this next position in accordance with corresponding rules. The carry, if any, generated by the addition of 1 in this second position has a weight of 4 and is added analogously into the third digit position, and so on. The weights of the digits in the various positions of a binary number can be expressed in terms of powers of 2 as 2^0 2^1 2^2, 2^3, and so on up to 2^{n-1}. Because every digit in a binary number can be 1, an n-digit binary number can be used to represent all integers from zero to $2^n - 1$. In slightly more general terms, an n-digit number can be used to represent any one of 2^n different numerical values.

As a more explicit example, the binary number 1101001 is equal to

$$(1)2^6 + (1)2^5 + (0)2^4 + (1)2^3 + (0)2^2 + (0)2^1 + (1)2^0$$
$$= 64 + 32 + 0 + 8 + 0 + 0 + 1$$
$$= 105$$

When working with binary numbers in any of the many applications in which they are encountered, the designer will find that he needs to determine or know various powers of 2 in countless instances. The smaller powers, up to about 2^6, are easily determined and memorized. For larger powers a convenient technique involves committing to memory the fact that $2^{10} = 1024$. With this power known, other large powers can be computed quickly. For example, $2^{12} = (2^{10})(2^2) = (1024)(4) = 4096$. For another example, $2^9 = (1024)/2 = 512$. For much larger binary exponents, an approximation to the magnitude is often all that is pertinent, in which case the convenient approximation to know is that $2^{10} \simeq 1000$. Then, for example, the fact that 2^{33} is approximately 8 billion can be ascertained almost instantly by observing that $2^{33} \simeq (2^{10})(2^{10})(2^{10})(2^3) = (1000)(1000)(1000)(8)$.

An appreciation of the opposite problem, that of determining the binary representation for a given decimal number, is not needed in the development of the present subject of implementing binary arithmetic—although decimal-to-binary conversion is covered in a later chapter. Nevertheless, the newcomer may understand binary numbers better by converting a few decimal numbers to binary form by an elementary process. For example, 37 may be converted by noting that $2^5 = 32$ is the largest power of 2 that is less than 37. Therefore, the digit in the 2^5 position of the binary number is 1, and 32 is then subtracted from 37 to yield a remainder of 5. The 2^4 and 2^3 digits of the binary number can then be determined to be 0's, but $2^2 = 4$ is less than 5 so the 2^2

digit is 1. When 4 is subtracted from 5, the remainder is 1. The 2^1 and 2^0 digits are then obviously 0 and 1, respectively. The resultant binary equivalent of 37 is therefore 100101.

Binary Fractions and the Binary Point

Fractions can be represented in the binary system in essentially the same manner as in the decimal system in that, for example, the binary notation $110\frac{11}{101}$ would mean $6\frac{3}{5}$. However, physical machines are virtually never built to handle fractions of this category (except through programming where the numerator and denominator are each treated as conventional numbers).

Instead, fractional parts are represented through the use of a binary point where digits to the right of the point have weights corresponding to negative powers of 2 in ascending sequence. That is, the first digit to the right has a weight of 2^{-1}, the second a weight of 2^{-2}, and so on. With this notation the binary number 0.01101 for example has the value

$$(0)2^{-1} + (1)2^{-2} + (1)2^{-3} + (0)2^{-4} + (1)2^{-5}$$
$$= 0 + \tfrac{1}{4} + \tfrac{1}{8} + 0 + \tfrac{1}{32}$$
$$= 0 + 0.25 + 0.125 + 0 + 0.03125$$
$$= 0.40625$$

Every binary fraction expressed in this way has an exact decimal equivalent. However, many decimal fractions of this form cannot be expressed exactly by a corresponding binary number no matter how many binary digits may be used. An elementary example is the decimal number 0.1 (one tenth). This number can be converted to binary form by first noting that 2^{-1}, 2^{-2}, and 2^{-3} are all greater than decimal 0.1, so the first three digits to the right of the binary point are 0's. The 2^{-4} digit is 1 because $2^{-4} = 0.0625$, which is less than 0.1. The 2^{-5} digit is found to be 1 because $2^{-5} = 0.03125$ is less than $0.1 - 0.0625 = 0.0375$. The difference $0.0375 - 0.03125 = 0.00625$ is then formed. Because this difference is less than 2^{-6} and 2^{-7}, the next two digits of the binary equivalent are 0's. By continuing this process (or any other decimal-to-binary conversion routine), the following result is obtained.

0.1   = 0.00011001100110011001100110011001100110 · · ·
(decimal)   (binary)

This lack of a one-to-one correspondence between all possible decimal numbers having a finite number of digits and all possible binary numbers having a finite number of digits is not particularly objectionable with some types of problems to be solved by a computer, but it can be a serious nuisance in other problems. When the data are approximations anyway, the nearest equivalent binary number is selected by means of round-off techniques. However, in some problems the original data are in the form of decimal fractions that must be represented exactly. A common example is in accounting problems where money is in terms of dollars and hundredths of dollars (cents). The monetary amounts must be transposed (at least mentally) to cents only, so that each dollar appears in the computations as 100 (one hundred) cents. Moreover, the machine must be built to treat numbers as integers instead of fractions—a topic to be discussed in more detail later in connection with multiplication and division.

"Counting" in fractional parts proceeds in essentially the same way as when counting integers. Thus, when counting by eighths, for example, a typical number sequence might be

$$101.100$$
$$101.101$$
$$101.110$$
$$101.111$$
$$110.000$$
$$110.001$$

Doubling and Halving by Shifting

In the decimal system the shifting of all digits to the left or right by one position has the effect of multiplying or dividing, respectively, the number by 10 (ten). In the binary system a corresponding shift of digits likewise has the effect of multiplying or dividing by 10, but the meaning of binary 10 is 2, so the shift has the effect of doubling or halving, respectively. In general, a shift of s positions to the left has the effect of multiplying the number by 2^s and a shift to the right has the effect of dividing by 2^s. For example, 1101 shifted to the right two positions produces 11.01, which amounts to dividing 13 by 4 to produce $3\frac{1}{4}$.

The shifting of binary numbers is a very commonly used operation in physical systems, especially as an intermediate operation in the development of more complex operations such as multiplication and division and many others.

Round-Off

When a quantity cannot be represented exactly by means of a single number (as the fraction $\frac{1}{3}$, for example, cannot in either the decimal or binary number systems) or when the quantity is not known exactly, the procedure used is to retain only a specified number of the relatively "high order" digits in the representation and to discard all "low order" digits that would otherwise be needed for a more accurate representation of the quantity. However, the retained digits should represent a quantity which, for that number of digits, is most nearly equal to the given quantity. The process of deleting the lower order digits and obtaining the desired resultant number is called "round-off."

In the decimal system, a well known round-off technique is to add the digit 5 to the highest order digit to be deleted. The carry, if any, resulting from this addition is added to the higher order digits in the conventional way. Thus, for example, when rounding 739.63 to the nearest unit, a 5 is added to the tenths digit to produce 740.13 which becomes the rounded number 740 after the tenths and hundredths digits are deleted. When rounding 739.63 to the nearest ten the result would still be 740, but when rounding to the nearest tenth, the result would be 739.6 as can be determined by adding 5 to the units and hundredths positions, respectively. ("Deleting" a digit means changing it to 0 in those instances where a symbol must be retained to indicate the positions of the remaining digits.)

In the binary system, the same round-off technique can be used except that a 1 instead of a 5 is added to the highest order digit to be deleted. Thus, for example, 1011.101 would be rounded to 1100 when rounding to the nearest unit, and this rounding would be accomplished by adding 1 to the 2^{-1} position. That 1100 is correct can be appreciated by observing that a 1 in the first position to the right of the binary point indicates that the fractional part is $\frac{1}{2}$ or greater. If all lower order digits were 0's, the fractional part would be exactly $\frac{1}{2}$ in which case a rounding to 1011 would be equally accurate, but the customary convention is to accept the next larger value in this instance—a convention which is conveniently adaptable to physical devices. Round-off in some other digit position such as to the nearest two, four, eight, and so on or to the nearest half, quarter, eighth, and so on is accomplished analogously.

Binary Addition

In principle, any number of numbers could be simultaneously added together in a physical system. For example, in a variable instruction length system (see Richards' book *Electronic Digital Systems* pp.

278 IMPLEMENTING BINARY ARITHMETIC

149–151) a triple addition instruction designated ADD-3 and having two address parts might be included in the instruction repertory. The two addresses would refer to separate but simultaneously operating storage units. Then the instruction ADD-3 0041 9072 could mean that the numbers in addresses 0041 and 9072 are simultaneously transmitted to an accumulator for addition to the number already contained there. The three-number addition could be performed in an appropriate switching network. As the interest in parallelism as a means for higher computing speed intensifies, multinumber addition may become important. However, virtually all present and foreseeable systems work with only two operands at a time, so the treatment here will be correspondingly limited.

To add two binary numbers the digits in corresponding orders (that is, having corresponding weights) are paired and added individually as in conventional decimal arithmetic. When two binary digits are added, the sum is 0 if both digits are 0's, and the sum is 1 if one but not both the digits is 1. If both digits being added are 1's, the sum is 10 (decimal 2) where the right-hand digit, a 0, is called the sum digit and the left-hand digit, a 1, is called the carry. The amount of equipment needed to add two binary digits to produce the sum and carry signals is called a "half adder." From the rules of addition just described the sum and carry signals for a half adder are easily seen to be, in Boolean notation:

$$\text{HALF-SUM} = A\bar{B} + \bar{A}B$$
$$\text{HALF-CARRY} = AB$$

A and B represent the two digits being added. The carry digit, whether it is a 0 or a 1, generated in any given order has the same weight as the digits of the next higher order, and to develop the final sum of two numbers, each HALF-CARRY must be added to the HALF-SUM of the next higher order. Because the same rules of addition apply, a second half adder may be used for the purpose. In any given order, a carry digit may be generated by this second half adder in the same manner as in the original half adder. Regardless of which half adder generates a carry, the carry must be transmitted to the next higher order for addition to the HALF-SUM of that order. However, regardless of the magnitudes of the two binary numbers being added, no more than one carry digit of 1 will appear from the two half adders in a given order. This result can be understood by observing that a carry digit of 1 can appear from the first half adder only if both digits to that adder are 1's, in which instance the HALF-SUM from that adder will be 0. Therefore, in transmitting a carry digit from one order to the next, an OR device may be used to combine the carry signals from the two half adders.

Figure 6-1 shows the kind of array needed to add two binary numbers in accordance with the scheme set forth in the previous paragraph. In this figure each block labeled H.A. signifies a half adder, and the output lines designated S and C refer to HALF-SUM and HALF-CARRY, respectively. Signal sets $\cdots A_8 A_4 A_2 A_1$ and $\cdots B_8 B_4 B_2 B_1$ represent the two binary numbers to be added, and these signals are simultaneously applied to the indicated input lines. When the transients of the switching devices have terminated, a signal set $\cdots S_8 S_4 S_2 S_1$ representing the binary sum appears on the output lines.

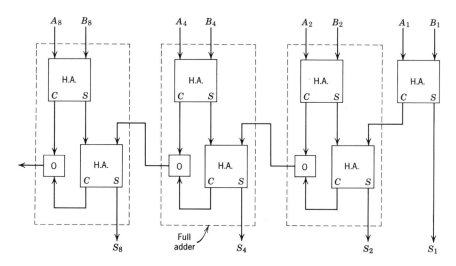

FIGURE 6-1. Binary addition by means of half adders.

A fundamentally important characteristic of the addition process is that the digits entering into the lowest order may affect the value of the sum digit from the highest order. Therefore, the time required for addition is equal to the time consumed for a signal to travel from the right-hand end of the network through the second half adder and the OR device in each order to the left-hand end of the network. This characteristic can be appreciated by considering the addition of the following two binary numbers.

```
  0 0 0 1 1 0 0 1 1 1
  0 1 1 0 0 1 1 0 0 1
  ───────────────────
  0 1 1 1 1 1 1 1 1 0   HALF-SUM digits
                  1     first CARRY digit
  ───────────────────
  1 0 0 0 0 0 0 0 0 0   sum
```

In this example the HALF-SUM digits generated by the first half adder in each order are all 1's except in the highest and lowest orders. Only one carry digit is generated by these half adders. This carry digit generates a carry in the second half adder of the next higher order where a new carry digit is generated. This new carry is then transmitted to the third order where still another carry is generated in succession, and so on. Much of the subject of implementing binary arithmetic is related to schemes for avoiding this time-consuming carry "rippling" process.

In addition (and also subtraction) the location of the binary point is irrelevant except that digits of corresponding weight must be paired in any two numbers to be added. That is, the addition scheme for digits to the right of the point is the same as for digits to the left of the point, and the carry, if any, is passed across the point in the same manner as between any other two digit positions. Although the complete system may include automatic means for shifting numbers as required to pair the digits properly in a manner called "floating point," the adding device itself need not contain any hardware that pertains to the location of the point. Floating point operation is covered in mote detail later.

Full Adders

An alternative way to view the binary addition process is to recognize that three binary digits are to be added together in each order. Two of the digits are from the corresponding positions of the numbers being added and the third digit is the carry from the next lower order. The sum is 00, 01, 10, or 11 (that is 0, 1, 2, or 3) in accordance with whether none, one, two, or all three of the digits are 1's. The right-hand digit of the pair represents the final sum digit in the corresponding position, and the left-hand digit is the carry to the next higher order. The equipment used to add three binary digits is called a full adder, and the SUM and CARRY signals may be expressed in Boolean notation as follows:

$$\text{SUM} = A\bar{B}\bar{C} + \bar{A}B\bar{C} + \bar{A}\bar{B}C + ABC$$
$$= (A + B + C)\overline{AB + BC + AC} + ABC$$
$$\text{CARRY} = AB\bar{C} + A\bar{B}C + \bar{A}BC + ABC$$
$$= AB + AC + BC$$

where A, B, and C may represent any three binary digits but where A and B usually represent the two digits from the corresponding orders of two binary numbers being added, and where C usually represents the carry signal from the next lower order. The first expressions for the SUM and CARRY specify in detail the combinations of 1-signals that cause the respective output signals from the full adder to be 1. The second expression for the CARRY is simply a minimized form of the first expression and indicates that the carry is 1 if any two of the input signals are 1. The second expression given for the SUM may be visualized as meaning that the SUM is 1 if any of the three input signals is 1 "and" at the same time no two of the input signals are 1 (i.e., "and" at the same time that the CARRY is 0) "or" if all three of the input signals are 1.

Figure 6-2 shows how binary addition is accomplished by means of full adders. Each block labeled F.A. represents a full adder switching network with outputs S and C to represent SUM and CARRY, respectively. A comparison of Figures 6-1 and 6-2 indicates that a full adder

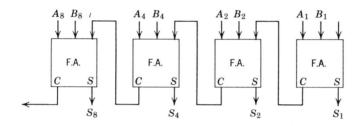

FIGURE 6-2. Binary addition by means of full adders.

performs the same switching functions as do two half adders plus an OR device. When adding two positive numbers in the conventional way, the third input to the lowest order is not needed, although for certain subtraction schemes the third input to the lowest order is needed for the "end-around" carry to be discussed later.

The design of a full adder is essentially nothing more than that of a three-input, two-output switching network described in some detail in more general terms in earlier chapters. Countless specific switching arrangements (specific solutions to the three-input, two-output problem) have been worked out to meet various specialized requirements pertaining to the characteristics of various switching components and circuits. Five different full adders employing either O-I or A-I modules are shown in Richards' book *Electronic Digital Components and Circuits,* pp. 85–87, and several other adders adapted to other specific module types or to specific components are shown elsewhere in the same book. More full adder designs are shown in *Arithmetic Operations in Digital Computers,* pp. 89–98, by Richards. However, rather than present an extensive exposition of the myriad of possible full adder designs, attention will be directed here to addition methods which illustrate the more important principles of the subject. Emphasis will be placed on module types and system design approaches which seem to predominate in current practice and which offer the most attractive possibilities for future systems.

An Accumulator Employing J-K Flip-Flops

In a digital system, two numbers to be added may first be transmitted to two registers, one of which is designated the "accumulator." Each register consists of a set of flip-flops, one for each binary digit of the number. The flip-flop outputs are connected to the adder inputs, and the adder outputs are returned to the inputs of the accumulator. Upon actuation of this array, the sum appears in the accumulator, and the number originally in the accumulator is lost. A new number may then be entered into the other register for addition to the number now in the accumulator. This process may be repeated for an indefinitely long series of numbers, and the accumulator "accumulates" the sum of the series. (Originally, the term "accumulator" meant a register equipped with carry-handling facilities as needed for the addition function—as contrasted with an "adder" which was strictly a combinatorial switching network. Now, "accumulator" is generally used to mean the register used to store the sum, regardless of how the addition is actually accomplished.)

When adding two numbers, a flip-flop in the accumulator need not change state if the corresponding sum digit happens to be the same as the digit already stored in that flip-flop. If the digit in the corresponding flip-flop of the other register is designated B and if the carry digit into the given position is designated C, the sum digit in the accumulator flip-flop A will not be affected if B and C are both 0. If B and C are both 1, the sum digit will likewise remain the same even though two 1's are, in effect, added into that position. The sum will, however, be changed to its opposite value if one, but not both, of B or C is 1. If the accumulator flip-flop is of the complementing form, the signal to the complementing input can be represented by the Boolean expression $B\bar{C} + \bar{B}C = (B + C)\overline{BC}$, which is the same as the exclusive-OR function.

With J-K flip-flops (see Chapter 4) in the accumulator, a complementing action is obtained when the signals to both the J and K input lines are 1's at the time of the clock or actuating pulse. An accumulator

AN ACCUMULATOR EMPLOYING J-K FLIP-FLOPS 283

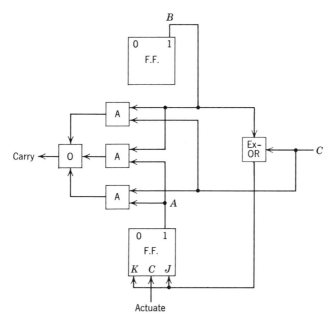

FIGURE 6-3. Switching functions required for binary addition when *J-K* flip-flops are used in an accumulator.

can then be formed as shown in Figure 6-3 for one order. The CARRY to the next higher order is formed in accordance with the expression $AB + AC + BC$ as explained previously. The block labeled "Ex-OR" can be a special exclusive-OR module or it may represent any set of other modules that produces the same net effect.

Although Figure 6-3 illustrates the switching functions needed for binary addition by means of an accumulator, the physical devices used to construct an accumulator would seldom be exactly of this form. For one thing, the carry signal when passing through the chain of AND and OR devices would need amplification, and in most instances the amplifiers would include inversion of the signal. Specifically, if the system is being assembled with A-I modules comprised of DTL or TTL transistor circuits, for example, the OR function can be obtained by connecting the collectors of two or more transistors together to a single load resistor. The resulting A-O-I modules can be used in the carry circuits as shown in Figure 6-4, which indicates the manner in which two successive orders of an accumulator would function. The signal transmitted from the right-hand to the left-hand order is the inverse of the carry because of the in-

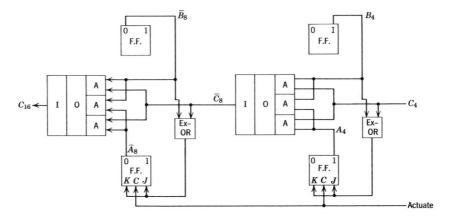

FIGURE 6-4. Two orders of an accumulator employing A-O-I modules and requiring only one inversion per order in the carry propagation path.

version in the A-O-I module. Rather than use a separate inverter to obtain the intended carry signal, the switching array of the left-hand order is designed to use the inverted carry as its input. Because of the symmetrical nature of the binary addition function, the circuit happens to be the same except that the $\bar{A}$ and $\bar{B}$ signals instead of the A and B signals are used. The carry output from the left-hand order is then the carry because, with the single inversion in the A-O-I module

$$\text{CARRY} = \overline{\bar{A}\bar{B} + \bar{B}\bar{C} + \bar{A}\bar{C}} = AB + BC + AC$$

as can readily be established by Boolean manipulation. The exclusive-OR device in the left-hand order generates a signal that causes the accumulator flip-flop to change state whenever one but not both of the corresponding $\bar{B}$ and $\bar{C}$ signals is 1, that is, when $\bar{B}(\bar{\bar{C}}) + (\bar{\bar{B}})\bar{C} = \bar{B}C + B\bar{C} = 1$, which is equivalent to the switching function in the right-hand order. In other words, in a complete accumulator, alternate orders are designed to function with inverse signals, and the purpose is to cause only one inverter per order to appear in the carry propagation path.

After a time sufficient to allow a signal to propagate through all orders in the carry path, a control signal is applied to the line marked "actuate" at which time all appropriate flip-flops in the accumulator register are changed to their opposite state of equilibrium to store the sum. A new number may then be transmitted from a storage unit or some other source to the B register through circuits not shown in Figure 6-4, and a new sum may be "accumulated" by a repetition of the process.

A Parallel Adder Employing Only O-I Modules or Only A-I Modules and Requiring Only One Module per Order in the Carry Path

With some types of transistor circuit modules, RTL modules for example, a common collector connection does not produce any new switching functions but only has the effect of increasing the fan-in capability. When such modules have been selected because of engineering considerations, the switching network designer is obliged to devise arrays that produce the desired results by employing modules that perform only the O-I function—or alternatively only the A-I function when the polarity convention with regard to 1's and 0's is inverted. With only O-I modules or only A-I modules available, the design of a full adder as defined previously is impossible unless the carry signal passes through at least two modules per order, even though the use of inverted carry signals in alternate orders is permitted.

However, by using two signal lines (wires) for each carry signal in a manner whereby a signal on either line represents a carry or a "not-carry," as the case may be, an addition scheme can be devised that employs modules of only one type and which provides a carry propagation path having only one module per order. The scheme is illustrated in Figure 6-5, which shows two orders of only the carry circuits. The two lines marked C_4 transmit the carry signal from the next lower order. For present purposes, the manner in which the carry signal is generated is irrelevant, except that the signals on the two lines must be such that C_4 is properly 1 if a signal appears on either line of the pair. The two

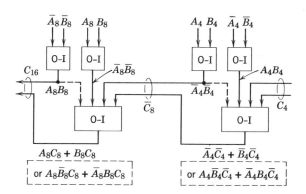

FIGURE 6-5. Carry propagation scheme requiring only one O-I device per order in the carry path.

O-I modules in the upper-right part of the figure generate the signals $\bar{A}_4\bar{B}_4$ and A_4B_4. Because

$$\bar{C}_8 = \overline{\text{CARRY}_8} = \bar{A}_4\bar{B}_4 + \bar{B}_4\bar{C}_4 + \bar{A}_4\bar{C}_4$$

the $\bar{A}_4\bar{B}_4$ signal may be used directly as the signal to appear on one of the two lines that transmit $\bar{C}_8$ to the next higher order. Without the dotted connection, the signal generated by the O-I module in the lower-right part of the figure is $\overline{A_4B_4 + C_4} = \bar{A}_4\bar{C}_4 + \bar{B}_4\bar{C}_4$, which is equal to the remaining two terms in the representation for $\bar{C}_8$, and this signal is therefore correct for use on the other of the two lines that transmit $\bar{C}_8$. If the dotted connection is made, the signal appearing at the output of this O-I module is

$$\overline{\bar{A}_4\bar{B}_4 + A_4B_4 + C_4} = A_4\bar{B}_4\bar{C}_4 + \bar{A}_4B_4\bar{C}_4$$

But

$$\bar{A}_4\bar{B}_4 + \bar{B}_4\bar{C}_4 + \bar{A}_4\bar{C}_4 = \bar{A}_4\bar{B}_4 + A_4\bar{B}_4\bar{C}_4 + \bar{A}_4B_4\bar{C}_4 = \bar{C}_8$$

even though

$$\bar{B}_4\bar{C}_4 + \bar{A}_4\bar{C}_4 \neq A_4\bar{B}_4\bar{C}_4 + \bar{A}_4B_4\bar{C}_4$$

Thus, a suitable two-line $\bar{C}_8$ signal is properly generated with or without the dotted connection. The connection happens to be useful when designing the complete array.

The operation of the carry generation circuit in the left-hand order in Figure 6-5 is analogous. Besides using $\bar{C}_8$ instead of C_8 as an input, the A_8 and B_8 signals are inverted in relationship to the corresponding input signals in the right-hand order. Also, the carry output signal is C_{16} instead of $\bar{C}_{16}$. That is, in a complete parallel adder, all signals are relatively inverted in alternate orders.

Two orders of a complete parallel adder employing the carry propagation scheme of Figure 6-5 are shown in Figure 6-6. Three additional O-I modules in each order are used to generate the SUM_4 and $\overline{\text{SUM}_8}$ signals, respectively. Note that two-wire $\overline{\text{SUM}_4}$ and SUM_8 signals appear on the indicated wire pairs. If devices to which the sum signals are transmitted have suitable built-in OR functions, the sum outputs can be taken directly from these wire pairs.

The carry scheme of Figure 6-5 and the complete adder scheme of Figure 6-6 function correctly without any network alterations when A-I modules are substituted for O-I modules throughout—see an exercise at the end of this chapter.

If the accumulator is composed of J-K flip-flops, an exclusive-OR module or its equivalent in O-I or A-I modules can be used in each order instead of the sum-generating circuit of Figure 6-6.

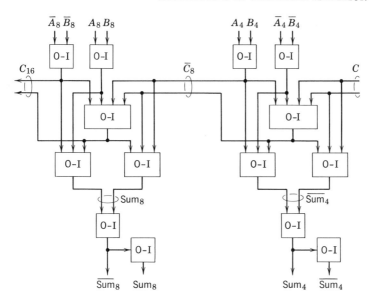

FIGURE 6-6. Parallel adder based on the carry scheme of Figure 6-5.

Simultaneous Parallel Addition

All digits of a binary sum can be generated essentially simultaneously without waiting for a ripple-type carry propagation by using the scheme illustrated for three orders in Figure 6-7a. The orders, designated 1, 2, and 3, can be any consecutive three orders in the accumulator, but for complete simultaneous operation the three orders as shown would be the lowest three orders, and the networks for higher orders would be obtained by a straightforward extension of the scheme.

If A_1 or B_1 (but not both) is 1, the CARRY$_{in}$ signal is propagated through the AND device in the lower half adder of the first order and then through the OR device of that order to the second order. (If both A_1 and B_1 are 1, a carry is initiated in the first order.) If A_2 or B_2 (but not both) is likewise 1, the CARRY$_{in}$ signal is propagated through the three-input AND device in the second order and then through the OR device of that order to the third order. If A_3 or B_3 (but not both) is additionally 1, the CARRY$_{in}$ signal is propagated through the four-input AND device of the third order and then through the OR device of that order to the CARRY$_{out}$ line. If a carry is initiated in any one of the three orders as a result of both A_i and B_i for that order being 1's, the carry is similarly propagated to the higher orders as appro-

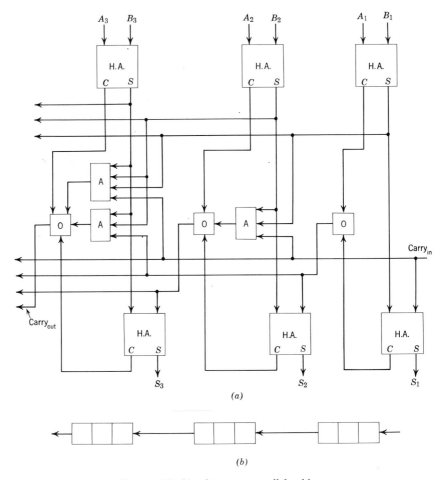

FIGURE 6-7. Simultaneous parallel adder.

priate for the digits of the two binary numbers being added. In any instance, each carry signal needs to pass through only one AND-OR sequence of switching to arrive at its destination. The added components in Figure 6-7a do not alter the signals appearing elsewhere within the otherwise basic adder. Instead, the added components merely provide a two-level AND-OR duplicate of the switching functions that otherwise involve multiple AND-OR switching levels.

Although the simultaneous addition scheme of Figure 6-7a provides the ultimate in high-speed operation, the scheme has obvious disadvantages for the addition of binary numbers having many digits. In particular, for n-digit numbers the highest order requires $n-1$ added AND devices with fan-in requirements up to $n+1$. Also, the fan-out requirements range up to n with the heaviest load appearing on the $CARRY_{in}$ line. Moreover, when the adder is used with a subtraction method involving an end-around carry (to be described later), the carry may be initiated in any order, and all orders, not just the highest order, would then need to be equipped with the full set of added equipment to achieve simultaneous generation of all sum digits with all possible number combinations.

A compromise between speed and equipment requirements can be made by grouping the orders of a parallel adder. Within each group, the generation of all sum digits is simultaneous and the $CARRY_{out}$ signal is generated with the propagation of any carry signal through at most one AND-OR sequence of switching. However, for those number pairs requiring it, the carry signal ripples through the groups sequentially. The arrangement is illustrated in Figure 6-7b for three groups each having three orders. Each rectangle represents the amount of equipment shown in Figure 6-7a, where the $CARRY_{out}$ of one group becomes the $CARRY_{in}$ of the next group to the left.

Carry Bypass

Another form of compromise between speed and equipment requirements is the carry bypass scheme illustrated in Figure 6-8.

Before explaining this scheme, the observation will be made that in the simultaneous adder of Figure 6-7a the input signals to the AND devices could have represented the OR functions $A_1 + B_1$, $A_2 + B_2$, and so on, instead of the "exclusive-OR" functions generated by the HALF-SUM signals from the upper row of half adders. In any instance where both the A_i and B_i signals to a given order are 1, the carry initiation function for that order would be duplicated, but no undesirable effects would occur.

Now in Figure 6-8, which shows five instead of just three orders, the OR devices along the top of the figure could be in addition to the conventional full adders, or each OR device could be a part of its associated full adder if the design of the full adders happens to be such that in the course of developing the sum and carry signals the $A_i + B_i$ signals are generated internally. Similarly, the OR device which generates the $CARRY_{out}$ signal need not be a separate module. Instead, the output from the AND device can be fed to an added input to the OR device that generates the carry signal within the left-hand full adder.

Alternatively, each OR device along the top of Figure 6-8 could be

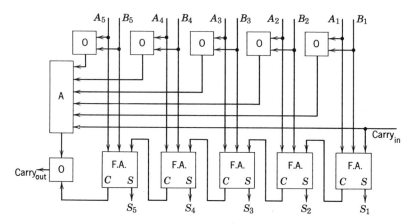

FIGURE 6-8. Carry bypass scheme.

replaced by an exclusive-OR device, either internal or external to the full adder, for reasons just explained. Thus, in effect, the network in Figure 6-8 is derived from the network in Figure 6-7a by deleting all of the AND devices except the highest-order one by which the CARRY$_{in}$ signal is bypassed around all of the adder circuits within the group. That is, if one (or both) of each A_i and B_i is 1 "and" if CARRY$_{in}$ is 1, CARRY$_{out}$ should be 1, and the network produces this result with a carry propagation path that includes only one AND-OR sequence.

Ordinarily, the carry bypass scheme of Figure 6-8 would not be useful for bypassing the carry the complete length of a parallel adder, because the bypass would then be effective only in the relatively few instances when a carry would otherwise propagate the full length of the parallel adder. Instead, the orders would be grouped, and the carry would be bypassed around the groups as appropriate for the binary number pairs being added. However, when a carry is initiated in the first order of a group and is to be propagated to the input of the last order in the next successive group, the bypass circuits afford no improvement in speed. The groups can be made smaller to improve this situation, but the increased number of groups thereby required increases the carry path length (in terms of the number of successive AND-OR functions through which the carry must pass) in those instances when the carry must be propagated through many bypass networks. An alternative to making the bypasses shorter and more numerous is to add bypass networks that overlap others, as illustrated in Figure 6-9.

In Figure 6-9a each block labeled "carry bypass" corresponds to a network of exclusive-OR devices and an AND device of the form illus-

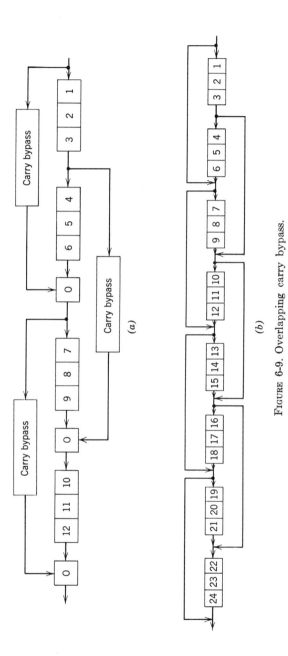

FIGURE 6-9. Overlapping carry bypass.

trated along the top of Figure 6-8. If only the two top bypass networks in Figure 6-9a were included, the carry propagation time would be reduced in any instance (any combination of binary numbers being added) for which the carry must otherwise propagate through orders 1 through 6 or through orders 7 through 12. On the other hand, if the carry is initiated in order 1 and propagates no further than the input to order 12 (whether or not a carry is initiated in order 12), the carry must be propagated through the full adders of orders 2 through 11 inclusive. This situation would represent the "worst" case. With the lower bypass path included in Figure 6-9a, the carry from order 1 would in this instance be propagated through the full adders of orders 2 and 3, then through the lower bypass network, and then through the full adders of orders 10 and 11. Because each full adder and each bypass network places one AND-OR sequence in the carry path, the lower bypass network has reduced the path length by five AND-OR sequences. However, the case just described is not the "worst" case with the lower bypass network added. The "worst" case is now represented by the instance where a carry is initiated in order 1 and affects the input to order 9. The carry must be propagated through orders 2 through 8, and none of the bypass networks offer any improvement. Nevertheless, in comparison with an adder that includes only the upper bypass networks, the lower bypass network has reduced by three the maximum number of full adders (AND-OR sequences) through which the carry must pass sequentially within two adjacent groups, and no deterioration in speed is caused when the carry must propagate through many groups.

The bypass scheme of Figure 6-9a extended to a 24-order adder is illustrated in a further simplified notation in Figure 6-9b. For adders of many orders the carry transmission path composed of the bypass networks can become of greater length (in terms of the number of successive AND-OR functions) than the path within two adjacent groups. In this particular example a "worst" instance is encountered when the carry input to order 1 affects the carry intput to order 24. The carry signal must propagate through eight successive AND-OR functions where one such function is found in each full adder or in each bypass network. One such path consists of the first three bypass networks in the top part of the figure followed by the carry networks in the full adders of orders 19 through 23 inclusive. An alternate (or in some respects a "duplicate") path of the same length is through the carry networks of orders 1, 2, and 3, then through the three bypass networks in the bottom part of the figure and finally through the carry networks of orders 22 and 23.

Many other variations in the carry bypassing pattern may be considered. Also, the simultaneous carry scheme of Figure 6-7 can be incorporated within the groups—although this amounts to bypassing all possible combinations of full adders within a group. That is, in a three-order group, for example, bypass networks can be provided for orders 1 and 2, orders 2 and 3, and orders 1, 2, and 3. The groups may be overlapped in various ways, and the groups may include differing numbers of orders.

Simultaneous Bypassing of Carriers

A variation of the bypassing scheme described in the previous section results in the simultaneous rather than sequential bypassing of carries across groups of orders. For each group a signal

$$D_i = (A_{i1} + B_{i1})(A_{i2} + B_{i2})(A_{i3} + B_{i3}) \cdots$$

is developed, where the various A_i and B_i represent the ith group digits of the two binary numbers being added. The signal D_i represents the conditions for which a carry signal should be bypassed over the ith group. As before, the exclusive-OR function may be substituted for the OR function without any alteration in the final output signals. Also as before, the OR device need not be a separate module but may actually consist only of added inputs to the OR device which generates the carry in the left-hand full adder in the group. The carry signal to the next group is designated C_{i+1} and is generated in accordance with the relationship

$$C_{i+1} = C'_{i+1} + D_i C_i + D_i D_{i-1} C_{i-1} + D_i D_{i-1} D_{i-2} C_{i-2} + \cdots$$

where $C'_{i+1} = C_{i+1}$ from a logic standpoint but where the generation of C'_{i+1} has not received any speed-up benefits from bypassing around the ith group. In other words, the remaining terms in the expression would be considered superfluous if minimization of the number of components were the only objective. In this application, if all of the conditions for bypassing in groups i, $i-1$, and $i-2$, for example, are fulfilled (that is, if D_i, D_{i-1}, and D_{i-2} are each 1), any carry input to group $i-2$ will be bypassed around all three groups by the propagation of the carry signal through a single AND-OR sequence. The equipment needed for the ith group is illustrated in Figure 6-10.

This simultaneous bypassing scheme can be combined with conventional bypassing in any of many different ways.

In general, in the design of carry speed-up networks for a parallel binary adder the objective is to obtain the highest speed possible at reasonable cost—but no advantages are gained in obtaining an addition time that is less than the cycle time of the storage unit which is supplying

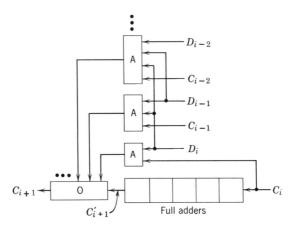

FIGURE 6-10. Simultaneous carry bypassing scheme.

successive numbers to the arithmetic unit. Therefore, the usual problem is to minimize cost (number of components or modules) for an adder that will achieve some specified time. The design must, of course, be such that all engineering requirements, notably fan-in and fan-out, of the available modules are properly observed. In arriving at a design, due recognition must be given to the fact that the switching time (and cost too, for that matter) of a module may be a function of fan-in and fan-out. Thus, a design which provides the minimum signal path length in terms of the number of successive modules may not provide the highest speed if the minimum-path design has large fan-in and fan-out requirements. In short, both the number of variations in the problem and the number of possible solutions are fantastically great.

A Carry Bypass Network for the Adder Composed of O-I Modules

Carry bypasses can be economically incorporated in the adder of Figure 6-6. Only one O-I module per bypass is required, and in a complete parallel adder the carry passes through only one O-I module for each full adder and each bypass through which it must propagate.

Because of the alternating C and $\bar{C}$ signals in successive orders, each bypass must be around an odd number of orders. The functioning of the bypass network is essentially as described for Figure 6-8 for the case where the bypass is from a point where the inverse carry appears to a point where the carry appears, but a peculiar complication is encountered where the bypass is from a carry point to an inverse-carry point.

The network and signals used when bypassing from an inverse-carry

CARRY BYPASS NETWORK FOR THE O-I MODULE ADDER   295

point to a carry point is illustrated in Figure 6-11. The single O-I module generates a bypass carry signal C_b expressed as follows.

$$C_b = \overline{\bar{C}_{in} + \bar{A}_1\bar{B}_1 + \bar{A}_2\bar{B}_2 + \bar{A}_3\bar{B}_3 + \cdots}$$
$$= C_{in}(A_1 + B_1)(A_2 + B_2)(A_3 + B_3) \cdots$$

The A_1B_1, A_2B_2, and so on, signals are already available at internal lines in the respective full adders—see Figures 6-5 and 6-6. Thus, C_b is 1 in exactly those instances where a bypass is desired. The complete carry output signal from the bypassed group is designated C_{out} in Figure 6-11 and it appears on three separate wires with C_b on one wire and the unbypassed carry, here designated C'_{out}, on the other two wires. A signal on any one of these wires represents a carry from the group, and the three-

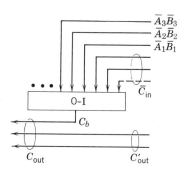

FIGURE 6-11. O-I module carry bypass network for the adder of Figure 6-6.

wire signal must therefore be combined in an "or" function. This "or" function is obtained in O-I modules of the next higher order as before, except that each module to which the carry signal is applied must have an extra input line to handle the third wire.

The opposite case, that is, when bypassing from a carry point to an inverse-carry point utilizes the same network as in Figure 6-11, but with all signals inverted. The output signal from the O-I module would then be $\bar{C}_b$, and the relationship would be as follows.

$$\bar{C}_b = \overline{C_{in} + A_1B_1 + A_2B_2 + A_3B_3 + \cdots}$$
$$= \bar{C}_{in}(\bar{A}_1 + \bar{B}_1)(\bar{A}_2 + \bar{B}_2)(\bar{A}_3 + \bar{B}_3) \cdots$$

That this function bypasses the carry is not obvious, but it can be understood by making the following observations. Signal $\bar{C}_b$ is 1 whenever C_{in} is 0 "and" at least one of each pair of A and B signals is 0. That is, $\bar{C}_b$ is 1 if no carry input signal is present to be bypassed "and" if no carry is initiated at any order within the group being bypassed. If C_{in} is 1, $\bar{C}_b$ will be 0 regardless of the binary values of the various A and B signals—as is most easily appreciated by noting the $\bar{C}_{in}$ factor in the right-hand expression for $\bar{C}_b$. Thus, the carry input signal is, in effect, bypassed to the output line in all instances including the particular instance where one and only one of each pair of A and B signals is 1 (which is the instance that the carry would otherwise need to be propagated through the full adders sequentially). The factors $(\bar{A}_1 + \bar{B}_1)$, $(\bar{A}_2 + \bar{B}_2)$, and so on, in the expression for $\bar{C}_b$ serve to ensure that $\bar{C}_b$ is 0 in all instances for which a carry is initiated in any of the orders in the group being bypassed.

The difficulty in the case of bypassing from a carry point to an inverse-carry point arises from the fact that to obtain a proper three-wire carry output signal, $\bar{C}_{out}$, the signal must be 0 on all three of the wires. A signal of 1 on any one of the wires will erroneously actuate the O-I modules in the next higher order. (This situation is to be contrasted with the previous case where a signal of 1 on any one—or more—of the three wires would constitute a proper C_{out}.) Unfortunately, the designer cannot in general be certain that the two-wire unbypassed carry signal $\bar{C}'_{out}$ will be 0 prior to the conventional propagation of the carry through the full adders. To insure that $\bar{C}'_{out}$ is 0, carries may be temporarily forced into alternate orders as shown in Figure 6-12 for a five-order adder that by-

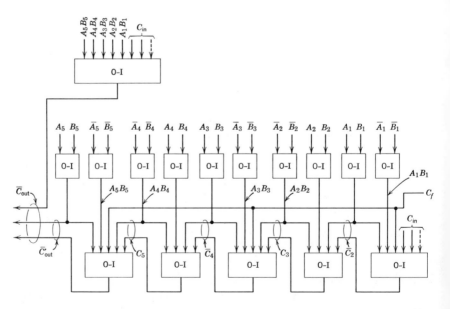

FIGURE 6-12. Bypassing from a carry point to an inverse carry point in the adder composed of O-I modules.

passes from a carry point to an inverse-carry point. The forcing signal C_f is caused to be 1 temporarily. If C_{in} is 1 and if the A and B signals are such that a carry would otherwise propagate through the adder, the changing of C_f to 0 will not cause any alterations of the signals elsewhere within the adder, and the desired bypassing will occur. In those instances where C_{in} is 0 or where the A and B signals are such that C_{out} should be 0 ($\bar{C}_{out}$ should be 1), a signal propagation action will occur when C_f is changed from 1 to 0, but the time consumed for this process is comparable with the time consumed in propagating a carry that originates within the group in conventional bypassing (or in bypassing from an inverse-carry point to a carry point in the 0-1 module adder).

The third line for $\bar{C}_{in}$ and C_{in}, shown dotted in Figures 6-11 and 6-12, respectively, indicates the carry input connections to be used when the carry input is derived from another bypassed group.

Restoring the Bypassed Carry Signals

Consider the addition of two binary numbers where a carry has been propagated over many orders. Now assume that the adder is to be used for the addition of a new pair of numbers where one number happens to be the same as before and the other number happens to differ only by a change from 1 to 0 in the digit position in which the carry originated. Although no carries are involved in developing the new sum, the correct new sum will not be generated until the previously propagated carry signals have "collapsed." In other words, a "no carry" signal must propagate through the carry network to nullify the previously generated carries. Except for presumably minor differences in the turn-on and turn-off times of individual switching devices in the carry network, the "no carry" propagation time would be the same as the carry propagation time. Thus, the time that must be allotted for the second addition is substantially the same as for the first addition.

However, the problem of restoring propagated carry signals is more complex in adders that utilize bypassed carries to obtain increased addition speed. With particular reference to Figures 6-8 and 6-9, note that even though the bypass carry signals may be changed from 1's to 0's very quickly, this changing accomplishes nothing toward resetting the carry signals generated within the full adders in the conventional carry propagation path. Therefore, the "no carry" requires the full propagation time.

One means for avoiding this problem is to cause one set of adder inputs (one of the numbers being added) to be all 0's temporarily prior to the transmission of any new number to the adder. The temporary resetting quickly causes all interorder carries to be 0's.

Alternatively, each interorder carry can be quickly forced to 0 by means of a control signal appropriately applied to the full adders. For example, in the adder composed of O-I modules as in Figures 6-5 and 6-6 a control signal similar to C_f in Figure 6-12 but applied at the inverse-carry points instead of the carry points will serve to restore all carry signals to 0's without the need for a "no carry" propagation time. (Actually, with the O-I module adder, this control signal is not needed when bypassing from a carry point to an inverse-carry point as in Figure 6-12, but the control signal is needed when bypassing from an inverse-carry point to a carry point or when no bypassing is used.)

A scheme which, in effect, bypasses "no carries" when changing adder input numbers is based on full adders which propagate both the carry and inverse-carry signals from one order to the next. Such an adder based on O-I modules is illustrated in Figure 6-13. In comparison with

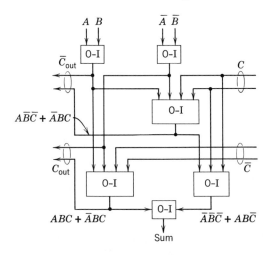

FIGURE 6-13. An O-I module full adder which utilizes both the carry and inverse carry signals.

the adder of Figure 6-6, one module per order is added to generate the inverse carry signal, but with this extra module the SUM signal can be generated with only two instead of three additional modules. Therefore, the number of modules needed to provide the inverse carry signal has not been increased.

For each full adder group for which "no carries" as well as carries are to be bypassed, an inverse carry bypass network as well as a carry bypass network is supplied. The design must be such that the bypassing signals override the propagated signals at the output of each group. The scheme is illustrated in Figure 6-14 for the O-I module adder. The network is an obvious elaboration of previously described bypass networks except for the added lines designated X and Y. When C_{out} is 1 as a result of the carry bypass signal C_b being 1, a signal is applied through line X to an added input to the O-I module which develops a signal for one of the three $\bar{C}_{out}$ lines so that this component of $\bar{C}_{out}$ is forced to 0. A second

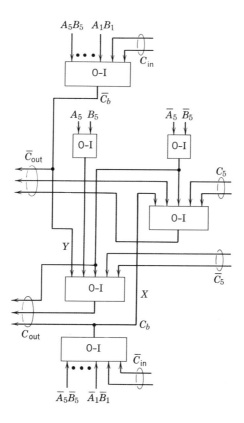

FIGURE 6-14. Carry network for one order of an O-I module adder that provides "no carry" bypass as well as carry bypass.

component of $\bar{C}_{out}$ is supplied by the O-I module to which A_5 and B_5 are supplied as inputs, and this component will be 0 in all instances for which the carry is to be bypassed. The third component is $\bar{C}_b$, and this component will be 0 for reasons explained before. Therefore, in developing $\bar{C}_{out}$ the need for the forcing signal C_f (as used in Figure 6-12) has been eliminated.

When an input number is changed so as to initiate the removal of a carry transmitted through many orders, C_{in} to the bypassed group will become 0 and will cause the inverse-carry bypass signal $\bar{C}_b$ to become 1 (except in those instances where a carry was initiated within the group). Signal $\bar{C}_b$ is transmitted along line Y to the O-I module which generates one of the three components of C_{out} and causes that component to be 0. The component of C_{out} which is equal to A_5B_5 will be 0 unless a carry originated at that point. The third component C_b will be changed to 0 directly as a result of the change in $\bar{C}_{in}$ from 0 to 1. Therefore, regardless of whether the carry input signal is changed from 0 to 1 or from 1 to 0, the effect is properly bypassed to the carry output lines without the need for separately controlled forcing signals.

The bypass scheme in Figure 6-14 has the disadvantage that the bypass signals must pass through two O-I modules in succession instead of just one module for each group bypassed. The X and Y connections have the effect of introducing the second module in the path. The SUM signals, not shown in Figure 6-14, may be generated as in Figure 6-13.

Asynchronous Operation of the Parallel Binary Adder

Thus far in the discussion of binary addition the viewpoint has been that the sum signals cannot be utilized until a fixed period of time has been allowed to elapse afer beginning the operation. This fixed time period must be great enough to allow carry generation in the "worst" case for the particular adder design chosen, and a suitable safety factor must be included to allow for variations in the operating speed of the individual circuits or modules. Speed improvements were obtained by manipulating the design so that the number of modules which need to be actuated in succession in the "worst" case (which, after manipulation, may not correspond to the same input numbers as originally) is reduced.

A quite different approach to speed improvement is to recognize that, for most number pairs to be added, the carry will not need to be propagated through all orders of the adder. In fact, for most number pairs, the number of orders through which any carry must be propagated is relatively few. Therefore, if networks are provided which will detect the completion of the carry propagation process, the sum can be utilized at carry completion time, and the system need not stand idle while waiting for a possible further propagation of carries. Of course, in those instances where a carry must be propagated through all orders of the parallel adder, the detection of carry completion provides no speed increase at all. Another point that needs emphasis is that in many applications any reduction in addition time to a duration less than the cycle time of the storage unit supplying numbers to the adder is of no value in increasing system speed.

Apart from speed considerations, some systems are through choice of the designers designed to operate asynchronously. That is, each individual operation is commenced upon the completion of the previous operation as opposed to performing operations at a more or less uniform rate as controlled by "clock" signals. In asynchronous systems a need exists to determine, in particular, when an addition operation has been completed. For most conventional full adders the need would be to sense a condition described in words such as: "no more carries are being generated." The development of networks to detect this condition is, at best, awkward.

Instead, the usual approach is to provide a carry and "no carry" signal (which is not quite the same as the "inverse of the carry," as will be explained) between each successive pair of orders and to design the networks so that, at the output of any given order, the carry and "no carry" signals are both 0 until one or the other of the carry and "no carry" signals is 1 at the input to that order. To initiate the operation of addition, the "no carry" input line to the lowest order is caused to be 1. Either the carry output or the "no carry" output of the lowest order then becomes 1 in accordance with the digits being added. The networks for the second and successive orders are then actuated analogously in succession. The addition is complete when the signal on either the carry output lines or the "no carry" output line from the highest order becomes 1, and an OR device is used to detect this condition (although a carry of 1 from the highest order may additionally indicate an overflow).

The switching functions required in the carry networks of a full adder operating asynchronously as described in the previous paragraph are illustrated in Figure 6-15. The carry and "no carry" signals (which in this type of adder are not the Boolean inverses of each other) for any one order may be expressed as

$$\text{CARRY} = AB\bar{C} + BC + AC$$
$$\text{NO CARRY} = \bar{A}\bar{B}C + \bar{B}\bar{C} + \bar{A}\bar{C}$$

where C and $\bar{C}$ here represent the carry and "no carry" input signals, respectively. This notation is a slight corruption of conventional Boolean notation, because C and $\bar{C}$ can be 0 simultaneously with this form of asynchronous operation, and this situation allows CARRY and NO CARRY to be 0 at the same time.

An adder having carry networks as in Figure 6-15 may be called a "full length" asynchronous adder, because a signal must be propagated

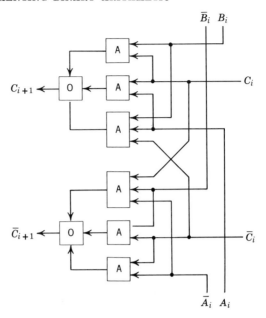

FIGURE 6-15. Carry technique for one order of a "full length" asynchronous parallel adder.

sequentially through all orders regardless of the carries encountered along the way. The signal propagation path shifts back and forth between the carry and "no carry" portions of the carry networks in accordance with the digits of the two binary numbers being added.

Carry Bypassing in the Asynchronous Adder

Carry bypassing networks can be installed in an asynchronous adder in essentially the same way as was described for synchronous adders, but the beneficial effects are much less even when both the carry and "no carry" signals are bypassed. For example, in the case where the numbers being added are such that carries are initiated in alternate orders but with no carry being propagated beyond the next successive order from which it originated, the propagating signal would shift back and forth between the carry and "no carry" lines in alternate orders throughout the length of the adder, and no reasonable pattern of carry bypassing would result in a speed improvement. Moreover, even in those instances where a speed improvement does result, means must be included in the network to indicate when all carry propagation within the bypassed groups of full adders is complete. Rather than pursue this subject as a problem in carry bypassing, a somewhat different method of improving the speed of asynchronous parallel adders will be described in the next section.

Asynchronous Adders Equipped with Completion Sensing Networks

Instead of propagating a signal the full length of an asynchronous adder, an alternative technique is to allow carry and "no carry" signals to be injected into the propagation path at whatever orders they originate in accordance with the binary numbers being added. Specifically, a carry signal is injected at each order where the A and B digits are both 1, and a "no carry" signal is injected where they are both 0. Because the numbers to be added are not known, means must be provided to sense that a signal is present on one or the other of the carry and "no carry" lines from each order. The operation is known to be complete when signals are sensed at all orders.

Before commencing the addition operation, carry and "no carry" signals that would result from the propagation of signals originating in lower orders must be made 0. Otherwise, extraneous signals on the various carry and "no carry" lines would generate a premature completion signal. To hold the propagated signals at 0 until the intended commencing of the addition operation, a control signal G is added to the carry and "no carry" functions as follows.

$$\text{CARRY} = AB + BCG + ACG$$
$$\text{NO CARRY} = \bar{A}\bar{B} + \bar{B}\bar{C}G + \bar{A}\bar{C}G$$

When G is 0, a CARRY or NO CARRY signal can appear on the corresponding output line of a full adder if the A and B digits for that order are both 1 or both 0, respectively, but if the output carry status is dependent on the input carry status, both the CARRY and NO CARRY output signals from that order will be 0. When G is changed from 0 to 1, signal propagation through the orders is allowed to proceed.

The switching functions needed in the carry networks for this form of asynchronous adder with completion detection are illustrated in Figure 6-16. Three orders, arbitrarily numbered 5, 6, and 7 are shown. The AND device at the bottom of the figure senses when one or the other of the carry and "no carry" signals from each order is 1. Thus, the output from this AND device is 1 when addition is complete.

The time required for an addition operation is determined by the longest "chain" of successive carry or "no carry" signals. For example, in the addition of two 40-digit binary numbers, a carry might be initiated

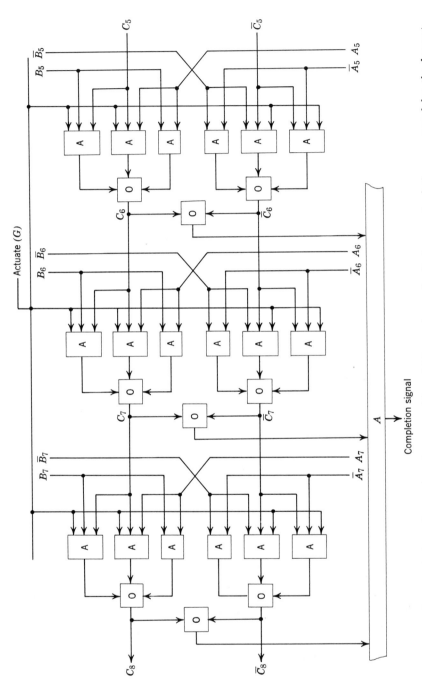

FIGURE 6-16. Carry technique for an asynchronous adder which requires only that amount of time consumed by the longest carry or "no carry" propagation sequence.

in the second order from which it is propagated to the sixth order (i.e., for the third, fourth, and fifth orders one of the A and B input digits would be 1 and the other would be 0, but for the sixth order both the A and B digits would be 0). Also, in the addition of the same two 40-digit numbers another carry might at the same time be initiated in the 20th order from which it would be propagated to the 28th order. If these two carries represented the only signals to be propagated, addition time would be dependent on the carry originating in the 20th order because the propagation path is longer for this carry. However, if in this same example one of the A and B signals is 1 with the other 0 for each of the seventh to 18th orders inclusive, the "no carry" signal originating in the sixth order would need to be propagated through the seventh to 18th orders inclusive, and this "no carry" path is longer than either of the carry paths. Therefore, in this example, addition time would be dependent on the propagation of the "no carry" signal.

Empirical studies have shown that for random n-digit binary numbers the maximum number of successive carry signals or successive "no carry" signals is (on the average for many number pairs) closely approximated by the expression

$$\log_2 (5n/4) \simeq 0.32 + \log_2 n$$

The "no carry" input signal is assumed to have been applied to the lowest order prior to commencing the addition operation—a reasonable assumption in most applications. For an example, when adding 32-digit numbers the average maximum sequence is about 5.3 carry or "no carry" digits. For 64-digit numbers the figure is about 6.3, which indicates a speed increase greater than a factor of ten in applications where speed is wholly dependent on carry path length. (Incidentally, the effect of the need to propagate "no carries" as well as carries is not great. With carries only, the number of successive signals would be slightly less than $\log_2 n$ instead of slightly greater than $\log_2 n$.)

Unfortunately from the standpoint of estimating average addition time, the numbers entering into the computations are not random in many applications. In some programs the accumulator may be frequently tested for zero contents by a process that involves transmitting a carry (or a "borrow" to be discussed later in connection with subtraction) through all orders. Detection of carry completion might then provide negligible speed improvement. In other programs most of the numbers may represent small quantities and therefore have relatively few 1's with a corresponding reduction in average carry propagation length and a corresponding great speed improvement in comparison with the "full length" asynchronous adder.

306 IMPLEMENTING BINARY ARITHMETIC

Bypassing of carries and "no carries" can be included as a feature in this form of asynchronous adder in essentially the same manner as described for synchronous adders. The bypssing would improve speed in those instances where the carry would otherwise need to propagate through many orders, but presumably in most applications these instances would be relatively few, so the average speed improvement would not be great. Note that this situation is quite different from that encountered with synchronous adders where all additions consume the amount of time needed for the "worst" case.

The Detection of Carry Completion in the O-I Module Adder

The technique of carry completion detection of Figure 6-16 can be applied to the O-I module adder as illustrated for one order in Figure 6-17. The inverse of the actuate signal, $\bar{G}$, is applied to two of the O-I modules as shown to hold 0 signals on the carry and "no carry" lines corresponding to propagated signals as long as G is 0. An O-I module which, in effect, generates $\overline{C_i + \bar{C}_i}$ is added to each order. When G is 0, the output from this module will be 0 if either a carry or a "no carry" is initiated in this order, but will be 1 otherwise. When G is changed to 1, the output from this module will become 1 (if not already 1) when a signal is received on the carry input or "no carry" input line to this order. The O-I module at the bottom of Figure 6-17 combines the signals from all orders to produce the signal

$$\overline{\overline{C_1 + \bar{C}_1} + \overline{C_2 + \bar{C}_2} + \overline{C_3 + \bar{C}_3} + \cdots}$$
$$= (C_1 + \bar{C}_1)(C_2 + \bar{C}_2)(C_3 + \bar{C}_3) \cdots$$

which is the desired carry completion signal. Recall that $\bar{C}_i$ is not the inverse of C_i in the modified Boolean notation used in this application.

Because the individual carry and "no carry" signals from the various full adders must pass through two O-I modules to reach the "completion signal" line, the generation of the completion signal will occur a short time after the actual completion of the addition operation when all full adders are equipped as illustrated in Figure 6-17. In other words, some time is consumed in doing nothing but generating the completion signal, and this time could conceivably be used for part of the addition operation. This fact can be utilized, not only to advance the generation of the completion signal slightly in some instances, but also to decrease the amount of equipment needed to generate the completion signal. The scheme is to equip only alternate orders or perhaps only every third order with the carry sensing networks and $\bar{G}$ input line of Figure 6-17.

DETECTION OF CARRY COMPLETION IN THE O-I MODULE ADDER 307

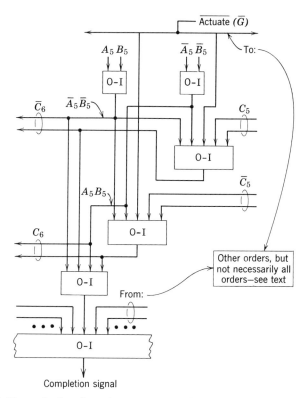

FIGURE 6-17. Network for detecting carry completion in the O-I module asynchronous adder.

Only every second one or every third one of the $(C_i + \bar{C}_i)$ signals would be utilized in sensing the completion of the addition operation, and the operation might not be quite complete when, of those signals utilized, all are 1's. However, no harm would be done because of the need to pass the $(C_i + \bar{C}_i)$ signals through the above-described O-I modules to generate the completion signal.

The discussion in the previous paragraph is over-simplified in that it ignores the O-I modules needed to develop the sum signals in each order, and it also ignores any added O-I modules the completion signal might be passed through in the utilization of that signal. Thus, an exact determination of the speed improvement and equipment saving must be made on the basis of the details of the peripheral networks pertaining to the parallel adder. The improvement is realizable only if the completion signal path is longer (in terms of the number of O-I modules through which it must pass) than the sum signal path. If equipment savings is the major consideration, the number of full adders to be equipped as illustrated in Figure 6-17 can be reduced to every fourth one or every fifth one or fewer by inserting an appropriate number of one-input O-I modules (inverters) in the completion signal path.

Binary Subtraction

In any given order in the subtraction of one binary number from another, $0 - 0 = 0$, $1 - 0 = 1$, and $1 - 1 = 0$ as is more or less obvious. For $0 - 1$, the usual viewpoint is that the difference is 1 but that a "borrow" digit is obtained from the next higher order to produce, in binary notation, $10 - 1 = 1$. The "borrow" has the effect of subtracting a 1 from the next higher order, where this 1 has twice the weight of a 1 in the given order. Although the borrowing seems to cause the transmission of a digit from the relatively higher order to the given order to allow the subtraction to take place without creating a negative quantity or a negative digit, the implementation of subtraction is such that the direction of signal transmission is toward the higher order as in addition. The borrow signal might appropriately be termed the "subtraction carry" or something similar.

With A the minuend and B the subtrahend as in $A - B$, the amount of equipment needed to subtract two digits as explained in the previous paragraph is called a "half subtractor." In Boolean notation the two signals to be generated are

$$\text{HALF-DIFFERENCE} = A\bar{B} + \bar{A}B$$
$$\text{HALF-BORROW} = \bar{A}B$$

The HALF-BORROW signal (digit) is subtracted from the A digit in the next higher order in the same manner as is the B digit of that order, and a second half subtractor may be used for the purpose. In any given order, the subtraction of the B digit or the subtraction of the borrow from the next lower order may produce a borrow to be transmitted to the next higher order. However, as may be visualized by considering all possible digit combinations, no more than one of the two HALF-BORROW signals in any given order can be 1 for a given pair of numbers A and B. Therefore, the two HALF-BORROW signals can be combined in an OR device to produce the borrow signal to be transmitted to the next higher order. The resulting array of equipment is shown in Figure 6-18a where H.S. stands for "half subtractor." The

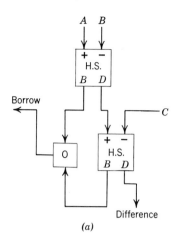
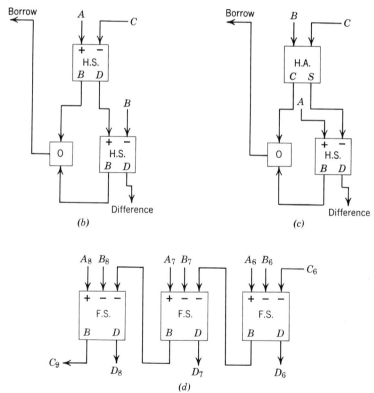

FIGURE 6-18. Binary subtraction arrangements.

output signals B and D here represent the borrow and difference, respectively. Contrary to a half adder, the two input signal terminals on a half subtractor are not interchangeable. In the figure they are designated $+$ and $-$ for the minuend and subtrahend digits, respectively.

Figure 6-18b shows the same network as in a, but the B and C input connections are interchanged, where C now represents the borrow from the next lower order. The net result is unchanged because the B and C digits have the same weight, and it is immaterial as to which digit is subtracted first. Figure 6-18c shows another subtraction scheme. Here, input digits B and C are added in a half adder with the half-sum subtracted from A. The borrow signal to the next higher order is formed by combining the half-carry with the half-borrow in an OR device. For the half-carry to be 1, both the B and C input digits must be 1, but in this instance the half-sum will be 0. Therefore, only one of the two signals to the OR device can be 1 for any given pair of input digits A and B.

Each of the networks in Figures 6-18a, b, and c is equivalent to a full subtractor, where a full subtractor is any switching network which has input signals corresponding to the A and B digits of one order and the borrow digit C from the next lower order and which generates output signals corresponding to the difference digit and the borrow to the next higher order. The difference is 1 if A is 1 "and" if the digits being subtracted (B and C) are both 0's "or" both 1's. Also, the difference is 1 if A is 0 "and" if one but not both of the digits being subtracted is 1. The borrow is 1 if A is 0 "and" if either or both of B and C is 1. Also, the borrow is 1 if both B "and" C are 1, regardless of A. In Boolean notation these signals are as follows.

$$\text{DIFFERENCE} = A\bar{B}\bar{C} + ABC + \bar{A}B\bar{C} + \bar{A}\bar{B}C$$
$$\text{BORROW} = \bar{A}B + \bar{A}C + BC$$

Note that the DIFFERENCE in a full subtractor is exactly the same as the SUM in a full adder. The BORROW differs from the CARRY only in the inversion of the A signal.

A parallel subtractor can be formed as shown in Figure 6-18d for three orders arbitrarily designated 6, 7, and 8. Each block labeled F.S. corresponds to a full subtractor which may be formed as in parts a, b, and c of this figure, or it may be formed as any three-input, two-output network which generates the switching functions specified in the previous paragraph. The input signals to the two negative ($-$) terminals on a full adder are interchangeable with each other but not with the signal to the positive ($+$) terminal.

A Parallel Binary Adder-Subtractor

Most computers must be capable of performing either addition or subtraction where the particular operation to take place at any given time is under the control of a special signal applied to a suitable input terminal to the arithmetic unit as a whole. Figure 6-19a illustrates one method of performing either operation in a controlled manner. The "sum-difference generator" generates an output signal of 1 when any one of the input signals is 1 or when all three of the input signals are 1's. The "carry-borrow generator" generates an output signal of 1 when any two or all three of the input signals are 1's. For addition or subtraction, control signal G_a or G_s, respectively, is made 1 with the other 0. By this means, either A or $\bar{A}$, respectively, reaches the "carry-borrow generator" so that the resultant carry-borrow signal to the next higher order is as desired for addition or subtraction, respectively—see previous section. In effect, the "sum-difference generator" and the "carry-borrow generator" together form either a full adder or a full subtractor in a selectively controlled manner.

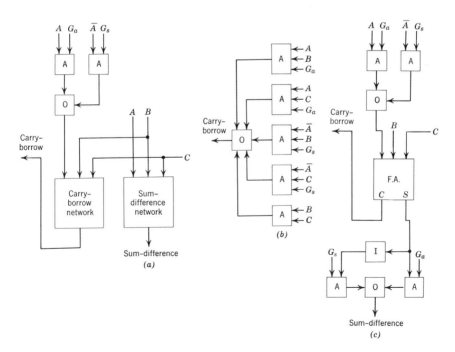

FIGURE 6-19. Binary addition-subtraction schemes: $A + B$ or $A - B$.

312 IMPLEMENTING BINARY ARITHMETIC

Figure 6-19b shows a variation of the "carry-borrow generator" portion of the network in a and illustrates the nature of the carry-borrow signal more straightforwardly. If $G_a = 1$ and $G_s = 0$, the signal is $AB + AC + BC$ as required for addition. If $G_a = 0$ and $G_s = 1$, the signal is $\bar{A}B + \bar{A}C + BC$ as required for subtraction.

Figure 6-19c illustrates how an unmodified adder can be used to achieve subtraction. When subtracting, $\bar{A}$ is applied to the full adder. The carry-borrow signal is affected as desired, but the sum-difference signal is affected in a manner which is not desired. As can be shown readily, the inversion of any one of the three input signals to a full adder causes the sum output signal to be inverted (but otherwise correct). Therefore, if A is inverted to obtain subtraction, the sum output signal must also be inverted to obtain the correct difference. The switching devices controlled by G_a and G_s at the bottom of Figure 6-19c achieve the necessary inversion of the sum signal to produce the correct difference digit when subtracting.

Although the subtraction method described in this section is important from the standpoint of binary addition-subtraction principles, the inversion of the subtrahend as to be described in a later section generally results in a somewhat more economical system than the inversion of the minuend as described here.

Negative Numbers

When manually subtracting B from A with $A < B$ and where A and B are multidigit numbers, a person would ordinarily notice that B is the larger and would then subtract A from B. The difference would be defined as a negative number and would be so designated by means of a minus sign. However, in a machine, facilities are not usually included for determining which of the two numbers is the larger before performing a subtraction. Instead, the usual machine procedure is to subtract without regard to relative magnitudes. If the subtrahend B happens to be larger than the minuend A, the result is a set of digits corresponding to the complement (commonly termed the "2's complement") of the magnitude of the difference. The complement of a number X may be defined as $2^N - X$, where N is an integer such that 2^{N-1} is equal to the weight of a 1 in the highest-order position in X. That is, for example, if $X = xxxx.xxx$, where each x may be either 0 or 1, the complement of X would be $10000.000 - xxxx.xxx$. Ordinarily, the complement of zero is assumed to be zero (not 2^N) because the digit in the 2^N position is physically nonexistent.

As an illustration, if $A = 1001010$ and $B = 1011100$ the subtraction of B from A produces

$$\begin{array}{r} 1001010 \\ \text{subtract:}\ \underline{1011100} \\ 1101110 \quad \text{difference} \end{array}$$

That the difference is negative (and therefore in complement form) will be signaled by the occurrence of a borrow from the highest order in the subtractor. In all instances where $A \geq B$, no such borrow will occur, and in these instances the difference is positive (or zero) and the result will be in "true" form.

To convert a number from complement form to true form the number may be subtracted from zero. In the illustration of the previous paragraph the negative difference is converted to true form as follows.

$$\begin{array}{r} 0000000 \\ \text{subtract:}\ \underline{1101110} \\ 0010010 \quad \text{difference} \end{array}$$

This difference represents the magnitude of $B - A$ in true form. Note that, except when a number being converted is zero, the subtraction from zero produces a borrow from the highest order. This borrow may be ignored because it would serve no purpose other than to change from 1 to 0 the digit in the 2^N position, which is physically nonexistent. In other words, subtracting from zero and ignoring the borrow has essentially the same effect as subtracting from 2^N.

If a number is initially in true form, a subtraction from zero will cause it to appear in complement form. Thus, repeated subtractions from zero will cause the representation to change back and forth between true and complement forms.

An alternative method of converting a number between true and complement forms is to sense the lowest-order 1 in the number being converted and then invert all higher-order digits. The lowest-order 1 and all lower-order digits (which will all be 0's) are left unchanged. Mathematically, this method is the same as above, but the networks for implementing it may be quite different.

In a machine, the signal that is used to indicate that a number is negative is a binary signal of exactly the same nature as any other binary signal. No inherent reason is encountered why this signal should not correspond to 1 and 0 symbols instead of $+$ and $-$ symbols. In fact, in a subtraction operation, the switching functions needed to determine the positive-negative status (sign) of the difference are the same as the switching functions needed for determining a difference digit. Therefore, a common practice is to add one digit to the left-hand end of each number to indicate sign. In "appearance" this added digit is the same as would be obtained by including a digit in the 2^N position and by assigning a weight of 2^N to the added digit, but here the added digit indicates sign and nothing else.

Thus, in the previous illustration the subtraction of B from A with both A and B positive would be as follows.

```
                    sign
                 0 1001010
      subtract:  0 1011100
                 1 1101110   difference
```

A full subtractor would be used for the sign position. In this instance, the A and B input signals to the sign position would both be 0's—but either A or B, or both, may be negative numbers in which case the corresponding digits in the sign position would be 1's. By experimenting with various magnitudes and signs for A and B, the reader may establish that this subtraction scheme functions correctly when either or both of A and B are negative (and in complement form).

No binary point was indicated in the above examples. The point can be at either end of the numbers for pure integers or pure fractions or it can be between any two adjacent orders. Alternatively, the point can be any number of orders to the right or to the left of the digits for each number, in which case a corresponding number of 0's is assumed to be present in each number although these 0's are not indicated physically in a machine. The subtraction method and the sign-indicating scheme are not dependent on the location of the point.

The Addition of Negative Numbers

Attention is returned temporarily to the operation of addition. Assume that negative numbers are represented in complement form as described. If one of the two numbers, say B, entering into addition operation is negative, the adder will receive A and $2^N - B$ as inputs. The sum will be $A + 2^N - B$. If $A > B$, the sum may be conveniently written as $2^N + (A - B)$ which is the correct positive difference in true form, because the 1 in the 2^N position is physically nonexistent and may be ignored. If $A < B$, the sum when written as $2^N - (B - A)$ is seen to represent the correct negative difference in complement form. These results are illustrated by the following examples.

$$\text{add:} \quad \begin{array}{c} \text{sign} \\ 0\ 1000 \\ \underline{1\ 1011} \\ 0\ 0011 \end{array} \quad \begin{array}{c} (+8) \\ + (-5) \\ \hline (+3) \end{array} \qquad \text{add:} \quad \begin{array}{c} \text{sign} \\ 1\ 1000 \\ \underline{0\ 0101} \\ 1\ 1101 \end{array} \quad \begin{array}{c} (-8) \\ + (+5) \\ \hline (-3) \end{array}$$

If A and B are both positive, a carry from the highest-order numerical position (not the sign position) does not indicate that a 1 should be added to the sign position to change sign. Instead, the indication is that the magnitude of the sum is 2^N or greater, which is an "overflow." When both A and B are negative, the absence of a carry indicates an overflow. Although the full adder for the sign position may be the same as the full adder for any other position, the adder as a whole must, in most applications, be additionally equipped to detect the condition $\bar{A}\bar{B}C + AB\bar{C}$ for the sign-position input signals so that an appropriate overflow signal may be generated.

Subtraction by the Addition of Complements

In manual arithmetic, the subtraction of a positive number and the addition of a negative number are virtually identical operations. However, in a machine the two operations are often distinctly different. A machine may add or subtract according to what it is instructed to do by the program. Although the signs of the numbers influence the results, the signs do not influence the switching functions performed by the machine. Thus, the addition of negative numbers as described in the previous section is not, by itself, a subtraction operation.

On the other hand, an actual subtraction operation may be derived easily from the scheme of using complements for the addition of negative numbers. The subtraction of a positive number is accomplished by converting it to complement form and adding. The subtraction of a negative number is accomplished by converting it to true form and adding. In either case the machine converts the number to opposite form and adds. Because the conversion process is the same in either case, the machine need not sense the sign of the number being subtracted.

Either of the two conversion methods described previously will produce the correct results, but a particularly convenient and economical scheme is to invert all digits of the number being converted (the subtrahend) and to add 1 in the lowest-order position of the adder. The inversion of all digits of a number X has the effect of subtracting X from

$111\cdots111 = 2^N - 1$. The addition of 1 in the lowest order produces a net result of $2^N - 1 - X + 1 = 2^N - 1$, the complement of X.

This subtraction method differs physically from the previously described method in that, instead of inverting the minued in only the carry part of each full adder, the subtrahend is inverted in both the carry and sum parts (of which the sum part now generates the difference), and a 1 is added in the lowest order.

Subtraction by the addition of complements is illustrated in Figure 6-20, which is to be compared with Figure 6-19a and c.

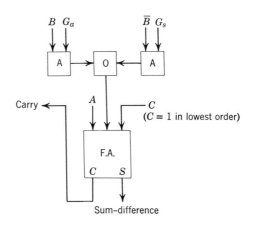

FIGURE 6-20. Subtraction by the addition of complements: $A + B$ or $A - B$.

1's Complements

The quantity $2^N - 1 - X$, obtained by inverting all digits of X as explained in the previous section, is called the "1's complement" of X. Negative numbers can be represented in 1's complement form rather than 2's complement form with the advantage being simplicity in the conversion process. Also, some variations of multiplication and division with negative factors tend to be somewhat simpler with 1's complements than with 2's complements.

When adding two numbers A and B where one, say B, is negative and in 1's complement form, the sum is $A + 2^N - 1 - B$. If $A < B$, the sum will be negative and should appear in 1's complement form, and this result will be obtained directly as seen by rewriting the sum as $2^N - 1 - (B - A)$. However, if $A > B$, the sum will be positive and should appear in true form. The true form can be obtained by subtracting 2^N and adding 1 to obtain $A - B$. Because $A - B$ is a positive quantity in this case, the addition process will produce a carry into the 2^N position, which is physically nonexistent. Therefore, the lack of a physical device to store the 1 in the 2^N position has the effect of subtracting 2^N, and by adding the carry in "end-around" fashion to the lowest-order position, the effect is to add 1. This and other aspects of the addition of negative numbers can be appreciated by studying the following examples, where each negative number appears in 1's complement form.

```
         sign                              sign
         0 1001   (+9)                     0 1001   ( +9)
add:     1 1010   (-5)            add:     1 0011   (-12)
         0 0011                            1 1100   ( -3)
add:          1
         0 0100   (+4)


         sign                              sign
         1 1001   (-6)                     0 1001   (  9)
add:     1 1100   (-3)            add:     1 0110   (-9)
         1 0101                            1 1111   (-0)
add:          1
         1 1010   (-9)
```

The end-around carry, when it occurs, need not be added in a separate step as implied in the above notation but it may be added in parallel fashion in the same manner as used for carries in other orders. The carry may originate in any order and may be propagated around the end to the next lower order so that, in effect, propagation through all orders of the parallel adder may be required in many more instances than when 2's complements are used. This fact must be given due recognition when carry bypass networks are incorporated in the adder. (No combination of numbers being added can cause the carry to propagate more than once around the loop formed by the end-around carry connection.) The end-around carry signal may be obtained from the sign position instead of the highest-order numerical position, and this variation may be a convenience in some applications although it lengthens the loop by one order.

A design nuisance encountered with the 1's complement scheme is illustrated by the last example shown above. When $A = B$ with one number negative, the indicated sum is a negative zero, whereas zero is usually viewed as being in the set of positive numbers. This situation can be avoided by representing positive numbers in 1's complement form and negative numbers in true form. Having positive numbers in complement form may seem "unnatural" and therefore objectionable, but when the numbers are transmitted to an output device for observation by a person the numbers are usually converted to the familiar magnitude-and-sign representation anyway, so the objection is not necessarily serious. Appropriate alterations must, of course, be made in the networks for other operations such as multiplication and division.

Subtraction with 1's complement notation is accomplished by inverting all digits (including the sign digit) of the number to be subtracted, and the inverted number is added with the same end-around carry.

Serial Addition-Subtraction

When the digits of the binary numbers are transmitted serially, low order first, the networks required for addition and subtraction reduce to a single full adder or subtractor, as the case may be, plus a device for storing the carry as required for combining it with the next pair of digits. The storage device can be a delay line, but is more likely to be a J-K flip-flop as illustrated in Figure 6-21 for addition. The C and $\bar{C}$ output signals from the full adder are applied to the J and K input terminals, respectively, of the flip-flop, and a clock pulse is applied to the C input terminal (where C here stands for "clock," not "carry") at substantially the same time that the binary number input signals are changed from one digit value to the next—or slightly prior to this time. The sum appears serially on the S output line from the full adder.

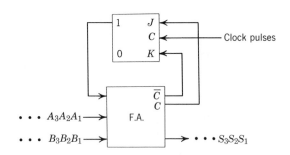

FIGURE 6-21. Serial addition network.

During the short periods of time when the input signals to the full adder are changing, the sum output signal may be temporarily incorrect. In applications where this situation is objectionable, the sum output line can be "sampled" at times when the full adder input signals are stable, and the sampled signals may be stored temporarily in a second flip-flop. A simple sampling technique is to employ a J-K flip-flop actuated by S and $\bar{S}$ signals applied to the J and K input lines, respectively, with the clock input actuated at the same times and by the same signals as used for the carry storage flip-flop.

Except for the slight delay in the full adder, the time required for an addition-subtraction operation is equal to the time required for the serial transmission of a number. Carry bypass or carry completion techniques are not applicable.

Compromises Between Serial and Parallel Operation

In some applications a compromise between the high speed but expensive parallel mode of operation and the economical but low speed serial mode may be appropriate. The digits of the numbers to be added or subtracted can be grouped with two, three, or any suitable number of digits in each group. The digits within each group are transmitted in parallel and the groups are transmitted serially. A parallel adder having a number of positions equal to the number of digits in each group is provided, and carry bypass or other features are included as necessary to generate a group of sum digits in the time that elapses between the transmission of successive groups. A carry storage device is included to allow the addition of the carry derived from the highest order of any given group into the lowest order of the next group.

Binary Multiplication

Binary multiplication can be accomplished in essentially the same manner as used for familiar manual methods of decimal multiplication, but the multiplication table for individual binary digits is so simple as to be almost trivial. For individual digits the table would consist of only four entries derived from the relationships $(0)(0) = 0$, $(0)(1) = 0$, $(1)(0) = 0$, and $(1)(1) = 1$. More commonly, binary multiplication is viewed as making each partial product equal to the multiplicand or zero according to whether the corresponding multiplier digit is 1 or 0, respectively. The partial products are positioned according

320   IMPLEMENTING BINARY ARITHMETIC

to the weights of the multiplier digits and added. For example, the multiplication of 1101 by 1011 would be accomplished as follows.

```
            1 1 0 1   multiplicand
            1 0 1 1   multiplier
            ───────
            1 1 0 1
          1 1 0 1
        0 0 0 0               partial products
      1 1 0 1
      ───────────────
      1 0 0 0 1 1 1 1   product
```

When implementing multiplication in a computer the addition of the partial products would not ordinarily be done simultaneously. Instead, the partial products are usually "accumulated" one at a time in essentially the same manner as the sum of any other set of numbers would be accumulated, as described previously for binary addition.

With n_1 digits in the multiplicand and n_2 digits in the multiplier, the product may have up to $n_1 + n_2$ digits—a result which is true for numbers in any radix. In most applications $n_1 = n_2 = n$, so the usual viewpoint is that the multiplication of n-digit numbers produces $2n$-digit products.

In applications where the numbers represent integers (binary point assumed to be at the right-hand end of each number) and where only one number may be used for the representation of an integer, a product having a nonzero digit in any of the n higher-order digits of the $2n$-digit product would be considered an overflow. Networks would need to be included to detect this condition. With or without an overflow indication, if some other number is subsequently added to the product through the action of a subsequent instruction, the number would be added to the n low-order digits of the product. A machine which functions in this way is said to be an "integral" machine.

On the other hand, if each number represents a fraction (binary point assumed to be at the left-hand end of each number), the product of 0.1101 and 0.1011, for example, would be 0.10001111. The concept of "overflow" would not be encountered. The n right-hand digits of the product would have less weight than any of the digits in either of the two factors being multiplied. If the two fractions being multiplied together are not exact but instead are approximations such as result from physical measurements or such as result from the rounding of previously obtained results, the significance of the right-hand n digits of the product is limited, and the usual procedure is to round the product to the nearest n digits to produce, in this example, a product of 0.1001.

The rounded product is most likely to be the most accurate n-digit representation of the exact product, but it is not necessarily the most accurate representation. The two original fractions of the previous example may have represented quantities ranging from 0.11001 to 0.11010111··· for the multiplicand and from 0.10101 to 0.10110111··· for the multiplier. If the true quantities happen to be near the lower end of the range for both factors, the most accurate four-digit product would be 0.1000, but if they happen to be near the higher end of the range, the most accurate four-digit product would be 0.1010. In short, the accuracy of an n-digit product is less than the accuracy of the n-digit factors. If the product is then used as a factor in a subsequent multiplication, the accuracy is further impaired.

Another accuracy loss is encountered when the factors in the multiplication are 0.1000 or greater but not great enough to cause the product to be greater than 0.01110111. In such instances the round-off procedure results in a product no greater than 0.0111, which in effect has only three ($n - 1$ in general) significant digits.

Regardless of accuracy loss, a number subsequently added to a product of fractions should be added to the n left-hand digits of the product. A machine which functions in this way is said to be a "fractional" machine.

Ordinarily, an integral machine is designed to "save" only the n right-hand digits of product—an automatic overflow procedure may be included. A fractional machine is designed to save only the n left-hand digits—an automatic round-off procedure may be included. However, many alternative designs are possible. The machine may be designed so that the programmer may (or must) also specify the overflow and round-off procedures to be used, all of which may be dependent, either through built-in equipment or through programming, upon the digits of the resulting product. Another alternative is to employ "floating point" techniques, to be described later.

Switching-Network Multiplier

An ordinary combinatorial switching network as discussed in a previous chapter can be used to form a multiplier. The two n-digit numbers A and B supply $2n$ input signals and the $2n$-digit product appears on $2n$ output lines. However, the conventional network design techniques for a $2n$-input, $2n$-output network are generally impractical for any value of n encountered in practical applications. On the other hand, a properly functioning network can be designed quite easily by using a set of parallel binary adders connected to add partial products in

322 IMPLEMENTING BINARY ARITHMETIC

the manner indicated for binary multiplication. The resulting switching-network multiplier is shown in Figure 6-22. The digits of the multiplicand are $A_4A_3A_2A_1$, and the various AND devices controlled by the multiplier digits B_1, B_2, B_3, and B_4 cause either the multiplicand or zero to be transmitted to the respective adder input terminals.

The time required for multiplication with this network is determined by the time required for a carry initiated in the right-hand end of the

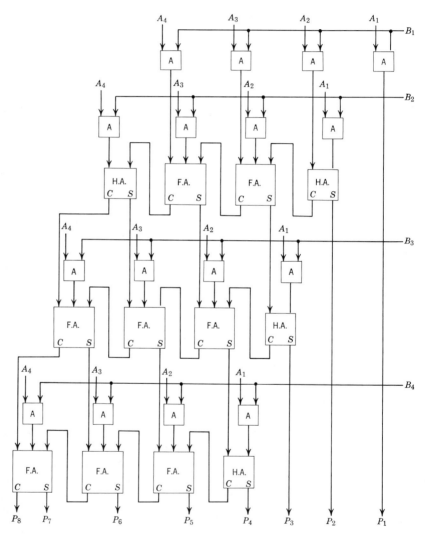

FIGURE 6-22. Switching network multiplier.

top adder to propagate to the P_8 output line. The path through which the carry must propagate depends on the digits of A and B. One such path is through the half and full adders of the top parallel adder and then through the left-hand full adder of each other parallel adder. Other paths may be determined readily by inspection of Figure 6-22. Although each possible path is of substantially the same length in terms of the number of switching modules through which the signal must propagate, the multiplicity of paths seems to cause the design of useful carry bypass networks to be quite impractical.

Some improvement in speed can be obtained by employing the "carry save" technique described in a later section. The array of half and full adders is modified so that the carries in any given row of half or full adders are not propagated to higher orders of that row but instead are added into the appropriate orders of full adders in the next lower row. A final parallel adder is included to add any remaining carries in a conventional fashion. Multiplication time is determined by the time required for a signal to propagate sequentially through a "column" of full and half adders plus the time required for operation of the final parallel adder, where the final parallel adder may be equipped with carry bypassing features such as have already been described.

Even without carry bypass networks the switching-network multiplier has extremely high speed in comparison with nearly all alternative multiplication implementations. Moreover, no sequencing or control networks are required. The obvious disadvantage of this multiplication method is the rather large amount of switching equipment required for n in the range of interest for most applications—n^2 AND devices and the equivalent of nearly $n(n-1)$ full adders. In fact, the equipment requirements have in the past proved to be a crucial deterrent to the adoption of this scheme of multiplication, although with large scale integration (LSI) of transistor circuits, the approach is reasonable.

Many variations in the network can be devised. The partial products can be added in accordance with different patterns, the carries can be intermingled among the adders in various ways, or the additions may be on columns of binary digits in the partial product array with carries handled appropriately. With some of the variations, carry bypassing can be combined with the carry save technique (to be described in a subsequent section) to obtain improved speed. A presentation of any of the countless variations that might be useful will not be attempted here, but reference is made to a paper by A. Habibi and P. A. Wintz in the *IEEE Transactions on Computers*, February 1970, pp. 153–157, which shows block diagrams of three significantly different variations of the switching-network multiplier.

324 IMPLEMENTING BINARY ARITHMETIC

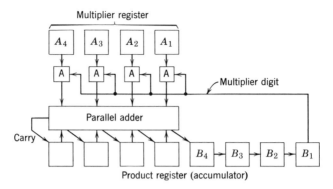

FIGURE 6-23. Arrangement for binary multiplication by accumulating partial products.

Multiplication by the Accumulation of Partial Products

The obvious scheme for multiplying two n-digit binary numbers through accumulating partial products is to employ a $2n$-digit accumulator. For each multiplier digit that is 1, the multiplicand is added into the appropriate positions of the accumulator. The positioning of the multiplicand may be achieved by placing it in a shift register and by shifting it one position for each multiplier digit. If the multiplier digits are sensed sequentially from the lowest-order digit to the highest-order digit, the multiplicand is first positioned to correspond to the n right-hand positions of the accumulator and is shifted to the left. However, the multiplication scheme would function substantially as well with the multiplier digits sensed in the opposite sequence with a corresponding reversal of shifting direction for the multiplicand.

A substantial saving in equipment is realized by using the scheme illustrated in Figure 6-23. The multiplicand A is initially stored in a static register, and the shifting capability is built into the accumulator which consists of an n-digit parallel adder and a $2n$-digit "product register," the storage elements of which are connected as shown in the figure. The n right-hand positions of the product register form a conventional shift register in which the multiplier B may be stored initially as indicated. The digit in the lowest-order position of the product register, which on the first step is the lowest-order multiplier digit B_1, controls a set of AND devices to allow the transmission of the multiplicand to the adder if that digit is a 1. Upon actuation of the accumulator as a whole (by means of a control network not shown), the output from the adder, including the carry digit, is transmitted to the $n + 1$ left-hand positions of the product register, and the multiplier digits are each shifted to the right-hand one position. Digit B_1 is lost, and B_2 is now stored in the lowest-order position. Note that B_1 is no longer needed in the multiplication process and that the lowest-order product digit, which is now stored in the position which initially contained B_4, will not be altered by subsequent accumulations of the multiplicand.

If B_2 is 1, a second actuation of the accumulator causes the multiplicand to be added to the contents of the n left-hand digits of the product register, and the sum including the carry is entered into the $n + 1$ left-hand positions. If B_2 is 0, the net effect is to shift all digits in the product register to the right-hand one position. In either case, B_2 is lost, B_3 appears in the lowest-order position, and the network is ready for a third actuation. This process is continued until all multiplier digits have been utilized, at which time the final product appears in the $2n$ positions of the product register.

The multiplication process can be made more rapid for some multipliers by including equipment which senses two or more successive 0's in the multiplier and which causes a shifting to the right by a corresponding number of positions in a single step.

Also, the parallel adder may contain carry bypass features or carry completion detection features as discussed previously. In fact, the designer may wish to consider operating the multiplication network asynchronously with carry completion detection even though the bulk of the machine operates synchronously, because once the multiplicand and multiplier have been entered into the network the multiplication process can be allowed to proceed at maximum possible speed without cycle time restrictions as imposed by a storage unit or other extraneous factors.

The "Carry Save" Technique

In the accumulation (addition) of any sequence of binary numbers, the carry digits resulting from the addition of any one number to the partial sum need not be accumulated prior to the addition of the next number. Instead, the carries can be "saved" and accumulated with the next number. Any new carries thus generated can likewise be saved for accumulation with the next subsequent number. In no instance will more than three digits (1's) need to be added in any one order on any one step. No carry propagation time at all is required for any of the numbers in the sequence until the last number has been added, at which time the carries may be accumulated in the usual way.

As an example, consider the addition of four numbers A, B, C, and D as follows.

```
                          A   000101110
                    add:  B   001010101
                              001111011
                                      1     carries
                    add:  C   000101100
                              001011111
                                    1 1     carries
                    add:  D   001110001
                              001111110
     add with carry propagation:    1 1 1   carries
                              100100000     sum
```

Although this "carry save" technique would be effective with any sequence of numbers when attempting to reduce the number of instances that carry propagation time need be provided, realistic applications where it is useful seem to be relatively few. However, the accumulation of partial products in an asynchronously operating binary multiplication unit is one application where the technique is highly useful in improving speed, because only one carry propagation operation is required for a complete multiplication.

A set of "carry save" binary storage elements with an appropriate control network to cause them to be effective on all steps except during the final accumulation of carries must be added to the equipment shown in Figure 6-23. The details of this added equipment would depend on many peripheral characteristics of the switching modules in use and of the multiplication unit as a whole, but in any case the details can be determined straightforwardly from the principles of the "carry save" technique.

Reducing the Number of Individual Operations
Required for Binary Multiplication

In binary multiplication techniques described so far the number of individual operations, all additions, required is equal to the number of 1's in the multiplier. By making use of the relationship

$$2^{k-1} + 2^{k-2} + 2^{k-3} + \cdots + 2^{k-a} = 2^k - 2^{k-a}$$

where k is any integer and a is any positive integer (ordinarily, $a \geq 3$), the number of operations can be reduced with many multipliers, although some of the operations become subtractions. In other words, when the multiplier happens to contain 1's in each of three or more consecutive orders, the corresponding additions of the multiplicand can be replaced by a subtraction of the multiplicand positioned to correspond to the lowest-order 1 in the set of consecutive 1's and an addition of the multiplicand positioned to correspond to the order one higher than the highest-order 1 in the set. For example, if the multiplier is

$$0111110001111010\,11100111$$
$$+\quad\ \ -\ +\ \ -++\ \ -\ +\ -$$

the 16 additions otherwise required by the sixteen 1's can be replaced by five additions and four subtractions with the various operations corresponding to the digit positions indicated by the plus and minus signs.

In instances where two adjacent 1's have a 0 in the next higher order and a 0 in the next lower order, no change would result in the number of operations required for that set. However, if a set of only two adjacent 1's is handled in the same manner as a longer set of 1's, a further reduction can be achieved in the number of required operations in some instances. For example, consider the multiplier digit sequence

$$\cdots 110101011\cdots$$
$$-\ +\ ++$$

where the plus and minus signs indicate the operations resulting from the technique just described. By a similar argument, the two adjacent additions can be replaced by an addition and a subtraction as follows.

$$\cdots 110101011\cdots$$
$$-\ ++\ -$$

Although the total number of operations remains the same, a different pair of adjacent addition operations appear, and these may similarly be replaced as follows.

$$\cdots 110101011\cdots$$
$$-+\ -\ -$$

Again the total number of operations has not been changed, but now an addition operation appears in an order which may be defined as having weight 2^j, where j is an integer, and a subtraction operation appears in the order of weight 2^{j+1}. Because $-2^{j+1} + 2^j = -2^j$, the two operations may be replaced by a single subtraction to produce

$$\cdots 110101011\cdots$$
$$-\ -\ -$$

An alternative method of arriving at the same result is to take as a set any sequence of consecutive multiplier digits wherein both the lowest-order digit and the highest-order digit are 1's. An addition and a subtraction are performed for the positions corresponding to weights 2^k and 2^{k-a}, respectively (where k and a are as defined previously), and for each 0 within the set a subtraction is performed with the multiplicand positioned accordingly.

However, in comparison with the version which permits no 0's within a set of consecutive 1's, a reduction in the total number of operations is realized only when each of the 0's within the set is isolated (that is, no two 0's can be in adjacent positions within the set) with 1's appearing in at least two nonadjacent groups of two (or more) consecutive positions—as in, for example, the set formed by the multiplier digit sequence 1011110110101. The simplest technique for minimizing the number of operations results from a requirement that the set be chosen so that a group of two (or more) consecutive 1's includes the two (or more) lowest-order positions of the set as a whole. An improvement occurs only in instances where at least one other group of two (or more) consecutive 1's appears in the set, but no harm is done when it does not appear.

All of the above factors can be collected together to form rules for determining the multiplicand positions for which the various additions and subtractions are to be performed. As before, the multiplier digits are sensed sequentially starting with the lowest-order (right-hand) digit and proceeding to the highest-order digit. When sensing any given digit, the action to be taken depends on whether the last previous operation was an addition or a subtraction and on the value of the next multiplier digit in accordance with Table 6-1. When the given multiplier digit is the lowest-order (first) digit, the last previous operation is assumed to have been an addition, and when the lowest-order digit is a 0, a skip to the "next" (first) 1 is made. When the given multiplier digit is the highest-order digit, the "next" multiplier digit is assumed to be a 0, and when the highest-order digit is 1 with the last previous operation being a subtraction, this "next" digit must be acted upon as though it were the $n+1$ digit in an otherwise n-digit multiplier. The skipping indicated in Table 6-1 refers to the next higher-order digit of the indicated value in the multiplier, and the skipping is accompanied by a shifting of the partially developed product to the right by a corresponding number of positions (see Figure 6-23).

When the last previous operation was a subtraction the accumulated balance of partial products will be negative, and the shifting action must be designed to take this fact into consideration. However, when

TABLE 6-1. Action to be Taken When Minimizing the Number of Operations Required for Binary Multiplication

Next Multiplier Digit	Given Multiplier Digit	Last Previous Operation	Action to be Taken
0	1	+	Add and skip to next 1
	0	−	
1	1	+	Subtract and skip to next 0
	0	−	
0	1	−	Because of the skipping, these combinations of conditions will not be encountered. (If skipping is not employed, neither add nor subtract—just shift one position and sense next multiplier digit.)
	0	+	
1	1	−	
	0	+	

either the 1's or 2's complements are used for negative numbers, the necessary added equipment consists only of a network to sense the sign of the balance and to shift 1's instead of 0's into the high-order end of the accumulator whenever the balance is negative.

The total number of operations required is equal to twice the number of sets thus formed plus the number of isolated 0's within the sets plus the number of 1's not in any set. In the first versions where the sets consisted strictly of consecutive 1's the number of operations is equal to twice the number of sets plus the number of 1's not in any set, but because the sets in that version tend to be shorter and more numerous the latter version results in the same or fewer operations for any combination of multiplier digits.

In any case, the maximum number of operations is $n/2$ with even n and is $1 + n/2$ with odd n, where n is the number of digits in the multiplier. This maximum is encountered when the multiplier consists of alternating 1's and 0's.

Again, the details of the equipment needed to control the indicated operations are strongly depending upon the characteristics of the switching modules and upon peripheral features of the arithmetic unit, but the details may be worked out straightforwardly in the manner suggested in earlier chapters.

Serial Techniques in Multiplication

The serial adder of Figure 6-21 can be used to accumulate the partial products in a multiplication provided the appropriate digits of each partially developed product are applied to one input at the correct times with respect to the digits of the multiplicand. Figure 6-24 illustrates in symbolic form the major details of one scheme of performing binary multiplication on a purely serial basis. Four storage registers are provided. These registers may be either delay lines or continuously operating shift registers, and the movement of digits is from left to right in each register. The digits of the multiplicand A and the multiplier B are initially placed in the indicated positions of two of the registers. The register for the multiplicand has $n + 1$ storage positions, and the right-hand position is initially caused to store a 0. The other two registers are initially reset to all 0's, and the n high-order digits of the product P

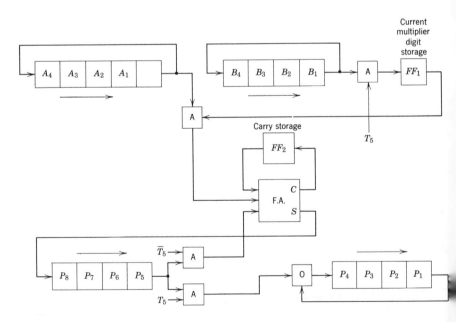

FIGURE 6-24. Serial multiplication scheme.

will eventually appear in one of them with the n low-order product digits appearing in the other.

Signal T_5 in Figure 6-24 is a control signal that is 1 during the first step and during each fifth following step (that is, T_5 is 1 during the 1st, 6th, 11th, 16th, and so on, steps) and is 0 during all other steps.

During the first step, multiplier digit B_1 is entered into the flip-flop FF_1, the "current multiplier digit" storage device, and is also shifted end-around fashion to the position initially containing B_4. All other multiplicand and multiplier digits are also shifted. During the second through the fifth steps, the digits of the multiplicand are (if B_1 is 1) transmitted serially through the full adder F.A. (the other input signals to the full adder being all 0's) to the high-order product digits register. At the beginning of the sixth step the digits of the multiplicand will again appear in the positions indicated in the figure, but digit B_2 of the multiplier will appear in the right-hand position of the multiplier register. During the sixth step, when T_5 is again 1, B_2 will be entered into FF_1, and the digit in the P_5 position, which is the lowest-order digit of the final product, will be transmitted through the OR device to the P_4 position of the low-order product digits register. Also during the sixth step the digits in the P_8, P_7, and P_6 positions will be shifted to the P_7, P_6, and P_5 positions, respectively. At the termination of the sixth step the multiplicand digits in the multiplicand register will have been returned to the same positions at which they appeared at the termination of the first step.

On the 7th through the 10th steps, when T_5 is 1, the multiplicand will be added (if B_2 is 1) to the digits from the high-order product digits register. At the end of the 10th step a carry digit may be stored in FF_2, the carry storage device, and on the 11th step this carry will be passed through the full adder to the P_8 position as all other digits in the high-order product digits register are shifted to the right. Note that on the 11th step the digit in the P_5 position will be transmitted through the OR device to the P_4 position. At the termination of the 11th step the two lowest-order product digits will appear in the P_4 and P_3 positions, with the lowest-order digit appearing in the P_3 position because of the continuously shifting end-around action of the register.

The process described above continues analogously until all multiplier digits have been acted upon, at which time all of the final product digits appear at the positions indicated in the figure. From a switching function standpoint the multiplier register and the low-order product digits register need not be actuated on every step, so that when the registers are shift registers (not delay lines) appropriate alterations may be made in the network.

A Serial-Parallel Multiplication Scheme

A serial-parallel binary multiplication scheme which produces a complete $2n$-digit product in $2n$ steps is illustrated in Figure 6-25. The digits of the multiplicand A are applied continuously in parallel fashion to a set of AND devices. The digits of the multiplier B are applied serially, low order first, to the other input of each AND device. The output signals from the AND devices are transmitted to a set of serial adders of the form illustrated in Figure 6-21, and these adders are interconnected chain-fashion with a digit storage flip-flop between each pair.

On the first step the multiplicand digits or 0's (according to whether B_1 is 1 or 0, respectively) are transmitted through the full adders to the network points designated X_4 through X_1—except that the highest-order digit happens to reach point X_4 directly without passing through a full adder. Point X_1 is the output line, and the digit appearing here on the first step is the lowest-order digit P_1 of the final product. At the end of the first step a clock pulse actuates all flip-flops, and B_2 is transmitted to the multiplier input line so that during the second step the digits previously appearing at points X_4, X_3, and X_2 now appear at points X'_4, and X'_3, X'_2, respectively. At the same time the multiplicand digits or 0's (now in accordance with the value of B_2) are added in the relative positions correct for the addition of partial products.

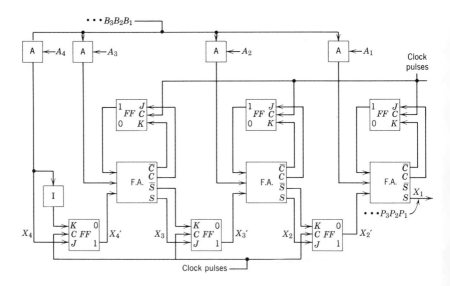

FIGURE 6-25. Serial-parallel multiplication network.

The second product digit P_2 now appears on the output line X_1, the first product digit presumably having been transmitted to some other part of the system as a whole.

At the end of the second step the clock pulse in Figure 6-25 causes the carry digits to be entered into the carry storage flip-flops, and the digits appearing at points X_4, X_3, and X_2 become stored in the flip-flops that interconnect the full adders. During the third step, product digit P_3 appears on the output line, and the multiplicand digits or 0's (now in accordance with the value of B_3) are added to the carries generated previously and to the digits currently appearing at points X'_4, X'_3, and X'_2. Thus, the digits are, in effect, handled in substantially the same manner as was described previously for the "carry save" technique. After n steps when all n of the multiplier digits have been transmitted to the unit, digits may remain in the various flip-flops. These digits are "run out" by actuating the unit for n additional steps during which time the n high-order digits of the $2n$-digit product appear in succession on the output line.

In applications where the multiplicand has n_1 digits and the multiplier has n_2 digits with $n_1 \neq n_2$, the number of steps required is $n_1 + n_2$. The amount of equipment required is dependent on n_1 but is independent of n_2.

Multiplication with the Multiplier Digits Handled in Groups

Increased multiplication speed can be achieved by sensing the multiplier digits in groups of two, three, four, or more and by taking action in accordance with the digit combination in each group rather than by acting upon the multiplier digits singly. This concept will be explained for the particular case of handling the multiplier digits in groups of three—the characteristics of the two-digit, four-digit, and so on, group schemes are reasonably obvious once the characteristics of the three-digit group scheme are appreciated. The groups are sensed in sequence, low order first (they could be sensed in the opposite sequence, but not so conveniently in general), and if the number of multiplier digits is not a multiple of three, one or two 0's are appended to the high-order end to make the number of digits a multiple of three.

Each group of three binary multiplier digits can be visualized as an octal (radix eight) digit, and the corresponding multiple of the multiplicand is taken as the partial product to be accumulated in forming the product. The various multiples of the multiplicand can be formed in any of several different ways, but a scheme which requires relatively little equipment and which is an elaboration of the previously described scheme for minimizing the number of addition-subtraction operations required for a multiplication is as follows. If the three multiplier digits in any given group are 000, no operation is performed except to shift the accumulated partial products to the right three positions (or the multiplicand to the left three positions) in preparation for the action to be taken on the next multiplier digit group. If the digits are 001, the multiplicand is added as a partial product in the conventional manner. If the digits are 010, the multiplicand is shifted to the left one position and added. However, the shifting in this instance is not necessarily by means of the shift register but is more likely to be by means of a combinatorial switching network controlled by the multiplier digits. Therefore, the shifted multiplicand is available in the same time as the multiplicand itself. (If the network is visualized as being an elaboration of the network in Figure 6-23, the portion of the network represented by the AND devices becomes a two-level AND-OR network controlled by signals that are active for certain combinations of multiplier digits within a group. Also, in the three-digit group example, the parallel adder and the product register must be extended by two positions at the high-order end to accommodate the temporary appearance of digits in these positions.) After adding the multiplicand shifted one position to the left, the digits in the product register are shifted, by a true shifting action, three positions to the right.

If the multiplier digits are 100, the action is similar except that the multiplicand is shifted by switching two positions to the left and then added.

If the multiplier digits are 111, the multiplicand is subtracted and 1 is added to the next higher order group of multiplier digits. That is, if the next higher order group is 000, it is made 001. If it is 001, it is made 010, and so on. If the next higher order group is 111, it becomes 000 with a "carry" to be added to the next group in succession. Thus, for a long string of 1's in the multiplier the net effect is the same as previously described except that the shifting is performed three positions at a time.

If the multiplier digits are 110, the multiplicand is subtracted after being shifted by switching one position to the left, and a 1 is added to the next higher order group of three multiplier digits.

For multiplier digit groups 011 and 101 two addition-subtraction operations are required, but two alternatives are possible in each instance. For 011, the multiplicand may be added and then added again after shifting by switching one position to the left. Alternatively, the multiplicand may be subtracted and then subtracted again after shifting by switching two positions to the left with a 1 being added to the next higher group. For 101, the multiplicand may be added and then added again after shifting by switching two positions to the left. Alternatively, the multiplicand may be subtracted and the subtracted again after shifting by switching one position to the left with a 1 being added to the next higher order group. In each instance the alternative corresponds to the use of a subtraction for isolated 0's in the scheme where the multiplier digits are sensed one at a time. Incidentally, an analogous alternative is available for 100 in that the multiplicand may be subtracted after shifting by switching two positions to the left with a 1 being added to the next higher order group.

To minimize the number of operations, the multiplier digit combinations 011 and 101 should be avoided as much as possible. Thus, for those digit combinations that offer alternative actions (namely 011, 100, and 101), the next higher order group is sensed. If that group is either 011 or 101, the alternative which causes the addition of 1 to that group is selected. The next higher order group is thereby caused to be 100 or 110, respectively, either of which can be handled with a single addition-subtraction operation. These actions are summerized in Table 6-2.

TABLE 6-2. Summary of Actions to be Taken on the Multiplicand When the Multiplier Digits are Sensed in Groups of Three

Multiplier Digits	Action on Multiplicand	Alternative Action on Multiplicand
000	...	...
001	Add	...
010	Add shifted left one position	...
011	Add, add shifted left one position	Subtract, subtract shifted left one position, and add 1 to next higher order group
100	Add shifted left two positions	Subtract shifted left two positions and add 1 to next higher order group
101	Add, add shifted left two positions	Subtract, subtract shifted left two positions, and add 1 to next higher order group
110	Subtract shifted left one position and add 1 to next higher order group	...
111	Subtract and add 1 to next higher order group	...

For n-digit multipliers having random digits the total number of addition-subtraction operations is slightly greater than $n/3$. The maximum possible number of operations is two in alternate groups and one in each remaining group plus a single addition in a position corresponding to the $n+1$ order in the multiplicand, which works out to $n/2+1$. The speed improvement results from the shifting by switching and the reduced time required to sense the multiplier digits, as the number of addition-subtraction operations is the same as when the multiplier digits are sensed one at a time.

Further improvements in speed can be obtained by supplying extra parallel adders to generate selected multiples of the multiplicand. In the example of sensing the multiplier digits in groups of three, a parallel adder which adds the multiplicand to twice the multiplicand (the multiplicand shifted left one position) to generate three times the multiplicand will allow multiplication with no more than one operation for each multiplier digit group. In particular, if the digits of the group are 011, switching networks cause the adder output to be accumulated in forming the product. If the multiplier digits are 101, the adder output is subtracted in the accumulation of partial products, and a 1 is added to next higher order group of multiplier digits. The number of addition-subtraction operations is then no greater than $n/3+1$ regardless of the multiplier digits.

Multiplication with the Digits of Both Factors Handled in Groups

The digits of both the multiplier and the multiplicand can be divided into groups to produce other types of multiplication schemes, some of which offer high speed and reasonable economy with appropriate switching modules. One such scheme is illustrated in Figure 6-26. The digits of both factors are grouped by threes and the complete product is developed in $(n_1+n_2)/3$ steps if the number of product digits is a multiple of three, where n_1 is the number of digits in the multiplier and n_2 is the number of digits in the multiplicand. If n_1+n_2, the number of product digits, doesn't happen to be a multiple of three, the necessary number of steps is $(n_1+n_2+1)/3$ or $(n_1+n_2+2)/3$ as required to make the numerator a multiple of three.

Figure 6-26a illustrates the equipment needed to generate the seven multiples A through $7A$ of the multiplicand A, with each number appearing three digits at a time on three parallel wires. The digits have relative weights of 1, 2, and 4 as indicated in the upper-left part of the figure. The doubled multiplicand $2A$ is generated by a sort of "shifting" action where the digits of relative weight 1 and 2 in A appear on the lines

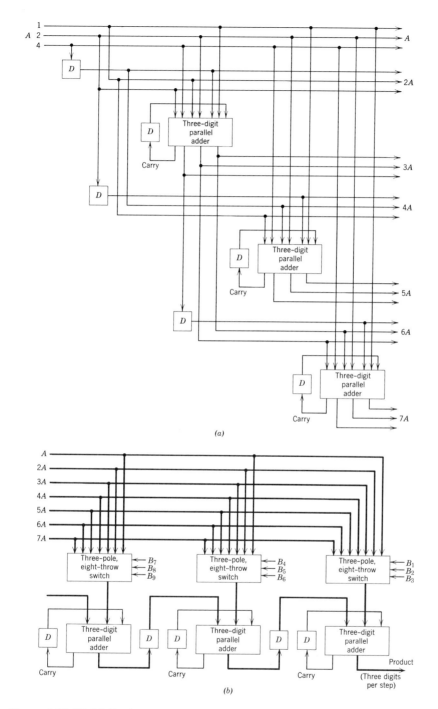

FIGURE 6-26. Multiplication scheme with digits of both factors handled in groups of threes. In *b* each heavy line represents three wires in parallel.

of relative weight 2 and 4, respectively, in $2A$. The digit of relative weight 4 in A is delayed one step by a delay device, indicated by D in the figure, and this digit becomes the digit of relative weight 1 in the next higher order three-digit group in $2A$. The delay device may actually be a flip-flop operated in the manner indicated in Figure 6-25. The tripled multiplicand $3A$ is obtained by adding A and $2A$, group by group, in a three-digit parallel binary adder, and the carry from the high-order end of this adder is returned through a similar delay device to the low-order carry input. Multiples $4A$ and $6A$ are obtained by "shifting" $2A$ and $3A$, respectively, and multiples $5A$ and $7A$ are obtained by adding appropriate lower multiples.

Multiplication by multiplier B is accomplished as indicated in Figure 6-26b. The three lowest-order muliplier digits B_1, B_2, and B_3 actuate a switching network which can be called a three-pole, eight-throw switch. The eight "throws" are for selecting the appropriate multiple of the multiplicand (with $0A$ being added to the set represented by $1A$ through $7A$), and the three 'poles" are for switching the three wires on which each multiple appears. Each heavy line in Figure 6-26b represents three wires in parallel. The next higher order multiplier digit group, as represented by B_4, B_5, and B_6, actuates a similar switch as does each succeeding higher order group. The various multiples thus selected are combined to form the product in the three-digit parallel adders along the bottom of the figure. The sum from each such adder is delayed by one step (where the block labeled D represents three flip-flops in parallel in this instance) before being added to the next lower-order three-digit adder. With this arrangement, three digits of the final product appear on the output lines after each step of operation.

Multiplication with Negative Factors Represented in 2's Complement Form—First Scheme

If negative numbers are represented in true form but with a sign digit appended, multiplication proceeds in the same manner as before when either or both of the factors are negative. However, equipment is included to sense the signs of the factors and to cause the sign of the product to be negative when one but not both of the factors is negative. If 1's complement representation is used for negative numbers, the negative numbers may be economically (because of the simplicity of the conversion process) converted to true form, and if the product is negative, it may be economically converted to 1's complement form.

With negative numbers represented in 2's complement form, the conversion to true form for multiplication is reasonable from an equipment standpoint, but schemes have been devised to eliminate the need for conversion. If only the multiplicand is negative, the multiplication can proceed as with positive numbers except that in shifting the partially developed product to the right, 1's instead of 0's are shifted into the left-hand position, as controlled by the multiplicand sign digit. For example, consider the multiplication of 10011 (decimal −13) by 01101 (decimal +13) with 2's complement notation being used for negative numbers. The steps are as follows, where each add-shift combination of operations corresponds to one step of the multiplication network of Figure 6-23 except that the addition networks have been extended by one order to accomodate the addition of the sign digits in the manner explained earlier for addition in general.

sign

0 0000	
1 0011	add multiplicand
1 0011	
1 10011	shift
0 0000	add zero
1 10011	
1 110011	shift
1 0011	add multiplicand
0 111111	
1 0111111	shift
1 0011	add multiplicand
0 1010111	
1 01010111	shift (final product: −169)

A complete full adder is not needed in the sign position of the multiplication unit. First, the carry output signal is not needed. Also, because the carry from the next lower order (that is, from the highest-order digit position) can be 1 only when the sign digits are equal (as can be appreciated by considering various examples in detail), the full adder can be reduced to a network which generates a "pseudo-sum" of the form $A\bar{B} + \bar{A}B + C$ in Boolean notation, where A is the sign digit of the multiplicand, B is the sign digit of the partially developed product, and C is the carry from the highest-order digit position.

If the multiplier is negative, a reasonably simple procedure is to sense that fact prior to the multiplication operation and to convert the multiplier to true form while the multiplication operation is in progress. Because the multiplier digits are acted upon one at a time, low order first, the conversion is accomplished easily by leaving all low-order 0's and the first 1 unchanged and by inverting all higher-order digits. For each 1 in the converted (true form) multiplier, the multiplicand is subtracted instead of added in forming the product. If the multiplicand is positive, the details turn out to be essentially the same as in the example of the previous paragraph. If the multiplicand is negative, the net result is essentially the same as the multiplication of two positive factors.

In some serial machines the sensing of the multiplier sign digit, which normally appears at the high-order end of the multiplier, prior to the multiplication operation is at best inconvenient. In such applications the multiplication unit can be designed to function by adding the multiplicand for each 1 in the multiplier regardless of the sign of either factor, except that when the multiplier is found to be negative when sensing the last (sign) digit, the multiplicand is subtracted from the corresponding position of the partially developed product. For example, the multiplication of 10011 by 10011 (decimal −13 by −13) would then proceed as follows.

```
sign
0 0000
1 0011      add multiplicand
─────
1 0011
1 10011     shift
─────
1 0011      add multiplicand
0 11001
─────
1 011001    shift
─────
0 0000      add zero
1 011001
─────
1 1011001   shift
─────
0 0000      add zero
1 1011001
─────
1 11011001  shift
─────
1 0011      subtract multiplicand
0 10101001  (final product: +169)
```

This procedure functions properly regardless of the sign of the multiplicand, although when the multiplicand is negative (as in the above example), 1's instead of 0's are shifted into the high-order end of the product register as before.

Multiplication with Negative Factors Represented in 2's Complement Form—Second Scheme

In the second scheme the multiplier digits are sensed one at a time, low order first, and for each given digit the action to be taken depends on the given digit and the next lower order digit. For the first (lowest order) digit the "next lower order" digit is assumed to be 0. The last digit, which represents the sign, is handled in the same manner as any other digit. If a given multiplier digit is 1 and the next lower order digit is 0, the multiplicand is subtracted; if the given digit is 0 with the next lower order digit 1, the multiplicand is added; and if the given digit and the next lower digit are of the same value, no action is taken. As with other multiplication procedures, a shift is made after each step. When shifting, a 0 or a 1 is shifted into the sign position according to whether the current balance of the partially developed product is positive or negative, respectively. With reference to the multiplication network of Figure 6-23, but with an added position for the sign digits, the digits encountered during the multiplication of 10011 by 10011 (decimal −13 by −13) are set forth as an example.

```
    sign
    0 0000
    1 0011      subtract multiplicand
    0 1101
    0 01101     shift
    0 001101    shift
    1 0011      add multiplicand
    1 011001
    1 1011001   shift
    1 11011001  shift
    1 0011      subtract multiplicand
    0 10101001  (final product: +169)
```

The similarity between this scheme as the scheme described earlier in the chapter for minimizing the number of addition-subtraction operations in a multiplication should be apparent. As described here, the number of operations is not minimized, although with suitable elaborations similar to those described earlier, the number of operations can be minimized.

Binary Division

Division is basically the inverse of multiplication with the dividend, divisor, and quotient corresponding to the product, multiplicand, and multiplier, respectively. However, some awkward problems are encountered with division that are not encountered with multiplication with the net result that implementing division is a computer tends to require a more elaborate control network, and for a given word length the division operation tends to consume more time than a correspondingly implemented multiplication operation.

First, the dividend is not restricted to those numbers which are exact products of other factors. That is, the quotient does not necessarily "come out even" but instead a nonzero remainder may result. In a few applications the value of the remainder may be important. For example, in a problem to determine the position of a shaft, its total motion in degrees may be divided by 360 (but expressed as a binary number), in which case the quotient represents the number of revolutions through which the shaft has turned, and the remainder represents the angular position in degrees. More commonly, however, the remainder is of no consequence except to determine whether or not the quotient should be rounded to the next larger number. Thus, to obtain an n-digit quotient, the divider unit may be designed to generate $n+1$ quotient digits but with a round-off procedure included to round the quotient to n digits. As an illustration, consider the division of 10010001 by 1011 (decimal 145 by decimal 11)—where, for the present, only positive dividends and divisors will be considered.

```
                           quotient
              10010001
   subtract:  1011
              ‾‾‾‾‾‾‾‾
              0111001      1???.?
   subtract:  1011
              ‾‾‾‾‾‾‾‾
              001101       11??.?
   subtract:  0000
              ‾‾‾‾‾‾‾
              01101        110?.?
   subtract:  1011
              ‾‾‾‾‾‾‾
              0010.0       1101.?
   subtract:  000.0
              ‾‾‾‾‾‾‾
              010.0        1101.0
```

The quotient digits are developed one at a time as indicated at the right. The final quotient 1101.0 would be rounded by adding 1 in the lowest-order position with carries, if any, to produce a rounded quotient of 1101 in this example.

A second problem is apparent from the example. On each step, the divider unit must subtract the divisor or zero in accordance with whether the corresponding quotient digit is 1 or 0, respectively. But, contrary to the analogous feature of a multiplication operation, the quotient digit is not known prior to the subtraction operation. One way to solve this problem is to subtract the divisor on every step. On each step where this subtraction produces a negative difference, the quotient digit is immediately known to be 0, and prior to the next step the divisor is added without shifting so that the net effect has been to subtract zero. A division operation which proceeds in this way is said to be of the "restoring" type, because the negative difference is restored to the positive value it would have had from the subtraction of zero. The number of addition-subtraction operations required for a division operation is then equal to the number of quotient digits plus the number of 0's in the quotient.

The number of addition-subtraction operations can be reduced to n through the use of a "nonrestoring" technique. With this technique, whenever a negative difference is encountered, the divider unit does not restore it but instead proceeds directly to the next step. The divisor is added instead of subtracted on this next step. The next difference is the same as it would be otherwise, because in general $-D + D/2 = -D/2$, where D is the divisor shifted to any given position. The factor of $\frac{1}{2}$ appears because of the shifting of the divisor to the right one position between steps. The quotient digit on this "next" step is 0 if the new difference is negative but is 1 if the addition produces a positive new difference. For each succeeding quotient digit, the divider subtracts or adds the divisor according to whether the sign of the current difference (intermediate remainder) is positive or negative, respectively.

When using the nonrestoring technique, the digits appearing in the division of 01111101 by 1101 (decimal 125 by 13), for example, would be as follows.

```
                              quotient
                 01111101
      subtract:     1101
                 0|0010101    1???.?
      subtract:     1101
                 1|100001     10??.?
      add:          1101
                 1|11011      100?.?
      add:          1101
                 0|1000.0     1001.?
      subtract:     110.1
                 0|001.1      1001.1
```

After rounding to the nearest integer, the final quotient would be 1010. In the example, a vertical bar is used to separate the sign digit from the other digits of each difference. The sign digit is generated as described previously for addition and subtraction, with negative numbers represented in 2's complement form.

The nonrestoring division technique is particularly appropriate for serial computers or for parallel computers where addition-subtraction is accomplished in an accumulator consisting of a set of complementing flip-flops equipped with carry-handling facilities.

However, most parallel computers perform the addition-subtraction operations in a switching network (the "adder-subtractor"), and the sum is placed in a set of flip-flops which may have been storing one of the numbers entering into the addition-subtraction operation. In these computers, neither the restoring nor the nonrestoring division technique is necessary. Instead, on each step of the division process, the divisor is subtracted from the appropriate relative positions of the dividend or intermediate remainder, but whenever a negative difference is indicated, the difference is not utilized. That is, the dividend or intermediate remainder digits are not altered. If the difference is positive, the dividend or intermediate remainder digits are replaced by the difference digits to form a new intermediate remainder. On each step the corresponding quotient digit is 0 or 1 according to whether the difference is negative or positive, respectively.

A Binary Division Network

A binary division network which is a close counterpart of the multiplication network of Figure 6-23 and which is based on the division method described at the end of the previous section is shown in Figure 6-27 for four-digit divisors and four-digit quotients. The dividend, which may have up to eight digits, is initially placed in the flip-flops along the top of the figure, and the divisor is placed in the flip-flops at the bottom. A five-digit parallel subtractor is used to subtract the divisor from the indicated dividend digits. The highest-order position in the subtractor has only two input signals, and one of these is the internal carry signal from the next lower order. If the borrow output from the parallel subtractor is 0, the output signal from the inverter is used to represent the quotient digit of 1. The inverter output signal is used to actuate certain of the AND devices to cause the difference from the subtractor to be entered into the top row of flip-flops with each digit being shifted to the left one position, and this signal is also entered into the right-hand flip-flop for temporary storage of the quotient digit Q_4. The actuation

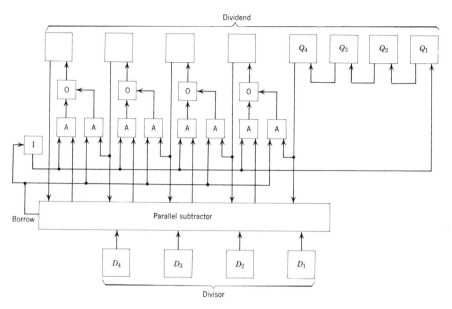

FIGURE 6-27. Binary division network.

of the flip-flops is accomplished by control signals not indicated in Figure 6-27. If the carry is 1, others of the AND devices are actuated so that the dividend digits are not altered but are shifted one position to the left. In either case, the digits in the low-order positions of the dividend register are shifted to the left.

Quotient digit Q_3 is developed on the second step in essentially the same manner except that the digits appearing in the seven left-hand flip-flops of the dividend register now represent the first intermediate remainder. Digit Q_4 is shifted to the flip-flop second from the right-hand end. After four such steps the quotient digits Q_4, Q_3, Q_2, and Q_1 appear in the flip-flops indicated in Figure 6-27.

Quotients having more than four digits may be generated by essentially the same network as in Figure 6-27, except that the quotient digits must be stored in a separate shift register instead of in the register initially used for storing the low-order dividend digits. As additional quotient digits are developed, 0's are entered into the right-hand end of the dividend register.

In Figure 6-27, the dividend and divisor registers may be one and the same as the multiplicand and product registers, respectively, in a machine capable of both multiplication and division. Also, the adder for multiplication and the subtractor for division may be formed by whatever adder-subtractor is used for ordinary addition and subtraction. A portion of the control network for division, notably the counter that counts the number of steps and terminates the operation, may likewise be the same as a portion of the control network used for multiplication, but most of the divider unit control network would, of course, be different.

Rounding the Quotient

For obtaining rounded quotients, elaborations must be added to the divider of Figure 6-27, but the nature of the elaborations may assume many different forms. Most straightforwardly, the control network is modified to cause the generation of five instead of just four quotient digits, but when generating the fifth quotient digit a special switching network causes a 0 (instead of Q_4) to appear as the remainder digit from which the D_1 divisor digit is subtracted in the parallel subtractor. The five-digit quotient, which now is stored in the right-hand five positions of the dividend register is then shifted to the left-hand five positions, and a 1 is caused to be added (with carries handled by means of the same parallel adder-subtractor used for other purposes) to the lowest-order divider digit which, after the shift, is stored in the flip-flop designated Q_4 in the figure. The resulting digit in this flip-flop is discarded, and the rounded four-digit quotient appears in the left-hand four positions of the dividend register.

An Alternative Quotient-Rounding Procedure

A quite different rounding procedure may be preferable when the dividend has the same number, n, of digits as the divisor—the dividend presumably being itself a rounded approximation, as the divisor may also be. With four-digit numbers and with the binary point assumed to be at the right-hand end, the division of 1011 by 1101, for example, would be viewed as dividing $1011xxxx$ by 1101 where the x's represent previously discarded digits that are presently assumed to be 0's. That is, the problem would be viewed as dividing 10110000 by 1101 (decimal 176 by 13). Instead of adding 1 to the quotient in the fifth-digit position, prior to the division operation the divisor is added to the dividend in the relative positions which produce the same net effect. The correct relative positions are obtained by shifting the divisor to the right one position so that, in effect, 110.1 is added to produce an initial dividend of 10110110.1. Fortunately, when developing the four-digit quotient (which will be properly rounded) the digit to the right of the point in the dividend will not affect the outcome, so the digit to the right of the point may be ignored.

Moreover, when adding the shifted divisor to the dividend, the corresponding dividend digits will all be 0's, so the addition is merely a matter of entering the three high-order divisor digits into the three low-order positions of the dividend register prior to commencing the division operation. Thus, the division of $1011xxxx$ by 1101 is accomplished by dividing 10110110 by 1101 to produce a rounded four-digit quotient of 1110 (decimal 14), and the network of Figure 6-27 may be used for this purpose.

Starting the Division Operation—Division Overflow

Another problem encountered with division that is not encountered with multiplication is that the dividend may be so large relative to the divisor that, for a given location of the binary point and for a given design of the equipment, the quotient will have 1's in positions that are of higher order than are represented by the physical equipment provided. This condition is called a "division overflow," and means must usually be included in the computer to detect this condition.

A straightforward scheme for detecting a division overflow is to subtract the divisor twice from the dividend positions corresponding to the first quotient digit. If the second subtraction leaves a positive intermediate remainder, the quotient digit would be at least 2—an impossibility in the binary number system. This second subtraction would not be needed in any instance where the first quotient digit has been found to be 0.

An alternative and more commonly used scheme is to attempt a subtraction of the divisor from the left-hand digits of the dividend. This first subtraction is to the left one position in comparison with the first subtraction in any of the division methods described previously. This subtraction has the same effect as the two subtractions of the previous paragraph. If the difference is positive, an overflow is detected. If the difference is negative, the dividend and the divisor have proper relative magnitudes, and the division operation can be allowed to proceed. This overflow-detecting subtraction can be corrected by either the restoring or the nonrestoring method as described previously for other subtractions in the division operation. Alternatively, and perhaps preferably, the storage, even temporarily, of the difference can be omitted as in the division method of Figure 6-27.

Another alternative method of detecting a division overflow is to sense the digits shifted out of the left-hand end of the dividend register in a divider such as the one in Figure 6-27. If any 1's appear at this point during the division operation, an overflow is indicated.

In all instances a division operation with a divisor of zero results in a division overflow (except perhaps in a few rare applications where a zero dividend should produce a zero quotient regardless of the divisor, and in these applications network means must be included to detect a zero dividend).

The action to be taken when a division overflow is detected depends on the application for which the computer is designed. In some applications the computer is to be stopped with a light or other indicator actuated to indicate the cause of the stoppage. In other applications, built-in equipment for an automatic corrective action of some sort is appropriate. In still other applications, the signal may be used to set a flip-flop to the "on" state, and this flip-flop may cause an automatic jump to a corrective program that is designed by the programmer. As a further elaboration, the computer may be designed so that the jump itself is under the control of the programmer, and the jump may be performed at whatever step desired by the programmer and to whatever subprogram desired by the programmer.

Division with the Binary Point at Positions Other Than the Right-Hand End

The mechanics of the division operation are independent of the locations of the binary points in the dividend and divisor. However, these point locations must be known because they obviously influence the point location in the quotient.

For each position the point is moved to the left of the right-hand end in the $2n$-digit dividend, the quotient point location must be moved one position to the left. The opposite relationship prevails for points that are to the left of the right-hand end of the divisor. If the point is at the left-hand end of both the dividend and the divisor (a common assumption), the net result is to cause the point to appear at the left-hand end of the quotient.

Also, in any given number the point location may be beyond either the right-hand end or the left-hand end. In either case, 0's are implied to exist between the end of the number and the point position. However, the mechanics of the division operation are still unchanged.

Of course, in a fixed-point computer as implied here, no physical devices are used to represent point locations. Instead, point locations are imaginary in the mind of the programmer, who must nontheless know the point locations so that by proper shifting of digits the points can be inserted in the proper positions when the final results are transmitted to an output device.

Determining the point location in the result of a single arithmetic operation is simple enough, but when many successive operations are applied to a given set of input data the problem of keeping track of point location can become not only tedious but really rather difficult. The nature of the difficulty will be described in some detail later in connection with a description of "floating-point" operation, which is a scheme for relieving the programmer of the need to keep track of the point locations.

Reducing the Number of Addition-Subtraction Operations in a Division Operation

Consider the position of the highest-order 1 in the divisor. At any step in a division operation, if the digit in the corresponding position of the dividend-remainder (i.e., the dividend on the first step and the intermediate remainder on subsequent steps) is a 0 and if all higher-order digits of that number are 0's, the quotient digit to be determined on that step will be 0. No subtraction need be performed.

Moreover, if in the dividend-remainder k digits in consecutive positions to the right of the position corresponding to the highest-order 1 are also 0's, then k additional quotient digits, all 0's, can be immediately determined on the same step.

To implement this technique for eliminating subtraction operations, a switching network is needed which will, in effect, count the high-order 0's in the two numbers that would otherwise enter into each subtraction and will subtract one total from the other. Such a network can be designed straightforwardly by using methods described earlier in this book. However, rather than give further attention to the general case, which is of limited practical interest, attention will be directed to divisors for which the highest-order digit is a 1. A number having a 1 as its highest-order digit is said to be "normalized," and normalized numbers will be described in more detail in a later section on normalized floating-point operation. With a normalized divisor the determination of quotient 0's without a subtraction is a matter of counting the 0's in the high-order positions of the dividend-remainder.

To illustrate, the 15-digit number 010111000000101 will be divided by the five-digit number 10101 to produce a ten-digit quotient in seven steps involving five subtractions (one of which is temporary) as follows. The initial check for overflow is omitted.

350 IMPLEMENTING BINARY ARITHMETIC

			quotient
		010111000000101	
Step 1	subtract:	10101	
		00010000000101	1?????????
Step 2	count 0's:	10101	
		010000000101	100???????
Step 3	temp. subtract:	10101	
		10000000101	1000??????
Step 4	subtract:	10101	
		0101100101	10001?????
Step 5	subtract:	10101	
		000010101	100011????
Step 6	count 0's:	10101	
		010101	100011000?
Step 7	subtract:	10101	
		00000	1000110001

On step 1 the first quotient digit, a 1, is determined in the usual manner. On step 2 the three 0's in the highest-order positions of the intermediate remainder signify that the next two quotient digits are 0's. No subtraction need be performed. On step 3 the fourth quotient digit is found to be 0, but a conventional subtraction is needed. However, the results of this subtraction are not stored, so the intermediate remainder after step 3 is the same as after step 2. Steps 4 and 5 determine the fifth and sixth quotient digits, both 1's, in the usual manner. One step 6 the four 0's in the highest-order positions of the intermediate remainder indicate directly that the next three quotient digits are 0's, and step 7 determines the last quotient digit, a 1, in the usual manner.

With the above division method, the required number of subtractions is a function of the digits which happen to appear in the various steps. An isolated 0 in the quotient (that is, a 0 which has 1's in both the next lower-order position and the next higher-order position) is determinable without a subtraction in some instances, but not in other instances. In general, for j consecutive 0's in the quotient, where $j \geq 1$, the number of subtractions is either none or one. A subtraction is required for each 1 in the quotient.

Of course, a suitable switching network is needed to cause shifting an appropriate number of positions when more than one quotient digit is determined on a single step.

The average number of subtractions required to determine 0's in the quotient can be further reduced by using a switching network which will compare two, three, or more digits in the high-order positions of the divisor with the corresponding digits of the dividend-remainder. Thus, in step 3 of the above example, if the three high-order divisor digits, 101, are compared with the corresponding intermediate remainder digits, 0100, the quotient digit can be determined to be 0. However, when this idea is carried to the extreme of comparing all divisor digits with the corresponding dividend-remainder digits, the necessary comparison network becomes impractically extensive unless it is in the form of a conventional subtractor, in which case the subtraction operation would not have been eliminated.

Addition-subtraction operations can be eliminated in certain instances where the quotient digit is 1 by using the following technique. Whenever a subtraction is indicated in determining a 0 quotient digit, the negative intermediate remainder is retained in 2's complement form. (This action is in contrast to step 3 of the above example, where the subtraction was temporary.) Then a nonrestoring technique is used, but instead of simply shifting one position and adding without regard to other considerations, the high-order digits of the intermediate remainder are examined. If the intermediate remainder digit in the position corresponding to the highest-order 1 in the divisor is 1 and if all higher-order digits in the intermediate remainder are also 1, the addition (if made) would certainly produce a positive result to indicate that the corresponding quotient digit is 1. Therefore, the addition need not be made. Instead, a second shift of one position may be performed before making the addition. If the subtraction which produced the negative intermediate remainder is called $-D$, the shift of two positions with an addition would be $+D/4$. Because $-D + D/4 = -D/2 - D/4$, the net effect has been to perform two subtractions in the positions appropriate for determining the two quotient digits after the 0 which resulted in a negative intermediate remainder. The second of these two quotient digits will be 1 or 0 according to whether the $+D/4$ addition produces a positive or negative remainder, respectively.

Moreover, if in a negative intermediate remainder one or more digits to the right of the highest-order divisor 1 are also 1's, a corresponding number of additional quotient digits, all 1's, can be determined directly without an addition operation. The reasoning is essentially an extension of that used in the previous paragraph.

In the examples where the highest-order quotient digits are 1's, the above technique for deliminating addition-subtraction operations will not work unless a higher-order quotient digit (beyond the intended capacity of the computer) is artificially determined to be 0. Fortunately, one of the division overflow detection methods described earlier is substantially the same as a determination of this higher-order quotient digit. Therefore, when this overflow detection method is included as a first step, all quotient digits offer the same possibilities with regard to addition-subtraction eliminations.

The scheme is illustrated by the following example of dividing the 15-digit number 101101001100100 by the five-digit number 11001 to produce a 10-digit quotient. Intermediate remainders are in 2's complement form, and the digits to the left of the vertical bars are the sign digits. Round-off is omitted.

```
                                          quotient
                      101101001100100
Step 1   subtract:    11001
                    1|111011001100100
Step 2   count 1's:    11001
                     1|1011001100100     11????????
Step 3   add:           11001
                      0|001011100100     111???????
Step 4   count 0's:     11001
                       0|01011100100     1110??????
Step 5   subtract:       11001
                        1|1111000100     11100?????
Step 6   count 1's:       11001
                         1|1000100       11100111??
Step 7   add:              11001
                           1|110110      111001110?
Step 8   count 1's:         11001
                              .....      1110011101
```

The subtraction performed on step 1 establishes, through the negative intermediate remainder, that no division overflow is present. On step 2, no addition-subtraction operation takes place. Instead, the divisor is merely shown "lined up" with the appropriate digits to illustrate the relative positions of the high-order 1's in the intermediate remainder resulting from step 1. The three high-order 1's to the right of the negative sign digit indicate that the first two quotient digits are 1's. The intermediate remainder resulting after step 2 is the same as after step 1 except that the two highest-order digits are omitted as would be the result of shifting the intermediate remainder two positions to the left (see Figure 6-27). On step 3 the divisor is added with the relative digit positions shifted as indicated. The result is positive, which is the indication that the third quotient digit is 1. On step 4, no addition-subtraction operation takes place. Because the present intermediate remainder is positive, a subtraction would normally take place but the high-order 0's indicate that the fourth quotient digit is 0. On step 5, the divisor digits have been shifted one position relative to the intermediate remainder digits, and a subtraction is performed. The negative result indicates that the sixth quotient digit is 0. No addition-subtraction takes place on step 6, but the high-order 1's in the current intermediate remainder indicate that the next three quotient digits, the sixth, seventh, and eighth, are all 1's. On step 7 an addition takes place after a shift of three positions, and the negative result indicates that the ninth quotient digit is 0. On step 8, where an addition would normally take place because of the negative intermediate remainder, no addition-subtraction is performed, because the high-order 1's indicate that the next quotient digit, the tenth, is 1. The final remainder is not computed in this example, but if there were a need to determine more quotient digits for round-off or other purposes, the above-described process could be extended without material alteration.

Further Reductions in the Number of Addition-Subtraction Operations in a Division Operation

The procedure described in the previous section results in the minimum possible number of addition-subtraction operations for most digit combinations that might be encountered in a division operation. Many and perhaps all of the remaining digit combinations can be handled with the minimum number of addition-subtractions by adding one elaboration to the procedure. This elaboration involves observing those instances in which the two highest-order divisor digits are 11 and the three highest-order digits of a positive dividend-remainder are 010, where the left-hand 0 in 010 is in a position corresponding to an order one higher than the highest-order divisor digit. Because $010xxx \cdots < 11xxx \cdots$ regardless of the digits represented by the x's, the corresponding quotient digit is always 0 in such instances. Therefore, the quotient digit can be set to 0 without performing the subtraction that would otherwise take place. All divisor digits are then shifted to the right one position (or the intermediate remainder digits to the left one position) in preparation for the determination of the next quotient digit.

In some instances determination of a 0 quotient digit by this elaboration immediately follows the determination of one or more 0's without subtractions in the manner described in the previous section. To combine

354 IMPLEMENTING BINARY ARITHMETIC

the determination of all adjacent quotient 0's into a single step, the division procedure can be modified to observing those instances where the highest-order 1 in the intermediate remainder is preceded by a 0 and followed by a 0 (with the two highest-order divisor digits being 11), and in this instance the number of quotient 0's and the number of positions to be shifted is one more than indicated in the previous section.

As an illustration of the elaboration, consider the division of 100010001 by 1101. If the division operation is carried out by the procedure described in the previous section, five addition-subtractions are found to be necessary if testing for overflow is included. However, with the present elaboration of the procedure, only three addition-subtractions (all subtractions) are needed, and the steps are as follows.

```
                                              quotient
                              100010001
Step 1        see text:       1101
                              ─────────
                              100010001
Step 2        subtract:       1101
                              ─────────
                              0|01000001    1????
Step 3        see text:          1101
                              ─────────
                              0|1000001     10???
Step 4        subtract:         1101
                              ─────────
                              0|001101      101??
Step 5        count 0's:         1101
                              ─────────
                              0|01101       1010?
Step 6        subtract:         1101
                              ─────────
                              0|0000        10101
```

As before, each digit to the left of a vertical bar is a sign digit, although all remainders happen to be positive in this illustration. On step 1 the two highest-order 1's of the divisor are lined up with 010 in the divdend (the left-hand 0 being in a position not physically represented). The digits indicate a 0 quotient digit, but because it is the first step, the indications is that no overflow is detected. No subtraction takes place on step 1. On step 2 the divisor is subtracted in the usual manner, and the positive intermediate remainder indicates that the first quotient digit is 1. On step 3, the high-order 11 in the divisor happens to be lined up with 010 in the high-order intermediate remainder. Thus, the second quotient digit is determined to be 0 without performing a subtraction. The third quotient digit, a 1, is determined by means of a subtraction on step 4. On step 5, the fourth quotient digit, a 0, is determined in the manner described in the previous section. The fifth quotient digit, a 1, is determined by a final subtraction on step 6.

The division method described in this and the previous section is roughly the inverse of the multiplication method which was described on pages 326 to 330. For all examples encountered in one particular study, the number of addition-subtractions required for a division having a given quotient was the same as for a multiplication in which the given quotient is used as the multiplier. However, this division method has apparently never been studied exhaustively, so instances could conceivably arise in which the number of addition-subtractions has not been minimized.

The maximum number of addition-subtractions required is $n/2$ for even n and is $1 + n/2$ for odd n, where n is the number of digits in the quotient. This maximum is encountered when the quotient consists of alternating 1's and 0's.

The design of a switching network to detect 11 in the highest-order divisor positions and 010 in the dividend-remainder and to cause appropriate shifting is a straightforward problem solvable by techniques described in earlier chapters.

Determining Three or More Quotient Digits per Step

The division method described in the preceding section has the effect of determining at least two quotient digits per step if the steps on which only shifting occurs are integrated with the steps on which an addition-subtraction operation occurs. The number of quotient digits determined per step is much greater than two when the quotient happens to contain long sequences of 1's or long sequences of 0's. In applications where the objective is to perform a complete division operation with a given small number of sequentially performed addition-subtraction operations regardless of the quotient digits, further elaborations in the division process must be employed.

A "brute force" but straightforward division method can be used to develop any number of quotient digits per step, and this method will be outlined for the determination of three quotient digits per step. (The practicality of the method decreases rapidly as the number of quotient digits per step is increased above three.) On any given step, the divisor is "lined up" with the dividend-remainder (i.e., the dividend on the first step and the intermediate remainder on each subsequent step) two positions to the right of where it would be in the division methods described previously. Then, with D being the divisor in this position, either $0D$, $1D$, $2D$, $3D$, $4D$, $5D$, 6, or $7D$ is subtracted—instead of just

$0D$ or $1D$ as previously. The particular multiple of the divisor which is chosen on any given step is the largest multiple which leaves a positive remainder. The three quotient digits determined on the given step are 000, 001, 010, 011, 100, 101, 110, or 111, respectively, in accordance with which multiple is chosen. The problem is in generating the multiples and in selecting the correct one at high speed and with a reasonable amount of equipment.

In a parallel machine the multiples of the divisor can be generated by a switching network that consists primarily of three parallel adders. One adder is used to add $1D$ to $2D$ to generate $3D$, where $2D$ is obtained merely by shifting $1D$ one position to the left. The shifting is not a physical shifting as in a shift register, but is merely a matter of connecting the $1D$ wires one position to the left. The second adder is used to add $1D$ and $4D$ to generate $5D$, where $4D$ has been obtained by a shift of two positions. A similar shift of $3D$ is used to obtain $6D$. The third adder is used to add $1D$ to $6D$ to obtain $7D$. All seven multiples then appear as steady-state signals on seven sets of wires.

The determination of the correct multiple to subtract on any given step can be made by using seven simultaneously operating parallel subtractors, one for each of the seven divisor multiples. The result obtained from the subtraction of the largest multiple which leaves a positive remainder is selected by means of a switching network and is entered in the dividend-remainder register in preparation for the determination of the next three quotient digits. Of course, if the subtraction of $1D$ would leave a negative remainder, the three quotient digits are 000, and the current contents of the dividend-remainder register are not altered except for a three-position shift to the left.

If the divisor has been normalized, the above "brute force" method can be improved upon somewhat by using a switching network to "examine" the high-order digits of the divisor and the high-order digits of the dividend-remainder to select two of the seven multiples for trial subtractions. The number of parallel subtractors is thereby reduced from seven to two. For nearly all digit patterns, the examination of the four highest-order divisor digits (of which the highest-order digit will be 1 with normalized divisors) and the four highest-order dividend-remainder digits is sufficient. For a few digit combinations, an examination of the fifth digit in either the divisor or the dividend-remainder is necessary. The essential information is portrayed in Table 6-3. For example, if the highest-order dividend-remainder digits are 0101 and the highest-order divisor digits are 1011, the corresponding table entry is 3, 4 which means that the correct divisor multiple to be subtracted is sometimes $3D$ and sometimes $4D$ in accordance with the lower-order dividend-

DETERMINING THREE OR MORE QUOTIENT DIGITS PER STEP

TABLE 6-3. Divisor Multiples Indicated by Examining the Four
High-Order Digits of the Divisor and the Four High-Order
Digits of the Dividend-Remainder

High-Order Divisor Digits

	1000	1001	1010	1011	1100	1101	1110	1111
0000	0	0	0	0	0	0	0	0
0001	0,1	0,1	0,1	0,1	0,1	0,1	0,1	0,1
0010	1,2	1,2	1,2	1,2	1	1	1	1
0011	2,3	2,3	2,3	2	1,2	1,2	1,2	1,2
0100	3,4	3,4	2,3	2,3	2,3	2,3	2	2
0101	4,5	4,5	3,4	3,4	3	2,3	2,3	2,3
0110	5,6	4,5,6	4,5	4,5	3,4	3,4	3	3
0111	6,7	5,6,7	5,6	4,5	4,5	4	3,4	3,4
1000	7,+	6,7	5,6,7	5,6	5,6	4,5	4,5	4
1001	+	7,+	6,7	6,7	5,6	5,6	4,5	4,5
1010	+	+	7,+	6,7	6,7	5,6	5,6	5
1011	+	+	+	7,+	6,7	6,7	5,6	5,6
1100	+	+	+	+	7,+	6,7	6,7	6
1101	+	+	+	+	+	7,+	6,7	6,7
1110	+	+	+	+	+	+	7,+	7
1111	+	+	+	+	+	+	+	7,+

High-Order Dividend-Remainder Digits

remainder and divisor digits, which are not examined. In this instance $3D$ would be switched to the input lines of one subtractor and $4D$ to the input lines of the other subtractor. The output from the first subtractor will always be positive, but the output from the second subtractor will be positive only when $4D$ happens to be the correct multiple. Thus, a negative or positive sign of the output number from the second subtractor provides the signal to utilize the $3D$ or $4D$ divisor multiple, respectively, and the corresponding quotient digits are either 011 or 100, respectively. For some digit patterns (some table entries) the correct divisor multiple is independent of the lower-order dividend-remainder and divisor digits.

For a few digit patterns, any one of three different divisor multiples might be correct insofar as can be determined by examining the four high-order digits. In these instances the fifth highest-order digit in either the dividend-remainder or the divisor must be examined. For example, the table entry for 0110 dividend-remainder digits and 1001 divisor digits is 4,5,6. If the next digit in the dividend-remainder is 0 (that is, if the first five digits are 01100), the number to be subtracted is either $4D$ or $5D$, but if this next digit is 1, the number to be subtracted is either $5D$ or $6D$.

The $+$ signs in the lower-left portion of Table 6-3 indicate overflow conditions. For those digit combinations corresponding to table entries of 7,+ the division method can be so designed that $7D$ is subtracted. If this action happens to be incorrect and an overflow is in fact encountered, the overflow can be detected by sensing any 1's that are subsequently shifted out of the left end of the dividend-remainder register. Alternatively, the division method can be modified to subtract $8D$ in the second subtractor in which case a positive output number from this subtractor is the indication that an overflow has occurred. The design of the physical switching network to achieve this and the above functions involves too much detail to include here, but the design is another straightforward problem in which the techniques described in earlier chapters may be utilized.

To illustrate the division method, consider the division of the 15-digit number 100010010000110 by the six-digit number 110001. The x's indicate the digit positions at which the divisor D is "lined up" with the dividend-remainder on each step.

```
                                                     quotient
                              100010010000110
                              xxxxxx
Step 1      subtract 5D:      11110101
                              ───────────────
                              011101000110    101??????
                              xxxxxx
Step 2      subtract 4D:      11000100
                              ───────────────
                              100100110       101100???
                              xxxxxx
Step 3      subtract 6D:      100100110
                              ───────────────
                              000000          101100110
```

On step 1 the four high-order divisor digits, 110, are compared with the four high-order dividend digits, 1000. From Table 6-3, the divisor multiple is found to be either $5D$ or $6D$. A subtraction (not shown) in one subtractor indicates that $6D$ is too great because it leaves a negative remainder, and $5D$ (11110101) is subtracted in the other subtractor. On step 2, the four high-order intermediate remainder digits are 0111, and the table indicates that the correct divisor multiple is either $4D$ or $5D$. A subtraction (not shown) indicates that $5D$ is too great, so $4D$ is subtracted. On step 3, the high-order intermediate remainder digits are 1001. The table indicates the multiple to be either $5D$ or $6D$, and in this instance $6D$ is correct and is subtracted.

Many variations and elaborations can be included in this division method. In particular, in each instance where the three quotient digits are 000, no subtractions need take place. In each instance where a given three quotient digits will be either 000 or 001, the first and second of these may be set to 0's, and the dividend-remainder may be shifted two instead of three positions to the left in preparation for determining the third digit along with the next two quotient digits as a three-digit group. Similarly, when a given three quotient digits will be either 010 or 011, the first may be set to 0 with the dividend-remainder being shifted one position in preparation for determining the second and third digits along with the next subsequent quotient digit as a three-digit group.

Other elaborations involve utilizing negative remainders. Then, certain of the 1's in the quotient can be determined without actual addition-subtraction operations taking place by using the scheme described in the preceding sections. Moreover, the need for two subtractors can be reduced to a need for only one subtractor. Of the two multiples indicated by the table, only the larger one is subtracted. If a positive remainder results, this multiple will have been correct. If a negative remainder results, the three quotient digits, considered as a three-digit binary number, are reduced by 1, and the next three digits are determined by an extension of the nonrestoring technique described earlier. A second table (and a corresponding second switching network) is devised and is employed in each instance that the remainder is negative.

In applications where speed is more important than economy of the divider unit, the designer may wish to have both the positive version of the remainder and the negative version of the remainder available on each step so that subsequent quotient digits, if they are either a string of 1's or a string of 0's, can be determined without addition-subtraction operations. For this purpose, three subtractors are required. Two are used for the two multiples specified by Table 6-3, and the third is used for the next higher multiple. For example, if the table indicates the correct multiple to be either $3D$ or $4D$, the third subtractor subtracts $5D$. In this way, at least one of the three subtractor output numbers will be positive and at least one will be negative.

With elaborations such as these, the number of quotient digits that can be determined per addition-subtraction is significantly greater than three and is probably about five for random digit patterns, but the exact number is not known.

The switching network needed for the development and manipulation of seven divisor multiples is more extensive than is generally considered economical for most applications, although seven multiples have been used in at least a few computers, some of which employed serial modes of operation where the adder-subtractors require relatively few components. More commonly, only three divisor multiples, $1D$, $2D$, and $3D$, are used in the manner suggested but with obvious simplifications. With three multiples and with suitable elaborations of the type described, the number of quotient digits determined per addition-subtraction can approach four for random digit patterns.

Switching-Network Divider

A switching-network divider analogous to the switching-network multiplier (Figure 6-22) can be designed. Of course, parallel subtractors instead of parallel adders would be used, but a more important difference results from the fact that the quotient digits, which correspond to the multiplier digits, are not known prior to the start of the operation. Therefore, the signals which "gate" the divisor, which corresponds to the multiplicand, are not input signals as were the multiplier input signals. Instead, each gating signal is derived from the highest-order carry signal from its respective subtractor. Moreover, the gating signal cannot be used to control whether zero or the dividend should be subtracted, because the subtraction of zero would destroy the gating signal. Accordingly, each gating signal must be used to control whether the "new" intermediate remainder or the "old" intermediate remainder is to be transmitted to the next parallel subtractor, and somewhat more switching equipment is required for this purpose than was required in the switching-network multiplier. On the other hand, the general concept of the switching-network divider is the same as for the multiplier in that, instead of performing the operation in a time sequence of steps with the various registers and other devices operating repetitively, a single large combinatorial switching array is designed so that when steady-state signals representing the dividend and divisor are applied at the input terminals steady-state signals representing the quotient will appear at the output terminals.

The time required to perform a division operation depends on the number of switching devices through which an input signal must pass in succession to reach an output terminal. A cursory consideration of the speed question might indicate that an input signal may need to pass as a carry signal through all orders of each parallel subtractor in succession to result in an undesirably long division time. However, when specific digit combinations are considered, an observation can be quickly made that the maximum signal path length is considerably less than cursory considerations would indicate, but no extensive studies of this subject are known to have been made.

Countless refinements can be incorporated in the divider as outlined in the first paragraph of this section. For example, instead of subtracting zero or the divisor in each portion of the switching network, parallel adder-subtractors may be used with a nonrestoring division method. For another example, the switching network may be designed to employ the division method described in the previous section whereby multiples of the divisor are used to determine two or more quotient digits for each subtraction from the dividend-remainder.

Switching-network dividers have been regarded as highly impractical because of the very large quantity of switching devices needed for division with numbers having enough digits to be useful. However, large scale integration (LSI) techniques now make the use of large quantities of switching devices less objectionable, and the high speed capabilities of switching-network dividers can be particularly attractive in very large systems where the entire system must otherwise stand idle while the arithmetic unit completes a division operation.

Other Considerations Pertaining to the Implementation of Binary Division

Negative numbers can be handled in binary division in much the same way that was described for multiplication. When the negative numbers are in 2's complement form, the roles of addition and subtraction are interchanged in an addition-subtraction operation if either the divisor or the dividend-remainder is negative. If both are negative, the roles are the same as when both are positive. The division method must be modified in minor details with respect to the action to be taken in response to an overflow from an addition-subtraction and with respect to the final remainder when rounding. These details depend on the particular version of the division method being used and can be worked out straightforwardly.

362 IMPLEMENTING BINARY ARITHMETIC

A serial division network analogous to the serial multiplication network of Figure 6-24 can be developed, but the resulting array of components tends to be somewhat more complicated. In serial machines the nonrestoring division method is virtually always adopted because each quotient digit can then be developed by passing the divisor and dividend-remainder through the full adder-subtractor just once instead of twice as would be required otherwise. Although the time consumed in a division operation is then essentially the same as in a corresponding serial multiplication operation, a flip-flop must be added to "remember" whether an addition or a subtraction is in progress and to actuate the full adder-subtractor accordingly.

Division networks analogous to the serial-parallel multiplication networks of Figures 6-25 and 6-26 apparently have not been invented and appear to be impossibilities in a practical sense.

Iterative Division

Totally different division methods, long known to mathematicians, employ iterative processes requiring only additions, subtractions, and multiplications to develop the reciprocal of the divisor. This reciprocal is then multiplied by the dividend to obtain the quotient. One such iterative process is expressed by the formula

$$b_{i+1} = b_i(2 - xb_i)$$

where x is the divisor and the b_i are successively closer approximations to the reciprocal, $1/x$. For the process to converge, the first approximation, b_0, must fall in the range $0 < b_0 < 2/x$.

If Δ is defined as the unknown error in a given b_i, then $b_i = 1/x + \Delta$. After making this substitution in the above formula, the relationship $b_{i+1} = 1/x - x\Delta^2$ may be easily obtained. If Δ is enough smaller than 1 to contain at least a few 0's between the point and the highest-order nonzero digit, Δ^2 will contain approximately twice as many 0's between the point and its highest-order nonzero digit. Then, if x is not greatly different from 1 (as it would not be when normalized fractions are being used), the number of significant digits in the b_i is roughly doubled for each iterative application of the formula. Actually, the same approximate doubling of the number of significant digits per iteration is obtained even when x is greatly different from 1, because Δ is correspondingly altered for a given proportionate error in b_i.

The formula given above can be modified to

$$b_{i+1} = b_i(2 - a_i)$$

where $a_0 = xb_0$ and

$$a_{i+1} = a_i(2 - a_i)$$

ITERATIVE DIVISION 363

The a_i approach 1 as i is increased. The resulting two-formula iterative process is mathematically identical to the one-formula process, but the various numbers enter into the computations in a different sequence so that the individual arithmetic operations performed in the two processes are quite different. In particular, in the present two-formula process, the number x for which the reciprocal is sought enters into the computations only as an initial step, whereas x is needed as a factor during each cycle of iteration in the one-formula process.

Division networks based on either of these iterative processes can be designed. One problem is to determine a reasonably accurate b_0 so that the required number of iterations is not excessive. The usual suggestion is to employ a switching network which has as its input signals the high-order digits (say, the five high-order digits) of x and which has as its output signals the digits of b_0 to a corresponding accuracy. Alternatively, a small bulk digital storage unit may be used. The high-order digits of x are used for the address, and the number stored at each address is the corresponding b_0. If b_0 is accurate to five digits, for example, the number of significant digits in b_1, b_2, b_3, and so on, will be approximately 10, 20, 40, and so on, respectively.

One way of determining how many iterations to perform to attain a specified accuracy in $1/x$ is to subtract each b_{i+1} from its corresponding b_i. If the magnitude of the difference is no greater than the expected round-off error, the process may be terminated. If the magnitude is greater than the round-off error but is not so great that a doubling of the number of significant digits would fail to produce the specified accuracy, the process should be carried through one more iteration. For larger differences, a correspondingly longer continuation of the process is indicated. Alternatively, the maximum number of iterations required can be predetermined for whatever manner b_0 is selected and for whatever range or ranges in which x is expected to lie, and the network can be designed to perform this number of iterations.

Another iterative formula for obtaining the reciprocal of x is

$$b_{i+1} = b_i[3(1 - xb_i) + (xb_i)^2]$$

where again the b_i are successive approximations to $1/x$ with $0 < b_0 < 2/x$. With this formula, if a given b_i is a reasonably close approximation, b_{i+1} will contain approximately three times as many significant digits as b_i.

Although only a few iterations are required to achieve high accuracy in the reciprocal, round-off errors can be troublesome in any iterative process if x is a full n-digit number and if a computer having an n-digit word length is being used to compute the reciprocal.

The total time consumed in computing a reciprocal is another troublesome design aspect of the iterative processes. By judicious attention to detail, the numbers to be used as multipliers can be caused to contain long strongs of all 0's or all 1's so that an appropriately designed multiplication network will function at high speed. Nevertheless, at least two multiplications per iteration are required so that the processes are not particularly attractive from a speed standpoint.

In applications where division is needed only rarely, the computer may preferably be designed with no built-in division facilities at all. Then the above iterative techniques may be used to effect division by means of programming where only multiplication and other more elementary operations are utilized.

The above iterative formulas were presented in decimal notation because of the familiarity of that notation, but the formulas are valid in any radix. In particular, for the binary system, each 2 would be replaced by 10 and each 3 by 11.

Iterative division techniques are discussed in greater detail by E. V. Krishnamurthy in the *IEEE Transactions on Computers*, March 1970, pp. 227–231.

Floating-Point Notation

In applications involving long sequences of arithmetic operations three quite serious problems arise. One problem is to maximize the number of significant digits in each final result. In general, the solution to this problem is to shift the digits in each intermediate result so that the highest-order 1 appears in the left-hand position, that is, to "normalize" the numbers. The second problem is to keep track of the binary point in the various intermediate and final results, and this problem is odious because the amount of shifting required to solve the first problem cannot be predicted and varies in accordance with the particular magnitudes of the numbers entering into the computations. The third problem is to determine the digits that are truly significant in the final results.

Whether or not the numbers are normalized, keeping track of the point can be accomplished by using "floating-point" notation. In this notation, each number is accompanied by a second set of digits that indicates the location of the point. By numbering the digit positions consecutively, this second set of digits is itself a number, and this second number can be visualized as being the exponent of a multiplier equal to the radix. That is, if a "zero position" of the point is defined, a given number N actually having its point at the zero position is equal to $Nr^0 = N$, where r is the radix. If the given number has its point m positions to the right or to the left of the zero position, the number is visualized as Nr^m or Nr^{-m}, respectively. Because r is a constant (equal to 10, meaning "two" in the binary system), it need not be recorded in the representation of each quantity. Each quantity is defined by its corresponding N and m.

Either N or m, or both, may be negative, and the number of digits in N and m need not be the same. Seven or perhaps up to nine binary digits are sufficient for m in nearly all applications. Most general purpose computers employ 35 to 60 digits for N, although many computers intended for control purposes where the original input information is in analog form may employ as few as 16 or even fewer digits for N. In floating-point notation, N is commonly called the "mantissa" because of its similarity to the mantissa of a logarithm, but floating-point numbers are definitely not logarithms.

Usually, the zero position of the point is at the left end of the mantissa so that, for example, the following numbers in floating-point notation would have the indicated values in conventional notation. The left-hand digit of each exponent and mantissa is a sign digit, and the 2's complement notation is used for negative quantities.

Floating-Point Notation		*Conventional Notation*
Exponent	Mantissa	
00000	011001110	+0.11001110
00000	100110010	−0.11001110
00010	011001110	+11.001110
01011	011001110	+11001110000.
11110	011001110	+0.0011001110
10101	100110010	−0.0000000000011001110

Floating-point Addition and Subtraction

When performing an addition or a subtraction operation in floating-point notation, the two numbers must be "lined up" with respect to the point position. The numbers will be lined up when their exponents are the same. If the exponents are not initially the same, one exponent can be altered to equal the other if the digits in the corresponding mantissa are shifted in the correct direction a number of positions equal to the change in the exponent. Ordinarily, a shift of the digits to the left is undesirable, because the most significant 1's in the mantissa may be lost. On the other hand, a shift to the right can generally be tolerated because the digits, if any, that are lost will be in the low-order positions and, in particular, will have less significance than the least significant digit of the other number entering into the addition-subtraction operation. Because a shift to the right of j positions corresponds to a multiplication by 10^{-j}, the floating-point number can be restored to its initial value (except for any loss of low-order digits) by adding j to the exponent. In other words, two floating-point numbers may be lined up by first determining which of the two numbers has the smaller exponent (where "smaller" means "more negative" when one or both the exponents is negative) and then by changing this exponent in a positive direction with a corresponding right shift in the digits of the mantissa.

As an example, consider the addition of floating-point numbers 00111,010100101 and 00100,011101111 where the digits before and after a comma represent the exponent and mantissa, respectively. In this example, the two numbers would be lined up by increasing the exponent of the second number by 00011 (decimal 3) and by shifting the digits of the mantissa to the right three positions to produce 00111,000011101. If the first number is an approximation as might have been obtained, for example, from the conversion of an analog signal, the significance of its lowest-order digit is questionable, and the implied 0's in still lower orders (not physically represented) would have no significance at all. Because the three low-order 1's which have been lost by shifting the second number would have been added to digits having no significance, the digit loss is of little consequence.

In the floating-point addition of two numbers of the same sign (or in a subtraction where the two numbers have opposite signs) a carry from the highest-order digit position does not indicate an overflow as it does in fixed-point operation. Instead, the carry signal is used to cause a 1 to be added to the exponent of the result and to cause the imaginary binary point to be moved to the left one position. As before, the moving of the point to the left is accomplished by physically shifting all digits of the mantissa to the right one position. The lowest-order (least significant) digit of the sum is lost. Thus, the accuracy of the result suffers slightly when the lost digit happens to be a 1, but rounding is of no help. An overflow occurs only when the exponent becomes greater than representable by the number digits in the exponent.

Round-Off in Floating-Point Addition-Subtraction

If one number must be shifted for "line up" purposes, a slight improvement in the accuracy of the sum or difference is obtained in most instances by rounding the shifted number. The rounding is accomplished by adding 1 to the highest-order digit to be lost and by handling carries in the usual manner. In the example of the previous section, the second number would become 00111,000011110 after shifting and rounding, and this number when added to the first number yields a sum of 00111,011000011 where only the mantissas enter into the actual addition.

Peculiarly, for some digit combinations rounding is detrimental to the accuracy of the sum or difference when a shift of only one position is to be made. For example, consider a four-digit number 0111 that is to be shifted to the right one position in preparation for an addition to some other number. The shift to the right with rounding would produce 0100. However, if the given 0111 had originally been obtained by rounding 0110.1 (where the point is used to indicate the relative significance of the digits in this example), 0011 would be a more accurate result than 0100. In other words, two rounding operations have taken place with a net addition of $\frac{3}{4}$ instead of $\frac{1}{2}$ (decimal notation) with respect to the lowest-order binary digit to be retained. For other digit combinations, rounding improves the accuracy even when the shift is only one position. For example, if the four-digit number to be shifted in 0111 where this number had been derived by rounding $0111.0xxx \cdot \cdot \cdot$, with at least one x being a 1, the most accurate shifted number would be 0100, and rounding must be employed to obtain this number. For random digits and for right-shifts of one position, the magnitude of the average error resulting from rounding is virtually the same as the corresponding error resulting from not rounding. In short, although rounding is effective in limiting the loss of accuracy in the presence of right-shifts of two or more positions, rounding is not effective for right-shifts of only one position.

Zero in Floating-Point Notation

In floating-point notation, the representation of zero is not necessarily unique. In general, all digits of the mantissa must be 0's, but the exponent may have any value representable by the digits being used for exponent representation. This situation causes a specialized problem in some applications.

If a number is exactly zero, it contains an infinite number of implied 0's in positions to the right of those physically represented. Then if this number is to enter into an addition-subtraction operation with another number, the digits of this other number should not be shifted even though the exponent of this other number may be smaller (more negative) than the exponent of the number which is exactly zero. To handle this situation, the computer may be designed to detect instances when the number having the larger exponent has all 0's in the mantissa and to cause the sum to be equal to the other number in these instances.

In other applications a number having an indicated value of zero may be only a rounded approximation the same as any other number. In these applications the exponent is significant in that it affords an indication of the maximum value of the number. When this approximate zero enters into an addition-subtraction operation with some other number having a smaller exponent, the digits of this other number should be shifted to the right with the addition-subtraction operation being performed in the conventional manner. If the digits are not shifted, the result will contain meaningless digits which can be harmful in that they create a false illusion of high accuracy.

Analogous remarks apply to the manner in which floating-point zero should be recorded on an output device. If the number is truly zero, the interests of clarity would be served by making the exponent zero regardless of the sequence of arithmetic operations that produced the result. On the other hand, if the zero is only an approximation, a nonzero exponent resulting from a particular sequence of operations should be recorded because it contains information about the maximum possible value of the quantity.

In summary, the manner in which floating-point zero should be handled is a function of the application.

Floating-Point Multiplication and Division

For floating-point multiplication, the mantissas of the two numbers are multiplied by any of the techniques described earlier for fixed-point multiplication. The exponents are added. The sign of the product is determined by the signs of the mantissas. If both exponents are positive, an overflow from their addition implies that the magnitude of the product is greater than representable by the number of digits physically available. If both exponents are negative, an overflow implies that the product magnitude is smaller than representable even though the magnitude is greater than zero and even though all digits may be significant. Minor exceptions to the overflow conditions arise when the product contains a series of 0's at the right- or left-hand ends, respectively, in which case a digit shift in the product mantissa with a corresponding alteration in the exponent may bring the result within range in some instances.

Floating-point division is analogous. The mantissas are divided, and the divisor exponent is subtracted from the dividend exponent to obtain the quotient exponent. Overflow conditions can arise when the mantissa signs are opposite to each other, rather than when they are the same as in multiplication.

Normalized Floating-Point Operation

A number is said to be "normalized" if its digits are shifted so that the highest-order 1 appears in the left-hand digit position. If floating-point notation is being used, the exponent is adjusted accordingly. Then, if the binary point is assumed to be at the left-hand end (the most common assumption), the mantissa, N, of each normalized number falls in the range $0.1 \leq N < 1$.

The automatic normalization of all numbers is useful in achieving the maximum attainable accuracy of the results (for a given word length) when the data undergo multiplication and division operations. In particular, when two n-digit numbers are multiplied together the product will contain up to n significant digits with the highest-order significant digit being the highest-order 1 in the product. (See page 321 for a more detailed discussion of significance.) If the factors in the multiplication are not normalized, the highest-order 1 in the product may be in any of the $2n$ positions of the product. But if the factors are normalized, the product mantissa N_p will fall in the range $0.01 \leq N_p < 1$ so that the n significant digits will lie either in the left-hand n digit positions or in the positions represented by a right shift of one position with respect to the left-hand n positions. The location of the n significant digits is easily determined by sensing the highest-order digit. If this digit is a 1, the product is already normalized. If this digit is a 0, the product can be normalized by shifting all digits to the left one position and by subtracting 1 from the product exponent. In either instance, product round-off can be produced in the usual manner by adding 1 to the highest-order product digit to be deleted.

For division, if the dividend and divisor are normalized, the quotient mantissa N_q will fall in the range $0.1 \leq N_q < 10$. If $N_q < 1$, the quotient will be normalized. If $N_q \geq 1$, the condition is not regarded as an overflow (contrary to fixed-point operation). Instead, the final quotient is normalized by shifting its mantissa digits to the right one position and by adding 1 to the quotient exponent. In either instance, $n + 1$ quotient digits are ordinarily determined with round-off being obtained by adding 1, with carries, to the last digit and then deleting the last digit.

A division with a zero divisor is generally unwanted regardless of whether the zero is exact or approximate. However, in normalized floating-point operation where the divisor is presumed to be normalized, the detection of a zero divisor can be accomplished quite simply by sensing the highest-order mantissa digit.

A subtraction operation with two nearly equal numbers of the same sign (or an addition of nearly equal numbers having opposite signs) may produce a result which, without normalization, will contain a string of 0's to the left of the highest-order 1 in the mantissa. When using normalized floating-point notation the procedure is to shift all digits to the left until the highest-order 1 is in the left-hand positions, and the exponent is reduced accordingly. If the result happens to be zero so that all mantissa digits are 0's, the shifting should not be allowed to continue indefinitely, and means must be included in the computer to count the number of positions shifted and terminate the process at $n - 1$ single-position shifts or the equivalent.

Placing the Binary Point to the Right of the Highest-Order Digit

Although the customary practice is to place the binary point to the left of the highest-order digit, the point may equally well be to the right of the highest-order digit. With normalized floating-point operation the magnitude of each mantissa N would lie in the range $1 \leq N < 10$. The product mantissa N_p resulting from the multiplication of two such numbers would lie in the range $1 \leq N_p < 11.111\cdots$ so that, if $N_p \geq 10$, the product would be normalized without shifting, but if $N_p < 10$, a normalized product could be obtained by a shift to the right of one position. For division, the quotient mantissa N_q would lie in the range $0.1 < N_q < 10$ so that, if $N_q \geq 1$, no shifting is required to achieve normalization, but if $N_q < 1$, a shift to the right of one position is required. Addition-subtraction operations are not affected by placing the point to the right instead of the left of the highest-order digit.

Although certain of the details of the physical design of a computer are dependent on whether the point is to the left or to the right of the highest-order digit, the net cost, speed, or other material aspects are virtually unaffected. The major reason for having the point to the right is that it corresponds with the nearly universal practice employed by persons when using decimal notation. That is, $(6.23)10^8$, not $(0.623)10^9$. The major reason for having the point to the left is that this convention somehow became adopted early in the development of electronic computers so that any change in the convention would cause confusion.

Unnormalized Floating-Point Operation—Determining the Number of Significant Digits in the Final Results

The third problem mentioned in the first paragraph of the section on floating-point notation, namely, the problem of determining the number of digits that are truly significant in the final results, is not solved at all through the use of normalized floating-point notation. In fact, the normalization of numbers can render the final results grossly misleading as to the number of digits that are significant. For an illustration, consider a sequence of operations that first involves a subtraction of two roughly equal numbers each having n significant digits. The difference may have only $n/2$ (or fewer) significant digits, as explained previously. If the difference is then multiplied by another number having only $n/2$ significant digits, the product will have only approximately $n/2$ significant digits, but the n highest-order digits may contain 1's and 0's in a more or less random pattern that give the appearance that n digits are significant. If this product is now subtracted from an approximately equal number having $n/2$ significant digits, the difference may have no significant digits at all, yet the pattern of 1's and 0's in the result may give the erroneous appearance that $n/2$ digits are significant. A subsequent multiplication with another number may then result in an n-digit number with the appearance of having n significant digits, none of which are actually significant.

In realistic problems the loss of digit significance seldom occurs quite as rapidly as suggested by the illustration of the previous paragraph. Nevertheless, in many applications the sequence of arithmetic operations is sufficiently long that the number of significant digits in the final result is substantially less than n, and examples are not particularly rare where all significance has been lost completely.

An indication of the number of significant digits in each intermediate and final result can be obtained by adopting an unnormalized floating-point scheme of operation. The details of the scheme are too numerous and subject to variations and elaborations in accordance with individual requirements to present fully here, but the details for any given application can be worked out straightforwardly once the basic concept is appreciated. The basic concept is simple enough and is as follows.

Instead of normalizing each floating-point number, the digits known or believed to be significant in the mantissa of each floating point number are shifted to the low-order positions. Therefore, in any given number the number of significant digits is inversely related to the number of 0's to the left of the highest-order 1 in the mantissa.

Unnormalized floating-point addition-subtraction operations are substantially the same as before. However, because of the unnormalized nature of the numbers, any carry which results from the position containing the highest-order 1 to the next higher order does not initiate any shifting (except when the highest-order 1 is in the highest-order position). If a carry occurs, the result contains one more significant digit than was contained in either of the operands. Similarly, no shifting occurs when, because of the subtraction of two nearly equal numbers of like sign (or the addition of nearly equal numbers of opposite sign), the highest-order 1 in the result is in a position to the right of the highest-order 1 in either of the operands. The position of the highest-order 1 in the result is an automatic indication of the number of significant digits in the result.

For unnormalized floating-point multiplication the computer must be equipped with means for determining which of the two factors has the more 0's to the left of the highest-order 1 (the fewer significant digits) and with means for shifting the product digits to have the same number of 0's to the left of the highest-order 1. The product exponent is adjusted accordingly. Highest multiplication speed is usually obtained by causing the number having the fewer significant digits to be the multiplier.

For unnormalized floating-point division the computer must similarly be equipped with means for determining whether the dividend or the divisor has the more 0's to the left of the highest-order 1, and means must be included for shifting the quotient digits to have this same number of 0's to the left of its highest-order 1. However, in the case of division an initial shifting of the divisor relative to the dividend is appropriate to cause the highest-order 1 in each number to be in the same position. In this way, either the first or the second quotient digit will be 1. If the first quotient digit is 0, it is not counted as a quotient digit. In either case, the division operation proceeds until the number of quotient digits (after rounding) is equal to the number of significant digits in which ever number, the dividend or the divisor, had the fewer significant digits. The quotient exponent is adjusted accordingly.

To assist in assuring that all apparently significant digits in the final results are truly significant, the number of product or quotient digits retained should be one less than indicated in the previous paragraph (see page 321). That is, each product or quotient mantissa is shifted to the right one position with respect to the positions indicated above. The exponent is caused to be one greater than it would otherwise be, and the round-off procedures are modified accordingly.

On the other hand, the desirable thing to do in some applications is to increase by one the number of digits retained in each product or quotient (except when the number of retained digits is otherwise n, in which case no increase can be made). The reason is derived from a refinement in the definition of "significance." For example, when multiplying 0.1101 by 0.1011, each of which is itself a rounded approximation, the unrounded product is 0.10001111 and the rounded four-digit product is 0.1001, but the true value of the product may be as small as 0.1000 or as great as 0.1010 according to whether the true values of the factors were at their lower or upper limits, respectively. Although the fourth digit to the right of the point is of questionable significance in indicating the true value of the product, it is of significance in indicating the most probable value of the product. That is, if each of the two original factors can lie at any point in their respective ranges with equal probability, the product does not lie at any point in its range with equal probability, but instead the product has a higher probability of lying near the midpoint of its range than near either extreme. In the sense of indicating the most probable value of the product, all eight of the product digits are significant.

The retention of all $2n$ digits in the product of two n-digit numbers is not only expensive but is generally undesirable for the reasons explained previously. However, some value can be realized in some applications by saving, in the above example, a fifth digit to produce a rounded product of 0.10010. Hopefully, over a long sequence of arithmetic operations, statistical considerations will cause the final computed results to have a very high probability of being accurate representations of the wanted results.

(In a special but widely encountered application the temporary retention of all $2n$ product digits is particularly desirable and is reasonably economical. This application involves the computation of

$$a_1 b_1 + a_2 b_2 + \cdots + a_j b_j$$

where j may be quite large—perhaps one hundred or more—and the individual factors may be either positive or negative. Rather than compute a series of rounded products to be summed, the preferred procedure is to accumulate the $2n$-digit products and perform a single rounding operation as a final step. The individual $2n$-digit products are not placed in a storage unit, however. Instead, the multiplier unit is modified so that the accumulator which develops the product will store the full $2n$ digits of the first product, $a_1 b_1$. Then the partial products of the second product, $a_2 b_2$ are accumulated "on top of" the first product. The accumulator portion of the multiplier then contains $a_1 b_1 + a_2 b_2$ without ever having formed $a_2 b_2$ as a separate number. Subsequent products in the series are handled analogously. The ability to compute sums of products in this way allows a substantial improvement in the accuracy of the final results, particularly when the individual factors are exact numbers rather than rounded approximations. This feature is equally applicable to fixed-point or floating-point operation.)

In general, the designer is faced with the need to make a compromise between attaining usually highly precise results that, with unfortunate number combinations and operation sequences, may be wildly inaccurate or attaining results that, although having less than achievable accuracy in most instances, assuredly have the accuracy indicated by the position of the highest-order 1 in each number.

One possibility for avoiding the need to make such a compromise is to employ a normalized floating-point scheme where each number is accompanied by a third set of digits that records an estimate of the number of significant digits in the mantissa. The nature of the estimate would be dependent upon the degree of certainty implied by the definition of "significance" and, in the four-digit example cited above, could reasonably be three, four, or five in accordance with the criteria desired for the product. Schemes such as this are too expensive and too specialized to be practical as built-in features in most computer applications, although the schemes may be practicably and usefully simulated by programming.

Exercises

1. Determine the decimal equivalents of 0.11100111101 and 11100111101.
2. Use the method suggested on page 274 to determine the approximate decimal equivalents of 2^{38}, 2^{22}, 2^{46}, and the binary number 1100000000000000000 (19 digits).
3. Determine the binary equivalents of 523, 9047, 66.23, and 0.905.
4. Prove by Boolean manipulation that a full adder is equivalent to two half adders plus a suitably connected OR device.
5. Prove by Boolean manipulation that the signals on the lines marked $\overline{\text{SUM}}_4$ and $\overline{\text{SUM}}_8$ in Figure 6-6 do in fact represent SUM_4 and $\overline{\text{SUM}}_8$.
6. In the carry bypass pattern of Figure 6-9b find at least one "worst" (that is, "equally bad") instance other than the one described in the text. Do the same for the pattern in Figure 6-9c.
7. Prove that the carry network of Figure 6-5 and the adder network of Figure 6-6 each function correctly when A-I modules are substituted for the O-I modules shown.
8. Show how A-I modules may be used in the carry bypass scheme of Figures 6-11 and 6-12.
9. In Figure 6-14 assume that signal C_b is additionally connected to an added (third) input to the O-I device to which A_5 and B_5 are input signals. Would the adder still function properly? Prove your answer.
10. In the carry completion detection network for the O-I module adder as shown in Figure 6-17, two minor modifications are needed when A-I modules are substituted for the O-I modules. Determine these modifications.
11. Subtract 011101 from 100010 (a) directly, (b) by adding the 2's complement of the subtrahend, and (c) by adding the 1's complement of the subtrahend.

12. By Boolean manipulation prove that the three networks in Figures 6-18a, b, and c are each equivalent to a full subtractor.

13. Draw the combinatorial switching network multiplier analogous to Figure 6-22 but employing the "carry save" technique described in page 325. (Hint: One more row of full or half adders is required, but each row requires one less full or half adder.)

14. Design a three-pole, eight-throw switch (with AND, OR, and NOT devices) actuated by three binary signals—as would be needed in the multiplication scheme of Figure 6-26.

15. Work out the steps in the multiplication of -13 by -13 in the manner of the example on page 339.

16. Work out the steps in the multiplication of $+13$ by -13 in the manner of the example on page 340.

17. Work out the steps in the multiplication of $+13$ by $+13$, -13 by $+13$, and $+13$ by -13 in the manner of the example on page 341.

18. Prove mathematically the validity of the multiplication scheme, intended for serial computers, which was described on page 340.

19. In the nonrestoring division example on page 343, continue the division process to generate at least three more quotient digits with round-off.

20. By employing the method illustrated on page 352, show the steps in the division of 1001100001011 by 11001 to obtain the quotient of 11000011 (zero remainder) with no more than four addition-subtraction operations. Use the same method to divide 101110011111 by 11001 to obtain 1110111 (zero remainder) with three addition-subtraction operations.

21. Using the technique illustrated on page 354, show the steps in the division of the 15-digit number 100110101000010 by the five-digit number 11101 to obtain the 10-digit quotient 1010101010 (zero remainder) with only five addition-subtraction operations.

22. Work out a simplified version of Table 6-3 as would be appropriate for using only three divisor multiples, $1D$, $2D$, and $3D$, and for determining only two instead of three quotient digits per step. Work out the analogous table for negative intermediate remainders used in the manner suggested in the text.

23. Draw a block diagram showing the switching functions required for a four-digit switching network divider as outlined on pages 360 and 361.

7
IMPLEMENTING DECIMAL ARITHMETIC

In practical digital systems, decimal digits are virtually always represented by sets of binary signals. The manner in which decimal arithmetic is implemented is to some extent dependent on the "code" which is used in employing the binary signals to represent the decimal digits. Here, "code" implies neither secrecy nor error control but simply constitutes a list of the ten selected binary signal sets as they are used to represent the ten decimal digits, 0 through 9.

In the immediately following sections brief descriptions are given of the more important codes which have been used or seriously considered for use in decimal machines, although the bulk of this chapter will be confined to the implementation of decimal arithmetic with the so-called 8-4-2-1 code where each decimal digit is individually represented in the binary system. This code is commonly called the "binary-coded decimal" system of notation. Although many alternative codes can be shown to have this or that advantage in various specialized aspects of decimal arithmetic implementation, the 8-4-2-1 code is probably the most straightforward of all. Moreover, when all aspects of a decimal machine are considered, the net advantages (if any) in overall speed, economy, or other factors that can be realized by employing alternative codes are generally found to be small. Certainly, of the many possible codes, the 8-4-2-1 code is by far the most commonly used.

Decimal Codes Employing Four Binary Digits

The minimum number of binary digits needed to represent a decimal digit is four, because three binary digits would be capable of representing only $2^3 = 8$ different characters, whereas up to $2^4 = 16$ different charac-

DECIMAL CODES EMPLOYING FOUR BINARY DIGITS 377

ters could be represented with four binary digits. However, in any decimal code utilizing four binary digits, six of the possible digit combinations (or corresponding signal combinations) are not used.

The unused signal combinations imply an inherent inefficiency of the decimal system in comparison with the binary system of notation (in the absence of error control features). However, the loss of efficiency is not as great as the apparent 60% suggested by the ten used and six unused signal combinations. A more meaningful efficiency comparison is obtained by considering the information content of the various digits where information content is proportional to the logarithm (to any base) of the number of different "messages" represented by the digits—see Chapter 5. The number of different messages representable by four binary digits is 16 in the binary system and 10 in the decimal system. The information content ratio is then

$$\frac{\log 16}{\log 10} \simeq 1.204$$

which means that approximately 20.4% more information can be represented by four binary digits in the binary system than by four binary digits being used to represent a decimal digit. An alternative viewpoint is that, in comparison with its use in a purely binary system, a binary digit has an average information content of $1/1.204 \simeq 0.83$ bit when used as one of the four binary digits in a decimal digit representation. That is, when changing from the binary to the decimal system of notation, the loss of information content of the binary digits is approximately 17%.

The preceding paragraph is significant when considering the storage or transmission of number of a given magnitude. If the numbers are coded with each decimal digit in a four-binary-digit code, a digital storage unit must have the capacity to store approximately 20.4% more binary digits than would be required if the same numbers were represented in the binary system of notation. When transmitting the numbers in parallel, the amount of transmission equipment (lines, line drivers, amplifiers, etc.) would need to be approximately 20.4% greater. If the numbers are transmitted serially at a given binary digit repetition rate, the amount of transmission equipment would not be altered, but the required transmission time would be increased approximately 20.4%.

As an example, a 33-digit binary number can represent $2^{33} = 8{,}589{,}934{,}592$ different magnitudes or messages. Roughly this same amount of information (but actually up to 10,000,000,000 different magnitudes or messages) can be represented by 10 decimal digits, which in a four-binary-digit code require 40 binary digits—a rough approximation to a 20.4% increase over the 33 binary digits.

Although the storage-transmission inefficiency of the decimal system is some deterrent to its use, the 20.4% is probably not crucial in most applications. Instead, more serious objections to the use of decimal notation is encountered in the design of the arithmetic portion of a computer where the complications introduced by the unused digit combinations require a switching equipment increase considerably greater than 20.4%. This problem is encountered not only in the main arithmetic unit but also in the addressing and address-modification networks. The net result is that the decimal system is selected only in applications where its use is dictated by a close association with input-output data that is handled by persons accustomed to using only the decimal system. Even in these applications the binary system is sometimes preferable for internal machine operation with appropriate decimal-binary conversions (see next chapter) being made for input-output data.

Actually, apart from the lack of familiarity with it, the binary system is objectionable for use by people because of the relatively long sequences of nothing but 0's and 1's required to represent quantities having a suitable degree of precision. Not only are these sequences time-consuming to handle, but they are quite difficult to handle in an error-free manner. A number system having a radix of either eight or sixteen would provide a good compromise between a reasonable number of different digits and a reasonable number of digits in precisely-represented quantities. Either radix eight or sixteen could be implemented in a machine quite straightforwardly. About the only requirement would be a binary machine with the binary digits grouped in three's or four's, respectively. Because the usage of the decimal system seems traceable to the fact that people have ten fingers, the advent of electronic computers has given cause to wish that people had only eight fingers or possibly sixteen fingers.

The Number of Four-Bit Decimal Codes

To represent the decimal digit 0, any one of the 16 combinations of four binary digits may be assigned. Then to represent the decimal digit 1, any one of the 15 remaining combinations may be assigned. Next, for decimal digit 2, the selection may be any one of the 14 remaining combinations, and so on. Thus, for the ten decimal digits 0 through 9 considered collectively, the number of different possible assignments or "codes" is

$$(16)(15)(14) \cdots (7) = \frac{16!}{6!} \simeq (2.9)10^{10}$$

Of the approximately 29 billion four-bit decimal codes (where "bit" is again used synonomously with "binary digit" as is common practice), only very few have been found to offer any particular advantages at all for use in a decimal computer.

Weighted Four-Bit Decimal Codes

A code is said to be of the "weighted" type if "weights" can be assigned to the four binary digit positions in such a manner that the sum of those weights corresponding to binary 1's equals the decimal digit being represented. An example is the following code having the weights 7, 2, 1, and —3, respectively.

Decimal Digit	Coded Representation			
0	0	0	0	0
1	0	0	1	0
2	0	1	0	0
3	0	1	1	0
4	1	0	0	1
5	1	0	1	1
6	1	1	0	1
7	1	0	0	0
8	1	0	1	0
9	1	1	0	0
weights:	7	2	1	—3

Note that a weighted code does not necessarily produce a unique binary digit combination for each decimal digit. For example, with this code, 0111 would be an alternative representation for decimal digit 0.

A total of 88 different weighted four-bit decimal codes exist—not counting relatively trivial variations attainable by alternative representations for individual digits or attainable by interchanging binary digit positions. Of the 88, 17 employ only positive weights, while the remaining 71 require at least one negative weight. The 88 four-bit weighted codes are listed in Table 7-1. Those codes marked with an asterisk(*) are "self-complementing" in that, for any decimal digit d in the code, the inversion of all 1's and 0's produces $9 - d$, called the "9's complement" of d. For example, in the 8-7-(—4)-(—2) code, decimal digits 0 and

380 IMPLEMENTING DECIMAL ARITHMETIC

9 are represented by 0000 and 1111, respectively, decimal digits 1 and 8 are represented by 0111 and 1000, respectively, and so on. The self-complementing feature is useful in subtraction.

Excess-Weighted Codes

A code is called an "excess-weighted" code if, when adding the weights corresponding to the 1's, the sum is $d + k$, where d is the decimal digit and k is a positive integer other than zero. One excess code which has been found to have practical utility in some applications is the so-called "excess-3" code, which is analogous to the conventional 8-4-2-1 code except that the weights produce a sum of $d + 3$ instead of d. Specifically, the excess-3 code is as follows.

	Decimal Digit	Coded Representation
	0	0011
	1	0100
	2	0101
	3	0110
Excess-3 code:	4	0111
	5	1000
	6	1001
	7	1010
	8	1011
	9	1100

(Actually, $k = 3$ for several excess codes other than the common excess-3 code listed here.)

In addition to being self-complementing as defined in the previous section, the excess-3 code offers a minor advantage in addition. When adding two decimal digits, binary adders may be used for the individual binary digits (as in the 8-4-2-1 code), but when the decimal sum is equal to or greater than decimal 10 (ten) the apparent binary sum will be equal to or greater than binary 1000 (sixteen). Therefore, the carry into the fifth-order binary position can be used directly as the decimal carry to the next higher order decimal position. Unfortunately, compen-

TABLE 7-1. A Listing of the 88 Four-Bit Weighted Decimal Codes

3321*	432-1	441-2	542-3	621-4	63-1-1
4221*	442-1*	443-2*	543-3*	632-4	
4311*	522-1	531-2	621-3	652-4*	54-2-1
4321	531-1	541-2	642-3*	653-4	63-2-1
4421	532-1*	543-2	651-3*	721-4	64-2-1
5211*	542-1	621-2	654-3	751-4*	73-2-1
5221	622-1*	631-2	721-3	821-4	74-2-1
5311	631-1*	632-2*	751-3	832-4*	84-2-1
5321	632-1	641-2*	842-3	852-4	
5421	642-1	643-2		861-4	73-3-1
6221	732-1	731-2*			75-3-1
6311	742-1	741-2		643-5	
6321	842-1	841-2		763-5	72-4-1
6421		843-2		842-5*	82-4-1
7321					86-4-1*
7421				543-6	
8421				753-6*	84-3-2
				841-6	
				843-6*	81-4-2
	* self-complementing				83-4-2
				653-7	85-4-2
					87-4-2*

sating disadvantages are encountered when considering all aspects of utilizing the excess-3 code in a computer. The topic will not be carried further here because it involves a large amount of detail with very little in the way of basically different principles in comparison with the 8-4-2-1 code.

A more generalized study of excess codes reveals many curious properties of them. For one thing, any of the codes having a negative weight, as listed in Table 7-1, can be transformed into an excess code by inverting all bits in positions corresponding to negative weights. The resulting code has the same weights as before, but all weights are positive, and k is equal to the sum of those weights that were negative in the original code. Any excess code can be converted to a code having negative weights by inverting the bits in those positions which have binary 1's in the decimal representation for 0. In the excess-3 code for example, decimal 0 is represented by 0011, so inverting the bits in the two right-hand positions for all ten decimal digits produces the 8-4-(—2)-(—1) code, as can be checked by writing out this code in detail. Two duplications happen to occur in the conversion process with the net result that only 86, not 88, excess four-bit decimal codes exist.

For a more complete exposition of excess codes, reference is made to a paper by B. Lippel in the *IEEE Transactions on Electronic Computers* June, 1964, pp. 304–306.

Four-Bit Shifting Decimal Codes

Four-bit decimal codes can be devised whereby the binary-digit representations for successive decimal digits are obtained by successive shifts of the binary digits to the right (or to the left). These codes are useful in certain types of counters based on shift registers. For a discussion of these codes, reference is made to Richards' book *Electronic Digital Components and Circuits* (1967), pp. 417–419.

Five-Bit Decimal Codes

Although four bits are sufficient for the representation of a decimal digit, the use of five or more bits is sometimes advantageous, particularly for error detection and correction. The principles covered in Chapter 5 apply directly here, but at least one five-bit code, called the "two-out-of-five" code deserves special attention. With five bits, exactly ten different bit combinations exist for which two of the five bits are 1's. These ten combinations can be assigned to the ten decimal digits in any pattern to produce an error-detecting code in that, if a five-bit group contains less than or more than two 1's, an error is known to have occurred. Two errors in a single group will escape detection if a 0-to-1 errors happens to be compensated by a 1-to-0 error in another digit of the same group, but in some applications error patterns of this form are highly improbable. Instead, in these applications the effects of a malfunction are more likely to be limited to the changing of 1's to 0's or limited to the changing of 0's to 1's, in which case the code provides for the detection of any number of errors.

If one of the four-bit weighted codes were such that each digit representation contains either one or two 1's, a two-out-of-five code could be generated merely by adding a parity check digit to cause the total number of 1's to be even (two). Unfortunately, no such weighted four-bit code exists. However, the 8-4-2-1 and 7-4-2-1 codes are each close to having the desired characteristics, and by employing a reasonably simple switching network either code can be transformed as required. Specifically, the two-out-of-five codes which resemble the 8-4-2-1 and 7-4-2-1 codes, respectively, are listed as follows.

	Decimal Digit	8421P	7421P
	0	10100*	11000*
	1	00011	00011
	2	00101	00101
	3	00110	00110
Two-out-of-five codes:	4	01001	01001
	5	01010	01010
	6	01100	01100
	7	11000*	10001
	8	10001	10010
	9	10010	10100

The digits in the columns labeled P are the parity check digits for codes that are of the weighted 8-4-2-1 and 7-4-2-1 forms, respectively, except for the digit combinations indicated by the asterisks (*). In applications where the digits will undergo arithmetic operations the 8-4-2-1-P two-out-of-five code tends to be preferable because of the relatively simple arithmetic switching networks required for the corresponding weighted 8-4-2-1 code. In other applications the 7-4-2-1-P two-out-of-five code is preferable because the switching networks needed for transformations between this code and the corresponding weighted 7-4-2-1 code are relatively simple.

The design of a switching network suitable for transforming a group of signals in a two-out-of-five code to a weighted code or from a weighted code to a two-out-of-five code (or from any code to any other code, for that matter) is a straightforward problem for which the techniques described in Chapters 2 and 3 are applicable.

Three-out-of-five codes (obtainable by inverting all binary digits in the two-out-of-five codes) have characteristics analogous to the two-out-of-five codes—there being exactly ten different bit combinations for which three of the five bits are 1's.

The Biquinary Code

The biquinary code is a seven-bit decimal code in which the binary digits are divided into a group of two and a group of five. One and only one digit in each group is 1 for each decimal digit, and the code thereby has error-detection properties in that the appearance of more or less than one 1 in either group indicates an error. As the biquinary concept is usually used, the various digit positions can be visualized has having weights of 5 and 0, respectively, in the "bi" part and weights of 4, 3, 2, 1, and 0, respectively, in the "quinary" part. The biquinary code is then as follows.

	Decimal Digit	Coded Representation
	0	0100001
	1	0100010
	2	0100100
	3	0101000
	4	0110000
Biquinary code:	5	1000001
	6	1000010
	7	1000100
	8	1001000
	9	1010000
	weights:	5043210

A major feature of the biquinary code is that arithmetic networks (not just storage and transmission units) employing this code can be designed to have certain error-detection properties that are difficult to realize with codes having fewer than seven bits. However, the code is generally limited to the detection of errors that originate in the arithmetic networks themselves (or in the transmission paths to and from these networks) and is not capable of detecting errors that originate in any of the many miscellaneous control networks that are a part of the arithmetic unit as a whole. Because the various networks may be composed of the same types of switching modules, an error is as likely to originate in a control network module as in an arithmetic network module, and an error-detection scheme that fails to check all modules is of limited value. For this reason and also for the reason that the biquinary arithmetic networks are not particularly economical in the first place, the preferable approach to error control is most often found to be the duplication or triplication of the entire arithmetic unit. For a description of the manner in which the biquinary code may be used in an error-detecting adder and an error-detecting multiplier, reference is made to the authors's book *Arithmetic Operations in Digital Computers* (1955), pp. 225–230 and 263–266. The use of the biquinary code in a decimal counter is described in the author's *Electronic Digital Components and Circuits* (1967), pp. 424–427.

The "quibinary" code has weights 8, 6, 4, 2, and 0 in the "qui" part and weights 1 and 0 in the "binary" part, but otherwise has characteristics similar to the biquinary code.

Decimal Addition—8-4-2-1 Code

Because of the similarity between the 8-4-2-1 decimal code and the pure binary number system, the addition of two decimal digits in this code is very similar to the addition of two binary numbers. However, a problem arises whenever the sum of the two decimal digits (plus possibly a carry from the next lower decimal order) is greater than 9. In particular, if the sum lies in the range 10 to 15, inclusive, the resulting binary number will be in the range 1010 to 1111, whereas the desired indication for the sum in the 8-4-2-1 decimal code would be 0000 to 0101, respectively, with a decimal carry to the next higher order. Moreover, if the sum lies in the range 16 to 19, the resulting binary number will be in the range 0000 to 0011 with an apparent carry to the "16's" binary position, whereas the desired indication for the sum in the 8-4-2-1 code would be 0110 to 1001, respectively, again with a decimal carry.

A comparison of the results obtained from a pure binary addition with the results wanted to produce decimal 8-4-2-1 addition reveals that the correction to be applied is to subtract binary 1010 (decimal 10) whenever the sum is greater than binary 1001 (decimal 9). The binary "16's" carry signal may be used as the decimal carry signal in those instances where a "16's" carry signal is generated (that is, when the sum is 16 to 19). A decimal carry signal must be generated artificially when the sum is 10 to 15.

A slight simplification in the network generally results when the subtraction of decimal 10 is accomplished by adding 6 and subtracting 16. The addition of 6 is performed by adding binary 0110 in the conventional manner, but the subtraction of 16 is accomplished merely by ignoring the "16's" carry signal which results from the addition of binary 0110.

A Decimal Adder

With any four-bit decimal code, an adder can be formed in the manner of a nine-input, five-output switching network as illustrated in Figure 7-1. Eight of the input lines are for the signals representing decimal digits A and B, and the ninth input line transmits the carry signal from the next lower order. Similarly, four of the output lines are for the sum decimal digit S, and the fifth output line transmits the carry signal to the next higher order. The design of such an adder is another straightforward problem of the category discussed in an earlier chapter.

386 IMPLEMENTING DECIMAL ARITHMETIC

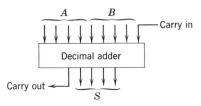

FIGURE 7-1. Decimal digit adder.

Actually, the digits A, B, and S need not be in the same code but may be in three different codes.

Although the visualization of a decimal adder as a conventional nine-input, five-output combinatorial switching network is useful in some applications, the design problem is often greatly simplified by making use of special code characteristics—as in the 8-4-2-1 code where the addition is performed in the manner suggested in the previous section. The resulting network is shown in Figure 7-2, where the "uncorrected" sum appears on lines U_8, U_4, U_2, and U_1, and the "16's" binary carry appears on line C_{16}. A cursory examination of the signal combinations corresponding to a sum greater than 9 reveals that the decimal carry

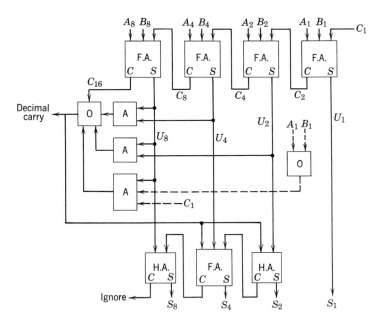

FIGURE 7-2. Decimal digit adder employing the 8-4-2-1 code.

should be 1 whenever $C_{16} + U_8 U_4 + U_8 U_2$ (Boolean notation) is 1, and a corresponding switching array is found in Figure 7-2 to develop the decimal carry. The corrective 6 should be added whenever the decimal carry is 1, and the carry signal itself can be used to form the 6 by transmitting it to the "4's" and "2's" positions at the half and full adders along the bottom of the figure.

Improving Carry Propagation Speed

The dotted lines in Figure 7-2 indicate one scheme by which a decimal carry input can be propagated through only one level of AND-OR switching to the carry output line in those instances when the carry output signal depends on the carry input. These instances are, of course, when the sum of A and B is 9, that is, when U_8 and U_1 are each 1. In the most straightforward version of the high-speed carry scheme, signals U_8, U_1 and C_1 would be applied as inputs to the bottom AND device, but in the versions shown, U_1 has been replaced by $A_1 + B_1$ as is possible because, for the decimal sum to be 9, either A_1 or B_1 must be 1. (If both A_1 and B_1 are 1, the functioning of the adder will not be impaired, although the carry output will not then be dependent on the carry input.) The OR device which generates $A_1 + B_1$ is shown as an added component in Figure 7-2 although it may actually be a part of the full adder which generates U_1. The purpose of using signal $A_1 + B_1$ instead of U_1 is that the former can be generated more quickly because the input signals need pass through fewer levels of switching.

Carry speed-up features can be added to the binary full adders along the top of Figure 7-2 in a manner described in the previous chapter on binary arithmetic. However, such features are largely superfluous because, fortuitously, the AND-OR network that generates the decimal carry eliminates much of the need for internal carry propagation although a binary carry is propagated from the "1's" position to the "8's" position (but not to the decimal carry output line) when adding, for example, 0111 to 0001 (decimal 7 to decimal 1).

When operating three or more decimal 8-4-2-1 adders in parallel, further improvement in decimal carry propagation speed can be achieved by adding carry bypass networks analogous to the binary bypass networks of Figure 6-7 except that the bypass networks serve to bypass the decimal carries around each decimal adder (rather than bypass the binary carries within a decimal adder as implied in the above paragraph). That is, for example, a decimal adder in the 10^2 (hundreds) position will not only have the AND device shown with dotted connections in Figure 7-2 but will also have a five-input AND device to which

388   IMPLEMENTING DECIMAL ARITHMETIC

signals U_8 and U_1 of the 10^2 position and U_8, U_1, and C_1 of the 10^1 (tens) position are applied. Signal C_1 of the 10^1 position is the carry output signal from the 10^0 (units) positions. The decimal adder in the 10^3 position will have this equipment plus a seven-input AND device for an analogous bypass of the decimal carry from the units position to the carry output of the 10^3 position when the uncorrected sum digits in the 10^3, 10^2, and 10^1 positions are all 9's. The decimal adders in higher order positions may be correspondingly more elaborate as necessary to achieve simultaneous generation of carry signals in all orders, and no allowance need be made for "propagation" time.

Simplifying the 8-4-2-1 Adder

If the decimal adder is to be formed with half-adder and full-adder switching modules, the arrangement in Figure 7-2 is probably satisfactory as it stands. However, if each half adder and each full adder is to be formed with more basic switching modules, a network requiring fewer of them would be desirable. Also, if the decimal adder is itself to be a single "module" built on a large scale integration (LSI) basis, an objective would be to minimize the number of transistors and other components within the module to improve reliability—even though total cost, size, and weight may be virtually independent of the number of components within the range under consideration here.

Apart from using the component minimization techniques described in an earlier chapter (which techniques tend to be unwieldy in a nine-input, five-output network), some component reduction can be achieved in the following manner. The uncorrected output signals U_8, U_4, U_2, and U_1 are listed as follows for the ten different sum digits and without a decimal carry.

Sum Digit	Decimal Carry = 0	Decimal Carry = 1
0	0000	1010
1	0001	1011
2	0010	1100
3	0011	1101
4	0100	1110
5	0101	1111
6	0110	0000 ⎫
7	0111	0001 ⎬ $C_{16} = 1$
8	1000	0010 ⎪
9	1001	0011 ⎭

Whenever the decimal carry is 0, the binary digits of the uncorrected sum are correct, but when the carry is 1 the uncorrected binary digits should be altered to equal the digits of the corresponding row (corresponding decimal sum digit) in the center column. An examination of the digit pattern reveals that the correct sum digit $S_1 = U_1$ in all cases as in Figure 7-2. Sum digit S_2 is 1 whenever one or the other, but not both, of U_2 and C_a is 1, where C_a is the decimal carry. That is,

$$S_2 = U_2 \bar{C}_a + \bar{U}_2 C_a$$

which is the same as the sum function in a half adder.

The correct sum digit S_4 is 1 whenever the decimal sum is 4, 5, 6, or 7 with or without a carry. Without a carry, S_4 can be generated by the switching function $U_4 \bar{C}_a$. The switching function $U_4 U_2$ is 1 when the decimal sum is 6 or 7 without a carry or when the decimal sum is 4 or 5 with a carry. The function $\bar{U}_2 C_{16}$ is 1 when the decimal sum is 6 or 7 with a carry. Thus, for all digit combinations

$$S_4 = U_4 \bar{C}_a + U_4 U_2 + \bar{U}_2 C_{16}$$

By similar reasoning, the expression $S_8 = U_8 C_a + U_2 C_{16}$ can be found. Further, by making the observation that signals A_8, B_8, and C_8 in Figure 7-2 will never all be 1 simultaneously (because having all equal to 1 would imply a decimal sum equal to or greater than 24), the sum part of the full adder which generates U_8 can be simplified to $U_8 = (A_8 + B_8 + C_8)\bar{C}_{16}$. When U_8 in the above expression for S_8 is replaced by this expression for U_8, the observation may be made that whenever $\bar{C}_a = 1$ then $\bar{C}_{16}$ must be 1, with the result that the appearance of $\bar{C}_{16}$ is found to be superfluous in the resulting expression for S_8. The net result then is

$$S_8 = (A_8 + B_8 + C_8)\bar{C}_a + U_2 C_{16}$$

In summary, with reference to Figure 7-2, the sum part of the full adder which generates U_8, and the half and full adders which are used to generate S_2, S_4, and S_8 may also be eliminated. In their place, switching networks which generate S_2, S_4, and S_8 in accordance with the above expressions are substituted.

Further simplifications are conceivable because, for example, no use has been made of the "don't care" conditions that arise from the fact that A_8 is never 1 when either A_4 or A_2 is 1. However, no further simplifications are known. The handling of "don't care" conditions is a multilevel network such as this is not as straightforward as when starting with a basic two-level network as in earlier chapters. Of course, in any case the actual detailed form of the adder would be strongly dependent on the nature of the switching modules available for use (AND, OR,

NOT, O-I, A-I, A-O-I, etc.) as explained previously, and this dependency would apply whether the individual modules are formed with discrete components or whether they are formed as parts of a functionally much larger LSI module.

Decimal Subtraction—8-4-2-1 Code

A decimal subtractor can be formed in the manner of a conventional nine-input, five-output combinatorial switching network in the same manner that was suggested for a decimal adder, although the fifth input line and the ninth output line would normally transmit "borrow" signals instead of carry signals. Again, any four-bit decimal code may be used, and the two input decimal digits and the output difference digit may even be in three different codes.

With the 8-4-2-1 decimal code, binary subtraction techniques may be used in analogy with the techniques described for addition, although the nature of the correction to be applied in certain instances is quite different. Specifically, for subtraction, in all cases where the difference digit is positive (including zero), no correction is required. In all cases where the difference digit is negative the binary output appears, without correction, as the 2's complement of the intended digit. For example, when subtracting 1000 from 0100 (decimal 8 from 4) with a borrow from the next lower order the result is

$$
\begin{array}{rl}
& 0100 \\
\text{subtract:} & 1000 \\ \hline
& 1100 \\
\text{subtract:} & 1 \quad \text{(borrow input)} \\ \hline
& 1011
\end{array}
$$

This result may be corrected by subtracting it from 0000 (actually 10000, although the binary digit in the "16's" position may be neglected). The corrected result then is

$$
\begin{array}{rl}
& 0000 \\
\text{subtract:} & 1011 \\ \hline
& 0101
\end{array}
$$

or decimal 5. Note that this difference digit, although apparently negative, may actually be a positive digit. For example, consider the subtraction of 183 from 641. In the tens position, the subtraction of 8 from 4 with a borrow from the units position is encountered, but the difference, 458, contains a positive 5 in the tens position.

The generation of the decimal borrow signal happens to be more simply

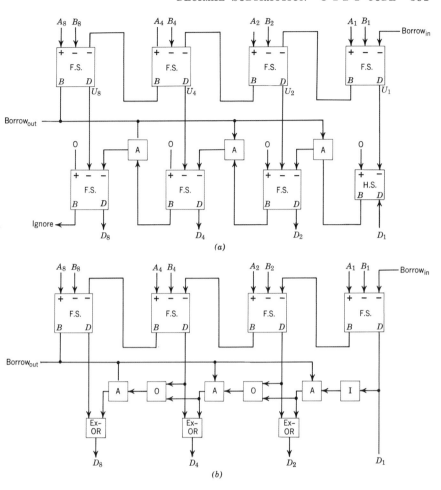

FIGURE 7-3. Decimal subtractor employing the 8-4-2-1 code.

accomplished than the generation of a decimal carry. The binary borrow output signal from the "8's" position may be used directly as the decimal borrow signal. Although this binary borrow signal appears to have a weight of 16, it may be regarded as having a weight of 10 (ten) because the binary input signal combinations are not unrestricted but are limited to those which represent decimal digits of 9 or less.

Figure 7-3a shows a decimal subtractor based on the above subtraction method, where H.S. and F.S. stand for half subtractor and full subtractor, respectively. When the decimal borrow output signal is 0, the AND devices in the borrow lines of the correction network have the effect of blocking any binary borrow signals that might occur there. Because the subtraction of a binary digit from 0 produces the same digit (no borrow signals), the corrected difference digits D_8, D_4, D_2, and D_1 are the same as the uncorrected digits U_8, U_4, U_2, and U_1 when the decimal borrow output signal is 0. When the decimal borrow output signal is 1, the subtraction of the uncorrected digits from 0000 (with binary borrows) takes place.

The correction portion of Figure 7-3a can be simplified to that shown in Figure 7-3b. Because D_1 is always subtracted from 0 without a binary borrow input, D_1 is always the same as U_1. The difference part of the half subtractor which generates D_1 may therefore be eliminated. The borrow part of this half subtractor may be replaced by an inverter, again because U_1 is always subtracted from 0. As for the three full subtractors which generate D_2, D_4, and D_8, the fact that the minuend input digits are all 0's causes the output digits to be related to the respective subtrahend and borrow digits in an exclusive-OR relationship. Moreover, with 0 minuend digits, the borrow signal in a full subtractor can be derived by a simple OR relationship between the other two input digits to that subtractor.

Decimal Subtraction by the Addition of 9's Complements

The 9's complement of a given decimal number is defined as the number obtained by subtracting each digit of the given number from 9. That is, for example, the 9's complement of 8206 is 9999 − 8206 = 1793. The 9's complement of a single digit d (a one-digit number) is, of course, merely $9 - d$. The 9's complement of an n-digit number B can alternatively be defined as $10^N - 1 - B$, where $N = n + 1$. If the lowest-order digit in B is not the units digit (that is, if the decimal point is not at the right-hand end), N should be adjusted accordingly, and the 1 should be multiplied by the power of 10 which will cause it to correspond to the weight of the lowest-order digit. Note that the 9's complement of the 9's complement of B is equal to B. That is,

$$10^N - 1 - (10^N - 1 - B) = B$$

Assume that negative numbers are expressed in 9's complement form so that, for example, "minus 2943" would appear in a machine as 17056 where the minus sign is represented by a single binary signal and the various decimal digits are represented by groups of binary signals in whatever code is being used.

With the above conventions, a scheme of decimal subtraction can be devised which has been found to be well adaptable to machine implementation. To subtract B from another n-digit number A, the basic step is to perform a 9's complementing action on the representation for B and then add with "end-around" carry. If A is negative, it will appear in 9's complement form. Likewise, B will initially be in 9's complement form if it is negative, but the complementing action will convert it to true form, so the subtraction of a negative number is equivalent to the addition of a positive number. The resulting "difference" (actually obtained by generating a sum) will appear in true or 9's complement form according to whether it is positive or negative, respectively. A zero difference will be associated with negative numbers and will appear in complement form—a minor problem to be discussed later.

The end-around carry mentioned in the previous paragraph is developed by transmitting the carry output signal from the highest-order decimal position to the lowest-order position (regardless of the location of the decimal point). To appreciate the role of the end-around carry, assume that A and B are both positive with $A > B$ and that the difference $A - B$, which will be positive, is generated by adding the 9's complement of B as follows.

$$A - B = A + (10^N - 1 - B) - 10^N + 1$$

In this case the sum of A and the 9's complement of B will always be at least as great as 10^N so that a decimal carry output will occur from the highest decimal order. If this carry is not used as having weight 10^N but instead is added "end around" fashion to the lowest-order position, the effect is to subtract 10^N and add 1, as indicated in the above equation. On the other hand, if $A < B$, then $A - B$ will be negative. In this case the sum of A and the 9's complement of B will always be less than 10^N and no carry from the highest order will occur. The sum can then be written as

$$10^N - 1 - (B - A)$$

which is the 9's complement of the desired difference. Thus, the presence or absence of an end-around carry can be used to indicate whether the difference is positive or negative, respectively, and whether the difference is in true or complement form, respectively.

Overflow situations can result when A or B, but not both, is negative. As explained above, if B is negative, the subtraction by the addition of the 9's complement has the effect of adding two positive numbers. In this case (which can be identified by sensing the signs of the numbers), a carry from the highest order indicates an overflow the same as in conventional addition. If A is negative, it will appear in 9's complement form so that when subtracting a positive B, the result will certainly be negative. However, in this case the effect is to add two 9's complement numbers to produce

$$(2)10^N - 2 - A - B$$

If an end-around carry occurs, the subtraction of 10^N and the addition of 1 yields a net result of

$$10^N - 1 - (A + B)$$

which is the 9's complement of the desired negative result. In this case the absence of an end-around carry indicates an overflow.

For example, assume that $A = -4758$, the digits of which appear in the machine in 9's complement form as 5241, and assume that $B = +2641$. The subtraction of B from A is accomplished by adding the 9's complement of B as follows.

```
            5241   (A in 9's complement form)
     add:   7358   (B in 9's complement form)
            ————
            2599
     add:      1   (end-around carry)
            ————
            2600
```

In this case the "difference" is -7399 because the presence of the end-around carry indicates no overflow, and the result must be negative and in 9's complement form.

If both A and B are negative, the subtraction of B causes, in effect, the addition of one 9's complement number and one true number so that the relationship between the end-around carry and the sign of the result is the same as when both A and B are positive.

In summary, when forming $A - B$ by adding the 9's complement of the representation for B, the character of the result is related to the sign of A, the sign of B, and the end-around carry as follows (with negative numbers being in 9's complement form).

Sign of A	Sign of B	End-Around Carry	Sign of Result
+	+	1	+
+	+	0	−
+	−	1	overflow
+	−	0	+
−	+	1	−
−	+	0	overflow
−	−	1	+
−	−	0	−

FALSE OVERFLOW WHEN THE DIFFERENCE IS $-99\cdots 99$ 395

If 1's and 0's are substituted for the $-$ and $+$ signs, respectively, in the above listing, the relationship between the indicated function and the sum part of a full adder becomes immediately apparent. In fact, a binary full adder may be used to generate the resultant sign signal in much the same way as was described in the previous chapter for binary arithmetic.

False Overflow When the Difference is $-99\cdots 99$

A peculiar problem arises in the particular instance when the result is $-99\cdots 99$. To illustrate, assume that $A = -4372$ and $B = +5626$. The formation of $A - B$ through the use of 9's complements would be as follows.

$$
\begin{array}{rl}
& 5627 \quad (A \text{ in 9's complement form}) \\
\text{add:} & \underline{4373} \quad (B \text{ in 9's complement form}) \\
& 0000 \\
\text{add:} & \underline{1} \quad (\text{end-around carry}) \\
& 0001
\end{array}
$$

The result is correctly indicated to be -9998 because, as in the previous example, no overflow occurs, and the result is negative and in 9's complement form. However, if $B = +5627$, the formation of the difference would be as follows.

$$
\begin{array}{rl}
& 5627 \quad (A \text{ in 9's complement form}) \\
\text{add:} & \underline{4372} \quad (B \text{ in 9's complement form}) \\
& 9999
\end{array}
$$

The correct result is -9999, but the absence of an end-around carry falsely indicates an overflow. The implication is that the maximum allowable magnitude of negative differences is $99\cdots 98$ instead of $99\cdots 99$.

The maximum negative magnitude can be increased to the expected $99\cdots 99$ by artificially adding 1 in the lowest-order position whenever a positive B is being subtracted from a negative A. This 1 takes the place of the end-around carry in this case, but the absence of a carry output from the highest-order position still indicates an overflow. Thus, when subtracting $B = +5627$ from $A = -4372$, the steps are as follows.

$$
\begin{array}{rl}
& 5627 \quad (A \text{ in 9's complement form}) \\
\text{add:} & \underline{4372} \quad (B \text{ in 9's complement form}) \\
& 9999 \\
\text{add:} & \underline{1} \quad (\text{artificially-introduced 1}) \\
& 0000
\end{array}
$$

396   IMPLEMENTING DECIMAL ARITHMETIC

A highest-order carry occurs, so no overflow is indicated. The negative difference, after conversion to true form, is —9999.

Whether the maximum allowable negative number is 99···98 or 99···99 is generally inconsequential from a machine capacity standpoint, but an overflow signal generated on a negative difference of 99···99 can cause vexing trouble on so-called "housekeeping" instructions in a program, especially if the programmer is unaware of the limitation.

Negative Zero

If $A = B$ and subtraction is performed by the addition of 9's complements, the result will be 99···99 with a negative sign indication. The equivalent negative zero, that is, —00···00, is mathematically acceptable in many applications, but zero is more commonly considered to be a positive number. A network may be included to perform an arbitrary 9's complementing action with an accompanying change in sign from — to + whenever the digits at the output of the adder are all 9's with a negative sign indication.

Alternatively, positive numbers can be represented in 9's complement form with negative numbers in true form. With this convention, a zero result still appears as 99···99 within the machine, but it is included in the set of positive numbers. When transmitting results to an output device where they will be observed by people, all positive numbers (instead of all negative numbers) will be converted to obtain the customary +00···00 when the result is zero. Minor modifications are required in the sign-overflow network, but no additional components are required when employing this convention as a means for avoiding negative zero.

A Subtraction Network Employing 9's Complements

Figure 7-4 illustrates the manner in which a combinatorial switching network may be arranged to perform decimal subtraction through the addition of 9's complements. The scheme is general in that any code may be used for the input numbers A and B and the output difference number D. Each heavy line represents as many wires in parallel as necessary (but normally just four) to represent decimal digits in the code selected. Each block labeled "Comp." is itself a small switching network which converts a decimal digit to its 9's complement form. Carry signals are transmitted from one decimal adder to the next in the usual fashion, where each decimal adder may be as indicated in Figure 7-1 or 7-2. End-around carry is included as described previously. The block labeled "Sign Net." is a switching network that produces output signals which indicate the sign of D and the presence or absence

NETWORK FOR DEVELOPING THE 9'S COMPLEMENT

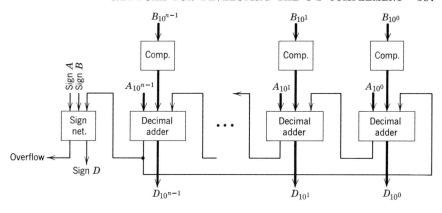

FIGURE 7-4. Parallel decimal subtraction by the addition of 9's complements. (Each heavy line represents four wires in parallel.)

of an overflow. The necessary switching functions in this network can be determined from the listing on page 394.

A characteristic to be observed in the network of Figure 7-4 is that a signal loop exists in the carry path. A carry may be initiated in any decimal adder, and this carry may be propagated through as many subsequent adders in the loop for which the digit sums are 9. However, the signal cannot circulate continuously nor can it cause the adder as a whole to "lock up" in a stable state in the manner of a flip-flop. Consider the example of subtracting $B = +8070215$ from $A = +8079715$

$$\begin{array}{rl} & 8079715 \quad (A \text{ in true form}) \\ \text{add:} & \underline{1929784} \quad (B \text{ in 9's complement form}) \\ & 9998999 \quad (\text{sum without carries}) \\ \text{add:} & \underline{1} \quad (\text{initial carry}) \\ & 0008999 \\ \text{add:} & \underline{1} \quad (\text{end-around carry}) \\ & 0009000 \end{array}$$

A carry signal originates in the fourth order and is propagated around the loop to this same order, but no further. When the input signals to the adder network of Figure 7-4 are returned to 0's or otherwise changed, the carry signals change accordingly.

Network for Developing the 9's Complement of a Digit in the 8-4-2-1 Code

The 9's complement of a decimal digit can, of course, be generated by means of a four-digit binary subtractor when the 8-4-2-1 decimal code is being used. Because some of the input signals (1001, representing

398 IMPLEMENTING DECIMAL ARITHMETIC

decimal 9) are constant, the binary full subtractors that would normally be needed for the subtraction of four-digit binary numbers can be replaced by a simpler network.

Alternatively, the formation of the 9's complement of a decimal digit can be treated as a conventional four-input, four-output combinatorial switching network problem, and this technique applied regardless of the code which may have been selected for the decimal digits. Moreover, the output digit need not be in the same code as the input digit.

However, in the case of the 8-4-2-1 code, little more than inspection of the output binary digits required for each set of input binary digits is necessary to derive a simple network for generating the 9's complement. Input digit combinations 1010 through 1111 (decimal 10 through 15) are "don't care" conditions. With d being the input digit and $9 - d = d_c$ being the 9's complement, the binary digits are as follows.

d 8421	d_c 8421
0000	1001
0001	1000
0010	0111
0011	0110
0100	0101
0101	0100
0110	0011
0111	0010
1000	0001
1001	0000

For any d, d_{c1} is the inverse of d_1. Output digit d_{c2} is always the same as d_2. Output digit d_{c4} is 1 only when one but not both of d_2 and d_4 is 1. Output digit d_{c8} is 1 only when none of the digits d_2, d_4, and d_8 is 1. That is,

$$d_{c1} = \bar{d}_1$$
$$d_{c2} = d_2$$
$$d_{c4} = d_2\bar{d}_4 + \bar{d}_2 d_4$$
$$d_{c8} = \overline{d_2 + d_4 + d_8}$$

With exclusive-OR and O-I modules available, the resulting complementing network would be as illustrated in Figure 7-5.

If the 9's complement of a digit is applied at the input to this network, the true digit appears at the output.

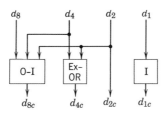

FIGURE 7-5. Network for developing the 9's complement of a digit in the 8-4-2-1 code.

Decimal Addition-Subtraction

In many applications the need is limited to addition or subtraction, but not both. In many other applications, notably the main arithmetic unit of a general purpose computer, the system must be capable of performing either operation. The details of the appropriate control signals can vary tremendously, but one of the simplest schemes of addition-subtraction is to select the operation by means of a single added signal which may be 1 for addition and 0 for subtraction. A multidigit parallel decimal adder-subtractor can then be viewed as one large combinatorial switching network as in Figure 7-4 but with an added binary input signal for control. If subtraction is to be performed by the addition of the 9's complements as in Figure 7-4, the added input signal can be used to cause the input number B to bypass the complementing networks when adding. Each individual complementing network then becomes a five-input, four-output network instead of the four-input, four-output network of Figure 7-5. Again, the techniques described in earlier chapters can be used to design the network, with the design depending on the nature of the switching modules available.

Note that even when adding, either A or B, or both, may be negative and in 9's complement form (or the opposite convention may be used with respect to signs and complements). In any case, the sign network in Figure 7-4 must be designed accordingly. If the sign network is designed as indicated previously for subtraction only, a sufficient alteration for addition is to invert the "sign B" input signal to the sign network.

Serial Decimal Addition-Subtraction

The previous discussion implied fully parallel transmission of the binary signals that represent a decimal number. Serial operation of three different types is possible. In one type the decimal digits are transmitted

400 IMPLEMENTING DECIMAL ARITHMETIC

one at a time, but the individual binary digits (bits) of any given decimal digit are transmitted simultaneously. In the second type the corresponding bits of each decimal digit are transmitted simultaneously, but the bits of each decimal digit are transmitted one at a time. (The effect of this second type is to transmit the decimal digits "in parallel" even though the individual bits within a decimal digit are transmitted serially.) The third type of transmission is fully serial in that only one bit is transmitted at any given time. The fully serial type of operation has several variations in accordance with the sequence in which the bits are transmitted, but most commonly all bits of one decimal digit are transmitted, then all bits of another decimal digit, and so on.

For the first type, which may be called serial-by-digit, parallel-by-bit, the considerations described in the previous chapter for serial binary operation apply directly. The single binary full adder is, of course, replaced by a single decimal adder of the type described earlier in this chapter, but the carry storage device functions in essentially the same way with a decimal carry as with a binary carry.

The second type, which may be called parallel-by-digit, serial-by-bit, requires networks quite different from those which have been described here. If the 8-4-2-1 code is used, each pair of decimal digits can be added in a manner analogous to serial binary addition except that a second cycle is needed to add the corrective 0110 (decimal 6) to those digits which require correction. More seriously, the handling of the decimal carries, although possible, tends to require a clumsy scheme having no particular advantages. For four-bit decimal codes that are not binary-like, the parallel-by-digit, serial-by-bit type of operation is highly disadvantageous. For one-out-of-ten codes, decimal counters can be employed to "accumulate" sums and differences, but this mode of operation is considered obsolete for nearly all applications. For details of carry schemes employing parallel decimal counters in this way, reference is made to Richards' book *Arithmetic Operations in Digital Computers* (1955), pp. 230–237.

The fully serial type of operation would appear to offer the advantage of requiring much less equipment—at the expense of an increase in overall addition-subtraction time. Although the equipment advantage may be realized in purely binary systems, the need to handle the binary digits in groups in accordance with the manner in which they represent decimal digits tends to make any significant equipment saving impossible of attainment. Even in the elementary operation of adding two decimal digits in the 8-4-2-1 code, temporary storage of certain of the sum bits is required for applying the corrective 0110 (decimal 6) for those digit combinations requiring a correction. Also, the 9's complementing of a decimal digit is found to require the temporary storage of some of the bits when this operation is attempted on a fully serial basis. Although a fully serial type of operation may be appropriate for transmitting numbers to and from the arithmetic unit, the preferable approach to the design of the arithmetic unit seems to be to convert each decimal digit to a parallel representation and to perform the arithmetic operations on a serial-by-digit, parallel-by-bit basis.

The above discussion does not exhaust the possibilities for utilizing a serial form of operation in some way. Many "trick" addition-subtraction schemes (too specialized to describe here) have been devised whereby serial and parallel operation have been combined in various manners. The availability of complementing or J-K flip-flops is particularly helpful in this regard. The schemes produce the effect of performing some of the switching functions within the flip-flops in contrast to the schemes described earlier in this chapter where all switching functions were performed by means of combinational switching networks with the flip-flops being used only for storage.

Decimal Multiplication—Simulating the Manual Method

The customary manual method of multiplication involves the memorization of a multiplication table. The multiplier digits are observed one at a time and are paired with each multiplicand digit. Each pair product, as derived from the memorized table, is then "accumulated" in some organized procedure to form the final product.

The usual procedure is to develop a set of partial products, one for each multiplier digit, where the "tens" digit of any given pair product is mentally added to the "units" digit of the next higher order pair product. After developing all partial products and placing them in appropriate relative positions to account for the different weights of the multiplier digits, the partial products are added to form the final product.

Machine multiplication methods can be devised which simulate the manual method quite closely—and in fact such methods have occasionally been used in machines. The role of the memorized multiplication table is played by a "read-only" (stored digits not alterable) digital storage unit. This unit has a capacity of 200 digits in a ten-by-ten array with two digits at each "address." One address is at each intersection of a row and a column. The selection of a row and a column is under the control of a multiplier digit and a multiplicand digit, respectively. The two digits stored at any given address correspond to the "units" and "tens" digits of the corresponding pair product. For an

M-digit multiplicand and an m-digit multiplier, the number of pair products is Mm and these pair products can be accumulated in any of many different fashions, one of which is essentially as described above for manual multiplication. Alternatively, as each pair product is determined, it may be accumulated in a "final product register" where the two digits of each pair product are entered into the product digit positions that are appropriate for the weights of the corresponding multiplicand and multiplier digits.

For a more parallel mode of operation, M separate but identical read-only storage units may be provided (i.e., M copies of the multiplication table) with each being controlled by a different multiplicand digit but all being controlled by the same multiplier digit on a given step in the multiplication process. The pair products thus generated for each multiplier digit may be added in parallel to form the partial products which in turn may be accumulated as they are developed.

In an alternative parallel mode of operation the M copies of the multiplication table are controlled in the same way, but for each multiplier digit the "units" digits of the M pair products are accumulated in the appropriate orders of one accumulator while at the same time the "tens" digits of the same pair products are accumulated in a second accumulator. As a last step, the two totals are then combined to form the final product.

Multiplication by Over-and-Over Addition

A multiplication method called over-and-over addition is probably the most elementary and most economical to implement, although the time required for a multiplication operation tends to be comparatively great. The multiplier digits are examined one at a time, usually from right to left. For each multiplier digit, the multiplicand is accumulated (added) a number of times equal to the value of the digit. After the additions for each given multiplier digit, either the multiplicand or the partially developed product is shifted one position in preparation for the additions corresponding to the next multiplier digit. The final product is obtained when the multiplicand has been accumulated in accordance with all multiplier digits.

The examining of any given multiplier digit is accomplished by placing that digit in a counter which counts in the reverse direction. Network means are included to sense the zero condition of the counter. If the counter is not at zero, the multiplicand is added and 1 is subtracted from the digit in the counter. This process is repeated until the counter does contain zero, at which time the next multiplier digit is entered, and either the multiplicand or the partially developed product is shifted one position as mentioned above.

In spite of the basic simplicity of the over-and-over multiplication method, a complete network for performing multiplication in this way requires a rather large number of components—more than is deemed worth describing in detail in a book of this scope. Also, the nature of any multiplication network would depend strongly on the characteristics of the switching modules available and would depend upon countless particular specifications as they might be encountered in any given application. However, once the basic scheme is appreciated, the designer may employ the switching network design techniques described in earlier chapters to devise an appropriate multiplication network as required for the specific application at hand.

The time required for a multiplication depends on the nature of the addition operation, and the additions may be performed either serially or in parallel, or some combination (such as adding three decimal digit pairs per step) may be employed. However, in any case the number of additions is a function of the magnitudes of the individual multiplier digits that are encountered in a given multiplication. With none through nine additions being required for multiplier digits 0 through 9, respectively, the average number of additions per multiplier digits over a large number of randomly selected multiplier digits is 4.5. Then the time required for a multiplication is $4.5nt + K$, where n is the number of multiplier digits, t is the time required for an addition and includes the time required to step the counter, and K is a constant that includes, primarily, the total time consumed in transmitting the multiplier digits to the counter but may also include an initial increment of time needed to transmit the operands to their intended registers.

Reducing the Number of Addition-Subtractions Required for Multiplication

For any given multiplier digit m, the addition operations described in the previous section for over-and-over multiplication can be replaced by $10 - m$ subtractions if a 1 is added to the next higher order multiplier digit. That is, if $m = 7$, for example, the seven additions of the multiplicand can be replaced by three subtractions, and the adding of 1 to the next higher order multiplier digit has the effect of adding the multiplicand ten times. This procedure reduces the total number of addition-subtraction operations only when $6 \leq m \leq 9$. For each m in the range $0 \leq m \leq 4$, over-and-over additions are performed as previously.

For a multidigit multiplicand such as 729638 for example, the lowest order digit, an 8, is sensed first. Two subtractions of the multiplicand are performed, and when 1 is added to the next multiplier digit, a 3, it becomes 4, and four additions (with the multiplicand appropriately shifted with respect to the partially developed product) are performed. For the third multiplier digit, a 6, four subtractions are performed and 1 is added to the next digit, a 9. This addition produces a 0, which requires no addition-subtraction operations, but a carry of 1 must be added to the next digit, a 2, to produce 3. The last multiplier digit, a 7, is handled by subtracting the multiplicand three times and by placing a 1 in an imaginary digit position of still higher order. Thus, in summary, a multiplier of 729638 is handled as though it were 1330442 where each underlined digit is regarded as being negative.

With the assumption that a multiplier digit of 5 requires five addition-subtraction operations, an average of approximately 2.5 addition-subtractions per multiplier digit are required for randomly chosen multiplier digits. (The average is not exactly 2.5 because of the 1 which may appear in the imaginary high-order digit position.)

Actually, the digit 5 can be handled in a special way to achieve a slight further reduction in the average number of addition-subtractions required for a multiplication. When a given multiplier digit is 5, the next higher order multiplier digit is sensed. If that digit is 4 or less, the given 5 is handled by adding the multiplicand five times, and no improvement is achieved. However, if the next digit is 5 or greater, the given 5 is handled by subtracting five times and adding 1 to the next multiplier digit to achieve a saving of one addition-subtraction in most such instances (although when the next digit is another 5 the realization of a saving will depend on the multiplier digits in still higher orders). As an example, consider the multiplier 25555 which would require twenty-two additions without this elaboration. If 5's are handled as described here, the multiplier would be treated as 3 4 4 4 5 and only 20 addition-subtractions would be required. A multiplier of 255 is an example where no saving would be realized.

Doubling

Further improvement in multiplication speed can be obtained by employing switching networks to form various multiples of the multiplicand. Of course, if all nine multiples of the multiplicand are formed, only one addition operation per multiplier digit is required. The complexity (cost) of the networks is strongly dependent on the code being used. With the 8-4-2-1 code, networks which will, in effect, multiply a number by two or five happen to be achievable relatively economically. Networks for forming other multiples can then be designed by adding the given number to its doubled and quintupled values in various ways to be described.

In particular, the doubling of a number can be accomplished by making a left shift of the bits in each decimal digit. For decimal digits in the range 5 through 9 this doubling produces quantities equal to or greater than 10 (ten) but expressed in a binary notation which must be corrected in essentially the same manner as was described for the addition of two decimal digits in the 8-4-2-1 code. For these digits, a decimal carry to the next higher order digit is produced (by means of switching devices that sense input bit combinations in the binary range 0101 through 1001). This decimal carry is added to the doubled digit in the next higher order, but because a doubled digit is always even, the 1-bit of the next higher order decimal digit is always 0. Therefore, the decimal carry from a given digit becomes the 1-bit of the next higher order decimal digit. The decimal carry, when it occurs, is also added to the 4-bit and 2-bit positions of the given doubled digit to have the effect of adding a corrective 0110 (decimal 6) as described previously.

An alternative procedure for designing a doubling network is to list the ten possible binary input signal combinations and the corresponding output signals for the respective doubled digits as follows.

| Digit | | | | Doubled Digit | | | | |
X_8	X_4	X_2	X_1	D_8	D_4	D_2	D_1	
0	0	0	0	0	0	0	0	
0	0	0	1	0	0	1	0	
0	0	1	0	0	1	0	0	
0	0	1	1	0	1	1	0	
0	1	0	0	1	0	0	0	
0	1	0	1	0	0	0	0	
0	1	1	0	0	0	1	0	plus
0	1	1	1	0	1	0	0	decimal
1	0	0	0	0	1	1	0	carry
1	0	0	1	1	0	0	0	

As always with the 8-4-2-1 code, input signal combinations 1010 through 1111 (decimal 10 through 15) will not occur and are therefore "don't care" conditions. Techniques described in earlier chapters can then be used to establish the following relationships, where C_{in} and C_{out} are the carry input from the next lower order and the carry output to the next higher order, respectively.

$$C_{out} = X_8 + X_4X_2 + X_4X_1$$
$$D_8 = X_4\bar{X}_2\bar{X}_1 + X_8X_1$$
$$D_4 = \bar{X}_4X_2 + X_2X_1 + \bar{X}_8\bar{X}_1$$
$$D_2 = \bar{X}_8\bar{X}_4X_1 + X_4X_2\bar{X}_1 + X_8\bar{X}_1$$
$$D_1 = C_{in}$$

This network is illustrated in Figure 7-6.

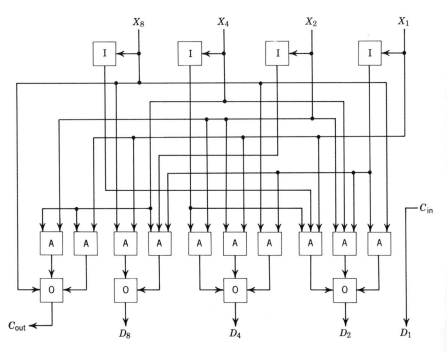

FIGURE 7-6. Digit doubler employing the 8-4-2-1 code.

For parallel doubling, an appropriate number of networks as in Figure 7-6 are used with the carry lines connected from one network to the next in sequence—although in contrast to a parallel adder, no carry "propagation" occurs beyond two adjacent networks. For serial operation, each carry digit must be stored temporarily as in serial addition.

By forming D_8, D_4, and D_2 as follows, an alternative decimal doubler can be developed which requires fewer switching devices in some applications (where inverted inputs signals are not available and where minimizing the number of inverters is important) but which requires more levels of switching.

$$D_8 = X_4 \bar{C}_{out} + X_8 X_1$$
$$D_4 = X_2 \bar{C}_{out} + X_2 X_1 + X_8 \bar{X}_1$$
$$D_2 = X_1 \bar{C}_{out} + \bar{X}_1 C_{out}$$

Quintupling

Switching networks for multiplying a number by 5 may be designed as follows. First, the observation is made that when multiplying a given digit by 5, the "units" digit of the product is either 0 or 5 according to whether the given digit is even or odd, respectively. Then, the observation is made that the "tens" digit of the product can be determined by dividing the given digit by 2 and discarding the remainder, if any. For example, if the given digit is 7, the tens digit 3 of the product 35 is determined by dividing 7 by 2 to obtain a quotient of 3 and a remainder of 1 which is discarded. With the given digit X in the 8-4-2-1 code the division by 2 is accomplished by an imaginary shifting of the X_8, X_4, and X_2 signals to the right one binary position.

When multiplying a multidigit number by 5, each "tens" decimal digit becomes a carry to be added to the "units" decimal digit of the next higher order. However, because a "tens" digit may have any value in the range 0 through 4, three signal lines are required to transmit this carry instead of the single line generally required when the carry can be only 0 or 1. The three lines carrying the input X_8, X_4, and X_2 signals become the three carry lines transmitting carry signals having weights of 4, 2, and 1, respectively. The line carrying input signal X_1 can be caused to represent 0101 (decimal 5) by connecting it to two appropriate destinations. The carry input signals from the next lower order may be designated C_{i4}, C_{i2}, and C_{i1}, and these three signals represent a three-bit binary number to be added to the "units" decimal digit of the product in any given order. This addition may be performed in two binary half adders and a binary full adder as illustrated in Figure 7-7a. The product digit Q, which may have a value as great as 9 (when X_1 is 1 and the input carry is 4), appears on the output lines indicated. The output carry signals C_{o4}, C_{o2}, and C_{o1} are developed as explained above and are transmitted to the quintupler network of the next higher order.

The array of full and half adders in Figure 7-7a can be viewed as a conventional four-input, four-output combinatorial switching network and may be designed accordingly. For each possible input signal combination the desired output signal combinations is as follows.

X_1	C_{i4}	C_{i2}	C_{i1}	Q_8	Q_4	Q_2	Q_1
0	0	0	0	0	0	0	0
0	0	0	1	0	0	0	1
0	0	1	0	0	0	1	0
0	0	1	1	0	0	1	1
0	1	0	0	0	1	0	0
1	0	0	0	0	1	0	1
1	0	0	1	0	1	1	0
1	0	1	0	0	1	1	1
1	0	1	1	1	0	0	0
1	1	0	0	1	0	0	1

After giving recognition to the fact that input signal combination 0101, 0110, 0111, 1101, 1110, and 1111 are "don't care" conditions, the output signals may be determined to be the following functions of the input signals.

$$Q_8 = X_1 C_{i2} C_{i1} + X_1 C_{i4}$$
$$Q_4 = \bar{X}_1 C_{i4} + X_1 \bar{C}_{i4} \bar{C}_{i1} + X_1 \bar{C}_{i4} \bar{C}_{i2}$$
$$Q_2 = \bar{X}_1 C_{i2} + X_1 \bar{C}_{i2} C_{i1} + C_{i2} \bar{C}_{i1} C_{i1}$$
$$Q_1 = \bar{X}_1 C_{i1} + X_1 \bar{C}_{i1}$$

The resulting quintupler is shown in Figure 7-7b.

Alternative expressions for Q_4 and Q_2 which reduce the number of switching devices with some types of modules but which require more levels of switching are as follows.

$$Q_4 = (X_1 + C_{i4}) \bar{Q}_8$$
$$Q_2 = (X_1 C_{i1} + C_{i2}) \bar{Q}_8$$

For serial operation with the quintupler, three binary storage elements are required for the three signals needed to represent the carry.

Multiplication With the Doubled and Quintupled Multiplicand Available

With the doubled and quintupled multiplicand available, the number of additions can be substantially less than is required in the basic over-and-over addition method of multiplication. For multiplier digits 1, 2, and 5, only one addition is required. For multiplier digit 3, two additions are required, one to add the multiplicand and another to add the doubled multiplicand. Multiplier digits 4, 6, and 7 may be handled as $2 + 2$, $5 + 1$, and $5 + 2$, respectively, so that two additions are similarly required. Multiplier digits 8 and 9 may be handled as $5 + 2 + 1$ and $5 + 2 + 2$, respectively, requiring three additions for each digit. On the

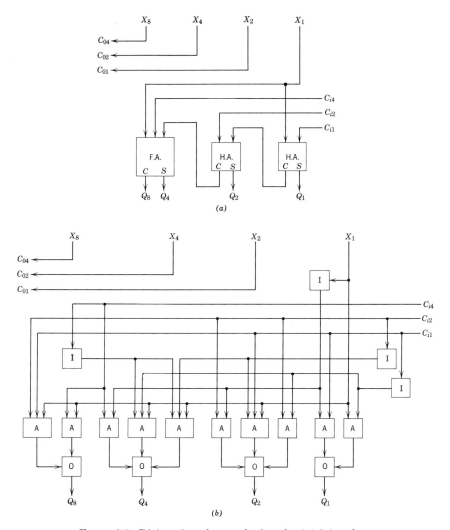

FIGURE 7-7. Digit quintupler employing the 8-4-2-1 code.

average, for random multiplier digits, 1.7 additions per multiplier digit are required.

By employing subtractions to handle multiplier digits 8 and 9 as suggested in an earlier section, only one operation is needed for either digit—a subtraction of the doubled multiplicand or the multiplicand, respectively, with 1 being added to the next higher order multiplier digit. An average of only 1.3 addition-subtraction operations per multiplier digit are then required.

A Switching Network for Generating All Multiples of the Multiplicand

Networks analogous to the doubler and quintupler can be designed for any of the other multiples of a number, but the number of components required tends to be as great if not greater than is required when generating other multiples by using only doublers, quintuplers, and adders in various combinations. Specifically, for example, in most applications the tripling of a number is most economically accomplished by employing an adder, as described previously, to add the number to its doubled value as obtained by a doubler. The resulting array for three orders is shown in Figure 7-8a, where each heavy line represents four

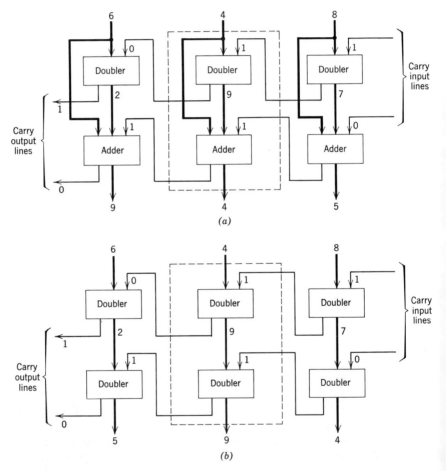

FIGURE 7-8. Tripler *a* and quadrupler *b*. (Each heavy line represents four wires in parallel.)

wires in parallel. The digits appearing at various points in the array are indicated for the tripling of the number $\cdots 648 \cdots$ with a carry input to the right-hand doubler. Note that a doubler-adder combination can be treated for design purposes as a six-input, six-output combinatorial switching network with one such network inclosed by a dotted line in the figure.

Quadrupling is most economically accomplished by employing two doublers in tandem as illustrated in Figure 7-8b. The two doublers in any order may be treated as a six-input, six-output combinatorial switching array, again as indicated by the dotted lines. Note that in this case the signals on the upper carry lines have a weight of 2 in contrast to the tripler where each carry signal had a weight of 1. The total carry may have a value as great as 3 (as when quadrupling 8 or 9, or when quadrupling 7 with a carry of 2 from the next lower order), in which case a carry signal will appear on both output lines of the corresponding order. The digits appearing at various points in the array are shown when quadrupling $\cdots 648 \cdots$ with a carry input signal of 1 (having a weight of 2 with respect to the quadrupled number) to the upper doubler in the right-hand position.

A Network for Obtaining All Nine Multiples

Figure 7-9 illustrates how adders, doublers, and a quintupler may be assembled to form an array which generates all nine multiples of a digit. For multidigit numbers, a corresponding number of such arrays would be utilized where the carry input signals for any given array are obtained from the carry output lines of the next lower order. With each heavy line representing four wires in parallel as would be used for any four-bit decimal code, the array amounts to a single 14-input, 36-output network. Because the value of the carry being transmitted from one order to the next cannot be greater than 8 (but can be 0), only four carry lines interconnecting the various orders would be required in principle. A reduction from the ten lines shown to eight lines can be achieved by replacing the quintupler with an adder which adds, for example, the doubled and tripled values to obtain the quintupled value. This reduction in carry lines would scarcely be consequential in a fully parallel system, but it might be worthwhile in a serial system where all carry signals must be stored temporarily.

Network for Generating Any One Multiple Under the Control of a Multiplier Digit

If the array of Figure 7-9 is to be used in the development of a multiplication unit, a switching network must be added to select the

412 IMPLEMENTING DECIMAL ARITHMETIC

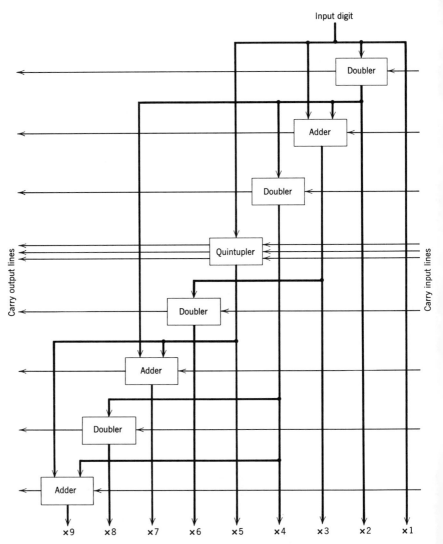

FIGURE 7-9. One order of a switching network for generating all nine multiples. (Each heavy line represents four wires in parallel.)

correct multiple under the control of the particular multiplier digit being sensed at any given time. Although straightforward in principle, such a network tends to require an undesirably large number of components. A much more economical approach to the selection of the desired multiple is to generate only the one multiple needed and to generate this one multiple by interconnecting the various doublers, quintuplers, and adders as needed. The interconnecting is accomplished by means of a switching network having the multiplier digit signals as inputs.

Many different schemes based on the suggestion of the previous paragraph can be devised with the one that is, so far as is known, the simplest and most economical being as follows. A doubler is employed to develop the doubled multiplicand, a second doubler is employed to develop the quadrupled multiplicand, and a quintupler is employed to develop the quintupled multiplicand. A single adder is then used to add none, one, two, four, and five times the multiplicand as needed to develop any of the nine multiples in accordance with the following pattern, where m_8, m_4, m_2, and m_1 are the bits of the multiplier digit in the 8-4-2-1 code.

Multiplier Digit	m_8	m_4	m_2	m_1	One Adder Input	Other Adder Input
0	0	0	0	0	—	—
1	0	0	0	1	×1	—
2	0	0	1	0	—	×2
3	0	0	1	1	×1	×2
4	0	1	0	0	×4	—
5	0	1	0	1	×5	—
6	0	1	1	0	×4	×2
7	0	1	1	1	×5	×2
8	1	0	0	0	×4	×4
9	1	0	0	1	×5	×4

After giving effect to the fact that multiplier bit combinations 1010 through 1111 are "don't care" conditions, the control signals for transmitting the multiplicand, the quadrupled multiplicand, or the quintupled multiplicand to one adder input are found to be $\bar{m}_8\bar{m}_4m_1$, $(m_8 + m_4)\bar{m}_1$, and $(m_8 + m_4)m_1$, respectively. For transmitting the doubled or the quadrupled multiplicand to the other adder input the control signals turn out to be simply m_2 and m_8, respectively. The resulting array is illustrated in Figure 7-10, where each heavy line represents four wires in parallel so that each AND and OR device illustrated is actually composed of four such devices.

The network in Figure 7-10 can be visualized as a single 14-input, 10-output combinatorial switching array, where four of the input signals are m_8, m_4, m_2, and m_1, another four are the four signals representing the multiplier digit, and the remaining six are the carry input signals. Conceivably, such a network could be designed by using the techniques described in earlier chapters, and the result might be reasonably economical because of the large number of "don't care" conditions that would

414 IMPLEMENTING DECIMAL ARITHMETIC

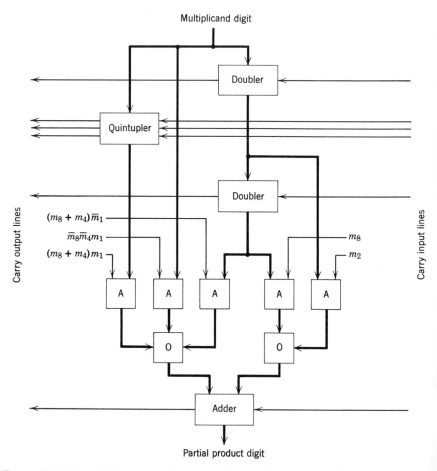

FIGURE 7-10. Network for generating any multiple under the control of a multiplier digit in the 8-4-2-1 code. (Each heavy line represents four wires in parallel.)

be encountered. (In fact, the number of carry input lines and carry output lines can each be reduced to four, and even then some signal combinations would not occur.) On the other hand, the network of Figure 7-10 contains at least four successive AND-OR levels of switching, and the design techniques of earlier chapters may be impractical with this many levels of switching. Thus, the number of components represented by the network of Figure 7-10 is probably close to, if not exactly at, the minimum achievable for the specified multiplication function. In other words, Figure 7-10 is a good example of the advantages of designing large switching networks in parts, with each part designed along the lines described in earlier chapters. Of course, if the resulting multiple levels of switching are objectionable for speed reasons, the designer would be forced to employ a design technique which treats the array as one large combinatorial network.

Alternative Schemes for Obtaining, In Effect, All Nine Multiples

An alternative scheme requires only one instead of two doublers for obtaining the effect of generating all nine multiples. Quadrupling is accomplished by subtracting the multiplicand digit from its quadrupled value. For a multiplicand digit of 9, the multiplicand is subtracted and then, in effect, multiplied by 10 and added. The multiplication by 10 is accomplished by shifting all digits to the left one decimal position, but this shifting can be merely the transmission of each digit to the left along four wires that are analogous to the carry transmission lines of Figure 7-10. The addition and the subtraction may take place simultaneously. For a multiplier digit of 8, the doubled multiplicand is subtracted along with the addition of 10 times the multiplicand. In an arithmetic unit which contains subtraction facilities for other reasons this alternative design may save a few components, but the complementer needed for subtraction and the switching devices needed to switch it "in" and "out" as appropriate for the various multiplier digits tends to be more expensive than the doubler which was eliminated.

In another alternative scheme only the first five multiples are developed. A doubler and a quintupler are used to obtain the doubled and quintupled multiplicand, respectively, and a decimal subtractor is used to subtract the doubled multiplicand and the multiplicand from the quintupled multiplicand to obtain the tripled and quadrupled multiplicand, respectively. For multiplier digits 6, 7, 8, or 9 the quadrupled multiplicand, the tripled multiplicand, the doubled multiplicand, or the multiplicand, respectively, is subtracted and a 1 is added to the next higher order multiplier digit. In comparison with Figure 7-10, one doubler is saved and the adder is replaced by a subtractor. However, the partial product accumulator must be capable of subtracting as well as adding, and the control network must include facilities for adding 1 to a multiplier digit.

A Multiplication Method That Develops One Product Digit Per Step

When an appropriate number of arrays as illustrated in Figure 7-10 are used in parallel, a complete partial product can be developed in a single step of operation, although suitable time must be allowed for

416   IMPLEMENTING DECIMAL ARITHMETIC

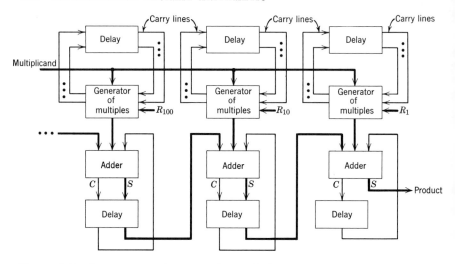

FIGURE 7-11. Arrangements for generating a decimal product, one digit per step. (Each heavy line represents four wires in parallel.)

carries to propagate through the various orders in succession. On this same step the partial product may be added in an accumulator, and the carry propagation time consumed by the accumulator can largely coincide with the carry propagation of the partial product generator. A multiplication scheme which handles carry propagation in a very different way and which generates a digit of the final product on each step is illustrated for three decimal orders in Figure 7-11.

In Figure 7-11 each block labeled "generator of multiples" represents an array as illustrated in Figure 7-10. However, instead of generating the partial products by a parallel process, each partial product is generated serially. The multiplicand is transmitted parallel-by-bit, serial-by-digit, low order first to each generator of multiples. The right-hand generator is controlled by the units multiplier digit R_1, and this generator forms the first partial product. The middle generator is controlled by the tens multiplier digit R_{10}, and it generates the second partial product, and so on. The various partial products are assembled by the decimal digit adders along the bottom of Figure 7-11. The blocks labeled "delay" may be composed of flip-flops in the manner described earlier, and they serve to store or "delay" the digits at the respective network points for one time step. In the case of the carries pertaining to the generators of multiples and to the adders (the latter being marked "C" in the figure), the delays serve to cause each carry signal to be added to the digits of the next higher order in each instance. In the case of the adder sum signals (marked "S") the delay has the effect of shifting the partial products relative to each other as appropriate for the weights of the various multiplier digits. The final product appears parallel-by-bit, serial-by-digit, low order first on the output line indicated.

A variation of the scheme of Figure 7-11 is to employ a single serially operating generator of all nine multiples as shown in Figure 7-9. A switching array forming essentially a four-throw, ten-position (including the 0 position) switch is used for each multiplier digit to select the corresponding multiple, which becomes the partial product corresponding to that multiplier digit. The accumulation of the partial products is as illustrated in Figure 7-11. This variation is preferable in applications where, because of one reason or another, the four-throw, ten-position switch is relatively inexpensive and the multiplier has a relatively large number of digits. However, with switches formed with electronic devices which perform the AND, OR, and NOT functions (or combinations thereof) the switch is comparable in cost to the generator of multiples as shown in Figure 7-10.

Decimal Division

In many respects division is just the inverse of multiplication with the dividend, divisor, and quotient corresponding to the product, multiplicand, and multiplier, respectively. However, division seems to be inherently more difficult (expensive) to implement. Two problems are encountered with division that are not encountered with multiplication. One problem is most easily explained if the factors are visualized as integers (although the decimal point may actually be in any fixed position). Assume that the dividend has $2n$ digits and that the divisor has n digits to correspond to a $2n$-digit product and an n-digit multiplicand, respectively. Because the dividend is not necessarily limited to numbers that are exact products of other numbers, the quotient is, first of all, not necessarily an integer, but more importantly is not necessarily limited to an n-digit number that would correspond with a multiplier in multiplication. Specifically, the quotient may have up to $2n$ digits (if the divisor is 1), and it may be infinite (if the divisor is 0). That is, a division overflow condition can be encountered which has no counterpart in multiplication. Ordinarily, the development of more than n quotient digits would be purposeless because, with a divisor having n digits, no more than n quotient digits would be significant. Therefore, division overflow can be expected to be encountered quite commonly, and means must ordinarily be included to detect it.

The other problem encountered in division can be appreciated by visualizing the inverse of elementary over-and-over addition. Over-and-over subtraction of the divisor from the dividend serves only to determine the remainder, whereas the objective is to determine the quotient. In contrast to multiplier digits, the quotient digits are, of course, not known in advance. Therefore, the number of subtractions to be performed cannot be determined by a simple process corresponding to "examining" a multiplier digit. Instead, subtractions must be performed until the intermediate remainder becomes negative, and then a subtraction must be nullified or some other corrective action must be taken.

Division by Over-and-Over Subtraction

A straightforward way to implement decimal division is to start by "lining up" the n digits of the divisor with the high-order $2n$ digits of the dividend. The divisor is then repeatedly subtracted until the intermediate remainder becomes negative, at which time the divisor is added once to restore a positive intermediate remainder. For each subtraction that leaves a positive remainder a 1 is entered into a counter which initially contains zero. The digit appearing in the counter after terminating the process just described is a quotient digit. After appropriately storing the quotient digit elsewhere, the divisor is then shifted to the right one position with respect to the intermediate remainder thus developed and the process is repeated to determine the next quotient digit.

The procedure described in the previous paragraph actually develops $n + 1$ quotient digits, whereas only n digits are desired. The first quotient digit is presumed to be 0. If it is not (that is, if the very first subtraction with the factors lined up as specified produces a positive intermediate remainder), an overflow has been detected, and an appropriate signal should be generated.

Ordinarily, the division network may be designed so that, when a given quotient digit is found to be 9, the next subtraction (and the corrective addition that would take place) may be eliminated, because a negative intermediate remainder would certainly result. However, in a variation of the division method, the factors are not initially aligned as specified above but the divisor is instead shifted to the right one position as appropriate for determining the first of the n quotient digits to be saved. In determining the first quotient digit, up to 10 subtractions are performed. If the 10th subtraction still leaves a positive remainder, an overflow has been detected essentially as before, because ten subtractions in this position are equivalent to one subtraction with the original alignment. The advantage of this variation is that no time is wasted in testing for an overflow in those instances where no overflow occurs.

Because two addition-subtraction operations are required for each 0 quotient digit, three for each 1 quotient digit, and so on, up to 11 for each 9 quotient digit, this division method requires an average of 6.5 addition-subtraction operations per quotient digit for random quotient digits.

Division by Nonrestoring Over-and-Over Addition-Subtraction

A modest improvement in division speed can be realized by employing a nonrestoring technique (as opposed to the "restoring" technique of the previous section) when the intermediate remainder becomes negative. Instead of adding the divisor to restore the last positive intermediate remainder, the divisor is shifted to the right and then added in over-and-over fashion until a positive intermediate remainder is obtained. If the next quotient digit is designated q, the number of additions thus performed is $10 - q$, and the quotient digit is developed by operating the quotient digit counter in the reverse direction. That the net result of this nonrestoring technique is the same as obtained by the restoring technique can be appreciated by observing that the restoring addition of the divisor in a given position can be designated $+D$ and the subsequent q subtractions which leave a positive remainder can be designated $-qD/10$. The two techniques are equivalent in net result because $D - qD/10 = (10 - q)D/10$ where the right-hand expression indicates the $10 - q$ additions without the previous restoring addition.

The number of additions needed to determine the second quotient digit ranges from one when the quotient digit is 9 to 10 when the quotient digit is 0. Therefore, the average number of addition-subtraction operations required per quotient digit has been reduced from 6.5 to 5.5.

A Division Method Which is Neither Restoring Nor Nonrestoring

The restoring and nonrestoring division methods described in the two previous sections are based on the assumption that the physical design of the addition-subtraction equipment is such that whenever a sum or difference (intermediate remainder) is generated it must be retained in replacement of the previous intermediate remainder. This assumption is valid in some computer designs, but it is not necessarily valid. When the addition-subtraction is performed in a combinatorial switching network that is separate from the various storage registers (as implied in most of the addition-subtraction implementations which were described), a negative intermediate remainder need not be retained. If

420   IMPLEMENTING DECIMAL ARITHMETIC

the carry-borrow signal from the highest order indicates that a given intermediate remainder is negative, the subtraction operation is nullified (by not transmitting the difference signals to the storage register), and the previous intermediate remainder is shifted in preparation for determining the next quotient digit.

With this scheme, all addition-subtraction operations are subtractions. When a quotient digit is found to be 0, one subtraction is necessary, although the difference (which will be negative) is not retained. For a quotient digit of 1, the first subtraction produces a positive difference, which is retained, and the second subtraction produces a negative difference, which is not retained. For random quotient digits, the average number of subtractions is 5.5 per quotient digit, the same as the average number of addition-subtraction operations required in the nonrestoring division method.

A Decimal Division Method Utilizing the Doubled and Quintupled Divisor

Further improvement in division speed can be obtained if various multiples of the divisor are available for subtraction from the dividend. As described earlier in this chapter, the doubled and quintupled values of a number can be developed reasonably economically, especially in serial systems. With the doubled and quintupled divisor available, a quotient digit is determined by first subtracting the quintupled divisor. If the difference is negative, the quotient digit is known to be in the range 0 through 4, and the next step is to add the doubled divisor to the difference so that, in effect, the divisor has been subtracted three times. If the new difference is still negative, the quotient is known to be in the range 0 to 2, and the doubled divisor is added a second time so that the net effect is to have subtracted the divisor once. If a negative difference remains, the quotient digit is known to be 0. If a positive difference remains, the divisor is subtracted to produce the equivalent of subtracting the divisor twice, and the quotient digit is known to be 1 or 2 according to whether this final difference is negative or positive, respectively. In instances where a positive difference resulted after the first addition of the doubled divisor, the divisor is added and the quotient digit can be determined to be 3 or 4 in accordance with the negative or positive, respectively, sign of the difference. Quotient digits 5 through 9 are analogously determined by subtractions of the doubled divisor and additions of the divisor.

Figure 7-12 illustrates the operations outlined in the previous paragraph for determining the ten possible quotient digits. Each block represents an operation followed in time by the block or digit indicated by the arrow corresponding to the resulting sign of the difference. (The

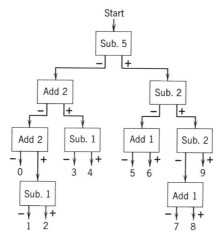

FIGURE 7-12. Addition-subtraction pattern in a division method employing the doubles and quintupled divisor.

blocks do not represent pieces of equipment interconnected by wires—although conceivably an appropriate number of doublers, quintuplers, adders, and subtractors could be connected in parallel much in the manner indicated by Figure 7-12.)

Control equipment which will cause the computer to proceed through the various steps can be organized in many ways. With any design, the control network must "remember" what has already taken place in the determination of any given quotient digit. This capability can be conveniently supplied by four flip-flops which are reset to 0 prior to the determination of each quotient digit. The first flip-flop may then be set to 1 (binary) if the sign of the first difference is positive. The second flip-flop is set to 1 if the sign of the difference after the second addition-subtraction is positive, and so on. Two additional flip-flops are employed in a four-counter to "remember" whether the first, second, third, or fourth addition-subtraction is being performed. The output signals from the six flip-flops are used as input signals to a combinatorial switching network which (a) determines the operation to be performed, that is, controls whether the divisor, its doubled value, or its quintupled value is to be used and also controls whether the operation is to be an addition or a subtraction and (b) determines the quotient digit. For example, if the sign-remembering flip-flops contain 1100 and the counter indicates that two addition-subtractions have been performed, the next operation is the subtraction of the doubled multiplicand, but if four addition-subtractions have been performed, the quotient digit is known to be 7.

If a positive intermediate remainder occurs after the determination of any given quotient digit, the next quotient digit may be determined in the same manner as just described. However, if a negative intermediate remainder occurs, the next quotient digit is determined by a nonrestoring process analogous to that described previously. Thus, in this scheme the initial addition or subtraction of the quintupled divisor for any given quotient digit is controlled by the sign of the previous intermediate remainder. Accordingly, the network described in the previous paragraph must contain an additional flip-flop which "remembers" that sign. The output signals from this flip-flop are used to invert the addition-subtraction operation and complement the quotient digit when appropriate.

The control networks of the two previous paragraphs were not completely described, and in particular the devices for "remembering" which quotient digit is being determined were not mentioned. A complete description, with diagrams, of all the switching devices needed to form a complete division unit operating by this or any other scheme would require more pages of text than is deemed appropriate to devote to this subject here. The division control network is nevertheless an excellent example of a "sequential circuit" as discussed in more general terms in a previous chapter.

As indicated in Figure 7-12, three addition-subtractions are required for quotient digits 0, 3, 4, 5, 6, or 9 and four for digits 1, 2, 7, or 8, and a corresponding situation exists when the initial intermediate remainder is negative for any given digit. Therefore, an average of 3.4 addition-subtractions per quotient digit are required for random quotient digits.

Division with All Nine Multiples of the Divisor Available

A decimal division scheme that consumes the time for only one addition-subtraction per quotient digit can be devised if all nine multiples of the divisor are available. However, in the most straightforward version, nine decimal subtractors operating simultaneously are required. Each of the nine multiples is subtracted from the dividend or partial remainder, as the case may be, and the quotient digit is equal to the largest multiple which leaves a positive remainder. Appropriate switching equipment must be included to select this remainder for the determination of the next quotient digit.

No division scheme practically analogous to the multiplication scheme of Figure 7-11 seems to exist.

A DIVISION METHOD   423

Another approach to high-speed division employing all nine multiples of the divisor involves examining the high-order digits of the divisor and dividend (or intermediate remainder) in much the manner a person would do so in manual division. A "guess" is then made as to the quotient digit, and the corresponding multiple of the divisor is subtracted. Further subtractions are made if the difference is positive, and subsequent additions are made if the difference is negative. By examining the two highest-order divisor digits and the two highest-order dividend (or intermediate remainder) digits and appropriately comparing them, a guess can always be made which will either be correct or too large by 1. For example, if the divisor is $21xxx\cdots$ and the dividend is $67xxx\cdots$, where the x's are any digits, the quotient digit will always be 3 regardless of the values of the x's. If with the same divisor the dividend is $63xxx\cdots$, the quotient digit may be 3 or it may be only 2 with the correct value being dependent on the values of the digits in the x positions. In either of these two instances a quotient digit of 3 is selected and the tripled value of the divisor is subtracted. If a negative difference is obtained in the latter instance (a negative difference being impossible in the former instance), the correct quotient digit is known to be 2, and the next successive quotient digit can be determined by a nonrestoring process.

For making the guess, a table of all possible combinations of the two high-order divisor digits and two high-order dividend (or intermediate remainder) digits may be constructed. The resulting table would contain 10,000 entries, a rather large number for mechanization. However, many of the entries are more or less trivial. In particular, the quotient digit will be 0 for the entire portion of the table for which the two high-order divisor digits represent a larger number than the two high-order dividend (or intermediate remainder) digits. By studying the complete table, many other hints for simplifications can be found. The appropriate guess can be generated by means of a 16-input, four-output combinatorial switching network (four lines per decimal digit). Instead of a switching network, a read-only storage unit having a capacity of 10,000 four-bit words may be used, as discussed in Chapter 3.

A Division Method Involving the Examination of the Highest-Order Divisor Digit and the Highest-Order Dividend Digit and Employing the Doubled and Quintupled Divisor

The features of the previously described decimal division methods can be combined to develop a scheme which, although neither the fastest nor the most economical, offers comparatively high speed characteristics at a reasonable cost.

The highest-order divisor digit and the highest-order dividend (or intermediate remainder) digit are examined. Then, either 0.2, 0.5, 1, 2, or 5 times the divisor is subtracted as set forth in Table 7-2, and a corresponding increment is made in the quotient digit. A doubler and a quintupler may be used for obtaining 2 and 5 times the divisor, respectively, and the doubled and quintupled values may be shifted to the right one decimal position to obtain 0.2 and 0.5 times the divisor, respectively. In the case of 0.2 and 0.5, the increment is not made on the given quotient digit, but is made on the next lower order quotient digit. In some instances the subtraction will produce a negative intermediate remainder. Whenever a negative intermediate remainder occurs, it is caused to appear in true (magnitude and sign—not complement) form. The next step is performed in the same way except that the appropriate multiple of the divisor is added instead of subtracted, and the corresponding quotient digit is decremented instead of incremented.

To illustrate the division method, the steps in the division of 9011360 by 49 shown below. Each underlined quotient digit is negative. Here,

TABLE 7-2. The Divisor Multiple to be Selected in Accordance with the Highest-Order Dividend and Divisor Digits

	Highest-Order Dividend Digit								
		1	2	3	4	5	6	7	8	9
Highest-Order Divisor Digit	1	1	2	2	2	5	5	5	5	5
	2	0.5	1	1	2	2	2	2	2	5
	3	0.5	0.5	1	1	2	2	2	2	2
	4	0.2	0.5	1	1	1	1	2	2	2
	5	0.2	0.5	0.5	1	1	1	1	2	2
	6	0.2	0.5	0.5	0.5	1	1	1	1	1
	7	0.2	0.2	0.5	0.5	0.5	1	1	1	1
	8	0.2	0.2	0.5	0.5	0.5	1	1	1	1
	9	0.2	0.2	0.5	0.5	0.5	0.5	1	1	1

$D = 49$ with the digits being positioned in accordance with the quotient digit being determined at any given step.

				Quotient
		9011360		
Step 1	subtract $2D$	98		2?????
		788640	(negative)	
Step 2	add $2D$	98		2_2_????
		191360	(positive)	
Step 3	subtract $0.2D$	98		22_2_???
		93360	(positive)	
Step 4	subtract $2D$	98		224???
		4640	(negative)	
Step 5	add D	49		224_1_??
		260	(positive)	
Step 6	subtract $0.5D$	245		224_1_05 = 183905
	final remainder:	15	(positive)	

For all steps the highest-order divisor digit is 4 in this example. On step 1, the highest-order dividend digit is 9, and Table 7-2 indicates that the doubled divisor should be subtracted. However, a negative intermediate remainder is produced, the highest-order digit of which (in true form) is 7. Again, the table indicates that the doubled divisor should be used, but instead of being subtracted, it is added to the corresponding positions, and the quotient is 2_2_????. Because $(20 - 2)10^k = (10 + 8)10^k$, where k is any integer, the quotient is equivalent to 18????, although this conversion is not shown until the last step. Steps 3 through 6 proceed analogously after giving due consideration to the statements in the preceeding paragraph. In examples where the final remainder is negative, 1 should be subtracted from the lowest-order position in the quotient.

Observe that in the example two steps were consumed in developing the third quotient digit (a 4 subsequently converted to a 3), but no steps were consumed in developing the fifth quotient digit (a 0). In total, only six addition-subtraction operations were required to develop a six-digit quotient. However, this result of one addition-subtraction per quotient digit is not necessarily typical or average. For other examples, especially when divisors of the form $10xxx\cdots$, where the x's are any digits, three addition-subtractions are required for many quotient digits. The required number of addition-subtractions is not a simple function of the magnitudes of the quotient digits as was the case for previously described division methods but is a function of the particular digits encountered in the divisor, the dividend, and the intermediate remainders. The average for random examples is not known, but it is probably slightly less than two addition-subtraction operations per quotient digit.

The mechanization of Table 7-2 can be accomplished by means of a combinatorial switching network. Eight input signals are required, four for each of the two decimal digits being examined. Most conveniently, five output signals are generated, one to control the selection of each of the multiples 0.2, 0.5, 1, 2, and 5. When the highest-order dividend (or intermediate remainder) digit is 0, all five control signals will be 0 (binary), and the next highest-order intermediate remainder digit is then examined with the divisor being shifted accordingly.

This division method is similar to, but not identical with, the inverse of the multiplication method described on page 409.

Iterative Division—Programmed Division

The formulas given in the previous chapter for iterative binary division are directly applicable to iterative decimal division. Other iterative procedures to effect division have been devised which are more directly applicable to decimal numbers. Although some of these procedures are highly ingenious, they have not been found to be particularly competitive with the division methods already described, either in terms of speed or cost. The reader who is interested in pursuing this subject further is referred to a paper by S. K. Nandi and E. V. Krishnamurthy in *Communications of the ACM*, May 1967, pp. 299–301. This paper gives additional references to the major prior publications on the subject.

Most interest in iterative division is related to performing division by means of programming a computer not equipped with the division operation as a built-in feature. However, any of the division methods which have been described in this chapter can be programmed on any general purpose computer. The number of instructions required and the amount of time consumed to perform a division would, of course, depend on the division method selected and would also depend on the instruction repertory built into the computer. A computer equipped with index registers (see Chapter 9) and an appropriate set of elaborations could utilize any of the division methods to perform division quite efficiently through programming.

Decimal Round-Off

Decimal round-off is accomplished in essentially the same manner as was described for binary round-off except that a 5 instead of a 1 is added (subtracted in the case of negative numbers in complement form) to the highest-order position to be discarded.

In the case of division where the quotient is developed in a register that is not equipped with addition and carry-handling facilities a convenient alternative round-off procedure is to add the quintupled divisor to those positions of the dividend which cause the quotient to be increased by 5 in the position to the right of the lowest-order position to be retained. This alternative procedure has the additional advantage of requiring the development of only n quotient digits (not $n + 1$ quotient digits) for an n-digit rounded quotient.

Decimal Floating-Point Operation

Decimal floating-point operation is likewise essentially the same as was described previously for binary floating-point operation. Many differences in detail are encountered, but these differences and their effect on network design can readily be determined from the information given.

Exercises

1. By relating the logarithm of the number of messages representable by a 10-digit decimal number to the logarithm of the number of messages representable by a binary number having the same number of bits (assuming four bits per decimal digit), confirm that approximately 20.4% more digit storage capacity is required for information in the decimal system in comparison with the binary system.

2. Of the weighted four-bit decimal codes (see Table 7-2) find at least two for which one and only one bit combination can be assigned to each decimal digit. Find at least two codes for which at least one decimal digit is representable by two different bit combinations.

3. Design an AND-OR-NOT switching network (that is, determine the required switching functions in Boolean notation) that will transform signals in the 8-4-2-1 weighted code to the 8-4-2-1-R two-out-of-five code, as these codes are defined in the text. Design another switching network which will perform the opposite transformation.

4. Assume that a decimal digit in the 8-4-2-1 code is stored in a set of four complementing flip-flops. Determine the signals, as functions of the flip-flop output signals d_8, d_4, d_2, and d_1, needed to transform the stored digit to its 9's complement.

5. Repeat the above exercise but with J-K flip-flops.

6. Determine the switching functions needed to develop the overflow and "Sign D" signals in the array of Figure 7-4. Repeat this exercise but with the convention that positive numbers are in 9's complement form and negative numbers are in true form. (The three input signals to the network may be designated A, B, and C.)

7. Prove that for the multiplication method described in page 404 the average number of addition-subtraction operations per multiplier digit is approximately 2.5.

8. Show by Boolean manipulation that the two sets of expressions for D_8, D_4, and D_2 in the doubled decimal digit (8-4-2-1 code) as given on pages 406 and 407 are equivalent.

9. Select a four-bit decimal code at random, and with this code design a doubler as a five-input, five-output combinatorial switching network.

10. Show by the use of diagrams (see Chapter 2) how the control functions $(m_8 + m_4)\bar{m}_1$, and so on, in Figure 7-10 can be derived.

11. For the multiplier variation (four carry input lines and four carry output lines) described on page 414, derive a switching array analogous to Figure 7-10.

12. Assume that the network shown in Figure 7-11 is being used to multiply 917 by 628. Determine the digits appearing at the output of each generator of multiplies and each adder on each of the six steps required to complete the multiplication.

13. Draw the chart corresponding to Figure 7-12 to be followed when the intermediate remainder resulting from the determination of the last previous quotient digit is negative.

14. With only a doubler (no quintupler) available, determine the sequence of addition-subtraction operations required to minimize the number of these operations required in a decimal division, and with this sequence determine the average number of addition-subtraction operations required for random quotient digits. Draw the chart corresponding to Figure 7-12. Repeat this exercise but with only a quintupler (no doubler) available.

15. Show the steps in the division of 9000138 by 107 when using the division method represented by Table 7-2.

16. Determine the switching functions, but not necessarily in minimum-component form, needed to control the selection of the multiples 0.2, 0.5, 1, 2, and 5 as prescribed by Table 7-1. (Assume that the input digits are in the 8-4-2-1 code. Designate the bits of divided digit by A, B, C, and D, respectively, and the bits of the divisor digit by E, F, G, and H, respectively. Suggestion: draw a chart as in Figure 2-4.)

17. Devise block diagrams analogous to Figure 7-10 for the two alternative multiplication schemes described on page 415.

8

IMPLEMENTING OTHER DIGITAL OPERATIONS

Many operations other than addition, subtraction, multiplication, and division, which were the topics of the two previous chapters, must be performed by a digital system in completing most information processing tasks. Although virtually all other operations can be performed by suitable sequences of the four basic arithmetic operations, the corresponding programs may be both expensive (in the amount of storage space required) and time-consuming. For any operation that is required only rarely, the programming approach tends to be preferable. On the other hand, built-in facilities for executing a given operation is preferable when that operation is known to be a frequent requirement in an application. This chapter will cover some of the more important of these more specialized operations, specifically: Gray-binary code conversions, decimal-binary conversion, comparison, square root extraction, one-variable function generation, random number generation, sorting, incremental computing (as in a digital differential analyzer), and checking of arithmetic operations.

Gray-Binary Code Conversions

In applications where an analog signal (which may be electrical or mechanical) is transformed to a digital signal, a desirable feature of the digital code is that successive digital representations differ in only one digit position. Many such codes can be devised, but one such code, commonly called the "reflected binary" or "Gray" code, is particularly straightforward and is widely used. The binary digit combinations for numbers 0 through 17 (decimal) are listed as follows.

Decimal	Binary	Gray
0	00000	00000
1	00001	00001
2	00010	00011
3	00011	00010
4	00100	00110
5	00101	00111
6	00110	00101
7	00111	00100
8	01000	01100
9	01001	01101
10	01010	01111
11	01011	01110
12	01100	01010
13	01101	01011
14	01110	01001
15	01111	01000
16	10000	11000
17	10001	11001

Observe that the Gray code digits are determined by changing a digit from 0 to 1 at each power of 2 (that is, at 2, 4, 8, 16, and so on) and by repeating the previously used combinations for the other digits, but in reverse sequence—hence the name "reflected."

A network for transforming any number in conventional binary code to Gray code can be devised by noting that each Gray code digit is the same as the corresponding binary digit except when the next higher order binary digit is 1, in which case the Gray code digit is the inverse of the corresponding binary digit. That is, in Boolean notation,

$$R_i = B_i \bar{B}_{i+1} + \bar{B}_i B_{i+1}$$

where R_i and B_i are the digits in the ith position of the reflected (Gray) and binary codes, respectively, and the $i + 1$ position is the next higher order position. The resulting network in terms of exclusive-OR modules is shown in Figure 8-1a for n-digit numbers.

For converting from the Gray code to the conventional binary code, the observation may be made that a given binary code digit is the same as the corresponding Gray code digit except when the next higher order binary digit (which must be determined prior to determining the given binary digit) is 1, in which case the binary digit is determined by inverting the given Gray code digit. In Boolean notation,

$$B_i = R_i \bar{B}_{i+1} + \bar{R}_i B_{i+1}$$

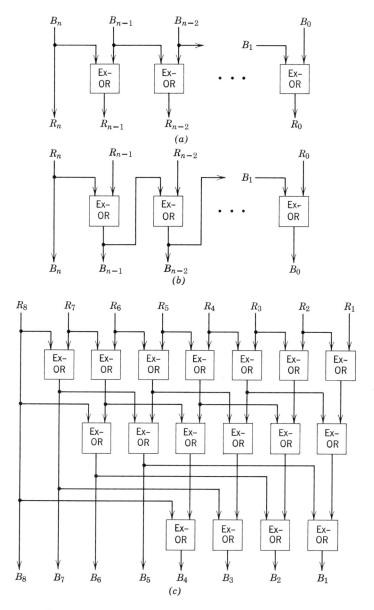

FIGURE 8-1. Gray-binary code conversion networks.

The resulting network is shown in Figure 8-1b. A characteristic of the code conversion process is that, although successive Gray-code numbers differ in only one digit, partial-amplitude signals are not usually tolerable at the input terminals of the conversion network. If one of the R_i signals is not decisively 0 or 1, the corresponding B_i and all lower-order binary code signals may be in error and may thereby cause a gross error in the binary representation of the quantity. Therefore, a network not shown in Figure 8-1b must usually be placed in the input lines to cause each R_i signal to be either 0 or 1 (even though at any given instant one such signal may be at some intermediate value) and to cause the input signals to remain static during the time that the B_i signals are being developed and utilized.

The array in Figure 8-1b has a speed disadvantage in that, for example, when R_n changes from one binary value to the other, the binary values of all digits in the binary number must change. The effect of the change in R_n must "ripple" through all exclusive-OR devices in succession. The problem is analogous to the ripple carry problem in binary addition, although in this case the signal travels from high to low orders instead of from low to high orders. Techniques analogous to the previously described high speed carry propagation methods may be used to improve the speed of the conversion process, but a somewhat simpler and yet effective speed improvement scheme is illustrated in Figure 8-1c for eight-digit words. The input signals must pass through at most three exclusive-OR devices in succession to reach the output lines. As explained in more detail in the next section for a corresponding step-by-step conversion method, the number of "levels" of devices is k for words having 2^k digits. At each level, the word which becomes available at the previous level is shifted to form the next word in succession, and the number of shifted positions is doubled after each level.

Other Gray-Binary Conversion Methods

Often a practical requirement is that the conversions be performed on a general purpose computer with a minimum of physical design features specifically intended for the task. Although general purpose computers are not described until the next chapter, the essential features of applicable conversion methods can be understood with only a knowledge of binary addition and subtraction as described in previous chapters. The adder is modified to suppress the carry signals so that each digit S_i is equal to $A_i \bar{B}_i + \bar{A}_i B_i$ in Boolean notation. In other words, the two operands A and B are combined in a manner whereby the digits of the resultant word are generated by the exclusive-OR function applied to the corresponding digits of A and B.

The binary-to-Gray conversion is then accomplished quite simply by shifting the binary word one position to the right (with the lowest order digit being discarded) and combining it with the unshifted binary word in an exclusive-OR fashion as described in the previous paragraph.

The Gray-to-binary conversion is somewhat more complex. The Gray word is shifted to the right one position (with the right-hand digit being discarded) and is combined with the unshifted Gray word in the same exclusive-OR fashion. The Gray word is shifted again (with the next digit on the right being discarded) and is combined in exclusive-OR fashion with the result of the previous step. This process is repeated $n - 1$ times, where n is the number of digits in the Gray word. If the n_h left-hand digits of the Gray word are 0's, the number of steps can be reduced to $n - n_h - 1$. As an example, consider the conversion of the Gray word 101101. The steps are as follows, where the word under each line is derived by combining the two words above the line on a digit-by-digit exclusive-OR basis.

```
              101101 Gray-code word for decimal 54
Step 1         10110
              ------
              111011
Step 2          1011
              ------
              110000
Step 3           101
              ------
              110101
Step 4            10
              ------
              110111
Step 5             1
              ------
              110110 binary-code word for decimal 54
```

The number of steps required to perform the Gray-to-binary conversion can be reduced to $n/2$ if n is even. If n is odd, it is in effect made even by appending a 0 to the left-hand end of the number. The first $n/2 - 1$ steps are as before, but then the word derived from the first steps (not the original Gray word) is shifted to the right $n/2$ positions and is combined in an exclusive-OR relationship with the word derived from the first steps. Thus, for example, the eight-digit Gray word 10101101 would be converted to its binary equivalent in four steps as follows, where the first three steps are accomplished as described in the previous paragraph and the fourth step is accomplished as described in this paragraph.

434 IMPLEMENTING OTHER DIGITAL OPERATIONS

```
           10101101 Gray-code word for decimal 201
Step 1      1010110
           ‾‾‾‾‾‾‾‾
           11111011
Step 2       101011
           ‾‾‾‾‾‾‾‾
           11010000
Step 3        10101
           ‾‾‾‾‾‾‾‾
           11000101
Step 4         1100
           ‾‾‾‾‾‾‾‾
           11001001 binary-code word for decimal 201
```

If $n = 2^k$, where k is an integer, the number of steps required to perform a Gray-to-binary conversion can be reduced to k. If k is not an integer, it can in effect be made an integer by appending a suitable number of 0's at the left end. The procedure is as follows. The first step is as before in that the Gray word is shifted to the right one position and is combined in an exclusive-OR relationship with the unshifted Gray word. On the second step, the result of the first step is shifted to the right two positions and is combined with the unshifted result of the first step. On the third step the result of the second step is shifted four positions to the right and is combined with the unshifted result of the second step. Succeeding steps are analogous, but the number of positions shifted is doubled on each step. As an example, the six-digit Gray word 101101 is treated as an eight-digit word 00101101 for which $k = 3$ and is converted to binary form in three steps as follows.

```
           00101101 Gray-code word for decimal 54
Step 1      0010110
           ‾‾‾‾‾‾‾‾
           00111011
Step 2       001110
           ‾‾‾‾‾‾‾‾
           00110101
Step 3         0011
           ‾‾‾‾‾‾‾‾
           00110110 binary-code word for decimal 54
```

In a quite different Gray-to-binary conversion method each digit position in the Gray word is assigned a weight $2^i - 1$, where i is obtained by numbering the digits, right to left, from 1 through n. That is, the weights, right to left, are 1, 3, 7, 15, 31, 63, 127, and so on. In the conversion process, the weight of the left-most 1 in the Gray word is taken as positive. The weight signs of the 1-digits to the right of the left-most 1 alternate between positive and negative. The weight of each 0 in the Gray word is zero. Thus, for example, the Gray word **10101101** would be equal to $255 - 63 + 15 - 7 + 1 = 201$. When performing the conversion in binary notation the process becomes quite simple because each weight is representable by a series of 1's, and no carries or borrows occur in any of the addition-subtraction operations. Therefore, no provision need be made for carry-borrow propagation time. In the above example the steps would be as follows.

		binary	decimal
Step 1	add:	11111111	255
Step 2	subtract:	111111	63
		11000000	192
Step 3	add:	1111	15
		11001111	207
Step 4	subtract:	111	7
		11001000	200
Step 5	add:	1	1
		11001001	201

Each series of 1's can be generated by starting with a word that consists entirely of 1's and by shifting this word to the right one position for each Gray word digit that is sensed. For each Gray word digit that is 0, no further action is taken, but for each Gray word digit that is 1, the alternating additions and subtractions are performed as described.

A Gray-to-binary conversion method that is based on binary multiplication is to multiply the Gray word by an equal length word containing all 1's, but the carries are suppressed when accumulating the partial products. The resultant binary word is found in the n high-order digits of the "product," and the $n - 1$ lower-order digits are discarded. (The "product" thus formed will have $2n - 1$, not $2n$, digits when the carries are suppressed.) If the Gray word is used as the multiplier, a step is saved for each 0 in the Gray word. Thus, for example, the conversion of the Gray word 101101 to its binary equivalent 110110 (decimal 54) would proceed as follows.

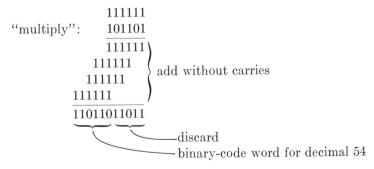

Decimal-Binary Conversion—General

Because of the previously discussed efficiency of machines employing the binary number system, binary machines are often used in applications for which the input and output information must be in the decimal system. In these applications an obvious need arises to convert the input decimal information to binary form and to convert the binary output information to decimal form. The conversions can be performed (a) by separate pieces of equipment specifically designed for the task, (b) by "pure programming" on the main information processing machine, which is assumed to of general purpose nature but which has no built-in features specifically intended for conversions, or (c) by programming the main information processing machine which has built-in facilities of some sort for improving the efficiency of the conversion process (i.e., for reducing the time and the number of instruction required). Method (c) usually involves adding at least one special conversion instruction to the instruction repertory, which is a topic to be included in the next chapter.

Regardless of whether method (a), (b), or (c) is employed, the designer may choose from several different procedures ("algorithms") for performing the conversions. In the following sections the more important of these procedures will be described and their general characteristics with respect to machine implementation will be pointed out.

In contrast to the four basic arithmetic operations, the location of the point is highly consequential in decimal-binary conversion. That is, for a given set of digits prior to conversion, the resultant set of digits is dependent on the location of the point. Two examples are as follows:

decimal		binary
23.	=	100111.
2.3	=	10.0100110011001100110011 $\cdots$
0.23	=	0.0011101011100001010001 $\cdots$

binary		decimal
1001.	=	9.
100.1	=	4.5
10.01	=	2.25
1.001	=	1.125
0.1001	=	0.50625

Moreover, as the examples indicate, a decimal fraction does not necessarily have an exact binary counterpart but instead may represent a transcendental number in the binary system. A binary fraction always has an exact decimal counterpart, although an objectionably large number of decimal digits is, in most instances, required to represent that counterpart. Thus, when performing a binary-decimal conversion on fractions the result must usually be only a rounded approximation (where rounding is accomplished as described in previous chapters). The subject of conversion error is treated more exhaustively in a paper by D. W. Matula in the *Proceedings of the Spring Joint Computer Conference*, April 1967, pp. 311–318.

The procedures for converting integers and fractions tend to be substantially different from each other. For quantities that have both an integral and a fraction part, the usual procedure is to treat the digits as though they represented either a pure integer or a pure fraction and to handle point adjustment in the manner to be described later for floating-point numbers.

Decimal-to-Binary Conversion—Integers

A straightforward manual method for converting decimal integers to binary form is to determine the largest power of 2 which is equal to or less than the given integer. This power of 2 is subtracted, and the largest power of 2 which is equal to or less than the remainder is then determined and subtracted. This process is continued until the remainder is zero. For each power of 2 that is subtracted, the corresponding binary digit is 1, and each other binary digit is 0. For example, 359 would then be converted to binary form by the following steps.

```
                    359
         subtract:  256    1--------
                    ---
                    103
         subtract:   64    101------
                    ---
                     39
         subtract:   32    1011-----
                    ---
                      7
         subtract:    4    1011001--
                    ---
                      3
         subtract:    2    10110011-
                    ---
                      1
         subtract:    1    101100111
                    ---
                      0
```

This conversion method is very simple for a person who has the various powers of 2 memorized, and it can be mechanized in several different ways. However, the method functions basically in the decimal number system, whereas the usual need is to perform the conversion in a binary computer.

A quite different conversion method generates the binary digits in the opposite sequence, that is, low order first. The decimal number is divided by 2, and the lowest-order binary digit is 1 or 0 according to whether the remainder is 1 or 0, respectively. The quotient is then divided by 2, and the second binary digit is equal to the new remainder. Subsequent binary digits are similarly determined by repeated divisions by 2 until the quotient is zero. The following steps illustrate the conversion of 359 to binary form by this method.

```
359
179    --------1
 89    -------11
 44    ------111
 22    -----0111
 11    ----00111
  5    ---100111
  2    --1100111
  1    -01100111
  0    101100111
```

The division by 2 can be mechanized by multiplying by 5 and then shifting to the right one position, that is, by multiplying by 0.5. The purpose of multiplying by 0.5 instead of dividing by 2 is that the conversion will, with most implementations, proceed at much higher speed, and the relatively economical quintuplers described in the preceding chapter may be used. In the parallel version, the decimal number to be converted is placed in a register, the outputs of which are used as the inputs to a set of quintuplers. After allowing for the small amount of delay required to generate the quintupler output signals (carry propagation being required through only one order), the quintupler output signals, properly shifted, are then entered into the same register to replace the digits originally stored there. This process is repeated until the register contains zero, and at each step the digit (which can only be 0 or 5—representing 0.0 or 0.5) which is shifted to the position to the right of the lowest-order position can be multiplied by 2 to obtain the binary digit currently being generated. However, an actual multiplication by 2 is unnecessary. For example, if the decimal digits are in the 8-4-2-1 code, the binary digit is simply 1 or 0 according to whether the 1's bit is 1 or 0, respectively.

The above conversion method can be modified to generate more than one binary digit per step. All that is necessary is to divide by an appropriate higher power of 2. One scheme for dividing by a higher power of 2 is to multiply by 0.5 a corresponding number of times, and this

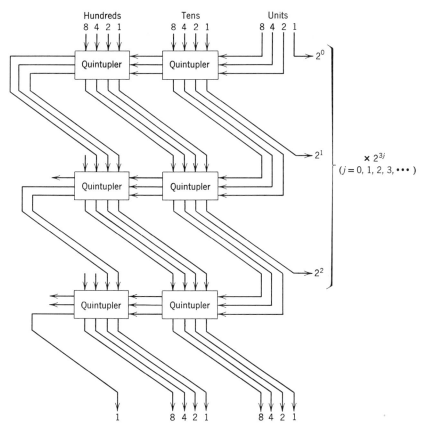

FIGURE 8-2. Divide-by-eight network for developing three binary digits per step in a decimal-to-binary conversion.

multiplication can be accomplished by added sets of quintuplers as illustrated in Figure 8-2 for multiplication by $(0.5)^3$, which is equivalent to a division by $2^3 = 8$. Three binary digits are generated by this array. The output signals along the bottom of the figure may be entered into the register (not shown) containing the original decimal number for the generation of the next three binary digits. Thus, only one third as many steps are required. Observe that for N-digit decimal numbers, at most $N - 1$ quintuplers are required for each multiplication by 0.5, and also observe that for some quintuplers some of the input and output signals are not utilized. This last situation results from the fact that the number being repeatedly halved gradually becomes smaller in magnitude.

440 IMPLEMENTING OTHER DIGITAL OPERATIONS

The conversion method illustrated in Figure 8-2 can be carried to the extreme of performing the complete conversion in a single step. The number of rows of quintuplers required is equal to n, the number of binary digits required for the largest number to be converted. However, because of the gradual reduction in the magnitude of the number being halved in the various rows, the total number of quintuplers required is considerably less than $n(N-1)$. Although the quintupler is a decimal network, it may without serious complications be included in an essentially binary computer. With large scale integrated (LSI) circuits, this method of conversion is practical and is attractive because of the manner in which it employs a single network form—the quintupler.

Another way in which a decimal number in the 8-4-2-1 code can be halved is to place the number in a register and shift all binary digits to the right one position with the individual decimal digits being corrected as necessary. The register is visualized as being divided into groups where each group contains four binary positions and stores a decimal digit. If after a shift but before a correction a given decimal digit contains a 1 in the "8's" position, that 1 must have been derived from a shift of a 1 from the "1's" position of the next higher decimal order. This 1 has the weight of an 8, but in the halving of a decimal number the weight of a signal transmitted from a given order to the next lower order should be only 5. Therefore, 3 should be subtracted in each instance that a 1 is shifted into the "8's" position of any group of four bits. To illustrate this scheme of repeatedly halving a decimal number by right shifts to obtain the equivalent binary number, the steps in the conversion of 359 are listed below. The binary digits are generated in sequence, low order first, by the digits shifted out of the lowest-order position of the register.

```
                0011 0101 1001 ← (359 in 8-4-2-1 code)
shift:          0001 1010 1100    --------1
subtract 3's:   0001 0111 1001
shift:          0000 1011 1100    -------11
subtract 3's:   0000 1000 1001
shift:          0000 0100 0100    ------111
shift:          0000 0010 0010    -----0111
shift:          0000 0001 0001    ----00111
shift:          0000 0000 1000    ---100111
subtract 3's:   0000 0000 0101
shift:          0000 0000 0010    --1100111
shift:          0000 0000 0001    -01100111
shift:          0000 0000 0000    101100111
```

As the binary digits are developed they may be entered into the high-order end of a second shift register the contents of which are shifted to the right one position at the time of each shift in the first register. At the termination of the process the digits will appear in the relative positions indicated in the last step above. Due allowance must be given to the fact that the number of binary digits in the 8-4-2-1 coded decimal number is greater than the number of binary digits in the purely binary equivalent.

Observe that after a shift in the decimal register, digit combinations 1101, 1110, and 1111 cannot occur in any group even before subtracting the corrective 0011 (decimal 3). This fact can be utilized to simplify the subtraction network. A combinatorial switching network is designed to transform digit combinations 1000, 1001, 1010, 1011, and 1100 (when they occur) to 0101, 0110, 0111, 1000, and 1001, respectively, and to leave digit combinations 0000 through 0111 unchanged. The resulting network is analogous to a quintupler as it appears in Figure 8-2. More than one row of corrective networks can be used to develop more than one binary digit per step in the same manner that quintuplers were used in that figure.

A conversion method which is more adaptable to programming on a purely binary computer is as follows. The decimal digits are examined one at a time, high order first. The binary equivalent of the first decimal digit is accumulated in a register initially containing zero. This quantity is then multiplied by binary 1010 (ten), and the next decimal digit is accumulated. This process is repeated until all decimal digits have been accumulated. By this method, the conversion of 359 proceeds as follows.

      add 0011:   000000011
      multiply by 1010: 000011110
      add 0101:   000100011
      multiply by 1010: 101011110
      add 1001:   101100111

With the decimal digits in the 8-4-2-1 code as implied, 359 will appear as the bit sequence 0011 0101 1001, which in the binary computer appears as single binary "pseudo-number" 001101011001. The various additions of the binary equivalents of the decimal digits may then be accomplished by selecting, by means of shifting, the appropriate group of four bits from the binary pseudo-number.

The above conversion method can be readily programmed without the use of a multiplication instruction, because each multiplication by 1010 amounts to nothing more than a single addition with appropriate shifts of the operand. After an initial accumulation of the highest-order decimal digit, the following binary operations are performed in a cyclic sequence until the units decimal digit has been accumulated: (a) shift the current operand to the left two positions, (b) add the current operand, (c) shift the new current operand to the left one position, and (d) add the next decimal digit. The particular instructions required to perform these operations depend on the instruction repertory available in the binary computer under consideration. If this conversion method is incorporated as a built-in feature, the "carry save" technique described for binary multiplication in an earlier chapter may be employed to avoid the carry propagation time otherwise required.

An alternative implementation of the conversion method utilizes a single large combinatorial switching network composed primarily of binary full adders. Only $N-1$ (not $n-1$) parallel adders are required. A multiplication by 1010 and the addition of a decimal digit in the 8-4-2-1 code may be accomplished in a single parallel adder plus an OR device as can be observed from the following digit array, where each x is a digit in the current operand and each y is one of the bits of the next successive digit in the 8-4-2-1 code.

$$\begin{array}{rl} & \cdots xxxxxx \\ \text{multiply:} & \underline{1010} \\ & \cdots xxxxxx0 \\ & \underline{\cdots xxxxxx} \\ & \cdots sssxx0 \\ \text{add:} & \underline{yyyy} \\ & \cdots sssssy \end{array}$$

Each digit designated by an s is irrelevant from the standpoint of illustrating the multiplication-addition scheme. The particular feature to be observed about the digit array is that, because of the assumed 8-4-2-1 code, if the y digit in the "8's" position is 1, then the y digits in the "4's" and "2's" positions must both be 0, in which case no binary carry will be transmitted from the "4's" position to the "8's" position. Therefore the "4's-to-8's" carry signal may be combined with the y digit in the "8's" position in an OR device to form the signal to be transmitted to one input line of the corresponding binary full adder.

The resulting network for the conversion of three-digit decimal integers is shown in Figure 8-3, where H and F designate half and full adders, respectively. The largest two-digit decimal number, 99, has the seven-digit binary equivalent of 1100011. Therefore, the eighth binary digit, which would be generated by the carry signal from the highest-order

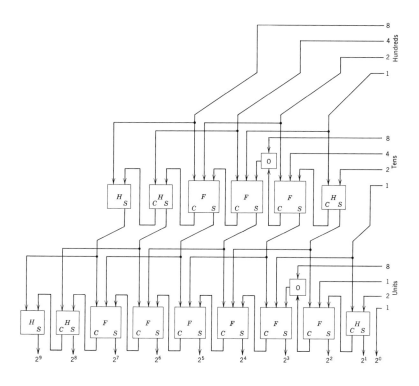

FIGURE 8-3. Another decimal-to-binary conversion network for integers.

half adder in the upper row in Figure 8-3 will always be 0, and this carry signal is not utilized. A similar situation arises in the bottom row where the largest three-digit decimal number, 999, has the ten-digit binary equivalent of 1111100111. The number of half and full adders required is obviously a rapidly increasing function of the number of decimal digits in the number to be converted, but with large scale integration (LSI) techniques the amount of equipment is within practical limits even for relatively large numbers.

Important variations in the combinatorial network for decimal-to-binary conversion can be obtained through a different visualization of the digit pattern. The units, tens, hundreds, thousands, and so on, decimal digits are converted to binary equivalents by multiplying by the corresponding powers of ten, that is, by 1 (one), 1010 (ten), 1100100 (one hundred), 1111101000 (one thousand), and so on, respectively, and then adding. The multiplications may each be performed by an array of full and half adders as described in Chapter 6, and the additions by other sets of full adders—all assembled in a single combinatorial switching network. For three decimal digits, the pattern of binary digits to be added is as follows where $z_8 z_4 z_2 z_1$, $y_8 y_4 y_2 y_1$; and $x_8 x_4 x_2 x_1$ represent the units, tens, and hundreds digits, respectively, and where the binary digits are to be summed column by column.

```
                                          z₈   z₄   z₂   z₁
                                    y₈   y₄   y₂   y₁
                          y₈   y₄   y₂   y₁
                                    x₈   x₄   x₂   x₁
                x₈   x₄   x₂   x₁
         x₈   x₄   x₂   x₁
```

Any array of binary full adders (half adders where only two input digits are encountered) that adds binary digits in accordance with the above pattern will perform the conversion correctly.

A decimal-to-binary conversion method that is suitable for programming on a binary computer not equipped with multiplication capability is as follows. The binary equivalents of decimal 1, 10, 100, and so on, are stored. In a ten-digit binary machine, the numbers would be 0000000001, 0000001010, 0001100100, and so on. These binary numbers are each added a number of times equal to the value of the corresponding decimal digit. In the example of converting 359, the binary number 0001100100 is added three times, 0000001010 is added five times, and 0000000001 is added nine times to produce 0101100111. The conversion process can be made more rapid if the binary equivalents of all nine multiples of each power of decimal 10 (ten) are stored.

Still another decimal-to-binary conversion method for integers is available. On the first step the first (highest-order) decimal digit is doubled, shifted to the right one decimal position, and added to the original number. On the second step the two highest-order decimal digits of the sum are doubled, shifted to the right one decimal position, and added to the above result. On the third step the three highest-order digits of the new sum are handled similarly, and so on. With the individual digits in the 8-4-2-1 code, each 1 in an "8's" position of a sum (but not in the relatively low-order decimal digits which did not enter into a given addition operation) is changed to 0, and a 1 is added to the next higher-order digit—with a corresponding "correction" in that digit if the "8's" bit becomes 1 as a result of binary carries. The process is complete after the step for which, for an N-digit decimal number, the number

DECIMAL-TO-BINARY CONVERSION—FRACTIONS 445

of digits doubled, shifted, and added is $N - 1$. The binary digits in all "8's" positions (all 0's) are discarded, and the binary equivalent of the original number is represented by the remaining binary digits.

The above conversion method will be illustrated by showing the steps in the conversion of 359 to its binary equivalent.

```
                              0011 0101 1001   359 in 8-4-2-1 code
add 3, doubled and shifted:        0110
                              ───────────────
                              0011 1011 1001
"correct" the "8's" bits:     0100 0011 1001
add 43, doubled and shifted:       1000 0110
                              ───────────────
                              0100 1011 1111
"correct" the "8's" bits:     0101 0100 0111
delete the "8's" bits:         101  100  111
regroup:                      101100111  binary-code word for 359
```

Although this particular example doesn't happen to illustrate the point, a carry may occur from one decimal digit to the next when doubling or adding. This carry, if it occurs, should be added to the "2's" digit instead of the "1's" digit of the next higher decimal order. Alternatively, prior to an addition, the doubled digits may be "corrected" in the same manner that a sum is "corrected," in which case no such carry from one decimal digit to the next can occur during the addition. It is instructive for the reader to prove that this procedure does in fact convert decimal numbers to binary form in the general case.

The above conversion method can be implemented in various ways, one of which employs a large combinatorial switching network composed of parallel binary adders. The total number of full and half adders required is roughly the same as for the conversion method illustrated in Figure 8-3.

Decimal-to-Binary Conversion—Fractions

In general, each conversion method applicable to integers has a counterpart that is applicable to fractions, but the counterpart may be materially different with respect to its adaptability to implementation by machine. In the method of subtracting the powers of 2, the principles are the same but the method generally becomes unwieldy because of the large number of decimal digits required to represent 2^{-n}, when n becomes large. As an example, the conversion of 0.446 to binary form proceeds according to the following steps.

```
                       0.446
    subtract 2⁻²:  0.25              0.01
                       0.196
    subtract 2⁻³:  0.125             0.011
                       0.071
    subtract 2⁻⁴:  0.0625            0.0111
                       0.0085
    subtract 2⁻⁷:  0.0078125         0.0111001
                       0.0006875
    subtract 2⁻¹¹: 0.00048828125     0.01110010001
                       0.00019921875
```

The process may be continued indefinitely until the binary equivalent has been generated to the desired accuracy.

The conversion method which, for integers, involved repeated halving of the decimal number becomes a repeated doubling of the decimal number when adapted to fractions. The binary digits are determined one at a time, high order first, by the carries which are transmitted from the tenths position to the units position. These carries may be deleted from the decimal number and shifted into the right-hand end of a shift register to assemble the binary equivalent. In the example of 0.446 the steps in the conversion process are then as follows.

```
          0.446
          0.892      0.0
          1.784      0.01
          1.568      0.011
          1.136      0.0111
          0.272      0.01110
          0.544      0.011100
          1.088      0.0111001
          0.176      0.01110010
          0.352      0.011100100
          0.704      0.0111001000
          1.408      0.01110010001
```

If the decimal fraction is repeatedly multiplied by 4 or 8, then two or three, respectively, binary digits may be developed per step, and if the 8-4-2-1 code is being used, the binary digits in the units decimal digit can be taken directly as the intended digits in the resultant binary number. Multiplications by 16, 32, 64, and so on, can be performed to develop a correspondingly larger number of binary digits per step, but then the integral part of each resultant decimal product must itself then be converted to binary form to obtain the wanted digits.

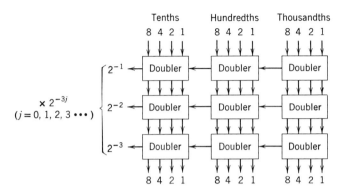

FIGURE 8-4. Network for developing three binary digits per step in a fractional decimal-to-binary conversion.

Alternatively, the multiplication by higher powers of 2 may be performed by sets of decimal doublers as illustrated for the development of three binary digits per step in Figure 8-4 (where a decimal doubler was described in the previous chapter). By this method a one-step conversion can be performed to any desired degree of accuracy by employing a corresponding number of rows of doublers. With N-digit decimal numbers a conversion of n-digit binary numbers requires less than nN doublers because the doublers in the lower-right part of an otherwise rectangular array do not affect the results and may be omitted. The time for conversion is determined primarily by the time required for signals to propagate through a column (not a row) of doublers, because in the doubling process carries do not propagate beyond the next higher decimal order.

The doubling of a decimal fraction through a shifting of the bits in the 8-4-2-1 code one binary position to the left and then applying an appropriate correction (in analogy with halving by a shift to the right, with a correction) results in a network that is the same as a row of doublers in Figure 8-4.

The fractional analogies of the other decimal-to-binary conversion methods tend to be quite awkward to implement. For example, when applied to fractions, the method that involves alternate additions of decimal digit equivalents and multiplications by 1010 becomes alternate additions (but with the decimal digits taken in oppositive sequence, that is, lowest-order digit first) and divisions by 1010. Division by 1010 is much more complex and time-consuming than multiplication by 1010. Moreover, in the directly analogous version the numbers added are the binary equivalents of 0.1, 0.2, 0.3, and so on, which (except for 0.5) do not have exact binary representations and must therefore be stored as binary numbers having sufficient digits to achieve the desired accuracy. Alternatively, the binary equivalents of 1, 2, 3, and so on, can be used, although an extra division by 1010 is required as a last step of conversion process.

When applied to fractions the decimal-to-binary conversion method that, for integers, involves additions of the binary equivalents of decimal 1, 10, 100, and so on, may be modified to perform additions of the binary equivalents of decimal 0.1, 0.01, 0.001 (one 10th, 100th, 1000th), and so on. The difficulty here is that, because decimal fractions do not have exact binary equivalents, substantial round-off errors will accumulate unless the computations are carried out with numbers having greater accuracy (having a greater number of digits) than is required in the final results.

Binary-to-Decimal Conversion—Integers

The various binary-to-decimal conversion methods are largely counterparts of corresponding decimal-to-binary methods, but the details tend to be so different that, from the standpoint of implementation, the methods are distinctly different.

One binary-to-decimal conversion method is to subtract powers of 10 (ten), but expressed in the binary number system, from the binary number to be converted. The decimal digits are generated one at a time, high order first, and each digit is determined by the number of times that the corresponding power of 10 can be subtracted without causing a negative reminder. Thus, for example, 10110111 would be converted by first subtracting the binary equivalent of 10^2, which is 1100100, three times. A difference of 111011 would remain. The binary equivalent of 10^1, which is 1010, would then be subtracted five times, after which a difference of 1001 would remain. Then 1 would be subtracted nine times to reduce the original number to zero. The several subtractions indicate the decimal equivalent of 359. This method can readily be programmed on virtually any binary computer.

The speed of the above conversion method can be improved substantially if, instead of over-and-over subtractions, a division operation is used to determine each decimal digit. In the example cited, the original binary number 10110111 would first be divided by 110100 to produce a quotient of 11 (decimal 3) and a remainder of 111011. This remainder would be divided by 1010 to produce a quotient of 101 (decimal 5) and a remainder of 1001. This remainder may then be divided by 1 to produce a quotient of 1001 (decimal 9) and a remainder of 0. Of course, this last division by 1 (or the over-and-over subtraction of 1 in the previous paragraph) is actually superfluous in determining the last (units) decimal digit, but the method is more straightforward to implement when all decimal digits are determined in the same basic way. Because each quotient will represent a decimal digit in the 8-4-2-1 code, each quotient will have at most three 1's, and therefore the division operations may be less time-consuming than would be the case if random n-digit quotients were required, where n is the number of binary digits in the number being converted.

The need for storing the binary equivalents of all powers of 10 (ten) can be avoided if, prior to the determination of each decimal digit after the first, the remainder is multiplied by binary 1010 (ten). The binary equivalent of only the highest-order power of decimal 10 then need be stored, and this number is used as the divisor in developing each quotient digit. In the example of the previous paragraph the first decimal digit (a 3) would be determined as before, but the remainder, 111011, would be multiplied by 1010 to produce 1001001110. This number would be divided by 1100100 (binary equivalent to decimal 10^2) to produce a quotient of 101 (decimal 5) and a remainder to 1011010. This new remainder would now be multiplied by 1010 to produce 1110000100 which would be divided by the same 1100100 to produce a quotient of 1001 (decimal 9) and a final remainder of 0. The numbers entering into the computations tend to have many more digits than do the numbers in the previous conversion methods. Therefore, this method is not attractive for manual use. However, the increased number of digits appears only in the divisor and dividend of each division operation with the number of quotient digits being unchanged. Therefore the time required for each division operation may be increased only slightly or none at all in some computers. Moreover, each multiplication by 1010 can, as before, be accomplished by a single addition with appropriate shifts of the operand.

In a different binary-to-decimal conversion method the decimal digits are determined in the opposite sequence, that is, lowest order first. The binary number to be converted is divided by 1010 (ten), and the remainder is equal to the lowest-order decimal digit. The quotient is then divided by 1010, and the remainder from this division is equal to the second decimal digit. For successive decimal digits, each new quotient is divided by 1010, and the remainder is equal to the decimal digit being determined. In the example of converting 101100111 to decimal form, the first division by 1010 produces a quotient of 100011 and a remainder of 1001 (decimal 9). The division of 100011 by 1010 produces a quotient of 11 and a remainder of 101 (decimal 5). Although the quotient 11 (decimal 3) actually indicates the last (highest-order) decimal digit, the conversion process is made uniform by a final division of 11 by 1010 to produce a quotient of 0 and a remainder of 11. Unfortunately, a divisor of 1010 results in a division operation that is, in general, as time-consuming as for any multidigit divisor and is, in particular, much more time-consuming than a division for which the quotient is known to be in the range 0000 through 1001. Therefore, this conversion method has relatively poor speed characteristics, although it is probably the simplest of any.

In another binary-to-decimal conversion method for integers the decimal representations of each power of 2 are stored. To perform a conversion, the digits of the binary number are examined one at a time, in any sequence, and the powers of 2 are accumulated or not according to whether their corresponding binary digits are 1 or 0, respectively. In the example of converting 101100111 the decimal numbers to be accumulated would be 256, 64, 32, 4, 2, and 1, which total 359. This conversion method can be implemented entirely by programming or entirely by built-in equipment, or by some combination of the two approaches, in many different ways with the details being dependent on the computer at hand. Although this conversion method is probably most directly applicable to decimal computers, it is usable on binary computers also. The decimal digits in the 8-4-2-1 code may be stored and accumulated separately. That is, for example, in the conversion of 101100111 the units digit is determined by adding 0110, 0100, 0010, 0100, 0010, and 0001 (decimal 6, 4, 2, 4, 2, and 1 as obtained from the decimal equivalents listed above) to result in a sum of 100011. This binary number is converted to 1001 (decimal 9) with a carry of 0001 by repeated subtractions of 1010, with only one subtraction being required in this instance. (If more than one subtraction is required, the carry is correspondingly increased.) The tens digit is determined by accumulating 0101, 0110, and 0011 to which the carry is added to produce 1111 which is similarly converted to 0101 (decimal 5) and a carry of 0001 by the repeated subtraction of 1010, with again only one subtraction being required in this instance. The hundreds digit is determined by adding this new carry to 0010 (the 2 in 256) to produce 0011 (decimal 3).

Another way to convert integers to decimal form is to enter the binary digits, high order first, into the lowest-order position of a decimal register, the contents of which are doubled after the entry of each binary digit except the last. In the conversion of 10110111 by this method the decimal number is built up as follows, where on each step the corresponding binary digit has been added to the doubled value of the previous contents of the decimal register.

decimal	binary
0	101100111
1	01100111
2	1100111
5	100111
11	00111
22	0111
44	111
89	11
179	1
359	

In effect, the decimal requivalent of each binary digit has been generated by doubling each binary digit a number of times corresponding to the weight of that binary digit as determined by its position in the binary number.

The above conversion method may appear to be most applicable to a decimal computer where the doubling can be accomplished in various ways such as through the use of an instruction which adds the number to itself. However, the method is also applicable to binary computers that are to be equipped with built-in facilities to aid in the conversion process. The storage elements in an otherwise binary register are grouped in fours and a decimal doubler is assigned to each group. After the entry of a binary digit in the lowest-order position, the contents of the register are replaced by the decimally doubled value of the contents. Actually, the entry of the next successive binary digit in the lowest-order position can take place simultaneously with the doubling, because this next binary digit will affect only the "1's" signal of the lowest-order decimal digit in the doubled number. (Although decimal doublers are covered in the previous chapter, it will be recalled here that decimal doubling in the 8-4-2-1 code is accomplished by shifting all bits one binary position to the left with the addition of a corrective 6 in each instance that a given decimal digit is greater than 9. Certain network simplifications can be made because some combinations of input signals cannot occur. A minor variation is to add a corrective 3 to each digit greater than 4 prior to the shift.)

By using two, three, or more sets of doublers, the conversion can proceed at the rate of two, three, or more, respectively, binary digits per step. When carried to its limit this idea results in a single large combinatorial switching network which will perform a complete binary-

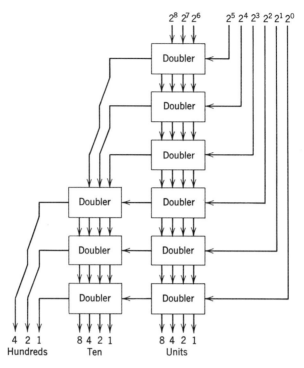

FIGURE 8-5. Binary-to-decimal conversion network employing doublers—for integers.

to-decimal conversion in a single step. However, as in decimal-to-binary conversion the number of doublers required is less than nN because the magnitude of the decimal number builds up gradually, and no doublers are required for positions where the decimal digits will always be 0's. Also, no doubler is needed at any point in the array where the magnitude of the doubled digit will never be greater than 9. Figure 8-5 illustrates the number of doublers needed and their interconnections for the conversion of nine-digit binary numbers to decimal form. Note that the largest possible decimal number in this case is 511. In the figure, the signals entering the right-hand end and leaving the left-hand end of each block labeled "doubler" are the carry input and carry output signals, respectively, for that doubler. The "8's" inputs for some doublers are not utilized.

Still another binary-to-decimal conversion method involves grouping the binary digits in threes (adding one or two high-order 0's as necessary when the number of binary digits is not a multiple of three) and considering each group of three as a decimal digit in the 8-4-2-1 code.

BINARY-TO-DECIMAL CONVERSION—FRACTIONS

The largest possible digit will be 7, but an "8's" bit (0 in each instance) is added to each group of three binary digits to create the effect of a four-bit code. Then as a first step the highest-order digit is doubled decimally, shifted to the right one position, and subtracted from the original number. On the second step the two highest-order digits of the difference are doubled, shifted to the right one position, and subtracted from the difference. On the third step the three highest-order digits of the new difference are similarly handled, and the process is continued analogously until after the step where the lowest-order digit of the doubled and shifted number is in line with the lowest-order digit of the previous difference. For example, the conversion of binary 101100111 to decimal 359 would proceed by first grouping the digits as 101 100 111 which appears as 0101 0100 0111 after the "8's" bits are included and which is regarded as 547. Then, the steps are as follows.

$$
\begin{array}{r}
547 \\
\text{subtract 5, doubled and shifted: } 10 \\
\hline
447 \\
\text{subtract 44, doubled and shifted: } 88 \\
\hline
359
\end{array}
$$

Binary-to-Decimal Conversion—Fractions

A direct method of converting binary fractions to decimal form is to accumulate the decimal equivalent of each 1 in the fraction to be converted. For example, the conversion of binary 0.101101 would be accomplished by adding 0.5, 0.125, 0.0625, and 0.015625 to produce 0.703125. One way to implement this process is to use a combinatorial switching network consisting of decimal digit adders. The adders may be visualized as being in rows where one decimal input number to the adders in a given row is derived from the sum signals of the decimal adders in the previous row, and the other input number is derived from the corresponding binary digit signal which is transmitted to the appropriate input terminals. That is, for example, the 2^{-4} binary digit signal, which has the decimal weight of 0.0625 would not be transmitted to an adder in the 10^{-1} position but would be transmitted (assuming the 8-4-2-1 code) to the "4's" and "2's" input lines of the adder in the 10^{-2} position, to the "2's" input line of the adder in the 10^{-3} position, and to the "4's" and "1's" input lines of the adder in the 10^{-4} position. Because many adder input terminals are not used, the network as just described can be simplified greatly, and an interesting problem is to find ways to eliminate as much hardware as possible. For example, only one decimal digit adder is needed in the 10^{-1} position, and neither of the "8's" input lines is utilized in this adder. Moreover, because no carries can occur to the 10^0 position, neither the carry output network nor the corrective-6 network is needed in this adder. For another example, the carry input to the 10^{-1} position can be derived from an OR device to which carry signals from the two decimal adders in the 10^{-2} position are transmitted. Although this conversion method is straightforward, an awkward round-off problem is encountered which is essentially the same as the round-off problem arising in the addition of any long series of numbers.

In another conversion method the binary digits are entered in sequence, lowest order first, into the units position of a decimal register, the contents of which are halved after each such entry. The conversion of binary 0.101101 to decimal 0.703125 would then proceed according to the following steps.

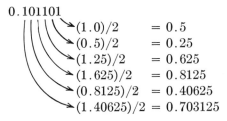

No serious round-off problem is encountered because at each step the decimal digits to the right of any given digit position do not affect the given digit (or higher-order digits) in the halving process. For round-off, a 5 is added (with carries, if any) to the highest-order decimal digit to be discarded.

Each halving in the above conversion method can be accomplished by a multiplication by 0.5, and this multiplication can be performed by quintuplers the outputs of which are shifted to the right one decimal position. A combinatorial switching network which performs a complete conversion in one step and which contains only quintuplers is illustrated in Figure 8-6. Some of the input and output lines for some of the quintuplers are not needed, and correspondingly simpler networks could be used at these points in the array. Note that after each halving (each row in the figure) the lowest-order digit is always a 5, and the carry from quintupling this digit is always a 2, so the quintuplers in the corresponding positions have been eliminated entirely.

If a decimal fraction such as $0.653\cdots$ is repeatedly multiplied by 10 (ten), the digits of the fraction are successively shifted to the left of the point. Analogously, if the binary equivalent of the same fraction is repeatedly multiplied by the binary equivalent of ten, 1010, the binary equivalents of the corresponding decimal digits are successively shifted to the left of the point. For example, the conversion of binary 0.101101 to decimal 0.703125 by this method is as follows, where the underlined groups of four digits are the successively generated decimal digits and are removed from the computations as they are generated. The conversion is complete when all digits to the right of the point are 0's.

```
                        0.101101
         multiply:          1010
                        1 011010        decimal
                        101 101         equivalent
                        0111.000010     0.7
         multiply:          1010
                        0 000100
                        000 010
                        0000.010100     0.70
         multiply:          1010
                        0 101000
                        010 100
                        0011.001000     0.703
         multiply:          1010
                        0 010000
                        001 000
                        0001.010000     0.7031
         multiply:          1010
                        0 100000
                        010 000
                        0010.100000     0.70312
         multiply:          1010
                        1 000000
                        100 000
                        0101.000000     0.703125
```

Because multiplication by 1010 is nothing more than an addition with appropriate shifts, each multiplication can be performed by a parallel binary adder as illustrated in Figure 8-7 for the generation of three decimal digits of the converted number. In this figure each block labeled H or F is a half adder or a full adder, respectively. Obvious extensions of each parallel adder are required when the input binary number has more than six digits. Round-off may be accomplished in the usual way

456 IMPLEMENTING OTHER DIGITAL OPERATIONS

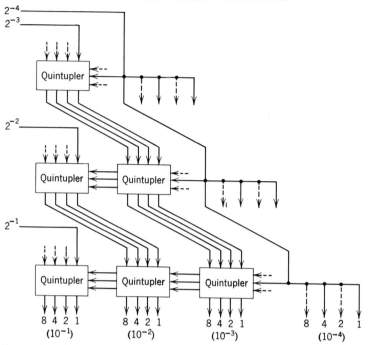

FIGURE 8-6. Quintupler array as employed to convert binary fractions to decimal form.

of generating an additional decimal digit to which a 5, with carries, if any, is added. Alternatively, the binary equivalent of the 5 can be added to the original binary number to be converted. Contrary to 0.5, the decimal fractions 0.05, 0.005, and so on, cannot be expressed exactly in the binary system, but only the high-order 1's in the binary equivalent affect the rounded results, so the addition can be accomplished in most cases with only small extensions of the parallel binary adders.

If the multiplications are performed sequentially (instead of by means of a combinatorial switching network as in Figure 8-7), a scheme for increasing the speed of the conversion process is to multiply by the binary equivalent of a higher power of 10 (ten) and thereby generate a correspondingly larger number of decimal digits per step. In particular, the binary equivalent of decimal 1000 (one thousand) is 1111101000. As explained in an earlier chapter, multiplication by this number can be accomplished by forming $(2^{10} - 2^5 + 2^3)M$, where M is the multiplicand. Thus, only two addition-subtraction operations are required, and if the previously described "carry save" technique is used, only a single carry propagation time is required. The binary digits appearing to the left of the point are the binary equivalent of the three decimal

BINARY-TO-DECIMAL CONVERSION—FRACTIONS 457

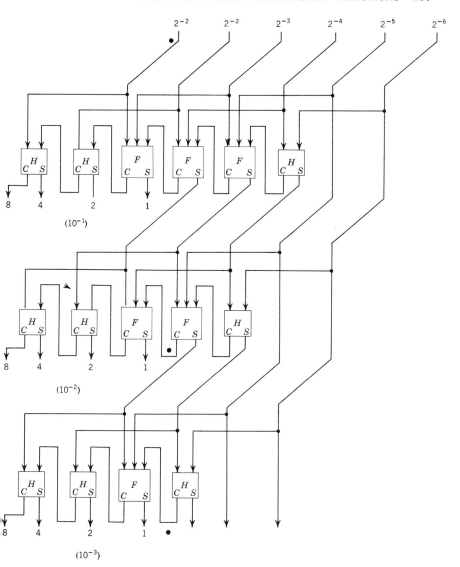

FIGURE 8-7. Combinatorial switching network for converting binary fractions to decimal form.

digits being generated by the multiplication by decimal 1000. This binary integer may be converted to a three-digit decimal integer by any of the methods already described. In the example of the conversion of binary 0.101101 the first multiplication by binary 1111101000 produces a product of 1010111111.001000, and the digits to the left of the point are the binary equivalent of 703. On the second step a multiplication of 0.001000 by 1111101000 produces a product of 1111101.000000, and the digits now to the left of the point are the binary equivalent of 125. The equivalent decimal fraction of 0.703125 is thereby produced. This conversion method is not particularly useful for short binary fractions as in this six-digit example, but it is useful when the binary fractions contain a large number of digits, say 40 or more.

Binary-Decimal Conversion—Floating-Point Numbers

The conversion of floating-point numbers between the binary and decimal number systems cannot be accomplished by simple independent conversions of the mantissa and exponent. Instead, the digits of the converted number must be determined as functions of both the mantissa and exponent considered collectively. The major features of the conversion process will be explained by means of an example. The decimal floating-point number $(0.2331)10^2$ is equal to 23.31 which can be readily converted to its binary equivalent $10111.010011110\cdots$ by converting the integral part 23 and the fractional part 0.31 independently by the methods described previously. The resulting binary number may then be "normalized" (as defined in a previous chapter) to $(0.1011101000111)2^5$ after rounding to thirteen significant binary digits. This result is the floating-point binary equivalent of the given floating-point decimal number.

Alternatively, the point in the original decimal floating-point number can be arbitrarily shifted to one end of the number with a corresponding alteration in the exponent. The number is then converted as a pure integer or a pure fraction, as the case may be, and the result must then be multiplied or divided as required by the binary equivalent of the corresponding power of ten. Because a multiplication is generally less time-consuming than a division, the preferable approach is to shift the point in the original number to the left to form a fraction. In most applications, as in the present example of the floating-point decimal number $(0.2331)10^2$, the mantissa appears as a fraction, which may be converted to $0.001110111010110001\cdots$. This binary version of the fraction must then be multiplied by 1100100 which is the binary equivalent of decimal 10^2. The product is $10111.010011110\cdots$ which is in agreement with the result given in the previous paragraph and which may be normalized as before.

If the original decimal floating-point number already has the point, in effect, appearing far to the left of the most significant digit, as for example $(0.2331)10^{-69}$, the conversion of $0.00 \cdots 002331$ could be performed as before, but the process would consume an excessive amount of time and would entail other difficulties. In such cases the preferable approach is to convert 0.2331 and multiply the result by the binary equivalent of 10^{-69}. The binary equivalents of the various powers of 10 may be precomputed to the necessary number of significant digits and stored as floating-point binary numbers.

The binary-to-decimal conversion of floating-point numbers may be performed analogously. That is, for example, $(0.10111010011110)2^5$ may be treated as 10111.010011110 and the integral and fractional parts converted independently to form 23.3085 which may be rounded to $23.31 = (0.2331)10^2$. Alternatively, the original binary floating-point number may, in effect, be divided by 2^5 by reducing the exponent to 0, and the fraction is converted to 0.7284 (after rounding to four decimal digits). This number is then decimally multiplied by $2^5 = 32$ to produce 23.3088 which may be rounded to $23.31 = (0.2331)10^2$.

A difficulty with the latter conversion method is that a decimal computer must be available to perform the final multiplication whereas the usual need is to perform the conversion on a binary computer. This problem can be solved if the decimal exponent d can be predetermined in some way. Then, prior to the conversion, the binary fraction B is multiplied by the binary equivalent of $(2^b)/(10)^d$, where b is the binary exponent. The multiplication by 2^b compensates for the reduction of b to 0, and the division by 10^d has the effect of normalizing the resulting decimal fraction D. With the definitions of symbols just given, $(B)2^b = (D)10^d$. If $b = d = 0$, then $B = D$, but in all other instances $B > D$ when both floating-point numbers have been normalized. The proof of this last statement will be omitted here—although it is an interesting little problem in mathematics. However, the relationship can be illustrated by the example of the floating-point binary and decimal representations of one $(1.00\cdots)$ which are $(0.1)2^1$ and $(0.1)10^1$, respectively, but in binary the 0.1 has the meaning of one half whereas in decimal the 0.1 has the meaning of one tenth so that, in this instance at least, $B > D$. Then, because $(2^b)/(10^d) = D/B < 1$, d is chosen to be the smallest integer which causes the relationship $(2^b)/(10^d) < 1$ to be satisfied.

In the example of converting $(0.10111010011110)2^5$ where $b = 5$, the smallest such d is 2, and $(2^5)/(10^2) = 32/100 = 0.32$, the binary equivalent of which is $0.010100011110101\cdots$. When this number is multiplied by the B of this example the product is binary $0.001110111010101\cdots$, which converts to decimal 0.2331 after rounding, and because $d = 2$, the decimal floating-point number is $(0.2331)\,10^2$.

Comparison

In the design of digital computers many applications are encountered where two numbers (or sometimes "words" containing alphabetic characters) are to be compared and where the subsequent action of the machine depends on the results of the comparison. In some applications the comparison is a part of the main problem being solved by the computer, but in other applications the comparison is a control function having only an indirect, if any, relationship to the main data processing task. If performed by programming as a part of the main problem, comparison may be accomplished by subtracting one number from the other. Under the assumption that zero is represented as a positive number (see previous two chapters), a negative difference after the subtraction of B from A is the indication that $A < B$, and a positive difference indicates that $A \geq B$. Of course, a subtraction in the opposite direction, that is, A from B, will serve to determine whether $A > B$ or $A \leq B$. One way to determine whether or not $A = B$ is to perform the subtraction in both directions. A positive result from each operation indicates that $A = B$, but if either difference is negative, then $A \neq B$.

An alternative method of determining whether or not $A = B$ is to subtract either number from the other number and to subtract 1 in the lowest-order position of the difference. If $A - B$ is positive and $A - B - 1$ is negative (where A and B are interchangeable), then $A = B$. If both $A - B$ and $A - B - 1$ are positive, then $A > B$, but if both are negative, then $A < B$.

When accomplished by programming, comparison by either of the above techniques is performed in the same adder-subtractor that is used for other purposes. When accomplished by built-in equipment, another adder-subtractor designed in essentially the same way may be used. However, a simpler method of generating the $A = B$ signal is to employ a zero-detection network which consists primarily of a multi-input AND device which detects the presence of 0 signals for all digits of the difference.

The above discussion applies either to binary computers or decimal computers, but binary subtraction is sufficient even for decimal numbers provided that the decimal code is such that the apparent binary equivalents of the ten decimal digits are in the proper arithmetic sequence. Specifically, the binary equivalent of decimal 1 must be greater than the binary equivalent of decimal 0, the binary equivalent of decimal 2 must be greater than the binary equivalent of decimal 1, and so on. The 8-4-2-1 code is one of many decimal codes that meet this requirement. Codes having more than four bits per decimal digit are permissible.

COMPARISON 461

In fact, the use of a binary subtractor for comparison is not limited to numbers but is applicable to any symbol system having a code of the characteristics just mentioned. In particular, alphanumeric words can be "compared" in this way, although the terms "greater than" and "less than" would, in most applications, be appropriately replaced by "precedes" and "follows," respectively. For example, letter A might be represented by 010000, letter B by 010001, C by 010010, and so on. Then word BCC could be determined to follow word BAB by performing a binary subtraction of 010001010010010010 from 010001010000010001. Such comparisons are useful in sorting, which will be discussed in more detail in a later section.

Comparison can alternatively and often more economically be accomplished by a network which, in effect, mechanizes the mental steps a person might use when comparing two numbers. Ordinarily, a person would not subtract one number from the other but instead would first look at the highest-order digit in each number. If these two digits are not equal, the larger of the two numbers would be determined by the larger of these two digits, regardless of the digits in all lower orders. If the highest-order digits are equal, the person would next look at the digits in the position next to the highest order. If these digits are unequal, the relative magnitudes of the two numbers would be indicated, but if they are equal, the digits in the next position would be correspondingly examined, and so on.

Figure 8-8a illustrates the equipment needed for generating an $A > B$ signal by this technique. The left-hand AND device generates a signal (a "1") if A_5 is 1 and B_5 is 0. This signal passes through the multioutput OR device to the output line to indicate that $A > B$. Here $A = A_5A_4A_3A_2A_1$ and $B = B_5B_4B_3B_2B_1$. Similar signals are generated at each of the five positions, but the added inputs to the various AND devices serve to block these signals when they would otherwise generate an erroneous $A > B$ output signal. That is, the third input to the AND device in the position for comparing A_4 and B_4 carries a signal which is 1 when $A_5 + \bar{B}_5$ (Boolean notation) is 1. $A_5 + \bar{B}_5$ will be 1 whenever $A_5 = B_5$ (both 0's or both 1's). Also, $A_5 + \bar{B}_5$ will be 1 when A_5 is 1 and B_5 is 0, and in this instance a redundant $A > B$ signal may be generated from the position which compares A_4 and B_4, but no harm is done. However, when A_5 is 0 and B_5 is 1, then A cannot be greater than B, and the $A_5 + \bar{B}_5$ signal (which is 0 in this instance) nullifies the effect of all lower-order signals. For numbers having a large number of digits, the fan-in requirements of the switching devices can become objectionably large, but at a moderate "price" in speed these requirements can be reduced as necessary by using extra levels of switching in the same manner as was described for high speed carry networks.

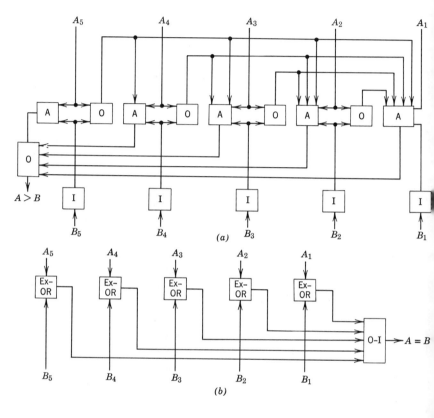

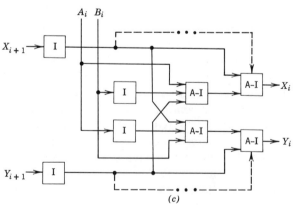

FIGURE 8-8. Networks for determining when $a: A > B$ and $b: A = B$.

The network of Figure 8-8a is applicable to decimal numbers and alphabetic characters in the same way as previously, where the individual binary input signals represent code elements instead of binary digits.

Figure 8-8b shows an arrangement, in terms of exclusive-OR devices and one O-I module, for generating a signal that is 1 when $A = B$. No subtraction is performed. Instead, the network detects directly that each $A_i = B_i$.

A more universally applicable comparison technique is illustrated in Figure 8-8c, where the network has been designed to use inverters and A-I modules. This network is for one pair of digits in the words to be compared, and a similar network is assumed for each other pair. With n digits in each word, $X_{n+1} = Y_{n+1} = 0$, and the results of the comparison are signaled by X_1 and Y_1. If $X_1 = Y_1 = 0$, then $A = B$. If $X_1 = 1$, then $A > B$, and if $Y_1 = 1$, then $A < B$. The network is such that X_i and Y_i cannot both be 1 simultaneously. The functions generated by the network for any given pair of digits are:

$$X_i = X_{i+1} + \bar{Y}_{i+1} A_i \bar{B}_i$$

$$Y_i = Y_{i+1} + \bar{X}_{i+1} \bar{A}_i B_i$$

These functions indicate that each X_i and Y_i is primarily dependent on X_{i+1} and Y_{i+1}, respectively, (not A_i and B_i) as is required in determining relative magnitudes. That is, only when $X_{i+1} = Y_{i+1} = 0$ does X_i and Y_i depend on A_i and B_i. The dotted lines in Figure 8-8c indicate a bypassing scheme that can be used to improve the speed of comparison when n is large. The connections are made from a relatively high-order network to a relatively low-order network, and these connections serve to avoid the time-consuming propagation of signals through intermediate-order networks when the comparison is dependent on the high-order digits. Overlapped bypass connections may be used in the manner described in Chapter 6 for carry bypassing.

Extraction of the Square Root—Integers

A simple method of extracting the square root is based on the fact that the squares of successive integers may be obtained by accumulating the odd integers. That is, $1^2 = 1$; $2^2 = 1 + 3$; $3^2 = 1 + 3 + 5$; $4^2 = 1 + 3 + 5 + 7$; and so on. To determine the square root of some number, A, the integers 1, 3, 5, 7, and so on, are sequentially subtracted, and the number of subtractions which leaves a positive remainder is the largest integer which is equal to or less than the exact square root.

Determining the rounded value of the square root happens to be possible by subtracting the even integers 2, 4, 6, 8, 10, 12, and so on, instead of the odd integers. The process is terminated when the difference is zero or negative, and the rounded square root is equal to one half the last integer subtracted. The need to sense a zero difference can be avoided by subtracting an initial 1 from A. For example, to determine the square root of 95, rounded to the nearest integer, 1 is subtracted to obtain 94, and then the following steps are taken.

$$
\begin{array}{rr}
 & 94 \\
\text{subtract:} & \underline{2} \\
 & 92 \\
\text{subtract:} & \underline{4} \\
 & 88 \\
\text{subtract:} & \underline{6} \\
 & 82 \\
\text{subtract:} & \underline{8} \\
 & 74 \\
\text{subtract:} & \underline{10} \\
 & 64 \\
\text{subtract:} & \underline{12} \\
 & 52 \\
\text{subtract:} & \underline{14} \\
 & 38 \\
\text{subtract:} & \underline{16} \\
 & 22 \\
\text{subtract:} & \underline{18} \\
 & 4 \\
\text{subtract:} & \underline{20} \qquad 20/2 = 10 \\
 & \text{(negative)}
\end{array}
$$

The square root of 95 is 9.74⋯ which becomes 10 when rounded to the nearest integer.

When implementing the above square root procedure, A may be placed in an accumulator capable of subtraction. After an initial subtraction of 1, a continuous series of clock pulses are counted in a counter, and on each step the doubled value of the counter contents (see "doublers" in the previous chapter) is subtracted from the contents of the accumulator. The process is terminated when the contents of the accumulator become negative, at which time the rounded square root appears in the counter.

EXTRACTION OF THE SQUARE ROOT—INTEGERS

The above method of extracting the square root works equally well with the binary number system. In fact, the network is somewhat simpler in that the doubling is achieved by a simple shift to the left by one position. The digits appearing in the extraction of the square root of 1011111 (decimal 95) are, after the initial subtraction of 1, as follows.

$$
\begin{array}{rr}
 & 1011110 \\
\text{subtract:} & 1 \\
\hline
 & 1011100 \\
\text{subtract:} & 10 \\
\hline
 & 1011000 \\
\text{subtract:} & 11 \\
\hline
 & 1010010 \\
\text{subtract:} & 100 \\
\hline
 & 1001010 \\
\text{subtract:} & 101 \\
\hline
 & 1000000 \\
\text{subtract:} & 110 \\
\hline
 & 110100 \\
\text{subtract:} & 111 \\
\hline
 & 100110 \\
\text{subtract:} & 1000 \\
\hline
 & 10110 \\
\text{subtract:} & 1001 \\
\hline
 & 100 \\
\text{subtract:} & 1010 \\
\hline
 & \text{(negative)}
\end{array}
$$

The last quantity to be subtracted, 1010 before shifting, is the binary equivalent of the rounded square root.

Because the number of steps required is a rapidly increasing function of the number of digits in the result, the range of practicability of the above square root extraction method is a function of the required accuracy of the results, the speed of the switching devices being used, and the time available for the process.

A square root extraction method that is somewhat more complex but which requires much fewer steps and is therefore much faster for large numbers follows the conventional manual method of obtaining the square root digits one at a time, highest order first. The digits of A are grouped in two's with a 0 appended to the high-order end if the number of digits is otherwise odd, and the highest-order group is handled first.

The largest integer which is equal to or less than the square root of this group is determined either by "inspection" or by successive subtractions of the odd integers. The square of this integer (which must be less than the radix of the number system in use) is subtracted from the highest-order group, and the difference is then placed with the next highest-order group of two digits from A to form a sort of "intermediate remainder" as in division.

The second digit of the square root is determined by finding the largest digit which, when appended to the right of the doubled first digit, forms a number which, when multiplied by this same largest digit, is equal to or less than the intermediate remainder. This determination is usually accomplished manually by a "trial-and-error" technique. However, for machine operation a preferable procedure is to increment the second square root digit by 1 on each step, and the digit thus developed when combined with the first square root may be designated S_i. The quantity $2S_i + 1$ is repeatedly subtracted as long as a positive remainder results. (In effect, this procedure forms the successive odd integers and adds them to the doubled value of the first square root digit.) The third group of two digits from A is then appended to the new difference to form the next intermediate remainder, and the determination of the third and subsequent square root digits is accomplished analogously.

To illustrate the method, the steps employed in the extraction of the square root of 106,929, which is 327, are set forth below.

		square root
	10 69 29	
subtract:	1	1--
	9	
subtract:	3	2--
	6	
subtract:	5	3--
	1 69	
subtract:	61	31-
	1 08	
subtract:	63	32-
	45 29	
subtract:	6 41	321
	38 88	
subtract:	6 43	322
	32 45	

			square root
subtract:	6 45		323
	26 00		
subtract:	6 47		324
	19 53		
subtract:	6 49		325
	13 04		
subtract:	6 51		326
	6 53		
subtract:	6 53		327
	0		

The details of these steps are based on the assumption that, prior to the entry of a difference in the accumulator, the computer is capable of determining each instance that a negative difference is being encountered. In each such instance the subtraction is voided. Most parallel computers have this capability, but for computers which do not have this capability (as in most serial computers), either a restoring technique or a nonrestoring technique must be used, as was described for division.

With the binary number system this square root extraction method reduces to a single comparison-subtraction operation per square root digit. The steps required to determine the square root of 10011111011001 (decimal 10,201), which is 1100101 (decimal 101), are set forth below for purposes of illustration. In the decimal example the comparison operations were omitted but are shown here.

			square root
		10 01 11 11 01 10 01	
Step 1	subtract:	1	1------
		01 01	
Step 2	subtract:	1 01	11-----
		00 00 11	
Step 3	compare:	11 01	110----
		00 11 11	
Step 4	compare:	1 10 01	1100---
		00 11 11 01	
Step 5	subtract:	11 00 01	11001--
		00 11 00 10	
Step 6	compare:	1 10 01 01	110010-
		00 11 00 10 01	
Step 7	subtract:	11 00 10 01	1100101
		0	

In the development of each S_i the "current" square root digit is taken as 0 so that in Step 2, for example, the quantity to be compared or subtracted is determined to be 101 by doubling (shifting to the left) 10 and appending 1. In Step 3 the quantity to be compared or subtracted is determined to be 1101 by doubling 110 and appending 1, and so on.

With this extraction method, round-off seems to be most expeditiously accomplished in the conventional manner by generating one additional square root digit (see next section) and then adding a decimal 5 or a binary 1, as the case may be, to this digit (with carries, if any, to higher-order digits). The resulting digit in the added position is then deleted.

Extraction of the Square Root—Fractions and Floating-Point Numbers

For fractions the square root extraction process is essentially the same as for integers except that, if the number of digits to the right of the point is odd, an extra 0 is added to the right-hand (low-order) end of the number instead of the left-hand end. If the number for which the square root is to be extracted has both an integral and a fractional part as, for example, the decimal number 320.603, the grouping by two's is done independently on the integral and fractional parts so that the number appears as 03 20.60 30, and the square root is extracted by the same process as described previously. For fractions for which the digits immediately to the right of the point are 0's, such as 0.0003075, for example, the 0's must be considered in grouping the digits to form 0.00 03 07 50 in this example, but of course the initial digit pairs which are 00 result in 0 square root digits. For any number, the location of the point in the square root corresponds exactly to the location of the point in the original number with the digits of that number grouped in two's as explained.

The square root extraction process gives the illusion that the number of digits in $A^{1/2}$ is one half the number of digits in A, but this illusion is misleading. Except in the relatively rare instances where A is the exact square of an integer, the usual requirement is to express $A^{1/2}$ with the same number of significant digits as was used in the expression for A. To achieve the required accuracy in $A^{1/2}$, an appropriate number of 0's are appended to the low-order end of A, and the square root computation is continued for the necessary number of steps by an extension of the process already described.

For floating-point numbers the square root extraction process is the same as for fixed-point numbers except that when the exponent is odd the digits of the mantissa must be shifted to cause the exponent to be even. The exponent of $A^{1/2}$ in floating-point notation is then equal to half the exponent of A.

Increasing the Speed of the Square Root Extraction Process

With the numbers in the decimal system, the speed of the square root extraction process can be increased by schemes analogous to those described for decimal division. With the numbers in the binary system, the same schemes can be used if the binary digits are grouped by two's, three's, or four's to simulate radices four, eight, and 16, respectively. However, in the case of square root extraction, accompanying increases in cost tend to render the schemes unattractive in nearly all applications.

In those applications where high speed is particularly important, a single reasonably economical combinatorial switching network can be designed to develop the square root if the required accuracy is moderate. For each value of the input number A, a specific value of $A^{1/2}$ is to be generated, and this fact may be used as a basis for the design of the combinatorial switching network. To illustrate, all possible values of A expressed to four significant binary digits are listed below with the corresponding rounded five-digit expressions for $A^{1/2}$.

A	$A^{1/2}$
$A\ B\ C\ D$	$a\ b\ c\ d\ e$
0.0 0 0 0	0.0 0 0 0 0
0.0 0 0 1	0.0 1 0 0 0
0.0 0 1 0	0.0 1 1 0 0
0.0 0 1 1	0.0 1 1 1 0
0.0 1 0 0	0.1 0 0 0 0
0.0 1 0 1	0.1 0 0 1 0
0.0 1 1 0	0.1 0 1 0 0
0.0 1 1 1	0.1 0 1 0 1
0.1 0 0 0	0.1 0 1 1 1
0.1 0 0 1	0.1 1 0 0 0
0.1 0 1 0	0.1 1 0 0 1
0.1 0 1 1	0.1 1 0 1 1
0.1 1 0 0	0.1 1 1 0 0
0.1 1 0 1	0.1 1 1 0 1
0.1 1 1 0	0.1 1 1 1 0
0.1 1 1 1	0.1 1 1 1 1

The fourth digit d, for example, in $A^{1/2}$ is 1 in each instance that input number A is 0.0011, 0.0101, 0.1000, 0.1011, 0.1110, or 0.1111. With the

470 IMPLEMENTING OTHER DIGITAL OPERATIONS

individual binary digits of A expressed as A, B, C, and D as listed, d in Boolean notation becomes

$$d = \bar{A}\bar{B}CD + \bar{A}B\bar{C}D + A\bar{B}\bar{C}\bar{D} + A\bar{B}CD + ABC\bar{D} + ABCD$$

which may be simplified and reduced to switching devices in the manner described in earlier chapters. The other digits of $A^{\frac{1}{2}}$ may be determined analogously, and a four-input, five-output network is indicated.

Of course, the number of digits employed in the above illustration is too few to be useful in many applications. (Nevertheless, one application is in the determination of b_0 as defined in the next section.) However, in many applications the required accuracy is no more than about 10 or 12 binary digits in which case a very high speed square root generation network can be designed at reasonable cost with large scale integrated (LSI) circuits, although the designer would doubtless wish to program the design problem on a computer rather than work out the details manually. Also, as in analogous problems described previously, a read-only storage unit may be employed instead of a switching network.

Extracting the Square Root—Iterative Methods

Iterative formulas, analogous to those described in previous chapters for division, are available for determining the square root of a number A. Four such formulas are

$$b_{k+1} = \frac{1}{2}\left(\frac{A}{b_k} + b_k\right) \tag{1}$$

$$b_{k+1} = \frac{b_k}{2}\left(3 - \frac{b_k^2}{A}\right) \tag{2}$$

$$b_{k+1} = \frac{1}{8}\left(3b_k + \frac{6A}{b_k} - \frac{A^2}{b_k^3}\right) \tag{3}$$

$$b_{k+1} = \frac{b_k}{8}\left(15 - \frac{10b_k^2}{A} + \frac{3b_k^4}{A^2}\right) \tag{4}$$

where the b_k are successively closer approximations to $A^{\frac{1}{2}}$. Formulas 1 and 2 are "second order" and formulas 3 and 4 are "third order," which means that, once a reasonably close approximation is obtained, the number of significant digits is doubled and tripled, respectively, upon each application of the formula.

For the formulas to function properly, the initial approximation b_0 must be chosen appropriately, although the restrictions are not severe. Specifically, in 1, b_0 may have any value except zero. In formula 3 the requirement is that $b_0 < (5A)^{1/2}$. Although formulas 2 and 4 appear to be more complex than formulas 1 and 3, respectively, the former require division at each iteration, whereas the latter require no divisions except for the determination of $1/A$ at the beginning of the process. With any of the formulas, one way to determine when the iterations may cease is to observe when the difference $b_k - b_{k+1}$ is less than the allowable error.

To obtain $A^{1/2}$ to the desired accuracy with a relatively few iterations, b_0 must itself be at least a crude approximation to $A^{1/2}$. Various mathematical procedures for selecting b_0, particularly those adaptable to the "pure programming" use of the iterative formulas, are discussed in a paper by C. T. Fike in *Communications of the ACM*, April 1966, pp. 297–299. However, when the iterative formulas are used as the basis of built-in square root extraction capability, a more attractive method of selecting b_0 is to employ a combinatorial switching network or a read-only storage unit as described in the previous section.

The Implementation of Other Functions, Including Very Complex Functions, of a Variable

The implementation of other functions such as raising a variable to some power other than one half (other than square root), determining the sine, tangent, and other trigonometric functions of a variable, determining logarithmic functions, Bessel functions, and countless others that might be mentioned is ordinarily accomplished as a part of the overall problem to be programmed on a computer. Built-in facilities are not usually included. No inherent reason is encountered for this situation other than the fact that the need for any one such function is generally so limited that the cost of the special equipment would out-weigh the disadvantages of the "pure programming" method of generating these functions. For any specific one of these other functions, a wide variety of arithmetic procedures and corresponding equipment could be designed to achieve the desired results more directly than through programming. However, because of the cost-usage situation just mentioned, descriptions of the possible techniques would be outside the intended scope of this book.

Nevertheless, one technique which is applicable to all functions just mentioned deserves at least a brief mention. This technique applied equally well even to very complex functions of A as might be suggested by

$$f(A) = A^{1.8} - k_1 \log_2 (\tan k_2 A) e^{-k_3 A}$$

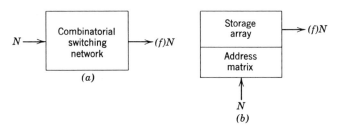

FIGURE 8-9. Illustrating the use of *a* a combinatorial switching network and *b* a storage unit for the generation of an arbitrary function of N.

as an illustration, where k_1, k_2, and k_3 are fixed constants. The technique is to precompute the function for all expected values of A and then provide either (a) a combinatorial switching network which develops the function as output signals for each value of A applied as input signals (as was described in some detail for the square root function) or (b) a storage unit in which the various addresses correspond to values of A and the respective stored words correspond to the values of the function—see Figure 8-9.

The precomputations may be carried out to any desired degree of accuracy to achieve the desired accuracy in the function—a point which is particularly important in instances such as a function that is a difference $f = f_1 - f_2$, where f_1 and f_2 are nearly equal. Either the combinatorial network or the storage unit can then produce highly accurate representations of the function at high speed.

For many functions, achieving a given number of significant digits in the results may require word lengths two or three or more times as great in the intermediate computations. However, regardless of the accuracy required in the precomputations, the development of binary 12-digit accuracy, for example, by means of a combinatorial switching network requires a network having no more than 12 output lines. The number of input lines required depends on the accuracy of the input number supplied. In the 12-digit example the input number would most likely be likewise expressed to an accuracy of 12 digits, but the number of input digits may be substantially more or less than 12. In applications where the range of the input variable is known to be restricted in some way, the resulting "don't care" conditions may be utilized to simplify the network.

In the storage-unit implementation of function generation, the length of the stored word is equal to the number of significant digits required in the output function. The number of addresses required in the storage unit is 2^n, where n is the number of binary digits in the input variable. In the 12-digit example this number is 4096. When the range of the input variable is known to be restricted in some way, the required number of addresses in the storage unit is correspondingly reduced (which is the counterpart of "don't care" conditions in the combinatorial switching network).

Either A or $f(A)$, or both, may be expressed in floating-point notation. With either a combinatorial network or a storage unit, the design technique is essentially the same as with fixed-point numbers. For each combination of input signals, with the mantissa and exponent bits considered collectively, a precomputed combination of output signals is to be generated, where some of the output signals represent the exponent of $f(A)$ and others represent the mantissa of $f(A)$ in floating-point notation.

A read-only storage unit may be used in applications where the generation of a specific function is to be built into the machine. Alternatively, read-write storage unit may be used when the unit is to be used for any function under the control of the programmer. In such designs the computer may be equipped with an instruction broadly designated FUNCTION (for readers not already familiar with computer instructions, see next chapter), and when this instruction is executed, the address part of the instruction is interpreted as referring to the function storage unit rather than the main storage unit. Two or more storage units with a separate instruction for each may be included to provide facilities for the high speed generation of a corresponding number of functions.

An obvious disadvantage of the above technique for function generation is the large number of components required. The combinatorial switching network can require roughly as many switching devices as required in a simple but complete computer. The storage unit may require as much storage capacity as is found in the main storage unit of many small computers. Nevertheless, the demands for high computational speed seems to increase continually, and the very high speed provided by this technique is certainly attractive in applications having a need for the generation of one or a few specific functions of this category.

Cryptographic Operations

Electronic cryptographic machines, that is, machines for the encoding and decoding of messages for secrecy purposes, employ switching and storage devices which may be and often are identical to those employed in conventional computers and other digital systems. The design procedures described in earlier chapters are directly applicable, although the techniques for manipulating the digital signals are, of course, a function of the code being used. Cryptographic techniques, especially those required for cryptanalysis (the "breaking" of unknown codes), form an extensive subject requiring perhaps several volumes for a reasonably complete exposition. Here, however, the objective will be limited to a very brief introduction to the subject with particular reference to the role of random numbers, the generation of which is discussed in the next section.

A secrecy code which is unbreakable (except through the theft of the code itself—a possibility which must always be considered) is based on a series of random numbers which are used to modify the message at the transmitting terminal. Prior to the decoding of the secret message, the series must be made available at the decoding terminal. Ordinarily, the series of random numbers must be recorded on paper, punched cards, magnetic tape, or some analogous medium and physically transported to the receiving terminal by trusted personnel. The general character of the code will be explained by the following example. Assume that only the 26 letters of the alphabet and spaces are used to form each message, and assume that each letter is represented by a number corresponding to its position in the alphabet ($A = 01$, $B = 02$, $\cdots$, $Z = 26$) and that a space is represented by the number 27. A series of random numbers, each in the range 01 to 27, inclusive, is used to encode the message by adding the character number to the corresponding random number. The addition is performed "modulo 27," which in this application means that, if the sum is greater than 27, only the excess above 27 is recorded. For example, the encoding of the message ATTACK NOW would be as follows for the random number sequence indicated.

message:	A	T	T	A	C	K		N	O	W
digital form:	01	20	20	01	03	11	27	14	15	23
random numbers:	18	07	26	01	11	14	08	19	22	12
encoded message:	19	27	19	02	14	25	08	06	10	08

At the receiving terminal the random number sequence 18 07 26 $\cdots$ is known, and the decoding is accomplished by subtracting these numbers on a character-by-character basis from 19 27 19 $\cdots$, which is the encoded message as received. When decoding, negative differences are avoided by adding 27 to each number in the random sequence that is smaller than the corresponding number in the encoded message. The decoded messages appear in numeric form but can be transposed in the output printer to conventional alphabetic characters.

Apart from keeping the random number sequence secret from the "enemy," a given random number sequence must never be used to encode more than one message if the unbreakable feature of the code is to be realized. Should a given random number sequence be used to encode two messages which are both intercepted by the enemy, the enemy is given the opportunity to decode both messages. The technique to be used by the enemy is to guess that one message contains a selected common or expected word or word sequence and to employ the indicated random number sequence in a trial decoding of the other message. If valid words appear in this trial decoding, the assumption is that the guess was correct. From this guess, more accurate guesses at other words in one or both messages may similarly be made and checked out. Of course, the enemy does not know the positions of the guessed words, especially at the beginning of the process, so the guess must be made for each possible position. Moreover, the enemy does not necessarily know which pair, if any, of messages was coded by the same random number sequence nor does he know how that pair, if any, is "lined up" with respect to the sequence. Thus, the code-breaking process is obviously laborious when performed manually, although computers can be used to speed the process in many ways.

The need to have, for each pair of transmitting-receiving terminals, a separate series of random numbers as long as the total length of all messages to be transmitted between those terminals is an intense disadvantage of the above coding method in spite of its virtually perfect secrecy qualities.

In alternative coding methods, instead of furnishing both terminals with the entire random number sequence, the terminals are furnished with a formula by which the numbers may be generated. The numbers are not then truly random but are called "pseudo-random," although they may pass all the conventional tests for randomness. Now the objective is to devise a formula which is simple enough to be implemented economically and at high speed but is sufficiently complicated that the enemy will not be able to deduce it even though he may have intercepted many messages encoded by the formula. The development of suitable formulas is where the subject of cryptography becomes extensive. Certain very simple formulas will be discussed in the next section.

To increase the difficulty in breaking a code based on pseudo-random numbers as generated by a relatively simple formula, the digit sequence as transmitted can be "messed up" in several different ways, some of which are as follows.

1. Each word in the message may be replaced by a code word. In the example, the words ATTACK and NOW might be replaced by FLOOR and TREE, respectively. In fact, this coding scheme is adequate by itself in some applications. However, when the scheme is used by itself for many long messages, the code words can usually be deduced by an enemy. The code words need not be recognizable as words, and in this example might be XTU and REYBV, respectively. A sort of dictionary is required to obtain the code words, and unfortunately the implementation of a dictionary tends to be awkward (high cost and low speed).

2. Each character (each letter or space in the above example) may be replaced by some other character. For example, each A in the message might be transposed to Q, each B to a space, each C to A, each D to H, and so on. This procedure is simple to implement—requiring only a difference in wiring connections in some applications—but it has the effect of modifying the pseudo-random numbers in such a way as to obscure the formula by which they were generated.

3. Each successive small group of digits may be replaced by some other small group. This technique is a compromise between the techniques of paragraphs 1 and 2 above. If, for example, each character is represented by five binary signals, each two-character group could be transposed to any selected other two-character group through the use of a 10-input, 10-output combinatorial switching network. At the receiving terminal a similar switching network may be provided to restore each two-digit group of the original message.

4. The characters in the uncoded message (or alternatively the numbers in the coded message) may be transposed in countless different ways. For example, each successive group of five characters may be modified by interchanging the second and fourth digits with the first digit being interchanged with the fifth digit of the following group. For some transposition patterns and some machine designs this technique can be implemented by nothing more than a difference in wiring connections, although in other instances an appropriate price must be paid in equipment and speed.

5. The formula for generating the pseudo-random numbers can be made message-dependent. For example, each random number may consist of two decimal digits, and each group of five successive numbers may have been computed, by some formula, from the previous group of five random numbers. However, in the computations, one or more of the numbers in the previous group may be modified by the number-equivalents of selected characters within the message. Elementary versions of this scheme are economical to implement with very little loss in speed,

and the code can be rendered very much more difficult to break. However, the technique has the disadvantage that, in the event of an error in the encoding or in the transmission, the remainder of the message may not be decodable, whereas in the previously described techniques such errors affect only the characters in question plus perhaps a few immediately adjacent characters.

6. The formula for generating the pseudo-random numbers can be made time-dependent. Again, countless variations are encountered, but to indicate the scheme, one possibility is to change the transposition pattern in accordance with the time of day that the various messages are transmitted.

7. The above schemes can be combined in many different ways.

The Generation of Genuinely Random Numbers

One method of generating random numbers manually is, for the decimal system, to employ a deck of 100 cards on which the numbers 00 through 99 are recorded with each card bearing a different number. The deck is shuffled. A few cards are drawn from the deck, and the numbers on the drawn cards are recorded. The cards are returned to the deck, which is shuffled again for a repetition of the process. For purely random numbers, only one card should be drawn after each shuffling, but the drawing of more than one speeds the process greatly with negligible impairment to the randomness of the numbers. In fact, the drawing of more than one card after each shuffle insures that long sequences of the same digit, such as long sequences of all 0's, will not occur. Such sequences are considered objectionable for some applications of random numbers even though a nonzero probability exists for their occasional occurrence in genuinely random numbers.

Random number generation methods analogous to card shuffling are not readily implemented by machines, at least not by digital electronic machines. In fact, digital electronic machines employing only switching devices (AND, OR, and NOT functions) and storage devices seem to be incapable of generating genuinely random numbers by any technique. Special equipment of some sort must be added.

Nearly all practical electronic random number generation schemes involve a device which generates pulses at random points in time. By relating each such pulse to some other pulse (which may or may not be randomly timed), time intervals of randomly varying duration are defined. A gating signal which is "on" or "1" for the corresponding intervals is formed. During each period of time that the gating signal is 1, a sequence of uniformly spaced clock pulses is allowed to pass

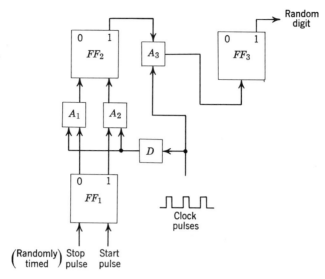

FIGURE 8-10. Network for generating a random binary digit.

to the complementing input of a flip-flop. If the duration of the randomly timed interval is long relative to the time interval between successive clock pulses, the probability of passing an odd number of pulses in virtually identical to the probability of passing an even number of pulses. Thus, regardless of the initial state of the flip-flop, the binary digit stored in the flip-flop at the termination of the clock pulse sequence is random.

To ensure proper operation of the scheme described in the previous paragraph, switching devices must be included to cause each clock pulse to be either blocked entirely or transmitted with full amplitude to the flip-flop. If the last clock pulse of a sequence is likely to be of partial amplitude, as might be caused by the arrival of the random pulse at essentially the same time as one of the clock pulses, any unbalance in the complementing flip-flop will tend to cause one binary digit to be generated with greater probability than the other.

Figure 8-10 illustrates how this difficulty may be overcome (and is also another illustration of a "sequential circuit"). Flip-flop FF_1 is set to the 1 state during the time interval between a "start" and the "stop" pulse which is randomly timed. Flip-flop FF_2 is initially in the 0 state as a result of the prior passage of clock pulses through AND device A_1. After FF_1 is set to 1, the clock pulses pass through A_2 instead of A_1 and cause FF_2 to be set to 1. Because of delay device D, which may consist of a suitable number of inverters connected in tandem, the particular clock pulse which sets FF_2 to 1 will have dissipated at the input to A_3 by the time FF_2 actually becomes in the 1 state. Therefore, this particular clock pulse does not reach the complementing flip-flop FF_3. However, FF_3 does change back and forth between the 0 and 1 states in response to the next and succeeding clock pulses until the randomly timed "stop" pulse occurs. Delay device D comes into play again at the termination of the clock pulse sequence in that the particular pulse which causes FF_2 to be returned to the 0 state will have passed through A_3 by the time it reaches A_1. Thus, the last clock pulse to reach FF_3 will arrive in full amplitude form.

Nondigital equipment is required for the generation of the randomly timed pulses. One scheme that can be implemented economically is to fire a silicon controlled rectifier (SCR) or a thyratron by means of a sine wave signal having a frequency that is not synchronized with the pulse repetition rate of the digital clock pulses. Further assurance of randomness can be obtained by modulating the frequency or amplitude, or both, of the SCR input signal or by forming the input signal with two or more sine waves of different and unsynchronized frequencies.

Another scheme for obtaining randomly timed pulses is to employ a radioactive substance. Radioactive particles are emitted at random times, and the pulses may be generated by a scintillation detector. A third scheme is to detect the zero-crossings of a noise signal as might be generated by thermal noise, suitably amplified, in a resistor.

Because of the need to have the random interval long compared with the clock pulse interval, the generation of digits having high randomness qualities tends to be slow relative to the pulse repetition rate adopted for a given machine. Many random digit generators can, of course, be operated in parallel.

All known schemes for generating genuinely random numbers have the characteristic that no given sequence of numbers is repeatable, except by chance. In many applications employing random numbers a need arises to employ the same sequence more than once—to check the computations and to diagnose error sources if for no other reason. Because the storage of long sequences of random numbers is often impractical at reasonable cost, the designer is commonly obliged to forego genuine randomness and employ pseudo-random numbers.

The Generation of Pseudo-Random Numbers

Countless formulas may be devised to produce very long sequences of numbers the digits of which appear to be random. Two general requirements arise. One is to obtain a formula which is as simple as possible so that the generation of the numbers (or the digits within the numbers) can be implemented at high speed and at low cost. The other requirement is to insure that the numbers do not contain patterns, possibly obscure and subtle, which will render the numbers unsatisfactory for a particular application. The two requirements are in opposition to each other, and the final selection of a formula is virtually always a compromise.

Most attention has been directed to formulas of the form

$$x_{j+k} = f(x_j, x_{j+1}, \cdots x_{j+k-1})$$

where x_j is the jth number in the series, f any suitably selected function, and k a constant. Even this formula is found to be more elaborate than necessary in nearly all applications (the cryptographic application being an exception). In particular, k may be set to 1. The formula then becomes $x_{j+1} = f(x_j)$ so that, in other words, each pseudo-random number is a function of only the previous pseudo-random number in the series. Although numbers having a "high quality of randomness" can be obtained in this way, the series for any given f is finite in length. The maximum length of the series is equal to the number of different values x_j may assume and is 2^n for n-digit binary numbers and 10^N for N-digit decimal numbers (or 2^{nk} and 10^{Nk}, respectively, for any k).

Unfortunately, for a given f and a given initial pseudo-random number x_0, the value of j for which $x_{j+1} = x_0$ may be very much less than the value corresponding to a series of maximum length. Moreover, for most f the length of the series which will be obtained cannot be predicted and must be determined by experiment. Situations as illustrated by way of example in Figure 8-11 for four-digit binary numbers are commonly encountered. If x_0 is 0010, the succeeding numbers will be as shown, but after 1001 is reached, the next number will be 0110, which had been generated previously. The system will then cycle through a loop having a "period" of eight numbers. If x_0 had been any number of the loop, the four initial numbers obtained previously would not have been obtained. With an x_0 of 0111, for example, the system would cycle in a different loop, which has a period of only three numbers. With an x_0 of 0000, no different numbers at all would be obtained with the f implied by this illustration.

Many different f's could be contrived to yield the number sequence of Figure 8-11, but because the number sequence was just randomly selected to illustrate the nature of the periods, the f's would, in general, be odd functions of no practical consequence. One such f can be generated through the use of Boolean techniques where the four binary digit positions are assigned the designations A, B, C, and D, respectively.

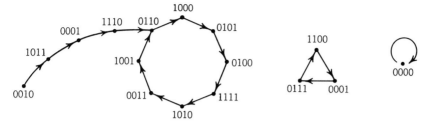

Figure 8-11. Illustrating "period" of pseudo-random numbers.

A_{j+1} is a Boolean function of A_j, B_j, C_j, and D_j, as is each of the other signals B_{j+1}, C_{j+1}, and D_{j+1}. The four Boolean functions can be derived straightforwardly by employing the techniques described in earlier chapters (doing this would be a good exercise), and the resulting network together with a storage device for containing the current number is an example of a "sequential circuit."

For many applications pseudo-random numbers generated by the simple formula

$$x_{j+1} = ax_j \quad (\text{mod } 2^n \text{ or } 10^N)$$

are satisfactory, where a is a constant. The term "mod" is an abbreviation for "modulo" and in this application can be interpreted to mean that in generating any x_j only the n or N (as the case may be for binary or decimal numbers, respectively) low-order digits of the product ax_j are retained.

When the above formula is used with binary numbers having $n \geq 3$, the maximum period is $(2^n)/4$ and is obtained when $a = 8t \pm 3$, where t is any integer 1, 2, 3, and so on. The initial number x_0 must be odd. For decimal numbers having $N \geq 4$, the maximum period is $(10^N)/20$ and is obtained when $a = 200t \pm r$, where t is as before and r is any of the integers 3, 11, 13, 19, 21, 27, 29, 37, 53, 59, 61, 67, 69, 77, 83, or 91. In addition to being odd, x_0 must not be a multiple of 5 when decimal numbers are used.

The period can be increased to the maximum possible, 2^n or 10^N for binary or decimal numbers, respectively, by adding a constant c to the formula to produce

$$x_{j+1} = ax_j + c \quad (\text{mod } 2^n \text{ or } 10^N)$$

With this formula and binary numbers, c may be any odd number and $a = 4t + 1$. With decimal numbers, c must be odd and not a multiple of 5, and $a = 20t + 1$. Because the period is the maximum possible, all numbers (integers) from zero to $2^n - 1$ (binary) or $10^N - 1$ (decimal) will appear in the series, and x_0 may therefore be any number. Unfortunately, the "quality of randomness" of the resulting pseudo-random number series tends to be impaired when $c \neq 0$ is added.

For those applications where the "quality of randomness" is adequate, the generation of pseudo-random numbers by the above formulas can be implemented economically and at high speed when a is chosen properly. In particular, with binary numbers and the latter formulas where $a = 4t + 1$, any t which is a power of 2 will produce a multiplier having only two 1's. Thus, the multiplication can be carried out by shifting x_j an appropriate number of positions to the left (with the corresponding number of high-order digits being discarded) and by adding the shifted number to x_j. Then, if an appropriate c is selected, a single parallel binary adder is sufficient to form $ax_j + c$ (mod 2^n). With the decimal system, t is chosen to be one of the numbers 5, 50, 500, and so on, in which case a similarly becomes a decimal number having only two 1's with all other digits being 0's (recall that $a = 20t + 1$), and a single parallel decimal adder is analogously sufficient.

When $c = 0$, all x_j in the series are odd. With $c \neq 0$, successive x_j's alternate between even and odd values. Clearly, the numbers are only pseudo-random and not genuinely random. Although this particular defect in randomness is not objectionable in some applications of random numbers, other nonrandom patterns are encountered which can be highly objectionable. These patterns tend to be complex functions of n (or N), a, and c, and in general cannot be predicted. In fact, many of the patterns tend to be undetectable in a practical sense except through testing the numbers in whatever application is at hand. A given series of pseudo-random numbers may be adequate for one application and totally unsatisfactory in another application, whereas the opposite results may be obtained for a differently derived series.

For further information on the character of the randomness of numbers derived by the above formulas, reference is made to a paper by R. R. B. Whittlesey in *Communications of the ACM*, September 1968, pp. 641–644, and to a paper by H. C. Ratz and J. V. Hildebrand in *IEEE Transactions on Electronic Computers*, December 1967, pp. 854–856. The references contained in these two papers will lead the reader to the bulk of the prior work on the subject.

Increasing the Length of the Period and Improving the "Quality of Randomness"

One reasonably economical scheme for obtaining both a great increase in the length of the period and a great improvement in the "quality of randomness" of pseudo-random numbers is to employ two generators as described in the previous section. The constants a and c may be different for the two generators. Each pseudo-random number is obtained by adding (modulo 2^n or 10^N as the case may be) the output numbers of the two generators, but on each step at which one generator returns to its initial x_0 the other generator is caused to omit a step (or alternatively caused to proceed through two steps) so that the two generators do not immediately repeat with the same number pairs. The period is thereby increased to $2^n(2^n - 1)$ or $10^N(10^N - 1)$ for binary or decimal numbers, respectively.

At little or no further increase in cost, the odd-even pattern in successive numbers can be destroyed and other improvements in randomness can be obtained by cyclically permuting or otherwise "scrambling" the digits of one number before adding it to the other number in obtaining the pseudo-random number to be used.

Sorting

Sorting, as the term is used in computer applications, means the arrangement of items in numerical or alphabetical sequence (as opposed to the "sorting" of objects by color or some other property). Usually, each item is a record containing at least two "fields," one of which is the "key" field upon which the sort is to be made. For example, a record might contain one field for a person's name, a second field for the license number of a car owned by that person, and various additional fields for information describing the car. The initial list of records may be in random sequence, and for some purposes the list may need to be sorted alphabetically by owner's name, but for other purposes the list may need to be sorted by license number (which may be alphanumerical) or by some characteristic, such as make, of the car.

When each record is stored on a separate punched card, the records may be sorted by physical rearrangement of the cards, and card sorters for this purpose are long established products of business machines manufacturers. Basically, a card sorter consists of an input hopper and at least ten (for decimal numerical sorting) output pockets. The cards are first passed through the machine and sorted in accordance with the least significant digit of the "key" field. A second pass is used to sort them in accordance with the next higher-order digit, and so on, until a final pass where the cards are sorted in accordance with the most significant digit of the "key" field. This sorting method, which has come to be known as "digit" sorting (or sometimes "radix" sorting) is illustrated by the following example.

484   IMPLEMENTING OTHER DIGITAL OPERATIONS

	After sorting on:		
Random Sequence	Units Digit	Tens Digit	Hundreds Digit
729	071	401	015
071	401	309	071
309	652	613	309
652	613	015	401
401	015	729	613
839	729	839	652
015	309	652	729
613	839	071	839

The time required to perform the sort is proportional to the product of the number of records and the number of digits in the "key."

In electronic systems where the records are stored on magnetic tape instead of punched cards, one tape unit may be used as the input hopper, and ten other tape units may be used for the pockets. After each pass, the 10 pocket tapes may be mounted, one at a time, at the hopper tape position with new tapes being used for pockets. Alternatively, all records on the pocket tapes may be electronically transmitted to the hopper tape in preparation for sorting on the next digit. Another alternative is to employ a total of 20 tape units. Although workable, each of these alternatives is inefficient with respect to the use of tape units in comparison with other sorting schemes which have been devised.

If in an electronic system the capacity of the high speed storage unit is sufficiently great to store the entire file to be sorted (plus a suitable sorting program), all sorting is accomplished within the machine. Only one tape unit is required. The randomly sequenced file is transmitted from the tape to the storage unit within the machine, and the sorted file may subsequently be transmitted to the same tape. Such sorting is called "internal" sorting, and several different internal sorting procedures (to be described in later sections) have been devised.

If the high speed storage unit is not large enough to store the entire file, but the file is nevertheless short enough to be stored on one reel of tape, the file may be sorted through the use of as few as three tape units by a process called "merging," where merging is the process of forming a single properly sequenced "string" of records from two or more strings, each of which is properly sequenced. Of course, in randomly sequenced records the individual strings may be very short and some strings may consist of only one record. (In a file which is in the exact reverse sequence, every string consists of only one record.) Several different merging procedures have been devised (also to be described in later sections).

If the file is not only too long to be stored in the high speed storage unit but is also too long to be stored on a single reel of tape, each tape full of information is sorted to form a single string, and then a different kind of merging procedure called "tape merging" is used to merge the string into a final sorted file.

Ordinarily a digital computer or other information processing machine is capable of storing at least a few records simultaneously in its high speed storage unit. In most applications the speed of the sorting process can be increased greatly if, prior to merging, the strings can be made as long as possible by internally sorting record batches as large as can be handled by the high speed storage unit. Thus, the complete sorting process generally breaks down into three major steps: first, an internal sort by batches is formed to generate relatively long initial strings; second, a merging of the strings on each reel of tape is performed; and third, the strings represented by complete reels of tape are merged.

As might be expected, the details of sorting requirements vary vastly from one job to the next. Consequently, in spite of the well established position of card sorters, electronic digital machines specifically intended for sorting are rare. Instead, sorting is accomplished by programming a general purpose system. Because the physical design of a digital system is therefore seldom closely related to sorting, only the basic principles of sorting are within the intended scope of this book.

Internal Sorting—General

Because of the nature of stored program computers, records to be sorted can be moved from position to position in the high speed storage unit with much the same flexibility available to a person who is manually sorting cards.

With any internal sorting procedure, the rearrangement of the complete records is unnecessary—and undesirable when the records are several times as long as the "key." Instead of rearranging the complete records, only the keys are rearranged, but included with each key is an address that "remembers" where the remainder of the associated record is stored.

Internal Sorting—Insertion

In one internal sorting procedure, variously called "inserting," "sorting by insertion," or "sifting," the records are placed in their proper position one at a time as they are entered into the high speed storage unit from the tape unit. For each record that is entered, the key of that record is compared with the key of each record already in the high speed storage unit, where the comparisons start at one end, say the "low" end, of the partially developed string and continue until a record is found which has a larger key than the key of the new record to be inserted. The new record is then stored in the position previously occupied by the first record having a larger key, and each record having a larger key is moved to the next higher position in the string. By this procedure the numbers 729, 071, 309, 652, 401, 839, and 613 would be sorted as follows, where each "new" record is underlined.

729	071	071	071	071	071	015	015
	729	309	309	309	309	071	071
		729	652	401	401	309	309
			729	652	652	401	401
				729	729	652	613
					839	729	652
						839	729
							839

The time required to perform the sort depends on the initial sequence of the records. If the records are initially in the correct sequence, much time is consumed in comparing keys, but no time is required for moving previously entered records. If the records are initially in the exact reverse sequence, only one comparison for each record determines the position of that record in the string, but much time is consumed in shifting previously entered records. For a random sequence of R records, an average of about $R^2/4$ comparisons and $R^2/4$ record shifts are required to obtain a properly sequenced string.

By incorporating elaborations in the sorting program, both the number of comparisons and the number of shifts can be greatly reduced. When a string contains more than a few records, each new record to be inserted may be first compared with a record near the middle of the string. The second comparison may be made with a record near the first half or near the second half of the string, as appropriate from the first comparison, and so on. When a shift is made, the shift may be by two or more positions under the assumption that the second shift will be required subsequently. As a final part of the sorting process, return shifts may be made in those instances that the shifts prove to be unnecessary.

Internal Sorting—Address Calculation

In "sorting by address calculation" an estimate of the final position for each new record is made by performing a computation on the key. As each new record is entered, it is stored in the position determined by the estimate. In many instances, the same estimate may be made for two or more records, and in these instances key comparisons must be made to determine the proper position of a given "new" record. Shifting of previously stored records may be required. Of course, the shifted records may then occupy positions corresponding to estimates made for subsequent "new" records, in which instances further shifting is required. This sorting procedure is obviously inferior when the keys tend to be "bunched up" around certain values, and it is also inferior when the number of positions available in the high speed storage unit is little or no greater than R, the number of records to be sorted internally. However, when the keys are randomly distributed over their possible range when R is reasonably large, this sorting procedure requires reasonably little comparing and shifting when the number of storage locations is as little as 20% greater than R.

Internal Sorting—Exchanging

In "exchanging" all records are assumed to be in the high speed storage unit prior to the start of the sorting operation. The key of each record is compared with the key of the record in the next successive position, and when a pair of out-of-sequence records is found, the pair is interchanged. This process is carried out repeatedly until a sort is achieved. In the eight-number examples used above, the numbers would appear as follows after each "pass" through the list.

Position	Original Sequence	First Pass	Second Pass	Third Pass	Fourth Pass	Fifth Pass	Sixth Pass
1	729	071	071	071	071	071	015
2	071	309	309	309	309	015	071
3	309	652	401	401	015	309	309
4	652	401	652	015	401	401	401
5	401	729	015	613	613	613	613
6	839	015	613	652	652	652	652
7	015	613	729	729	729	729	729
8	613	839	839	839	839	839	839

On the first step of the first pass, the numbers **729** and **071** in positions 1 and 2 are compared and interchanged. On the second step, **729**, the number currently appearing in position 2 is compared with 309, the number in position 3, and this pair is interchanged. The number 729 is analogously compared with 309, 652, and 401 on succeeding steps of the first pass and is likewise interchanged with them. The number 729 now appears in position 5 and is then compared with 839 in position, and this pair is not interchanged. On subsequent steps of the first pass, 839 becomes analogously moved to position 8. Subsequent passes are carried out in the same way.

The number of passes required to perform the sort is determined by the maximum number of positions that any one number must be moved to lower numbered positions (not to higher numbered positions) to reach its final position in the sorted sequence. In the above example, the number 015 must be moved six positions (from position 7 to position 1) so that six passes are required. The time required to perform the sort depends, not only on the number of passes required, but also on the number of interchanges that must be performed. The required number of interchanges is a complex function of the initial sequence. So far as is known, this number has never been determined for random initial sequences.

When performed in detail as in the example, sorting by "exchanging" is not attractive from a speed standpoint, but the accompanying program is relatively simple. Elaborations can be incorporated in the sorting program to detect and "remember" properly sequenced groups so that many comparisons may be eliminated and many interchanges can be with nonadjacent numbers, and sorting speed may thereby be increased.

Internal Sorting—Merging

In one version of sorting by "internal merging" the initial unsorted number sequence of R records is viewed as consisting of R properly sequenced strings of one record each. Pairs of these strings are then merged to form $R/2$ strings of two records each. Next, pairs of the two-record strings are merged to form $R/4$ strings of four records each, and so on, until the R records are sorted. (Reasonably obvious modifications to the procedure may be made when R is not an integral power of 2.) In the first merging operation, the procedure consists of nothing more than comparing the numbers in positions 1 and 2, positions 3 and 4, positions 5 and 6, and so on, and interchanging the numbers when they are not in proper sequence. On second and succeeding merging operations, the merging is still accomplished by interchanging, but the pattern of pair comparisons is slightly more complex. When using the same example as before, the numbers after the various merging operations appear as follows.

INTERNAL SORTING—MERGING

Position	Original Sequence	After First Merge	After Second Merge	After Third Merge
1	⌊729	⌊071	⌊071	015
2	⌈071	⌈729	⌈309	071
3	⌊309	⌊309	⌊652	309
4	⌈652	⌈652	⌈729	401
5	⌊401	⌊401	⌈015	613
6	⌈839	⌈839	⌊401	652
7	⌊015	⌈015	⌊613	729
8	⌈613	⌊613	⌈839	839

Each line indicates a string pair to be merged in the next successive merging operation. In the first merging operation of this particular example, only the numbers in positions 1 and 2 are interchanged during the first merge.

The generalized merging process will be explained by defining any two strings to be merged as string A and string B where string A appears in the relatively lower numbered positions. On the first step of the merge, the lowest number in string A is compared with the lowest number of string B, and this pair of numbers is interchanged if they are not in proper sequence. Whether or not an interchange is made, the number now appearing in the lowest position of string A is in its final position of the merged string and is not considered further. If an interchange is made, the number then appearing in the lowest position of string B may not be in proper position in that string, but it may be shifted to its proper position by a single pass of interchanging as described for the "exchanging" method of sorting. On the second step, the second number in string A is compared with the number currently appearing in the first (lowest numbered) position of string B. The pair is interchanged if not in proper relative sequence. The number now appearing in the second position of string A is in its final position in the merged string and is not considered further. Again, if an interchange was made, the number moved to string B is shifted, if necessary, to its proper position in that string. The number of such steps required to complete the merge is equal to the number of numbers (records) in string A.

An important variation of sorting by internal merging is to merge three or more strings during each merge operation. Further improvements in speed can be obtained by employing analogies of the tape merging methods to be described later, although for a given number of records R, the storage unit must be capable of storing substantially more than R records, whereas a capacity of R records (plus program) is sufficient for the sorting method described in this section.

Internal Sorting—Selection

In sorting by "selection" all records are again assumed to be in the high speed storage unit prior to the start of the sorting process. The number (that is, the key of the record) in the first position is compared with the number in the second position. The smaller of these two is compared with the number in the third position, and the smaller of these is then compared with the number in the fourth position, and so on. After all numbers have entered into the comparison, the smallest of all the numbers is known and this number is "selected" and may then be transmitted to an output device as the first number in the sorted sequence. The position originally containing this number is filled with 9's (or 1's in a binary computer), and the process is repeated to determine the second number in the sorted sequence by a similar process. For R records, $R(R-1)$ comparisons are required.

If registers suitable for the temporary storage of certain numbers are available, the number of comparisons can be approximately halved. Before returning the larger number to storage after a given comparison, the next number in the random sequence is obtained from the storage unit. The number to be returned is then returned to the position from which the new number was obtained rather than to the returned number's original position. The file is thereby shortened by one number after each number in the sorted sequence is determined. The number of comparisons then becomes $(R-1) + (R-2) + (R-3 + \cdots + 1$.

A scheme which produces a more dramatic reduction in the number of required comparisons, especially for large R, is to divide the R numbers into $R^{1/2}$ groups with $R^{1/2}$ numbers in each group. (These quantities may be approximate when R is not the square of an integer.) In the first part of the process the smallest number in each group is selected and transmitted to a portion of the high speed storage until reserved for the purpose. Along with each number, an indicator is stored which "remembers" the group from which that number came. The smallest number from this new group is then selected analogously and is ready for transmission to an output device as the first number in the sorted sequence. The savings in comparisons is realized in the selection of the second and successive numbers in the sorted sequence. To select the second number, the next smallest number in the group from which the first number case is selected and transmitted to the new group to replace the first number there. The smallest number in the new group is then selected to determine the second number in the sorted sequence. Successive numbers are selected in the same way. Thus, without the refinements of the previous paragraph, only $2R(R^{1/2} - 1)$ comparisons are required to sort the sequence (an approximation when R is not the square of an integer).

INTERNAL SORTING—COUNTING 491

Further decreases in the number of comparisons required can be obtained by increasing the number of "levels" of selection. For three levels, the R numbers are divided into $R^{1/3}$ groups each of which will contain $R^{2/3}$ numbers. Each group is divided into $(R^{2/3})^{1/2} = R^{1/3}$ subgroups each of which will contain $R^{1/3}$ numbers. The smallest number in each subgroup is selected as described in the previous paragraph, and then the smallest number in each group is analogously selected. From these numbers, the smallest is selected to produce an output number in the sorted sequence. When R is the exact cube of an integer, the total number of comparisons is then $2R(R^{1/3} - 1)$. For L levels, the numbers of comparisons can be as small as $LR(R^{1/N} - 1)$. This quantity continues to decrease as L is increased, but an increase in L beyond that value which corresponds to only two numbers in the smallest subgroups is pointless.

For most applications, multilevel selecting is probably the preferred sorting method, although in comparison with other sorting methods it does have the disadvantage that the accompanying program is not among the simplest. Also, for a given R, the storage capacity requirements are not minimized.

Internal Sorting—Replacement-Selection

"Replacement-selection" is the same as "selection" except that when a number is selected and transmitted to an output device it is replaced by a new number from the input device. If at any given step of the internal sorting process the new number happens to be smaller than the selected number, the order of selection will not be altered, and the sort will continue as though the number had been replaced by all 9's (or all 1's in a binary machine). However, if the new number is greater than the selected number, this new number will eventually be selected in the course of forming the output string. Thus, the internal sorting operation will produce a longer string than obtained without the replacement feature.

Near the beginning of an internal sorting operation, the probability is high that the new replacement number will be larger and will thereby be effective in increasing the string length. However, near the end of the string formation, the probability is low that the new number will be effective. For random numbers and for a given R, the average string length is roughly doubled by adding the replacement feature to the sorting-by-selection method.

Internal Sorting—Counting

In internal sorting by "counting" each item to be sorted has another number called a "counter" associated with it. The counting may be performed as the items are entered, one at a time, into the high speed storage unit. As each item is entered, its key is compared with the keys of all other items already in high speed storage. For each comparison for which the new item has a larger key than a previously entered item, the counter of the new item is increased by one. For each comparison producing an opposite result, the counter associated with the previously stored item is increased by one. The number of comparisons required is $1 + 2 + 3 + \cdots + (R-1) = R(R-1)/2$ for R items. After all items are entered, each counter will indicate the proper position of the corresponding item in the sorted sequence.

However, to perform the actual sort, the items must be rearranged, and this operation is accomplished by transmitting each item to an address $K_1 + K_2 C$, where K_1 is a constant determined by the particular "region" available for the purpose in the high speed storage unit, K_2 is a constant determined by the number of addresses required to store each item, and C is the number in the counter. As before, instead of rearranging the complete items (records), each item may be represented by only its key, its counter, and an address number that "remembers" where the remainder of the item is stored.

Internal Sorting—Partitioning

In sorting by "partitioning" all items to be sorted are again assumed to be initially in the high speed storage unit. The random sequence is partitioned into two sequences where every key in one sequence is greater than every key in the other sequence. The two sequences are then similarly partitioned into four sequences, and this process is repeated over and over until each sequence consists of only one item, at which time a sort will have been effected.

A "partitioning" of a given sequence into two sequences is accomplished by determining a key value near the expected median for the sequence (not practical unless the keys are at least approximately randomly distributed over their possible values). The keys of the items at the presumed "low" end of the given sequence are examined, one at a time, starting from the first item in the unsorted sequence. When a key is found that is larger than the expected median value, its position is "remembered." The keys of the items at the "high" end of the random sequence are then examined, one at a time, starting with the last item. When a key is found that is smaller than the expected median value, the corresponding item is interchanged with the high-key item found previously. Keys of items progressively toward (and past) the center of the given random sequence are then analogously examined with interchanges being made as before. This process stops when all keys have been considered. The last key to be considered may not be at the center of the given sequence.

At the end of the process some high-key items from the "low" portion of the original sequence may be left without any low-key items from the "high" portion with which to be paired and interchanged. Or, alternatively, extra low-key items from the "high" portion of the original sequence may be encountered. In either case, the situation is corrected merely by adjusting the partition point so that the left-over items appear in the proper one of the two random sequences which have been generated. The two resulting sequences may not, of course, be of equal length.

Each of the two resulting two random sequences is then further partitioned in the same way but with a new median value assumed for each.

When a given sequence contains only two items, the partitioning procedure can be replaced by a simple comparison of keys with an interchange of the two items if they are found to be out of order. In fact, when the partitioning process has reduced the lengths of the random sequences to three, four, five, or perhaps a few more items, procedures other than further partitioning may be preferable for achieving the final sorted sequence.

Sorting by partitioning requires only R storage spaces for the sorting of R items and is competitive from a speed standpoint with other sorting procedures that require only this amount of storage. Partitioning is particularly attractive in computers having an "exchange" instruction. However, the exact time required to perform a sort is a function of both the detailed characteristics of the computer and the degree of randomness in the initial sequence.

Sorting the Items on One Reel of Tape—the Balanced Merge

When sorting the items on a single reel of tape only a few of the items need be stored in the high speed storage unit of the computer at any given time. The simplest sorting scheme is called "balanced merging" and any even number (except 2) of tape units may be used. The tape units are divided into two equal groups, and the items are transmitted back and forth between the two groups in a manner whereby the individual sorted strings are merged into longer strings. This procedure continues until all items appear in a single sorted string on one of the tapes.

With four tape units the sorting procedure is as follows. One item from each of two tape units is entered into the computer. The item having the smaller key, here designated L, is then transmitted to the output tape unit, and L is "remembered." Another item is then entered into the computer from whichever input tape unit supplied the item having L as a key. The keys of the two items now in the computer may be designated X and Y. If X and Y are each larger than L, the item having the smaller of X and Y is transmitted to the same output tape unit containing the item with key L. If only one of X and Y is larger than L, the item having the larger of the two keys is transmitted to the same output tape unit. If X and Y are each smaller than L, the item having the smaller of X and Y is transmitted to the opposite output tape unit with respect to the tape unit to which the item having L as the key was transmitted. Again, a new item is entered from whichever input tape unit supplied the item that was transmitted to an output tape unit. If one of the two input tape units is empty (as one will be during the first pass and as one may be at subsequent points in the sorting process), the next item is taken from whichever tape unit is not empty. The transfer process continues until all items have been transmitted to output tape units. The roles of the input and output tape units are then reversed and the process is repeated analogously.

An example of this form of a merging sort is as follows, where the items are initially all on tape unit A, and tape unit B is empty. The items are transferred to tape units C and D and then back to A and B, at which point a sort is achieved.

A	B	C	D	A	B
{729		⌈071	015⌉	⌈015	
⌈071		|309	401|	|071	
{309		{652	613⌋	|309	
⌊652		|729		|401	
⌈401		⌊839		|613	
⌊839				|652	
⌈015				|729	
⌊613				⌊839	

The various strings of properly sequenced items are indicated by brackets. The first pass, from A to C and D, does not necessarily produce any merging (example: initial sequence of 7, 8, 5, 6, 3, 4, 1, 2). On each subsequent pass the number of strings is approximately halved so that the maximum number of passes required is $1 + (\log_2 S)_{\text{int}}$, where S is the number of strings in the initial sequence and where the subscript means that, when the logarithm is not an integer, the next larger integral value is taken. The maximum value of S is equal to R, the number of items, and this maximum is obtained when the initial sequence is in the exact reverse order from that intended in the sorted sequence. However, when S is at a maximum, the first pass does produce merging, so an alternative expression for the maximum number of passes is $(\log_2 R)_{\text{int}}$.

With six tape units, a three-way merge is performed as the items are entered from the three input tape units. At any given step, three items are stored in the computer. Also, the key of the last previous item transmitted to an output tape unit is stored (that is, "remembered"). The output strings, as they are developed, are distributed in rotational sequence to the three output tape units on each pass. Extensions of the sorting method for eight, ten, and so on, tape units are straightforward. In general, to perform M-way merging, $2M$ tape units are required, and the maximum number of passes required is $(\log_M R)_{\text{int}}$, but in many applications the number of passes may be significantly fewer because of the manner in which strings are encountered on the initial unsorted sequence.

Because a general purpose computer is seldom utilized to its full capability in performing the merges described in this section, the usual procedure is to perform internal sorting by batches on the first pass and thereby obtain a substantial increase in string length on the first pass. However, subsequent merging passes are performed as described in this section.

The tape reels need not be rewound between passes, although appropriate modifications must be made in the program when operating the reels in the backward direction.

M-way merging may alternatively be accomplished with only $M + 1$ instead of $2M$ tape units by initially distributing the strings equally among M tape units and then merging them onto a single tape unit. This procedure is repeated on each pass until a sort is achieved. For a given value of M, more passes are required than with the previous procedure. However, for a given number of tape units, M is greater so that a net increase in sorting speed can be obtained in some applications.

Sorting the Items on a Reel of Tape—Unbalanced Merging

Further improvement in sorting techniques can be obtained by distributing the strings unequally, but in accordance with a certain pattern, among the tape units. Specifically, in the case of sorting through the use of only three tape units, the initial string distribution is made in accordance with the number series

$$0, 1, 1, 2, 3, 5, 8, 13, 21, \cdots k_j, \cdots$$

where the first two numbers are defined to be 0 and 1, respectively, and where each other number is derived by adding the two previous numbers in the series. If prior to the start of the sorting process the number of strings is not equal to k_j for some j, a sufficient number of dummy strings may be added to cause the number of strings to be equal to the next larger k_j.

The three tape units may be designated A, B, and C, and all strings may be assumed to be on A initially. The first pass is to distribute the strings on B and C so that the numbers of strings on these two tape units is k_{j-1} and j_{j-2}, respectively. On the next pass. k_{j-2} pairs of strings from B and C are merged, and the merged strings are transmitted to A, after which the remaining $k_{j-1} - k_{j-2}$ strings on B are transmitted without alteration to C (which would otherwise be empty at the completion of the pass). Observe that $k_{j-2} > k_{j-1} - k_{j-2}$ for any j. On the third pass the strings currently on C are merged with an equal number of strings from A, and the merged strings are transmitted to B, after which the strings remaining on A are transmitted without alteration to C (which again would otherwise be empty at the completion of the pass). Successive steps are carried out analogously and repetitively until a sorted sequence (a single string) appears on either A or B.

When the items to be sorted initially appear in eight strings for example, the strings are distributed on the tape units as follows, where each digit represents a string number (and not a number to be sorted and not the number of items in a string) and where a notation such as 1,6 means a string formed by the merging of string 1 and string 6.

Initial	After First Pass:		After Second Pass:		After Third Pass:		After Fourth Pass:		After Fifth Pass:	
A	B	C	A	C	B	C	A	C	B	C
1	1	6	1,6	4	1,4,6	3,8	1,3,4,6,8	2,5,7	1, $\cdots$ 8	
2	2	7	2,7	5	2,5,7					
3	3	8	3,8							
4	4									
5	5									
6										
7										
8										

As with the balanced tape sorting methods, no rewinding of tape is required between passes, although minor modifications are required in the above procedure when operating the tape units backwards to avoid rewinding.

For M-way merging, $M+1$ tape units are required. In general, for any $M \geq 2$, the first pass is used to distribute the strings in accordance with a number series formed by starting with a number of 0's equal to $M-1$ and then a 1 with each succeeding number being obtained by adding the M previous numbers. For example, with $M = 4$, the number series is

$$0, 0, 0, 1, 1, 2, 4, 8, 15, 29, 56, \cdots$$

Five tape units are required. On the second pass a four-way merge is performed as long as possible, then a three-way merge as long as possible, and then a two-way merge as long as possible. In the course of merging the strings on each given step, the merged strings are distributed on the available tape reels so that the number of strings on each reel is one of the numbers of the series. (This distribution does not come about quite as naturally as when $M = 2$, but is not complicated.)

The unbalanced merging technique can be modified to provide a M-way merging action throughout the process—at the expense of tape rewinding or alternatively at the expense of providing means for skipping over some strings. With rewinding, the three-tape $M = 2$ merging example given above would be modified as follows. The first pass would be the same as before. The first part of the second pass would also be the same after which string 1, 6, string 2, 7, and string 3, 8 would appear on tape unit A. Tape unit A is then rewound, and the first two of these strings is merged with strings 4 and 5, respectively, from B to produce strings 1, 4, 6 and 2, 5, 7 on C. String 3, 8 remains on A at this point. Tape unit C is then rewound, after which string 1, 4, 6 from C is merged with string 3, 8 from A, and the merged string 1, 3, 4, 6, 8 is placed on B. After rewinding B, this string may then be merged with string 2, 5, 7 from C with the completely sorted string 1, $\cdots$ 8 being transmitted to A. (This sorting process does not break down into neatly defined "passes.")

Extensions of this example to larger numbers of strings and to $M > 2$ are straightforward, although a substantial amount of detail is involved.

In comparison with the balanced merge method of tape sorting requiring $2M$ tape units, the unbalanced method offers the advantage of requiring fewer tape units to achieve M-way merging, for a given M. However, the time required to complete the sort is greater (even without considering the rewind time). In comparison with the balanced method requiring $M+1$ tape units, the unbalanced method is faster (except possibly for rewind time) because the M-way merging process may proceed continuously without any time wasted in the transmission of unaltered strings from one tape unit to another.

Sorting the Items on a Reel of Tape—Interspersing Merging and Internal Sorting

In a tape sorting procedure commonly called the "oscillating sort" the merging and internal sorting operations are interspersed. Because the procedure requires no rewinding of tape and because the replacement-selection technique (described in a previous section) may be used, the sorting procedure can be quite efficient in many applications.

For M-way merging, $M + 2$ tape units are required. One of these tape units serves only as an input device (and actually need not be a tape unit) with the merging functions being accomplished on $M + 1$ of the tape units. During the first part of the procedure, M batches of input items are sorted by internal sorting with the resulting strings transmitted to any M of the $M + 1$ tape units with one string on each unit. Here, a "batch" means that number of items which can be sorted internally with the computer at hand and with the internal sorting program at hand. A tape unit is assumed to be capable of storing several batches.

The next step is a merging operation (where the tapes may be read backwards), and the M strings are merged to form a single string which is transmitted to the remaining tape unit. An appropriate digit combination is then recorded at the end of this merged string to identify the end of this string and the beginning of the next string to be transmitted to this tape unit.

Again, M batches of input items are sorted by internal merging, and the resulting strings are transmitted to M tape units, but the unused tape unit must not be the same as the previously unused tape unit for this part of the process. Thus, one of the tape units will contain the merged string from the previous operation followed by one of the strings just formed. The M strings just formed are then merged with the merged string being transmitted to the tape unit which was unused during the current internal sorting, and an identifying signal combination is recorded on the tape to mark the end of the string.

The operations described in the three previous paragraphs are repeated M times, at which point all but one of the $M + 1$ tape units used for merging will contain a merged string of M strings as produced by internal sorting. The merged strings are then entered into another M-way merging operation that produces a single string of M^2 strings as produced by internal sorting. This string is transmitted to that tape unit which was empty at the beginning of this part of the process. An identifying signal combination is recorded on the tape to mark the end of the string.

Further batches of M^2 strings may then be generated and merged in an analogous way, but each resulting string is transmitted to a different tape unit. When all but one tape unit contains such a string, the resulting strings are similarly merged to form a single string containing M^3 strings as formed by internal sorting. This entire process may be similarly repeated to form M strings of M^3 internal-sort strings, and the results may be merged to form a single string of M^4 internal-sort strings, and so on.

To illustrate the procedure, the numbers of internal-sort strings which appear on the various tapes for $M = 3$ (input tape unit not shown) are as follows after the indicated parts of the overall sorting process.

A	B	C	D	
1	1	1	0	after M internal sorting operations
0	0	0	3	after the first merging operation
1	1	0	3-1	after the second M internal sorting operations
0	0	3	3	after the second merging operation
1	0	3-1	3-1	after the third M internal sorting operations
0	3	3	3	after the third merging operation
9	0	0	0	after merging the strings formed above
9-1	1	1	0	after the next M internal sorting operations
9	0	0	3	after the next merging operation
9-1	1	0	3-1	
9	0	3	3	
9	1	3-1	3-1	(note the procedure alterations needed to get the
9-3	0	3	3	next 9-string on a different tape unit)
9	9	0	0	
.		.		
9-3	9-3	0	3	
9	9	9	0	
0	0	0	27	
.				

The above process continues until all items have been entered from the input tape unit. As a final part of the sort-by-merge procedure, the strings then on the $M + 1$ merging tape units may be merged, for which at most two merging passes are required.

Merging the Items on Two or More Tape Reels, Each of Which May Be Filled

When the number of items to be sorted is more than can be stored on a single reel of tape, each reel may be sorted as described previously, and then the sort may be completed by merging the various reels. If the number of tape units on the computer is at least one greater than the number of tape reels to be merged, the merging can be accomplished in an obvious way with a single pass. The output reels must, of course, be removed one at a time from the output tape unit as they become filled with sorted items.

When the number of reels to be merged is equal to or greater than the number of tape units, more than one merging pass is required, and the usual problem is to determine the merging scheme which provides for the highest speed. In most applications, highest speed is obtained with the scheme which minimizes the number of "tapes full" of items that must be passed through the computer.

For example, a computer having four tape units can perform three-way merges, and if sixteen reels of tape are to be merged, the reels could be divided into five groups of three with one reel left over. Five passes may be used to merge the five groups. The result would be five strings of items with each string occupying three reels. A sixth pass may be used to merge three of these three-reel strings to form a single string distributed over nine reels. A seventh pass may be used to merge the remaining two three-reel strings with the remaining one-reel string to produce a string distributed over seven reels. An eighth and final merging pass may then be used to merge the nine-reel string and the seven-reel string to form a sorted file distributed over 16 reels. By this scheme, which is straightforward, a total of **47** "tapes full" of items have been passed through the computer. The merging pattern is illustrated in Figure 8-12 where each number indicates the length of the string, in terms of reels, at the corresponding point in the diagram.

The straightforward scheme illustrated by example in the previous paragraph seldom produces the highest possible speed, however. Highest possible speed is obtained by a scheme as follows. With P reels of tape to be merged and with M-way merging ($M + 1$ tape units), the remainder, r, from the quotient $(P-1)/(M-1)$ is determined. If $r = 0$, then M reels are merged as a first step. If $r > 0$, then $r + 1$ reels (one-reel strings) are merged on the first pass. On each succeeding pass, M strings are merged. The particular strings to be merged on each pass are determined as follows. The strings existing prior to each pass are written in the form of a list of numbers where each number represents the length, in terms of reels, of the corresponding string. The strings to be merged are taken from the top of the list, and the merged string is placed at the bottom of the list. The process is repeated until a single string (a sorted file) results.

In the example of 16 reels with three-way merging, the lists prior

MERGING THE ITEMS ON TWO OR MORE TAPE REELS   501

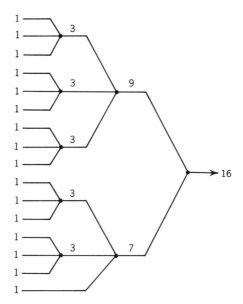

FIGURE 8-12. Straightforward, but not optimum, merging scheme for 16 reels and three-way merges.

to each pass are as follows, where the brackets indicate the strings to be merged. In this example, $r = 1$ as determined by dividing 15 by 2, so $1 + 1 = 2$ reels are merged on the first pass.

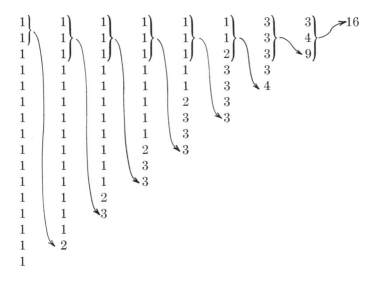

502 IMPLEMENTING OTHER DIGITAL OPERATIONS

With this scheme the total number of "reels full" of items to be passed through the computer has been reduced to 43 in this example.

Evaluation of Sorting Techniques

Analyses and comparisons of the various sorting techniques could be made in an effort to illustrate their relative merits. Although some useful generalized conclusions could possibly be reached in this way, the relative merits are so strongly dependent on the details of the application at hand and upon the details features available in the computer at hand that such evaluations and comparisons tend to be distinguished more by their limited range of applicability than by their generality. Often trial runs on sample random number sequences are the only practicable means of evaluation (an application for random number generators).

Integration and Integrators

In digital systems the forming of the integral

$$z = \int_{x_1}^{x_2} y \, dx$$

where $y = f(x)$ is a matter of accumulating quantities which may be visualized as representing increments of area as formed by a plot of the function. A basic version of digital integration is illustrated in Figure 8-13a, where y is assumed to have been "scaled" so as to have a maximum value less than 1.0 (decimal notation). A series of equally spaced points along the x axis is visualized where the distance between adjacent points is defined to be Δx, and x is assumed to have been scaled so that each $\Delta x = 1$. Then the value of y at each point on the x axis is determined by some manner (usually by computing it from the proscribed function or by adding increments in succession as in the digital differential analyzer to be described in the next section). The area represented by a rectangle in Figure 8-13a is equal to $y\Delta x$, where the value of y to be taken is the value at the left-hand edge, that is, at the beginning of the increment in x. However, because $\Delta x = 1$, the area is equal to y, and the integral from x_1 to x_2 may be obtained simply by accumulating the corresponding values of y. If in the problem being solved $y = f(x)$ has been multiplied by some scale factor F_y to cause it to have a maximum value of 1, and x has been multiplied by a scale factor F_x to cause the increments in x to be 1, the value of the integral as obtained above must be multiplied by $1/(F_yF_x)$ to obtain the desired final result.

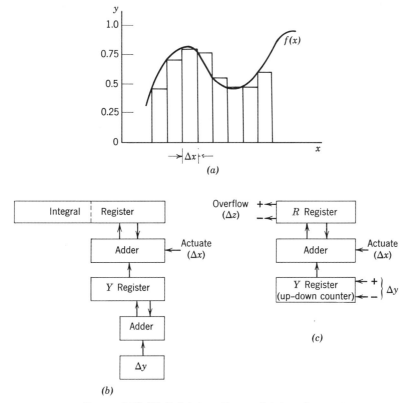

FIGURE 8-13. Digital integration and integrators.

To obtain high accuracy in the integral, the areas between the rectangles and the actual curve for $y = f(x)$ must be small, and this objective is achieved by making Δx small, that is, by accumulating y values at closely spaced intervals. However, regardless of the magnitude of Δx as it may appear on the graph, it may be visualized as being equal to 1 by using a suitable F_x.

Figure 8-13b illustrates the basic elements needed for one technique of forming the integral. The initial value of y is placed in the Y register, and the integral register is initially set at zero. The output signals from these two registers are transmitted to an adder the sum from which is returned to the integral register to replace the number there (the operation may be either serial or parallel). This action takes place each time an "actuate" signal is received. The "actuate" signals represent successive increments in x, and these signals may be applied uniformly

504 IMPLEMENTING OTHER DIGITAL OPERATIONS

in time in the manner of clock pulses, or they may be applied irregularly in time.

Prior to each "actuate" pulse after the first one, the value of y in the Y register is altered to correspond to the next point on the x axis. One way to accomplish this alteration is to compute the necessary increment, Δy, in y and then add Δy to y by means of a second adder as shown in Figure 8-13b. Ordinarily the number of digits needed to represent Δy would be much less than needed for y.

The integral register consists of two parts which will be called the right-hand and left-hand parts. The right-hand part contains the fractional portion of the integral z as it is developed—recall that y is scaled to be a fraction. The left-hand and part contains the integral part of z (that "integral" has two meanings is unfortunate). When the integral z is computed over a suitably long interval or (which amounts to the same thing, when the increments in x are suitably small), the significant digits in z will all appear to the left of the point, that is, in the left-hand part of the integral register. Although the digits in the right-hand part of the register must be retained during the summing of the successive y values, the digits in the right-hand part will be low-order digits of no significance in indicating the value of z, and these digits may be disregarded at the termination of an integration process.

The left-hand part of the integral register may be constructed in the form of an up-down counter with the input signals being obtained from the overflow signals generated by the addition process. The counter is caused to count "up" or "down" in accordance whether an overflow occurs when adding a positive y or a negative y, respectively.

Figure 8-13b has been described to illustrate the basic integration process, although this particular array of equipment does not necessarily have wide utility. A more widely applicable assemblage of equipment, called an "integrator," is shown in Figure 8-13c. The register is in the form of an up-down counter, and y is altered by applying pulse signals representing increments Δy. These signals are transmitted to the "up" or "down" counter input line according to whether Δy is positive or negative, respectively. The left-hand part of the integral register is omitted, and the output signals from the integrator are limited to the previously described overflow signals which represent positive and negative, respectively, increments Δz.

Because the form of the integrator output signals representing Δz is the same as the form of the input signals representing Δy, the output of one integrator may be used as the input of another integrator. By another viewpoint, the integral z_1 of one integrator can be formed in the Y register of a second integrator where it may be designated as

y_2 and may be used as the function to be integrated in the second integrator.

As an integrator has been described above, z_1 contains only digits to the left of the point whereas y_2 should contain only digits to the right of the point. Compensating for this discrepancy is a matter of scaling, to be discussed in more detail later.

A More Accurate Integrator—Trapezoidal Integration

A great increase in the accuracy of integration is realized if, on each step, the number added to the number in the integral register is $y + (\Delta y)/2$ instead of y. Such additions have the effect of accumulating trapezoidal areas, one of which is illustrated in Figure 8-14. The area of the trapezoid is $(y + y + \Delta y)(\Delta x)/2$, which is the same as adding the triangular area $(\Delta y)(\Delta x)/2$ to the rectangular area as determined previously.

The details required to implement this refinement may assume any of several different forms. Most straightforwardly, Δy is formed in a separate register as in Figure 8-13b, and the number in this register is halved (merely by shifting to the right one position when binary numbers are being used), and then on an intermediate step the halved number is added to the integral register through the adder shown but with control equipment to switch one set of adder input lines from the Y register to the lines on which $(\Delta y)/2$ appears.

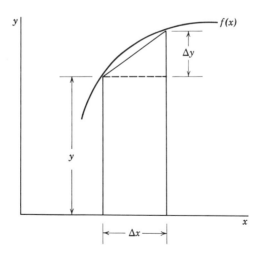

FIGURE 8-14. Trapezoidal integration.

Digital Differential Analyzers

Integrators of the form illustrated in Figure 8-13c can be interconnected in various ways to generate solutions of differential equations, and the resulting system is called a "digital differential analyzer" (DDA). A digital differential analyzer is sometimes called an "incremental" computer because the signals transmitted among the units represent increments in the variables rather than the variables themselves as in general purpose digital computers. Strictly speaking, a DDA is only a specialized version of the incremental computer concept, but in practice the terms incremental computer and DDA are virtually synonymous inasmuch as few, if any, incremental computers of consequence have been built which were not basically DDA's.

Consider two integrators interconnected as shown in Figure 8-15. Each Δz output line and each Δy input line may actually consist of two wires, one for positive increments and one for negative increments. The minus sign with an arrow pointing to one interconnection means that the two wires are transposed so that the sign of the corresponding Δz signal from the bottom integrator is inverted before entering the signal as the Δy input of the top integrator. Assume that all numbers are fractions and that the Y register in the top integrator initially contains the largest possible number 0.999 · · · and that the bottom integrator is initially set to zero. The continuous series of pulses designated dt in the figure applied to the Δx inputs of both integrators creates, for a number of steps, a corresponding continuous series of overflow (Δz) pulses from the top integrator, and these pulses are accumulated in the Y register of the bottom integrator. Eventually the successive additions in the bottom integrator will cause overflow pulses from that integrator, and these

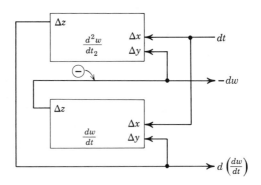

FIGURE 8-15. Integrator interconnections for generating the sine and cosine functions.

pulses will cause the number in the top integrator to be diminished so that the rate of overflow pulses from there will be diminished. The number in the Y register of the bottom register will nevertheless continue to increase (even though at a slower rate) so that the rate of overflow pulses from the bottom integrator will increase. This process will continue until the number in the top integrator is zero and the number in the bottom integrator will be 0.999 · · · (although it is not obvious that these two conditions will be reached at the same time). A continuation of the dt pulses will cause the number in the top integrator to build up in the negative direction so that overflow signals, when they again commence, will be of negative sign and will cause the number in the bottom integrator to start to diminish toward zero. The process continues indefinitely in a cyclic manner.

The action described in the previous paragraph can be better appreciated by considering the differential equation

$$\frac{d^2w}{dt^2} = -w$$

for which the solution is

$$w = A \sin t + B \cos t$$

where A and B are arbitrary constants. A number representing the initial value of $(d^2w)/(dt^2)$, which is the same as $-w$, is placed in the Y register of the top integrator. If this number is properly incremented by signals approximating $d(-w) = -dw$, the output will approximate increments in the integral of $(d^2w)/(dt^2)$, that is, the signals will approximate $d(dw/dt)$. These signals are correct for transmission to the Δy input terminals of the bottom integrator when that integrator contains dw/dt. The Δz output from the bottom integrator then approximates increments in the integral of dw/dt, that is, the output approximates dw, which with an inversion of sign is the proper signal to be entered into the Δy input of the top integrator. Thus, with proper choices for A and B, the numbers in the two integrators are seen to become $\sin t$ and $\cos t$, respectively.

Numbers other than zero and 0.999 · · · can be used for the two initial values to be placed in the Y registers of the two integrators, and the numbers will vary in the manner of sine and cosine functions, although visualization of the action is less straightforward. For any given pair of initial conditions, a particular solution of the differential equation is generated, that is, A and B become defined. However, the initial conditions must ordinarily be such that A and B will both be less than 1.

Observe that the accuracy of the differential equation solution is dependent upon the number of digits employed in the integrators, but as the number of digits is increased the time required to generate the solution (time required per cycle of the sine function in this example) increases greatly. Therefore, in any given application a compromise between speed and accuracy must be made.

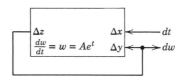

FIGURE 8-16. Integrator connections for generating an exponential function.

Figure 8-16 shows how a single integrator may be used to develop an exponential function. In effect, the differential equation $dw/dt = w$ is solved. The solution is $w = Ae^t$, where A is an arbitrary constant. Some initial value for dw/dt (which is the same as some initial value for w) is placed in the Y register. As dt pulses are applied, the rate at which overflow signals appear increases exponentially until w becomes equal to $0.999 \cdots$, the maximum capacity of the Y register. The Δz output signals are increments of the integral of dw/dt which, because $dw/dt = w$, are the proper increments to transmit to the Δx input terminal in the manner shown in the figure.

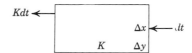

FIGURE 8-17. Using an integrator to multiply by a constant.

For multiplication by a constant, an integrator as shown in Figure 8-17 may be used. The constant K is placed in the Y register, and the Δy input terminals are not used. Ordinarily, K must be less than 1 because the rate of output pulses cannot be greater than the rate of input pulses (although the effect of a $K > 1$ can be obtained in a manner to be described later).

Division in a DDA

Integrators may be used to perform multiplication and division, although the nature of these operations is not quite the same as implied in conventional computers as described in earlier chapters. Division will be described first because integrators without modifications or elaborations may be used.

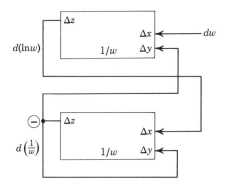

FIGURE 8-18. Maintaining the reciprocal in a DDA.

To effect division, the reciprocal of a number is formed, and this reciprocal may be multiplied by another number as described in the next section. For some initial value of w, the reciprocal $1/w$ is computed by external means and is entered into the Y registers of two integrators connected as shown in Figure 8-18. Thus, the network does not really compute $1/w$ but only maintains it as increments in w are applied.

Pulses representing increments in w are applied to the Δx input terminals of the top integrator which, in effect, sums areas formed by multiplying $1/w$ and dw. Thus, the Δz output pulses from the top integrator are designated $d(\ln w)$ and represent increments in the integral of $(1/w)dw$, that is, increments of $\ln w$. These increments are applied to the Δy input terminals of the bottom integrator where the output represents increments in the integral of $(1/w)d(\ln w) = (1/w^2)dw$, that is, increments of $-(1/w)$. As indicated in Figure 8-18, the resulting signal is inverted with respect to sign, and the result, $d(1/w)$, is applied as the input to the Δy terminals of both integrators to maintain the reciprocal of $1/w$.

Multiplication in a DDA

Figure 8-19 illustrates one scheme by which multiplication may be accomplished in a DDA. The scheme is most easily understood by assuming that, for given nonzero initial values of u and v, the initial value of the product uv is computed by external means and placed in the bottom integrator. The product is then maintained as the magnitudes of u and v are altered in analogy with maintaining the reciprocal as described in the previous section. However, in some applications u and v may both practicably be zero initially, in which case the arrangement in Figure 8-19a actually determines uv.

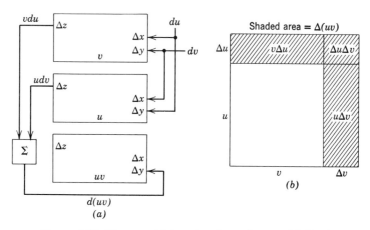

FIGURE 8-19. The use of integrators to perform multiplication.

The differential relationship used in maintaining the product is $d\,(uv) = u\,dv + v\,du$. Initial values of u and v are placed in the Y registers of the center and top integrators, respectively, as these integrators appear in Figure 8-19a. Increments in u and v are then applied as shown to the Δx and Δy input lines. The top and center integrators generate signals representing increments corresponding to $v\,du$ and $v\,du$, respectively. These signals are then "added" or "summed" in the block labeled Σ and entered into the Δy input of the bottom integrator, which merely acts as a counter to accumulate the increments in the product uv. The physical nature of the Σ block depends on the nature of the machine as a whole. In a parallel DDA it would be a binary half adder, and the sum and carry signals would be added to the two lowest orders, respectively, of the Y register. In a serial DDA, the Σ block would be a device that "remembers" whether none, one, or two pulses had been received at its input terminals and then supplies this number of pulses to the bottom integrator at the time that this integrator is actuated.

For obtaining highest possible accuracy in the product, recognition must be given to the fact that the signals transmitted among integrators represent increments and not differentials. In particular, $\Delta(uv) = u\,\Delta v + v\,\Delta u + \Delta u\,\Delta v$, where the third term, $\Delta u\,\Delta v$, has no counterpart in the differential expression. Graphically, the increment in the product is shown as the shaded area, consisting of three parts, in Figure 8-19b. The correct increment in the area (product) is obtained by causing the increment in either u or v (but not both) to be accumulated prior to the formation of the product increment. For example, if u is incremented, the quantity produced in forming the product increment is

$$(u + \Delta u) \Delta v + v \Delta u = u \Delta v + v \Delta u + \Delta u \Delta v$$

which is as desired.

With reference to Figure 8-19a this desired result is obtained by applying the du and dv signals (actually representing Δu and Δv, respectively) alternately in time. For example, as the first part of a given step the Δu signal is applied to form $v \Delta u$ at the output of the top integrator, and at the same time the Δu signal is applied at the Δy input of the middle integrator to form $u + \Delta u$ in the Y register there. During the second part of the given step, the Δv is applied to form $(u + \Delta u) \Delta v$, and at the same time v is incremented in the top integrator in preparation for the next step. In a serial DDA where the integrators are actuated one at a time, the process is better visualized by dividing each step into four parts as follows: (1) the signal representing Δu for the present step is generated, (2) the top integrator is actuated with the Δv (from the previous step) being accumulated in the Y register to form the present value of v, and then the addition, Y to R, in this integrator is made to generate $v \Delta u$, (3) the signal representing Δv for the present step is generated—most likely as the output signal from an integrator (not shown in Figure 8-19) that is actuated between (in time) the actuation of the top and middle integrators, and (4) the middle integrator is actuated with the quantity $u + \Delta u$ being formed first, after which the Y to R addition is made to generate $(u + \Delta u) \Delta v$.

Components can be saved by replacing the two integrators, two adders, and the summing device with a single network specifically intended for multiplication. The principle of the multiplication process is the same, but instead of the two R registers required by the two integrators, the v and $u + \Delta u$ quantities are accumulated in a single R register, and the overflows from this register represent increments of the product. A switching network is needed to select the appropriate Y register (the one containing u or the one containing v) at each point in the process, but this network is not extensive, especially in serial systems.

Another Example of Integrator Interconnections

As another example of integrator interconnections, consider the differential equation

$$\frac{d^2w}{dt^2} - w\frac{dw}{dt} - \sin w = 0$$

512 IMPLEMENTING OTHER DIGITAL OPERATIONS

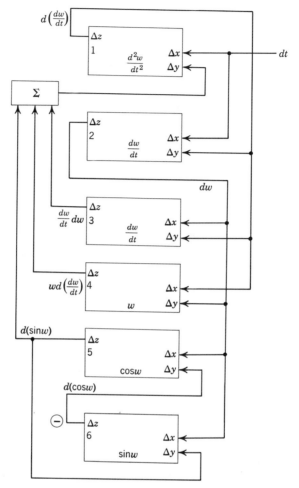

FIGURE 8-20. Another example of DDA integrator interconnections.

which can be solved by the integrator arrangement illustrated in Figure 8-20. The solution is not analytical, but for given initial conditions the value of w is maintained in integrator 4 as t is incremented. That is, successive values of w can be plotted to show its functional relationship to the independent variable t.

The integrator interconnections can be determined by transposing all terms except the highest-order derivative to the right-hand side of the equation and then determining the differentials to produce

$$d\left(\frac{d^2w}{dt^2}\right) = d\left(w\frac{dw}{dt}\right) + d(\sin w)$$

$$= wd\left(\frac{dw}{dt}\right) + \frac{dw}{dt}\,dw + d(\sin w)$$

so that increments in the second derivative are seen to be obtained by summing three other increments, where the multiplication of w by dw/dt is effected in the manner described in the previous section. The increments in $\sin w$ are obtained by solving a hypothetical differential equation $d^2s/dw + s = 0$, where s is an imaginary variable. Integrators 5 and 6 in Figure 8-20 are used for this purpose and are connected in the manner shown in Figure 8-15. Other aspects of the functioning of the network in Figure 8-20 can be appreciated by studying the differentials shown to appear at various points in the network.

A Serial Implementation of a DDA

In implementing the digital differential analyzer concept the designer has a choice ranging from an all-parallel machine with all integrators operating simultaneously and with all digits of each number being handled simultaneously to an all-serial machine with the integrators being actuated one at a time and with the digits in each number being handled one at a time. The all-parallel approach offers high speed but, of course, requires a relatively large number of components whereas the all-serial approach offers greater economy but, for a given pulse repetition rate, requires a much greater time for problem solution. The details of DDA design depend greatly on the extent to which parallelism is employed but for the most part the switching and arithmetic networks described in earlier chapters are directly applicable.

A major problem in DDA design is to provide a flexible means for interconnecting the various integrators as required for the solutions of different equations. One approach to this problem in a machine where all integrators operate simultaneously is through the use of a plugboard where the interconnections are made by manually inserted wires.

In an all-serial machine an ingenious fully electronic interconnection scheme is available, and this scheme will be outlined here. The Y and R registers, each n digits in length, consist of storage positions in two serial storage devices of some sort, and for purposes of illustration the storage devices may be assumed to be two tracks on a magnetic drum.

The Y registers are on one track and the R registers are on the other track, and the registers are positioned so that corresponding Y and R digits appear simultaneously at the read heads of the two tracks. All Y-to-R additions may then be performed by a single serial adder, and the details of the storage unit operation must be such that a number in any register can be replaced, digit by digit, with a new number.

In the simplest version the number of integrators around the periphery of the drum must be equal to n. Then, a third track, designated the Z track, may be used to store the Δz output signals from the various integrators. However, the read and write heads on the Z track are separated by only $n-1$ digit positions (with there being n^2 digit positions around the periphery of the drum). At the completion of the Y-to-R addition for any given integrator, the resulting overflow (Δz) signal, if any, is entered into the Z track for storage. (The Z track may actually consist of two tracks, one for positive and one for negative overflow signals, as described previously in connection with integrator operation.) During the Y-to-R addition for any given integrator, the Z track is storing the Δz signals from the other $n-1$ integrators. Moreover, during each Y-to-R addition, the Δz signals from all of the other $n-1$ integrators appear at the read head of the Z track.

With the arrangement described in the previous section, the Δy quantity to be added to the contents of the Y register of a given integrator is accumulated by means of a counter which counts (with due consideration given to negative Δz signals) the number of Δz signals originating in those integrators which should have their Δz output terminals "connected" to the Δy input terminal of the given integrator. These "connections" are made by means of a fourth track, commonly called the L track, which extends completely around the periphery of the drum. For the solution of any given problem, signals (1's) are placed in those positions of the L track which correspond to the Δz signals which are to be "connected" to the Δy input of the next Y-R register pair to appear under the read heads. An AND device then allows only the corresponding signals to be transmitted to the counter. The addition of the net Δy quantity to the Y register of a given integrator is performed in a second serial adder. The details of the two addition operations may be adjusted so that the net effect is to perform trapezoidal integration, as described in a previous section.

As indicated by the example of Figure 8-20, a Y-to-R addition does not necessarily occur at every integrator during every cycle (drum revolution) of operation. Only those integrators for which the Δx input signal is derived from the independent variable (t in the example) are so actuated. A second L track may be used to "connect" the various Δz

output terminals to an add-subtract control network that causes addition, subtraction, or no operation at all, as required in each instance. The operation of this second L track is similar to that described in the previous paragraph except that a counter is not necessary in most applications, because a Δx input signal is not usually obtained from the sum of two or more other signals.

Some DDA Elaborations

For the construction of a DDA with sufficient flexibility to be useful over a significant range of problems, many points must be taken into consideration in addition to the basic principles covered in the preceding sections. In particular, having exactly n digits in the registers of all integrators is a stringent limitation. For those registers for which n-digit accuracy is greater than necessary and for which a higher solution speed is desired, the register length can, in effect, be shortened by shifting the Δy quantity to appropriate higher orders before adding it to the Y register. In the serial DDA, an added track containing one pulse per integrator may be included to select the digit positions to which Δy is to be added in each integrator.

Another elaboration is to provide means for multiplying each Δz signal by some predetermined quantity. The details of the equipment required for accomplishing this function may vary widely, but one approach is to employ additional tracks to store a number for each integrator. Then when accumulating the Δz signals to form the Δy input signal for some given integrator, each control pulse in the L track initiates the addition of the corresponding number to an accumulator provided for the purpose (to replace the elementary counter described previously).

Although elaborations such as those just mentioned tend to be essential for practical DDA usage, the initial "scaling" of a problem to be solved becomes an extensive problem in itself. Not only must all Δy shifts, all Δz multiplicative factors, and all initial quantities be correct to produce the desired problem solution, they must be so chosen as to produce an acceptable balance between solution accuracy and solution time. The scaling problem can be difficult, but it is too dependent upon specific applications to be covered here. See the paper by R. J. Leake and H. L. Althaus in the *IEEE Transactions on Computers*, January 1968, pp. 81–84, for one approach to the problem and for further references on the subject.

Other DDA elaborations include means for limiting the magnitudes of the quantities in some integrators or for detecting when the quantities are outside of predetermined magnitude ranges. The detection signals may be used to affect the integrator interconnections or otherwise control the progress of the problem solution. Another elaboration is to provide means for delaying (by a predetermined amount of time or a predetermined number of cycles) the entry of some signals to some integrators as might be required for problems having "delayed action" or "hysteresis" effects.

Any problem solvable by a DDA can alternatively be solved by conventional programming on a general purpose computer (as described in the next chapter) or by employing a computer "language" which causes the general purpose computer to simulate a DDA. However, within the range of problems for which DDA's are applicable, a suitably elaborate all-parallel DDA can produce solutions at higher speed than can a general purpose computer employing comparable switching and storage components, and an all-serial DDA can be built at lower cost. For these reasons, DDA's play a relatively small but continuing role in digital technology.

The Checking of Arithmetic Operations—Codes for the Purpose

Although the codes described in Chapter 5 can be effective for the detection and correction of errors that occur during the transmission or storage of information, those codes are not necessarily effective when the information undergoes "processing" of some form. In fact, for most information processing functions (for which the interchanging of the third and fifth digits in a word might be an arbitrarily selected examples), the codes are utterly ineffective in the detection and correction of errors. Even when the information processing function is something conventional, such as binary addition, the previously described codes are largely inapplicable.

Consider conventional n-digit parallel binary addition as accomplished by a set of full adders without carry "skip" features or other elaborations. Assume that a single malfunction occurs in the full adder in the ith position where i ranges from 0 to $n - 1$ with the numbering from low-order to high-order positions. This malfunction may cause either the sum digit S_i to be in error or it may cause the carry digit C_{i+1} to the next higher order position to be in error—or some types of single malfunctions may cause both S_i and C_{i+1} to be in error. If only S_i is in error, the effect is to increase or decrease the number of 1's in the sum by one, and this effect could be detected by the previously described codes. However, if C_{i+1} is in error, the effect of the error will be propagated through as many successively higher orders as may be occasioned by the previously described carry propagation process in accordance with the digit pattern of the two numbers currently being added. Thus, an unknown number (up to n) of 1's may be added to or deleted from the sum with the result that the previously described coding techniques are quite impractical for detecting the error.

However, the effect of the error can be viewed in a different way. If S_i is in error, the effect is to add $+2^i$ or -2^i to the sum according to whether S_i has been erroneously changed from 0 to 1 or from 1 to 0, respectively. If C_{i+1} is in error, the effect is to add $+2^{i+1}$ or -2^{i+1} to the sum according to whether C_{i+1} has been erroneously changed from 0 to 1 or from 1 to 0, respectively. If both S_i and C_{i+1} are in error but in the opposite manner (as is possible with some full adder designs), the effect is to add $\pm(2^{i+1} - 2^i) = \pm 2^i$, which is the same net effect as when only S_i is in error. On the other hand, if both S_i and C_{i+1} are in error in the same manner, the effect on the sum is to add $\pm(2^{i+1} + 2^i) = \pm(3)2^i$. In summary, the net effect of a single malfunction in a binary adder is to add $\pm 2^i$, $\pm(2)2^i$, or $\pm(3)2^i$ to the sum.

Observe that, regardless of the value of i, the malfunction in a parallel adder never causes the increment in the sum to be a multiple of 5.

Therefore, a malfunction can be detected if the numbers A and B to be added are encoded by multiplying them by 5 (binary 101) prior to addition. The sum will be $5A + 5B = 5(A + B)$. For n-digit binary numbers the multiplication by binary 101 produces numbers having up to $n + 3$ digits, and the length of the parallel binary adder must be correspondingly increased. The desired sum is obtained by dividing the above result, $5(A + B)$, by 5. The occurrence of a malfunction is detected by sensing the remainder (often called the "residue" in this application), which will be zero or not according to whether a malfunction has not or has, respectively, occurred.

Note that the encoding operation, a multiplication by binary 101, is essentially an addition operation where the number to be encoded is shifted two positions to the left and added to itself. If a malfunction occurs in this encoding operation, the encoded number will not be a multiple of 5, and the resultant sum will likewise not be a multiple of 5—provided the encoding and adding are performed in separate adders and provided that only one malfunction occurs in the system as a whole. Therefore, only one error-detection network is needed to check both the encoding and the addition.

Unfortunately, the decoding operation, which involves a division by binary 101, is more complex, although it can be accomplished by any of the binary division techniques described in a previous chapter. However, various simplifications are attainable because the divisor is a constant.

The above encoding and checking method works equally well for subtraction. For multiplication and division performed by essentially over-and-over addition or over-and-over subtraction processes, respectively, the method likewise works, although in general the results of each individual addition-subtraction operation must be checked. If the individual operations are not checked, the multiple use of a malfunctioning adder-subtractor may (with certain designs, certain malfunctions, and certain digit patterns) result in compensating effects on the coding remainder and may thereby cause errors in the final product or quotient to be undetectable.

Checking by Using a Multiplier of 7

Although 5 is the smallest multiplier which will result in a code capable of detecting errors in binary addition, any odd multiplier greater than 3 will work. In particular, certain decoding advantages can be realized with multipliers that are equal to $2^k - 1$, where k is a positive integer ($k \geq 3$). Specifically, consider the case where $k = 3$ so that the multiplier is 7 (binary 111). The encoding becomes slightly more complex in that two additions are required to effect the multiplication, but if separate adders are used for the purpose, any single malfunction in either one is detectable as described in the previous section. Alternatively, the multiplication by binary 111 can be accomplished with a single addition-subtraction operation by shifting three positions to the left (multiplying by binary 1000) and subtracting, as is valid because $1000 - 1 = 111$.

In applications where the numbers are to remain in their coded form most of the time (including successive arithmetic operations possible intermixed with storage operations and transmissions to remove terminals and machines), the need to decode the numbers will be encountered only when the final results are to be utilized—and even then complete decoding may not be necessary, as when the results are transmitted to a digital-to-analog converter, for example. Nevertheless, the checking for errors may be required frequently as after each storage or transmission of a number or after each instance that a number enters an arithmetic operation. In these applications, an actual division by 7 need be performed only rarely, but the determination of the remainder (residue) that would result from such a division is required frequently.

A simple scheme for determining the remainder without actually performing the division can be understood by studying the following relationships where a notation such as $R \bmod_a X = Z$ means "the residue modulo a of X is equal to Z," or in other words, "the remainder obtained by dividing X by a is Z."

$$R \bmod_{111} 1 = R \bmod_{111} 1{,}000 = R \bmod_{111} 1{,}000{,}000 = 1$$
$$R \bmod_{111} 1{,}001 = R \bmod_{111} 1{,}001{,}000 = 10$$
$$R \bmod_{111} 1{,}001{,}001 = 11$$

By continuing with additional examples, the reader may quickly observe that the residue $\bmod_{111}$ of any binary number may be obtained by grouping the digits in three's and by adding those groups in a $\bmod_{111}$ fashion. That is, if the sum of any two groups of three digits is 111 or greater, the "count" is artificially advanced by 1. The additions of the groups may be accomplished in a parallel binary adder to which a minor elaboration has been made. When the sum of any two groups is 1000 or greater, the left-hand 1 may be added to the low-order digits of another group in the manner of an ordinary carry. Because the sum may be exactly 111, a network must be added to detect this condition and to generate a carry artificially.

The resulting array for generating the residue $\bmod_{111}$ of 12-digit binary numbers is shown in Figure 8-21, where X_1 through X_{12} represent the digits of the number for which the residue is desired. Extensions to numbers having more than 12 digits are straightforward.

Arithmetic Error-Detecting Codes Having the Check Digits Separate from the Information Digits

An alternative coding scheme which provides for the detection of errors in a parallel binary adder has the feature that the digits representing the numbers are unchanged. Check digits (but not parity check digits as with most of the codes described in an earlier chapter) are appended to each number, and these check digits undergo an arithmetic operation in such a way that any single malfunction can be detected. The check digits for a given number are determined by taking the remainder (residue) which results after dividing that number by a suitably selected factor ("modulus"). That is, the check digits which accompany some number N are $R \bmod_a N$ for a modulus of a. However, here the notation for the residue of N to an implied modulus of a will be abbreviated R_N. Observe that $R_N < a$ in all cases.

When adding two numbers A and B, the addition $A + B$ is performed in the conventional way, but the residues R_A and R_B are added in "modulo a" fashion, which for present practical purposes means that, if $R_A + R_B < a$, the addition is conventional, but if $R_A + R_B \geq a$, then a is subtracted from $R_A + R_B$. The result, $R_A + R_B$ or $R_A + R_B - a$, as the case may be, is equivalent to the $\bmod_a$ residue of $R_A + R_B$ and may

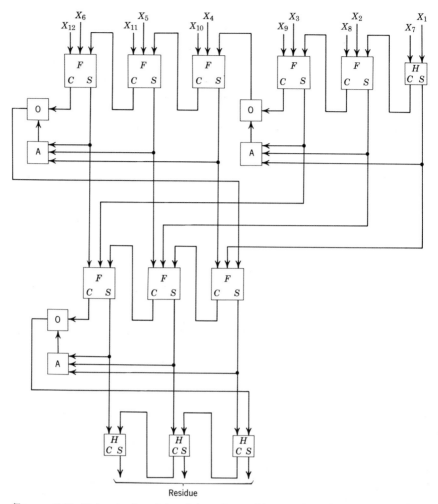

FIGURE 8-21. Network for determining the residue mod_{111} of a 12-digit binary number X.

be designated $R_{(R_A+R_B)}$. To check the addition, the residue R_S of the sum S is determined and is compared with $R_{(R_A+R_B)}$. In the absence of any malfunctions, $R_S = R_{(R_A+R_B)}$. This equality can be appreciated by observing the following relationships (where all numbers may be assumed to be integers).

$$\frac{A+B}{a} = \frac{A}{a} + \frac{B}{a} = K_A + \frac{R_A}{a} + K_B + \frac{R_B}{a}$$

where K_A and K_B are the integral parts of the respective quotients. If $R_A + R_B < a$, then $K_A + K_B$ must be equal to K_S, the integral part of the quotient $(A + B)/a$, and $R_A + R_B$ would be equal to R_S. If $R_A + R_B \geq a$, then K_S would be equal to $K_A + K_B + 1$ (although K_S need not actually be computed), and R_S would be $R_A + R_B - a$.

The next problem is to find a value of a that will cause R_S to be not equal to $R_{(R_A+R_B)}$ under the condition of any single malfunction in either part of the adder which adds the given numbers or in that part of the adder which adds the residues (the check digits). This problem may be analyzed in essentially the same manner as with the previous adder checking scheme, and it similarly turns out that any odd value of a greater than 3 will work for malfunctions in that part of the adder which adds the given numbers. The same range in a also seems to be satisfactory for that part which adds the residues provided the subtraction (when required) of a is mechanized in a manner which avoids compensating effects that erroneously cause R_S to be equal to $R_{(R_A+R_B)}$ and thereby renders the malfunction undetectable. A simple and effective value of a to use is 7 (binary 111) because, not only can the residues R_A, R_B, and R_S be computed economically as illustrated in Figure 8-21, but also the subtraction of a when required from $R_A + R_B$ (that is, the determination of the residue $R_{(R_A+R_B)}$ can likewise be accomplished simply. The subtraction of a is accomplished by an "end-around" carry signal which is generated in those instances that $R_A + R_B > a$. When $a = 7$ the end-around carry has the effect of subtracting $a + 1$ and adding 1.

The resulting array of equipment required for performing only the addition is illustrated in Figure 8-22. Besides this equipment a residue-determining network such as shown in Figure 8-21 is needed for each of the input numbers A and B (if these numbers do not arrive already supplied with check digits), and a similar residue-determining network is needed for the sum. Also, a comparison network is needed to compare the check digits as generated by the network in Figure 8-22 with the residue of the sum. Of course, malfunctions in these auxiliary networks must be considered as well as malfunctions in the main processing networks, but apparently any single defective switching device or interconnection will, in fact, have an effect on the final signal which indicates the presence or absence of an error.

Decimal Arithmetic Error-Detecting Codes

Analogous techniques may be applied to computers operating with the decimal number system, but the general problem tends to be substantially more complex. In particular, a single malfunction can conceivably

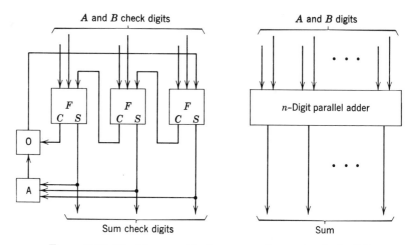

FIGURE 8-22. Checking an adder by means of separate check digits.

create (with some adder designs) an error of any amount up to ±9 in the sum output of that adder plus an error in the decimal carry to the next higher order. Because of this situation and because the necessary residue-determining networks tend to be so much more complex with decimal numbers than with binary numbers, no attempt will be made here to cover the subject. (The use of modulus 9 in the manner described for binary 7 in the previous section would constitute the so-called "casting out 9's" checking scheme which has long been used for checking manual computations.)

A decimal error-detecting arithmetic unit can be designed with the bi-quinary code as a basis, as described in Richards' *Arithmetic Operations in Digital Computers* (1955), pp. 225–230 and 263–266. In this scheme the check is accomplished on a digit-by-digit basis rather than on a number-by-number basis. Another digit-by-digit decimal checking scheme is reported in a paper by M. Y. Hsiao and F. F. Sellers in the *IEEE Transactions on Electronic Computers*, June 1969, pp. 265–268.

Some Disadvantages in the Various Arithmetic Checking Schemes

When provision is made for all necessary details, each of the known checking schemes causes the amount of equipment required for the arithmetic operations to be roughly doubled. This increase in the amount of equipment is a disadvantage in itself, but a more serious problem is encountered in the fact that the various schemes which check the arithmetic operations do not necessarily check the various control networks within a computer. Because the control networks are generally composed of exactly the same types of devices (such as AND, OR, O-I, A-I, etc., modules and flip-flops) as are the actual arithmetic networks, a failure in a control network would, in general, be equally probable and the final output of the computer would be equally unacceptable. Schemes for checking most or perhaps even all of the control networks can be devised, but computers vary so vastly in their design that generalized approaches to control network checking are elusive at best.

Another serious problem encountered with the arithmetic checking schemes is their limited applicability. For example, the various coding schemes do not necessarily work for adders equipped with carry skip features, because malfunctions in certain switching devices may cause the sum to be in error by an amount other than the $\pm 2^i$, $\pm(2)2^i$, or $\pm(3)2^i$ encountered in simple adders. Even if a code were found which worked for a given carry-skip adder design, difficulties may be encountered because of timing considerations. That is, for example, the carry-skip network may temporarily cause a sum to be correct, but a late-arriving erroneous conventional carry may introduce an error in such a manner that the time that must be allotted to detect the error may be undesirably great.

Moreover, when the information words undergo processing other than conventional arithmetic operations, the direct detection of errors can become virtually impossible in a practical sense. Even an operation as elementary as shifting introduces awkward checking problems.

Substantially all of the above problems (except for the undesirable increase in the amount of equipment required) can be solved by duplicating the complete arithmetic unit, control networks and all, and by comparing the final output signals from the two units. Further, any single malfunction can be corrected, as well as detected, by triplicating the complete arithmetic unit and by incorporating a network which rejects output signals from any one of the units when those signals do not agree with the output signals from the other two units. Although duplicating and triplicating the units will produce a large reduction in the probability of an undetected error when the error probability is reasonably low for one unit, even this technique is not particularly effective in increasing the average time between failures—all as discussed at some length in Chapter 10 on reliability in Richards' book *Electronic Digital Systems* (1966).

Exercises

1. Convert the Gray-code number 1001010 to conventional binary form. Convert the conventional binary number 1001010 to its Gray-code equivalent.

2. Determine the number of quintuplers required to convert, in a single step, four-digit decimal numbers to binary form in the manner of Figure 8-2.

3. With the decimal-to-binary conversion method described on pages 440 and 441, determine the switching functions needed to form the corrective network for each decimal digit.

4. How many half and full adders are required to convert four-digit decimal numbers to binary form by using the method of Figure 8-3?

5. Work out an array of half and full adders which will convert three-digit decimal numbers to binary form by the method suggested by the array on page 444. Do the same for the method explained on page 445.

6. Show the steps in the conversion of 723 and 0.297 to binary form by each of the methods described in the text for integers and fractions, respectively.

7. Show the steps in the conversion of 1110011011 and 0.1100001101 to decimal form by each of the methods described in the text for integers and fractions, respectively.

8. Determine the number of doublers needed to convert 12-digit binary numbers to decimal form by the method of Figure 8-5.

9. Devise a combinatorial switching network employing only doublers and subtractors which will convert nine-digit binary numbers to decimal form by the method described on page 453.

10. In the network of Figure 8-8, show how a rounded three-digit decimal fraction can be generated by the addition of one full adder and a minor rearrangement of the connections. (Hint: Add the binary equivalent of 0.0005 to the original binary number and observe that with six-digit binary fractions the rounding process happens never to affect the tenths and hundredths decimal digits.)

11. Show the steps in the extraction of the square roots of decimal 85.3 and binary 1011101.010 by each of the procedures described in the text.

12. In the generation of pseudo-random decimal numbers by the formula $x_{j+1} = ax_j + c \pmod{10^N}$ determine the number series generated by the smallest possible values of a, c, and N for which the period is 10^N. Point out any nonrandom patterns observable.

13. Select a sequence of about 10 random numbers and show the detailed steps involved in the sorting of these numbers by each of the internal sorting methods described in the text.

14. Determine the maximum number of useful levels in the sorting of 256 items by the multilevel selection technique.

15. For the 16-reel, three-way merge example set forth on page 501, draw the string-flow diagram corresponding to Figure 8-12. (Remark: Do not jump to the conclusion that this exercise is obvious.)

16. For 16 sorted reels and five-way merging, determine the minimum number of "tapes full" of information that must be passed through the computer to complete the sort, and draw the corresponding string-flow diagram.

17. With dx as an input, show how two integrators may be interconnected to generate x^2. (Hint: Integrate $x\ dx$.)

18. Explain why the Gray-to-binary conversion method exemplified on page 433 accomplishes the same result as the network in Figure 8-1b. Show the steps in the conversion of the Gray-code word 10101101 to its binary equivalent by the method exemplified on page 433.

9
AN INTRODUCTION TO THE GENERAL PURPOSE COMPUTER AND ITS PROGRAMMING

In designing a complete digital machine to perform some prescribed task, all manner of different design approaches may be taken and all manner of different machine designs may result with each design having its own peculiar set of advantages and disadvantages. However, one particular machine type, commonly referred to as the "general purpose stored-program computer," has emerged as the basis for design in an ever widening list of applications, including virtually all of the most important applications. The term "computer," incidentally, is used in the broad sense of an "information processing machine"—actual arithmetic computations may or may not play a significant role in any specific application.

In any digital information processing application, digital input signals are transmitted to the machine, and the machine generates digital output signals which are functions of the input signals. In a given application, if the function is fixed and known in advance, a so-called "special purpose" machine can be designed for the purpose. Many special purpose machines have been built, although they are usually for relatively elementary purposes as in the case of a counter where the input signals are pulses originating from objects to be counted (e.g., articles on an assembly line) and where the output signals represent the total count. Miscellaneous other input-output signals may be added, such as an input signal to reset the counter to zero and an output signal to indicate when the counter exceeds a predetermined quantity. Other special purpose (or at least non-stored-program) machines of importance include any of the so-called "desk calculators" where the operations to be performed are controlled by the manual depression of buttons on a keyboard.

In the more complex applications the function relating the output signals to the input signals is not fixed but must be alterable in accordance with the details of the problem at hand. Moreover, in these applications the function is not generally specificable in the same sense that a mathematical (possibly Boolean) function is specificable by an equation or a set of equations. Instead, the more usual situation is that a long list of "instructions" (often numbering in the tens of thousands or more) is required to specify the function in a practical manner.

The list of instructions is called a "program." For a given function, the character of the program depends upon the details of the particular instructions the machine is capable of executing. Thus, the design of the machine and the manner in which the information processing function is specified are closely interrelated. The repertory of instructions a given machine is capable of executing forms a "machine language." An information processing function need not be specified in the language of a particular machine, however. Some abstract language specifically designed for the purpose may be used, where the purpose of the abstract language is to provide a notation that is unambiguous and machine-adaptable (qualities not found in conventional spoken and written languages) but which persons can use with a minimum of confusion and drudgery. Of course, before the computer can perform its task, the program must be translated from whatever language in which it was originally written to the machine language of the particular computer to be used—although the translation is itself an information processing function that can be performed by a machine. The translation function employs a program which may, in turn, have been prepared with the aid of abstract languages with subsequent translations.

Thus, the design of a digital machine can involve a fantastically complex interplay of the details of the problems to be solved by the machine, the details of the machine language, and the details of the languages to be used in the preparation of programs to express the problems. The design process is often complicated by the fact that the machine designer has no opportunity to know the details of the problems for which the machine is to be used. In fact, in view of the general purpose nature of many computers, the designer often has no hint at all of even the nature of many of the problems which the machine will subsequently be called upon to solve.

In short, the design of a complete digital machine may well be regarded as more of an art than a science, especially when considering factors such as the relative "ease" by which a given problem may be programmed for different machine designs. Nevertheless, certain fundamental considerations apply to all computers of the stored-program type,

528 THE GENERAL PURPOSE COMPUTER AND ITS PROGRAMMING

and these will be explained by describing the organization of an elementary version of such a machine and the manner by which such a machine would be programmed to solve typical elementary problems. A few elaborations for improving computer "efficiency" (a subject to be discussed in more detail) will be mentioned. Finally, the breadth of applicability of the general purpose stored-program computers will be illustrated by discussing the manner by which they may be employed in selected widely differing applications.

The Organization of an Elementary General Purpose Stored-Program Computer

Figure 9-1 illustrates the major elements of a general purpose stored-program computer. Countless variations have been devised even for the most basic of the stored-program computer features, but rather than attempt an exposition of the variations the approach here will be to describe a rather specific machine organization which illustrates the principles common to all. First of all, the machine will be assumed to be decimal in character with each "word" having six decimal digits. When a word represents a number, a sign bit is assumed to be appended. The sign bit may also be useful when a word is used for other purposes, but in the example machine this elaboration is unnecessary. The digital storage unit, shown by a block in the central part of the figure, is assumed to have a capacity of 10,000 words with each word stored at a separate "address" and with the individual addresses being numbered from 0000 through 9999. A four-digit number placed in the "address register" controls the particular address position to be actuated when transmitting a number to the storage unit for storage or when "sensing" a number that has previously been stored in the storage unit. The numbers shown accompanying the transmission paths in Figure 9-1 indicate the groupings of the digits as they are used in the various parts of the machine. Thus, for example, the 6's to the right and left of the digital storage unit indicate that the digit groups ("words") each contain six digits at the input and output terminals of this unit, whereas the digit 4 at the input line to the address register indicates that an address number as transmitted to and as stored in this register contains four digits.

The instruction register consists of an "operation" part and an "address" part as indicated in Figure 9-1. The instruction register stores a word which represent an "instruction." The six digits of an instruction are divided into two groups: a two-digit group called the "operation part" and a four-digit group called the "address." As the name implies, the operation part controls the operation to be performed by the instruc-

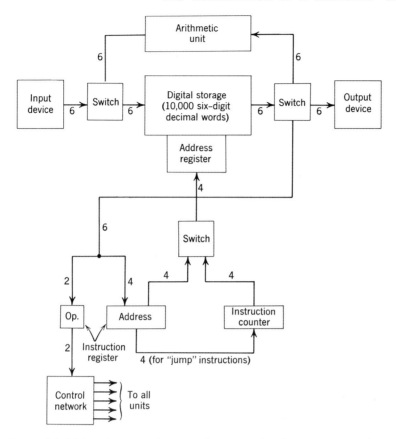

FIGURE 9-1. Major elements of a general purpose stored-program computer.

tion, and with two decimal digits any one of one hundred different operations may be selected. Signals from the operation part of the instruction register are transmitted to a control network which, in turn, generates signals for the appropriate actuation of all parts of the computer. At appropriate times, signals from the address part of the instruction register are transmitted through a switch to the address register in the storage unit.

The "instruction counter" in Figure 9-1 is a four-digit register. An appropriate control signal when transmitted to the instruction counter will cause the number stored there to be increased by 1 (hence the name "counter"), but other control signals may cause the counter contents to be completely replaced by some other number. In particular, in the computer of this example, suitable control equipment will cause the number in the address part of the instruction register to be transmitted to the instruction counter, and the number previously contained in the instruction counter will be lost (although this number may have been otherwise preserved by programming—as will be described later—or by incorporating special equipment for the purpose). The increasing of the counter contents by 1 is done for the routine selection of successive instructions for execution, and the replacing of the contents by some other number is done for the purpose of program "jumping," also to be described later.

The "arithmetic unit" in Figure 9-1 consists primarily of an accumulator and a multiplier-quotient register, all as described at some length in earlier chapters, but it may include a wide assortment of other networks. Numbers transmitted to or from the arithmetic unit may be transmitted to or from any register within that unit, but most commonly the transmission is to or from the accumulator.

The input device in Figure 9-1 can be almost any device capable of supplying input words to the computer as a whole. However, for purposes of this illustration the input device may be visualized as a punched card or punched tape which, when actuated by an appropriate control signal from the computer itself, causes the next successive six-digit input word to be sensed. This word must, of course, be stored at least temporarily somewhere within the computer, and the assumed mode of operation is that the word is transmitted through a switch to the digital storage unit. The address at which it is stored is controlled by the contents of the address register. Similarly, the output device can be almost any device capable of accepting output words and may be visualized as a printer. For output the storage unit is actuated for word recovery, and a control signal to the printer causes the printer to accept this word and print it.

Basic Operation of the Stored-Program Computer

In the operation of the computer as described in the previous section, a suitable sequence of instructions or "program" is first entered into selected addresses of the digital storage unit. The entry of the program can be from the same input device as used for the entry of data words, although some special features (not indicated in Figure 9-1) are required for the purpose. After the entry of the program, the instruction counter is caused to contain the address at which the first instruction of the program is stored. The main functioning of the computer may then begin.

Computer operation proceeds in a series of steps with each step having two parts: an instruction-obtaining part and an instruction-execution part. During the instruction-obtaining part, the contents of the instruction counter are transmitted to the address register, and the word (assumed to be an instruction) stored at the corresponding address is transmitted to the instruction register. During the instruction-execution part of the step, the operation part of the instruction, which is stored in the "operation" part of the instruction register, causes the intended instruction to be executed. For most instruction types the address part of the instruction, which is stored in the "address" part of the instruction register, is transmitted to the address register to control the storage unit location to come into play during the instruction-execution part of the step. However, for some instruction types, the address part of the instruction may serve totally different purposes, in which case the digital storage unit is dormant during this part of the step.

The time duration of the instruction-obtaining part of each step is ordinarily fixed by the physical characteristics of the storage unit. The time is generally equal to one storage unit cycle, where a cycle includes primarily the functions of setting the address register and recovering a word from the indicated address, but the cycle may also include (as in certain magnetic core storage units) the function of storing a new word (or "restoring" the same word) at the same address.

The time duration of the instruction-execution part of each step is not necessarily fixed but may be as long as required to execute the instruction. Of course, for those instructions for which the address part actually pertains to a storage unit address, a minimum time is set by the storage unit cycle time.

The speed of the computer can be roughly doubled by providing two digital storage units, one for instructions and one for data. This elaboration is practical in many applications of computers of more advanced design. However, a need arises to perform at least simple arithmetic operations on the instructions themselves (examples to be given in subsequent sections), and in the relatively elementary stored-program computers contemplated here this need can be satisfied in an economical and flexible manner by storing the instructions and data interchangeably in a single storage unit.

A Further Description of the Control Network

Figure 9-2 illustrates a basic approach to the design of the control network for a stored-program computer. A multivibrator (designated

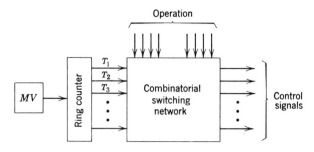

FIGURE 9-2. Control network.

MV) generates a continuous series of pulses which are applied to a ring counter which may consist of a shift register having only one stage in the 1 state at any given time and having the output of the last stage connected to the input of the first stage. The output signals from the ring counter are designated T_1, T_2, and so on, and these signals are cyclically generated. The multivibrator and ring counter, together with various switching elements not shown in the figure, are collectively called the "clock." A given clock output signal may be referred to as the "T_1 clock pulse," the "T_2 clock pulse," and so on.

The "operation" signal lines shown at the top of Figure 9-2 are obtained from the operation part of the instruction register in Figure 9-1. Eight such binary signals are shown, and these represent the two decimal digits of the operation part of an instruction. The operation signals and the clock pulse signals are combined in a combinatorial switching network designed to produce the various control signals as needed throughout the computer. For example, for each instruction type for which the digital storage unit is to be actuated, the corresponding operation signal combinations are caused to be combined with the appropriate clock pulse to generate a correctly time control signal for the storage unit.

The details of the control network are generally dependent on the physical characteristics of the elements of the computer as well as dependent upon the overall organization of the computer. For example, in the physical operation of the storage unit, a timed pulse may be inadequate. Instead, the need may be for a control signal that is 1 over a substantial portion of a clock sycle. Such a signal may be obtained by including a flip-flop within the control network, where the flip-flop is set to 1 by a selected clock pulse and is returned to 0 by a different clock pulse.

Another elaboration in the control network is means for causing the ring counter to skip over those parts of the cycle which are not needed for certain operations. Alternatively, the clock may be interrupted at some intermediate point in the cycle to permit the completion of the functioning of some portion of the computer where that functioning requires a variable and generally unknown amount of time. A completion signal from that portion will cause resumption of the clock operation. Thus, the concepts of synchronous and asynchronous operation (discussed in a previous chapter) may be combined in various ways.

An Illustrative Instruction Repertory

To illustrate the manner in which a general purpose stored-program computer processes digital information, the computer will be assumed capable of executing the following repertory of instructions. The instructions selected for this repertory are typical of those found in most computers, although the repertory is much more limited than would be considered desirable in nearly all practical applications. A few additional computer features with their accompanying additions to the instruction repertory will be described later. Nevertheless, the repertory to be presented is adequate to accomplish virtually any digital information processing task—although undesirably long programs would be required for many kinds of tasks. Each instruction type is given a name as indicated by capital letters, and its abbreviation is given in parentheses. To aid in understanding the following sections, it is suggested that the reader memorize the abbreviations and the computer action taken during the execution of each instruction type.

RESET (RE). The accumulator in the arithmetic unit (see Figure 9-1) is reset to zero. The address part of this instruction is not used. (Because the RESET instruction is most commonly followed in a program by an ADD or a SUBTRACT instruction, a preferable design variation may be to provide RESET AND ADD and RESET AND SUBTRACT instructions in the repertory. However, having a separate RESET instruction seems to add clarity to the elementary example programs to be described. In some computer designs and with some computing tasks the availability of a separate RESET instruction results in a program requiring the execution of fewer instructions. Moreover, the separate RESET instruction illustrates the point that not all instructions need have an address part.)

ADD (AD). The number stored at the storage unit address designated by the address part of the instruction is added to the contents of the accumulator in the arithmetic unit, and the sum remains in the accumulator. The previous contents of the accumulator are lost (unless previously transmitted elsewhere by a STORE instruction—see below).

SUBTRACT (SU). The number stored at the storage unit address designated by the address part of the instruction is subtracted from the contents of the accumulator in the arithmetic unit, and the difference remains in the accumulator. The previous contents of the accumulator are lost (unless previously transmitted elsewhere by a STORE instruction—see below).

MULTIPLY (MU). The number stored at the storage unit address designated by the address part of the instruction is multiplied by the contents of the accumulator, and the product replaces the original contents of the accumulator. (Here, an integral instead of a fractional computer will be assumed, and only the six low-order digits of the product are retained. A product having a nonzero digit in a higher-order position will be regarded as an overflow. The manner in which the multiplier register or other miscellaneous switching devices in the arithmetic unit come into play is not of consequence here.)

STORE (ST). The number in the accumulator is transmitted to the storage unit and stored at the address designated by the address part of the instruction. The number is also retained in the accumulator, but the word previously stored at the designated address is lost.

SHIFT RIGHT (SR). The digits of the word in the accumulator are shifted to the right (to relatively lower-order positions) a number of positions equal to the number in the address part of the instruction. Digits shifted from the lowest-order accumulator positions are lost, and 0's are shifted into the highest-order position. (Address numbers greater than 0005 have the effect of resetting the accumulator to zero, but such address numbers may actually be encountered in some programs. The designer may wish to incorporate computer facilities which will detect large address numbers in SHIFT instructions and cause response as desired for the application.)

SHIFT LEFT (SL). This instruction is the same as the SR instruction except that the shifting is to the left.

JUMP (JU). The computer is caused to take its next instruction from the storage unit address indicated by the address part of this instruction.

JUMP IF NEGATIVE (JN). If the number currently stored in the accumulator is negative, the computer is caused to take its next instruction from the storage unit address indicated by the address part of this instruction in the same manner as for the JU instruction. Otherwise, the next instruction is taken from the next higher numbered address position as in conventional instruction sequencing.

INPUT (IN). The input unit is actuated, and the word received from it is stored in the storage unit at the address designated by the address part of the instruction.

OUTPUT (OU). The word stored in the storage unit at the address designated by the address part of the instruction is transmitted to the output unit. The word in the storage unit is retained ("restored") there.

TERMINATE (TE). The computer terminates its functioning (stops) and generates a signal indicating that is has done so.

In the design of the computer, each instruction type is assigned one of the possible digit combinations representable by the operation part of an instruction word. With two decimal digits being used for the operation part as assumed here, 100 different instruction types may be incorporated in the computer, although only 11 are employed in the illustrative repertory. For example, 04 in the operation part of an instruction may represent MULTIPLY, and a six-digit word such as 040213 would mean "multiply by the number stored at address 0213" when this word is used as an instruction. However, the person preparing the program ordinarily need not be aware of the particular numeric code used for the various instruction types.

As will be illustrated, the use of arithmetic instructions to alter the address parts of other instructions is a highly useful technique. An alteration in the operation part of an instruction produces a vastly different result, because it transforms an instruction of one type to an instruction of a totally different type. If the programmer is aware of the numeric code used for the various instruction types, he can sometimes devise "trick" programs which advantageously employ the computer's capability of transforming instruction types, but the usefulness of this feature is not general and is dependent on the minute design details of a particular machine. Nevertheless, an example of using arithmetic operations to modify the operation parts of instructions will be given—see pages 542 and 543.

An Example Which Illustrates Basic Programming

To illustrate basic programming, assume that the numbers x, y, and z are available at the input unit in that sequence. The problem is to form $xy + z^2$. A program which performs this computation and transmits the result to the output device is as follows. Each instruction word and each input, intermediate, and final number is stored at the address indicated on page 536.

As indicated, the program is stored at addresses 0000 through 0013 inclusive. Input data are stored at addresses 0014, 0015, and 0016. An

Address	Word
0000	IN0014 ⎫
0001	IN0015 ⎬ enters data
0002	IN0016 ⎭
0003	RE ⎫
0004	AD0014 ⎪
0005	MU0015 ⎬ forms xy and stores it
0006	ST0017 ⎭
0007	RE ⎫
0008	AD0016 ⎬ forms z^2
0009	MU0016 ⎭
0010	AD0017 ⎫ forms $xy + z^2$ and stores it
0011	ST0018 ⎭
0012	OU0018 ⎫ transmits result to output device and stops
0013	TE ⎭
0014	x
0015	y
0016	z
0017	xy
0018	$xy + z^2$

intermediate result is stored at address 0017, and the final result is stored at address 0018. With the notations accompanying the various instructions, the functioning of the program is reasonably self-evident. One point that might be mentioned is that the addresses available for data storage are not necessarily evident prior to writing the program. The programmer may equally well store the data at some remote portion of the storage unit, say at addresses starting with address number 9000, but for problems where the total number of instructions and data approaches the capacity of the storage unit, the programmer must generally prepare the program with symbols for addresses and then, as a final step in preparing the program, substitute available address numbers for the symbols. This final step, although lengthy and tedious in complex programs, is routine and can itself be reduced to a computer program specifically intended for the purpose.

Countless variations and improvements in the above program can be made, especially if certain elaborations can be incorporated in the physical design of the computer. One variation requiring no physical changes results in only two, instead of five, addresses being required for data storage. During the first part of the program only two numbers, x and y, are entered into the computer. After determining xy this intermediate result may be stored at the address originally used for the storage of x. Then z is entered and stored at the address originally used for y. Then, after computing z^2 the final result $xy + z^2$ may be stored at the address used for xy.

Flow Charts and the JUMP Instructions

For most problems to be solved by computers effective programs cannot be prepared in the form of a list of instructions to be executed in a simple sequence as in the previous example. Instead, the program must be such that the instruction sequence is selected in accordance with the nature of intermediate results, which are not known at the time of preparing the program. As an example, assume that three unequal but otherwise unknown numbers x, y, and z are stored at addresses 0143, 0144, and 0145, respectively, and that the problem is to transmit these numbers in numeric sequence to the output device. In other words, the computer is to sort the numbers.

A straightforward approach (but not necessarily the "best" approach) to this problem is, first, to compare x and y. If y is smaller, the next step is to compare y and z. If z is the smaller of these two, the correct numeric sequence is then known to be z-y-x. If on the second comparison y is found to be the smaller, the smallest of the three is then known to be y, and y may then be transmitted to the output unit. However, a comparison of x and z is then needed to determine the second number in the correct sequence. If on the first comparison x was found to be the smaller, a different but analogous sequence of operations is followed to determine the correct sequence.

The various steps are illustrated by means of a diagram called a "flow chart" in Figure 9-3. The first comparison is expressed in the form of a question: "Is $x > y$?" The next step to be taken depends on whether the answer to this question is "yes" or "no." The pattern of steps, each represented by a block in the figure, follows the pattern described in the previous paragraph. Observe that not all steps are different from each other. For example, the "Is $x > z$?" step appears at two places in the figure. With elementary stored-program computers and with elementary programming techniques, instructions for performing this step must correspondingly be duplicated in the program, although with the subprogramming technique to be described later, a single sequence of instructions can be utilized to perform a given function regardless of the number of points in a program that sequence is to be executed. In Figure 9-3 some of the OUTPUT steps appear more than once, but in other instances the diagram illustrates that the program can arrive

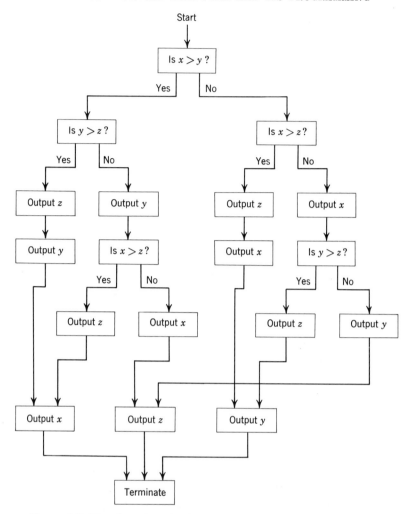

FIGURE 9-3. Flowchart for selecting x, y, and z in numerical sequence.

at a given step from two or more other steps. This latter feature is possible when the next succeeding step to be taken (TERMINATE in this example) is independent of the previous step.

The preparation of a flow chart prior to the preparation of a program is not exactly essential, but without a flow chart the writing of error-free programs for seemingly simple problems can be vexingly difficult, whereas relatively complex problems can often be accurately programmed with ease after the preparation of a flow chart.

PROGRAM LOOPS AND INSTRUCTION MODIFICATION   539

The program for sorting and printing x, y, and z as developed from the flow chart of Figure 9-3 is on page 540, where the first instruction is assumed to be stored at address 0100. The determination of whether or not y is smaller than x, for example, is accomplished by subtracting y from x. If y is the larger, the difference will be negative, and the JUMP IF NEGATIVE (JN) instruction may be used to jump to the instruction to be executed in the event of a "no" answer. However, if the answer is "yes," the difference will be positive and no jump will take place with this instruction. Thus, the instruction to be executed in the event of a "yes" answer may be placed immediately following the JN instruction. (Of course, the roles of "yes" and "no" can be reversed by interchanging the operands in the subtraction operation.)

In the program the dashed lines indicate the jumps that are made conditionally in the event of a "no" answer. The solid lines show jumps that are made unconditionally. In this program each unconditioned JUMP (JU) instruction could be eliminated and replaced by an added OUTPUT (OU) and an added TERMINATE (TE) instruction. Although the total number of instructions in the program would be increased by three, the number of instructions executed during any particular "run" of the program would be decreased by one—a situation which illustrates that writing a program is often a compromise between the length (storage requirements) of the program and the speed of computation.

If the sorting program is a part of some larger program, each TE instruction would be replaced by a JU instruction having as its address the location of the next instruction to be executed in the larger program.

Program Loops and Instruction Modification

In numerous problems where essentially the same function is to be performed many times, the number instructions required for the program can be reduced greatly by preparing the program or some part of it in the form of a "loop," that is, a sequence of instructions to be executed repeatedly as many times as necessary. The function to be performed may be a single-instruction function such as represented by the INPUT or the ADD instructions (with the latter being used to add a series of numbers as with an adding machine) or the function may itself be complex and require many instructions.

Usually, when proceeding through the loop repeatedly, the address parts of one or more instructions in the loop must be modified so that new data enter into the computations each time that the function is performed. Because instructions and data are interchangeable and indis-

Address	Word
0100	RE ⎫
0101	AD0139 ⎬ Is $x > y$?
0102	SU0140 ⎭
0103	JN0122
0104	RE ⎫
0105	AD0140 ⎬ Is $y > z$?
0106	SU0141 ⎭
0107	JN0112
0108	OU0141
0109	OU0140
0110	OU0139
0111	TE
0112	OU0140
0113	RE ⎫
0114	AD0139 ⎬ Is $x > z$?
0115	SU0141 ⎭
0116	JN0119
0117	OU0141
0118	JU0110
0119	OU0139
0120	OU0141
0121	TE
0122	RE ⎫
0123	AD0139 ⎬ Is $x > z$?
0124	SU0141 ⎭
0125	JN0130
0126	OU0141
0127	OU0139
0128	OU0140
0129	TE
0130	OU0139
0131	RE ⎫
0132	AD0140 ⎬ Is $y > z$?
0133	SU0141 ⎭
0134	JN0137
0135	OU0141
0136	JU0128
0137	OU0140
0138	JU0120
0139	x
0140	y
0141	z

PROGRAM LOOPS AND INSTRUCTION MODIFICATION 541

tinguishable in the storage unit of a general purpose stored-program computer, the arithmetic unit used for conventional mathematical operations may be used to alter the instructions as needed for each traversal of the loop.

The above points will be illustrated by the example of computing and transmitting to the output unit the sum of the squares of 50 numbers stored at consecutively numbered addresses 0200 through 0249. The address, 0200, of the first number and the number 50, of the numbers are treated as data and are stored as 000200 and 000050 at addresses 0121 and 0122, respectively, the latter of which acts as a "counter" to control the number of traversals of the loop. The number 1 is stored as 000001 at address 0123, and the sum of the squares is stored, as this sum is accumulated, at address 0124. In the program, each address or number which changes in the course of executing the program is indicated by brackets, with the indicated values being only the initial values. The first instruction is arbitrarily placed at address 0100.

Address	Word	
0100	RE	
0101	AD0122	The "counter" is reduced by 1 and sensed for a negative condition
0102	SU0123	
0103	ST0122	
0104	JN0119	
0105	RE	
0106	AD[0200]	x_i^2 is computed and added to the accumulated sum
0107	MU[0200]	
0108	AD0124	
0109	ST0124	
0110	RE	
0111	AD0106	
0112	AD0123	
0113	ST0106	The address parts of the instructions stored at 0106 and 0107 are increased by 1
0114	RE	
0115	AD0107	
0116	AD0123	
0117	ST0107	
0118	JU0100	
0119	OU0124	
0120	TE	
0121	000200	
0122	[000050]	the "counter"
0123	000001	
0124	[Σx_i^2]	

If the program as a whole is to be executed only once, the initial values of the various changeable quantities may be supplied in the same manner as for the other parts of the program. However, if the program is to be used more than once, program means for setting the initial values are desired. The details of the necessary program depend upon the application, but the following instructions, to be executed prior to the program above, illustrate some aspects of the problem and also illustrate programming techniques of more general application.

Address	Word	
0087	RE	} $x_i{}^2$ is set to zero
0088	ST0124	
0089	AD0106	} The address of the instruction at 0106 is set to 0200
0090	SR0004	
0091	SL0004	
0092	AD0121	
0093	ST0106	
0094	RE	} The address of the instruction at 0107 is set to 0200
0095	AD0107	
0096	SR0004	
0097	SL0004	
0098	AD0121	
0099	ST0107	

Observe that after a RESET (RE) instruction any number of storage addresses may be reset to zero by using one instruction per address. The zero contents of the accumulator are merely stored in the desired addresses by the STORE (ST) instruction.

The above program is an example of an opportunity to save instructions by modifying the operation part of an instruction. In particular, assume that the digits in the operation part of ADD (AD) and MULTIPLY (MU) instructions are 02 and 04, respectively. The number 020000 is stored at some available address, say 0138. The five instructions at addresses 0094 through 0098 may then be replaced by the single instruction AD0138. The addition of 020000 to an AD instruction transforms it to a MU instruction having the same address. Thus, after executing the instruction at address 0093 in the aborev listing, the accumulator still contains digits which represent the instruction AD0200, that is, 020200. The execution of the instruction AD0138 transforms the accumulator contents to 040200 which has the meaning of MU0200 and which may then be stored by the instruction at 0099. (Actually, the program should be rewritten to close the "gap" left by the elimination of four instructions.)

A greater reduction in the number of instructions required for the above program can be achieved by incorporating a STORE ADDRESS (SA) instruction in the computer. As the name implies, this added instruction type causes only the four low-order (address) digits in the accumulator to be transmitted to the corresponding positions in the address designated by the address part of the instruction. The two high-order digits, which are used to represent the operation, are not altered either in the accumulator or at the designated address. With the SA instruction available, the above program can be reduced to the following five instructions (after relocating the otherwise unchanged instructions).

Address	Word	
0095	RE	$\left.\begin{matrix} \\ \end{matrix}\right\} x_i^2$ is set to zero
0096	ST0124	
0097	AD0121	The number 000200 is placed in the accumulator
0098	SA0106	The addresses of the instructions at 0106 and
0099	SA0107	0107 are set to 0200

Subprograms

For many problems, numerous instances arise where a given sequence of instructions is to be executed at more than one point in the program, but not in an orderly repetitive manner as in the example of a program loop. Even though modifications may be required in the given sequence each time it is executed, the total number of instructions can often be greatly reduced by treating the given sequence as a "subprogram." The subprograms may be stored at any available portion of the storage unit, and the main program may be designed so that a jump to the subprogram is made each time the given sequence is to be executed. After completing that sequence, a jump back to the main program is made.

The program (which contains a loop) described in the previous section may itself be used as a subprogram. To illustrate some of the subprogramming features better, the function will be generalized from the addition of the squares of 50 numbers with the first stored at address 0200 to the addition of the squares of N numbers with the first stored at address M. In the example below, the assumption will be that the computer has arrived at the instruction stored at address 0410 and that the next function to be performed is the addition of the squares of N numbers with the first stored at address M, where the actual magnitudes of N and M are not known. In fact, even $L(N)$ and $L(M)$ may be unknown to the programmer, where a notation such as $L(X)$ means the address or "location" where X is stored. However, prior to the execution of the instruction at 0410, the actual numbers representing $L(N)$ and $L(M)$ must be inserted (possibly by programming) in the address parts of the instructions at addresses 0411 and 0414, respectively, in the program below.

Address	Word	
0100	RE	⎫
.	.	
.	.	Same as on page 541
.	.	
0119	OU0124	⎭
0120	[JU0421]	The "link"—see text
0121	[M]	
0122	[N]	
0123	000001	
0124	[Σx_i^2]	
0410	RE	⎫
0411	AD $L(N)$	
0412	ST0122	N and M are placed where the subpro-
0413	RE	gram can find them
0414	AD $L(M)$	
0415	ST0121	⎭
0416	RE	⎫
0417	AD0420	The "link" is prepared
0418	ST0120	⎭
0419	JU0100	} Jump to subprogram
0420	JU0421	} The computer never arrives at this step
0421	.	⎫
	.	Continuation of main program
	.	⎭

Prior to jumping to the subprogram, the main program places N and M at addresses 0121 and 0122, respectively, because the subprogram has been so written as to use these two addresses for the purpose. Alternatively, the main program could be prepared so that the address parts of the particular subprogram instructions which make use of N and M (i.e., the instructions at addresses 0101 and 0103 for N and addresses 0106 and 0107 for M—see page 541) could be set to $L(N)$ and the address parts of the instructions at addresses 0106 and 0107 could be set to M. The STORE ADDRESS (SA) instruction, if built into the computer at hand, would be helpful for this purpose, as explained previously.

Also prior to jumping to the subprogram, the main program prepares a jump instruction, commonly called the "link," in the subprogram so that the subprogram will cause a jump back to the main program upon completion of the function to be performed by the subprogram. In the above example, this jump instruction is stored at address 0420, and the instructions at addresses 0416, 0417, and 0418 serve to transmit this instruction to address 0120. This instruction, JU0421, is executed when the computer arrives at 0120 upon completion of the subprogram function. The computer never executes the instruction at address 0420 except in the indirect manner just indicated.

The scheme which has been described in this section for entering subprograms and returning from them is basic and quite practical. On the other hand, in most applications an overall improvement in computer performance can be realized by installing elaborations which will allow jumping to and from subprograms at greater speed as a result of reducing the number of instructions that must be executed. One such elaboration is a special register which "remembers" the address of the main program instruction which caused the jump to the subprogram. The instructions SUBPROGRAM JUMP and RETURN JUMP are added to the instruction repertory. Upon execution of the SUBPROGRAM JUMP instruction, the current contents of the instruction counter are increased by 1 and transmitted to the special register. Upon execution of the RETURN JUMP instruction (which should appear only in the subprogram—not in the main program), the contents of the special register are returned to the instruction counter to control the address from which the next main program instruction is taken. Multiple levels of subprogramming can be handled if the added equipment includes a corresponding number of special registers operated in a "last-in-first-out" (LIFO) basis. A SUBPROGRAM JUMP instruction to be executed at a time when all special registers are filled would constitute an attempt to employ more levels of subprogramming than the special facilities are capable of handling, and a programming error would be implied. Further equipment elaborations which will detect and indicate such errors are desirable.

Apart from suggesting specific means for handling subprograms, the above discussion should be construed as an illustration of the fact that the nature of a computer program is strongly dependent upon the physical design of the computer. Even with computers limited to the general purpose stored-program variety, the design variations can be so great that the program to solve a given problem on one computer may bear little resemblance to the program to solve the same problem on a different computer.

Library of Subprograms

Often the need to perform a given function will be encountered in many different problems. Much programming effort can be saved if the function is represented in the form of a subprogram which may be retained in a "library" of subprograms. However, as the main program for each problem is prepared, the addresses available for the storage of a given subprogram may change from one problem to the next. When using a subprogram from a library, the address parts of those instructions which refer to other parts of the same subprogram must be adjusted in accordance with the particular portion of the storage unit in which the subprogram is to be stored.

This readjustment of addresses is a routine task, but it can be tedious and a serious source of errors if performed manually. Because the task is routine, a computer program can be prepared to perform it. Many approaches can be taken, but basically the affected addresses are recorded˙ (as the subroutine is retained in the library) with reference to the address at which the first instruction of the subprogram is to be stored. For example, when the program appearing on page 541 is used as a subprogram the instruction shown as being stored at address 0101 would be stored in the library as AD0022 instead of AD0122. Then, when the subprogram is used, it assigned to a particular portion of the storage unit, and the address at which the first instruction is stored (address 0100 in the example) is added to the address recorded for the instruction in the library.

Some means must be included to distinguish those subprogram instructions which have addresses to be modified from those instructions (e.g., SHIFT instructions) which have fixed address parts. One straightforward scheme is to accompany each subprogram, as it is retained in the library, with a list of those instructions which have addresses that must be modified, where each instruction is identified by the address at which it would appear if the first instruction in the subprogram were to be stored at address 0000.

For maximum effectiveness of the subprogram library, each subprogram should be accompanied by a detailed description of what the subprogram does. Other information such as the expected accuracy of the results and the time required for the subprogram to accomplish its function (both of which may be dependent on the magnitudes of various parameters supplied to the subprogram) is also virtually essential in some applications. In fact, this so-called "documentation" of all programs, not just subprograms, is extremely important in all cases where a given program is to be used by persons other than the one who prepared the program—or even by the same person at a later time when he may have forgotten the details of what the program accomplishes.

Interpretive Programs

Interpretive programs constitute an important extension of the subprogramming concept. However, instead of preparing the main program to cause jumps to the subprograms, the interpretive program "interprets" each instruction in the main program and causes the appropriate jump. The computer does not directly execute the instructions as they appear in the interpreted program. The instruction types to be interpreted can be similar to, or even identical to, the instruction types built into the computer as a part of its repertory, or they may be totally different.

One important application of interpretive programs is the simulation of one computer by another. Such simulation is particularly useful during the design of a new computer when programs for it are to be tested. An existing computer may be used to run the programs prior to the completion of the new computer. Also, for one reason or another, programs for a given computer may have been prepared under circumstances where some users of the programs may have a different kind of computer available to them.

Interpretive programs exist in countless forms with many of them bearing little resemblance to each other. However, the basic principles of interpretive programming can be appreciated by studying the example to be presented. The heart of the interpretive program consists of the words to be stored at addresses 0000 through 0013, but the interpretive program also includes a "jump table" and the interpreting subprograms. The jump table in the example consists of 100 jump instructions, one for each of the instruction types in the interpretive program repertory. The jump table is stored at addresses 0014 through 0113. The interpreting subprograms, although a part of the interpretive program, are not fixed but may be designed in accordance with the desired repertory of instruction types to be interpreted. The subprograms may be stored at any conveniently available addresses, but in the example, addresses 0114 through 1999 are reserved for this purpose. (This number of addresses may be less than would be required for the storage of 100 usefully different subprograms, but if so, the number of addresses reserved for this purpose may be increased as required—at the expense of the storage space available for the program to be interpreted and its accompanying data.)

The program to be interpreted and its accompanying data may then be stored at addresses 2000 through 9999. Actually, in most applications at least a few addresses are needed for special purposes and are likewise unavailable to the programmer. In the example, an imaginary accumulator will be needed, and address 9999 will be used for this purpose. When studying the interpretive program, observe that the computer directly executes only those instructions that are in the interpretive program and never directly executes the instructions to be interpreted. In other words, the real instruction counter (as it appears in Figure 9-1) never contains a number greater than 1999. The imaginary instruction counter, identified by A in the address part of the instruction stored at address 0011, can contain only numbers that lie in the range 2000 through 9998 (in the absence of a programming error).

The means for initially entering the interpretive program and the interpreted program with its data will be omitted from the discussion, but the assumption may be made that the computer first executes the instruction at 0000 with the first interpreted instruction being at address 2000. Therefore, A in the address part of the instruction at 0011 must be set to 2000 initially. The example interpretive program is then as shown on page 549.

For convenience, the operation part of the word at 0011 has the digits corresponding to the instruction type ADD (AD). Therefore, this word may be visualized as an instruction although its major function is to "remember" the particular instruction to be executed next in the interpreted program, and the term "imaginary instruction counter" is derived from this function.

The first part of the interpretive program (the instructions at addresses 0000 through 0004) places the contents of the imaginary instruction counter, including the operation part, at address 0006 where it will actually be executed as an instruction. Also, these first instructions increase A by 1 so that when the computer arrives at address 0006 for its next instruction the contents of the address part of the word at address 0011

Address	Word	
0000	RE	⎫ Places the contents of the imaginary instruction counter in accumulator and at 0006
0001	AD0011	
0002	ST0006	⎭
0003	AD0012	⎫ Increases the contents of the imaginary instruction counter by 1
0004	ST0011	⎭
0005	RE	⎫ Places the operation part of the instruction to be interpreted in the two low-order accumulator positions
0006	AD[A]	
0007	SR0004	⎭
0008	AD0013	⎫ Prepares for jump to jump table
0009	ST0010	⎭
0010	JU[ ]	} Jumps to jump table
0011	AD[A]	} The imaginary instruction counter
0012	000001	⎫ Auxiliary words
0013	JU0014	⎭
0014	JU I_1	
0015	JU I_2	
.	.	⎫ Jump table
.	.	
.	.	
0113	JU I_{100}	⎭

To subprograms

will be $A + 1$ in preparation for the interpretation of the next succeeding interpreted instruction, although the interpretation and execution (by means of a subprogram) of the first interpreted instruction has not yet commenced.

When the computer executes the instruction at address 0006 the instruction to be interpreted, which is at address A, is placed in the accumulator. The form of this instruction may be the same as the form of a conventional instruction in that the two left-hand digits represent the operation or instruction type, and the four right-hand digits represent an address. However, the specific meaning of the operation digits need not bear any resemblance to the meaning of the operation digits in a conventional instruction. Instead, each digit combination refers to a specific subprogram, where each subprogram may be designed as desired for the application at hand. In this illustration of interpretive programming the digits $67xxxx$ will be assumed to mean a SUBTRACT instruction essentially the same as a built-in SUBTRACT instruction. The four digits indicated by x's represent a conventional address, and the address must normally be in the range 2000 through 9999, although various programming "tricks" might be possible by allowing the address to refer to one of the words in the interpretive program. (As will become apparent later, the use of address 9999 in this instruction will have the effect of subtracting the contents of the imaginary accumulator from itself.)

The instruction at 0007 in the interpretive program shifts the operation digits to the two right-hand positions so that these digits become, in effect, an address. The instruction at 0008 transforms the accumulator contents to a JUMP instruction and it also increments the address part so that one of the jump instructions in the jump table is indicated. The STORE (ST) instruction at 0009 stores the resulting JUMP instruction at 0010 where it is executed on the next step. If the interpreted instruction is a SUBTRACT instruction, the jump is to address $0067 + 0014 = 0081$ where the instruction JUMP I_{67} is stored. I_{67} is the address at which the first instruction of the subprogram for SUBTRACT is stored, and this address will be assumed to be 0340. A jump is then made to 0340.

The following subprogram interprets and executes the SUBTRACT instruction whenever such an instruction is encountered at an address A. The real accumulator is used in executing the instructions of the subprogram, but the interpreted SUBTRACT operation is with respect to the imaginary accumulator at address 9999. That is, the interpreted

Address	Word	
0340	RE	} The computer arrives here from the jump table
0341	AD0006	
0342	ST0344	The operand is located; after completing these instructions the accumulator contains 674082—see text
0343	RE	
0344	AD [A]	
0345	SL0002	} The accumulator contents are changed to 004082
0346	SR0002	
0347	AD0354	The accumulator contents are changed to SU4082 and positioned for execution
0348	ST0351	
0349	RE	
0350	AD9999	The subtraction called for by the interpreted instruction is performed
0351	SU4082	
0352	ST9999	
0353	JU0000	} The computer jumps back to the interpretive program
0354	SU0000	} Auxiliary word

SUBTRACT operation causes the number stored at the address designated by the address part of the instruction stored at A to be subtracted from the number at 9999, and the difference is left at 9999.

In the above example the SUBTRACT instruction at address A is assumed to have 4082 as an address. Recall that the interpretive program had stored this instruction at address 0006 where the subprogram can find it. Other aspects of the subprogram are indicated by the notes accompanying the various instructions.

Although the example chosen to illustrate subprograms as used in an interpretive program consists of an instruction which is essentially the same as one of the built-in instructions, the power and usefulness of interpretive programming arises from the possibility of designing the subprogram to perform instructions of a totally different nature. In particular, each interpreted instruction may have more than one address part, with the additional address parts being stored at consecutively higher numbered addresses with respect to the location of the operation part of a given instruction. However, the interpretive program must then be modified to increase A by 2, 3, or more as appropriate instead of by 1 as previously indicated. Then, for example, one instruction type to be interpreted might be SUBTRACT X, Y, Z with the meaning that the number at address X is to be subtracted from the number at address Y with the difference being stored at address Z. Each of the addresses X, Y, and Z could lie in the range 2000 to 9999 but should not ordinarily be addresses at which any part of the interpretive program is stored.

The interpreted instructions may also be of the JUMP form, although the jumping is with respect to the interpreted program, and the jump is accomplished by modifying A (by means of a subprogram) as A is stored in the address part of the word at address 0011, the imaginary instruction counter.

When using an interpretive program, the programmer need not have any knowledge of the manner in which the interpretive program functions. In fact, he need not even be aware that the computer is using an interpretive program to solve his problem. Instead, the person who has prepared the interpretive program may supply the eventual computer user with an instruction repertory in essentially the same manner that the designer of the physical computer supplies the user with an instruction repertory. In either case, when preparing a program, the user works with the repertory supplied regardless of the procedure employed within the computer to execute the various instructions.

Apart from the applications indicated at the beginning of this section, interpretive programs are useful for simplifying programming. The general manner by which simplification is accomplished is to furnish the programmer with instruction types which are extremely powerful. One instruction in an interpreted program may take the place of hundreds, thousands, or even more instructions in a conventional "machine language" program. One simple example is an interpreted INPUT instruction that will cause N words, not just one word, to be entered from the input device to the computer. For another example, the extraction of the square root may be accomplished by means of a single interpreted SQUARE ROOT instruction, whereas the programmer would otherwise need to bother with finding a suitable subprogram in a library or actually work out the details of the square root extraction process in the absence of such a library. Matrix inversion, Bessel function computations, sorting, or countless other operations might similarly be performed by means of a single instruction with an interpretive program.

A serious disadvantage of interpretive programming is in the time required to complete the computations. This point is illustrated by the basic interpreted SUBTRACT instruction where nearly thirty machine instructions must be executed in the various parts of the interpretive program to produce the same net effect as a single SUBTRACT (SU) machine instruction. The speed disadvantage is not quite so great with complex interpreted instruction types where the time spent in the interpreting process is proportionately less. Also, the speed disadvantage can be further lessened by employing a computer of more advanced design than the elementary computer described in this chapter. Nevertheless, in some practically encountered examples the computation time may still be 10 or so times as great with interpretive programming in comparison with machine language programming.

Program Tracing and the Detection of Program Errors

Although the bulk of the interest in interpretive programs is with respect to providing the programmer with more powerful (or at least different) instructions than are available with the built-in instruction repertory of some computer at hand, one important application is found where the interpred instruction repertory is essentially the same as the built-in repertory. This application is the so-called "tracing" program.

Programs often contain errors, especially when first prepared. Using only the final output data to deduce the particular one or more erroneous instructions in a long program is often utterly inadequate. The computer can be equipped with means for causing it to procede through one instruction at a time, and lights and other indicators may be used to indicate the contents of various registers after the execution of each instruction. This technique is in fact commonly used to obtain the information necessary to locate program errors, but the technique is objectionable in that during the time the computer is being used for this purpose it is not accomplishing its intended purpose.

The objective of determining the response of the computer to each individual instruction can be achieved by running the program on an interpretive basis. The interpretive program is supplied with subprograms that duplicate the operations of the built-in instructions but with the elaboration that the contents of selected registers (notably the imaginary instruction counter and the imaginary accumulator in the example previously presented) are printed after the execution of each instruction. To make the printed information more useful, the interpretive program can be refined in many ways, such as by designing it to print the operation and address parts of each instruction as it is executed. Alternatively, each subprogram can be refined to cause the printing of only that information which is likely to be useful for locating errors and to suppress the printing of similar information which will only clutter the report. For example, the contents of the accumulator need be printed only after those instructions which cause the contents to change.

Interpretive programs of this form have been called "tracing" programs because their use is to "trace" the action of the computer in proceeding through a program.

If a problem has been programmed in terms of the instruction repertory of some given interpretive program which has been prepared for other reasons, the tracing requires a different interpretive program which performs all the functions of the given interpretive program plus the tracing functions (but two "levels" of interpretation are not required).

If an error is suspected in the given interpretive program itself, a complication arises. A second interpretive program, the tracing program, may be adequate to provide information that will allow the programmer to deduce the source of the error. However, while the interpretive program is operating under the conditions of being traced, a practical requirement may be to supply the given interpretive program with a sample program to be interpreted. In such a case, the computer would, in effect, be proceeding under two "levels" of interpretation in that the interpretive tracing program would be interpreting the instructions of an interpretive program that is in the process of interpreting some program prepared to solve a problem. Needless to say, the finding of errors can be extremely difficult in complex programs. Although the above discussion does not begin to exhaust the programming aids that may be available, some programming errors may be virtually undetectable (not to mention correctable) for long periods of time, because the input data may be such that certain parts of the program come into play only rarely—in some actual instances not until years after the program is first put to use.

Housekeeping Instructions and Their Elimination

As is apparent from the example programs which have been presented, many of the instructions in a program do not perform any of the information processing functions required by the original problem to be solved by the computer. Instead, up to 90% and even more of the instructions perform only "housekeeping" tasks such as moving words from one place to another, modifying instructions, jumping, and controlling the progress of computations.

An obviously desirable objective is to minimize the number of housekeeping instructions that must be executed in the course of performing the main data processing problem. Of course, a part of the objective is to accomplish the minimization without abandoning program loops, subprograms, interpretive programs, and other programming concepts that are desirable for other reasons, which have been explained. Note that there is a difference between minimizing the number of instructions in a program as written and minimizing the number of instructions that must be executed when executing the program as a whole, because in the course of executing the program the computer may be called upon to execute some individual instructions many times.

General cleverness on the part of the programmer is a useful attribute in minimizing housekeeping instructions. Except for the simplest of problems, no one sequence of instructions is ever apparent for producing the required output information. Instead, for a given machine design and a given reasonably complex problem the program as might be written by one programmer is likely to be vastly different from a program as written by another programmer for the same problem and same machine. Not only would the programs be different on an instruction-by-instruction basis but the total number of instructions and the required computing time may likewise be greatly different. To a great extent, reducing the number of instructions or reducing the computing time may be achieved at the expense of the other, but a clever programmer can often achieve both objectives simultaneously to a far greater degree than can a programmer of lesser talents. Moreover, the difference is not usually just a matter of a few percent but may commonly be expected to be a factor or two or even much more.

For specific computer designs, many of the commonly encountered functions have been the subject of much thought and study with the resulting "best" programming techniques made available, usually through the instruction manuals prepared by the computer manufacturers. Also, many books on programming have been written in which clever and otherwise advantageous programming techniques are described for various problem types. However, again the techniques are generally applicable only to specific computer designs which are often hypothetical designs specifically prepared to aid in the teaching of programming principles—much in the same vein as the elementary computer described at the beginning of this chapter.

The above discussion is intended to indicate that an extended presentation of techniques for obtaining efficient (relatively few instructions and a relatively short computing time) programs would be of value for the specific computer design assumed but would be an elusive subject when considering the general purpose stored-program computer concept in a more general way.

Built-in Equipment, Notably Index Registers, for Eliminating Housekeeping Instructions

Although a general purpose computer with an instruction repertory as described earlier in this chapter is capable, in principle, of handling virtually any digital information processing problem, the storage capacity required for the program and the time required to execute the program would be undesirably great and perhaps totally impractical in many applications, even after all imaginable ingenuity had been applied to the programming task. The difficulty lies in the large number of housekeeping instructions which remain in the best of programs. However, many of the housekeeping instructions can be eliminated by incorporating various elaborations in the physical design of the computer, but then, of course, the nature of the program required to solve a given program depends on the particular elaborations which have been included.

Two such elaborations, the STORE ADDRESS and SUBPROGRAM JUMP instructions have already been mentioned along with the manners by which these added instruction types can reduce the number instructions required in a program.

Added registers of the form commonly called "index registers" constitute another elaboration that, with many problems, can provide a means for greatly reducing the number of instructions required in a program, although the number of different instruction types required in the repertory is increased. The function of an index register is to store a number which will automatically be added to the address part of an instruction to generate the actual address to come into play during the execution of that instruction. One simple but useful version of index register operation will be described as an example.

One index register will be assumed. Three instructions are added to the repertory, and modified versions of various other instructions are provided. One added instruction may be called STORE IN INDEX REGISTER, and when this instruction is executed the number currently stored at the address designated by address part of the instruction is transmitted to the index register. The second added instruction may be called INCREMENT INDEX REGISTER, and this instruction causes the number (often -1) at the designated address to be added to the current contents of the index register with the sum being retained in the index register. The third added instruction may be called JUMP IF INDEX REGISTER CONTENTS ARE NEGATIVE and this instruction causes the next instruction to be taken from the address indicated by the address part of the instruction if the contents of index register are negative; otherwise, the next instruction is taken from the next higher numbered address in the usual manner.

The modified instructions include ADD, SUBTRACT, MULTIPLY, STORE, INPUT, and OUTPUT, and these instructions are modified only in that, prior to execution, the address part is incremented by the number currently stored in the index register. The unmodified ADD, and so on, instructions are not necessary in the repertory, because at any point in the program the contents of the index register can be set to zero by using the STORE IN INDEX REGISTER instruction where the address part designates an address known to contain zero. However, such an instruction would be a housekeeping instruction that could be eliminated by providing both the modified and unmodified instruction types in the repertory.

Program loops represent a commonly encountered example where an index register would be useful. First, the number representing the number of times the loop is to be traversed is sent to the index register. Then prior to the traversal of the loop, the contents of the index register are reduced by 1 by means of the INCREMENT INDEX REGISTER instruction where the address part designates an address at which -1 is stored. The JUMP IF INDEX REGISTER CONTENTS ARE NEGATIVE instruction is used to cause traversal of the loop except when the index register contents have become negative, at which point a jump away from the loop is made. Thus, after initially setting the index register, only two housekeeping instructions per loop traversal are required at the beginning of each loop (plus a jump instruction at the end of the loop to jump back to the beginning of the loop).

The two housekeeping instructions per loop traversal can be reduced to one by incorporating an added instruction type called REDUCE INDEX REGISTER CONTENTS BY 1 AND JUMP IF NEGATIVE. This instruction type combines the functions of the second and third added instructions described above, and it may be incorporated instead of those two instructions. Although it would reduce both the number of instructions appearing in a program as written and the number of instructions executed by the machine for many commonly encountered problems, this instruction is more specialized and would be less effective in certain other programming situations.

The above example further illustrates that any "best" instruction repertory is nebulous and depends not only on the details of the problem to be solved but also on the details of the general approach which may have been taken to solve that problem—not to mention economic considerations, programming ease, and other ill-defined factors.

If the computer is equipped with two or more index registers, some means must be included to designate, for each instruction, which register is to come into play. One practical scheme is to expand the format of the instructions to include an index register designator in addition to the operation and address parts described previously. The number of digits required for this purpose depends on the number of index registers, and in a decimal machine any one of ten different registers can be designated by a single digit. A common practice is to include one less index register than can be designated by the number of digits assigned for the purpose. Then, for those instructions for which indexing is not needed, the nonexistent register is specified by the index designator part of the instruction.

To achieve highest possible speed of operation, the index registers must be composed of storage elements, such as flip-flops, which are not a part of the main storage unit. However, addresses within the storage unit may function as index registers, in which case the functioning is achieved by control network elaborations designed so as to produce the operations described earlier in this section. This design approach is not advantageous from a speed standpoint because more than one cycle of storage unit operation is required to execute most instruction types. However, a very large number of index registers can then be included at very little incremental cost. In fact, in the extreme, the address of any instruction can be incremented by the number currently stored at any address within the main storage unit, in which case the index register designator part of the instruction becomes an address having the same number of digits as the regular address part of the instruction.

An alternative way to use more than one index register (but a limited number of them in practical applications) is to increment the address of an instruction by the sum of the contents of a selected two or more index registers. This capability is useful in computations involving matrices and in other applications.

Further elaborations involving index registers, as well as many other computer features which help minimize the number of housekeeping instructions, are described in Richards' book *Electronic Digital Systems* (1966), especially in Chapter 3.

Program Improvements Attainable by Including a SUBTRACT-TO-ZERO Instruction

In the straightforward design of adder-subtractors a negative result is obtained whenever the subtrahend is larger than the minuend. In most mathematical problems this result is as desired. On the other hand, in many other problems negative numbers are actually rather artificial, and important programming improvements can be realized by avoiding negative numbers. The avoidance, however, requires a physical modification in the machine. This modification may assume the form of an added instruction in the repertory. Specifically, this added instruction is a SUBTRACT-TO-ZERO (SZ) instruction which causes the result (not always a true "difference") to be zero in all instances where the subtrahend is equal to or greater than the minuend. To obtain this result the signal which normally indicates a negative sign may be used to "force" all digits in the result to be 0's whenever the sign would otherwise be negative.

The improvements to be realized are a saving in storage space through a reduction in the number of instructions in the program and an increase in computing speed through a reduction in the number of instructions that must be executed to obtain a given result. (Recall that the number of instructions in a program and the number of instructions that must be executed are not necessarily proportional to each other.) Also, the need for jumping is eliminated in many instances with the result that, for some special purpose machines, the JUMP facilities can be entirely eliminated from the machine design.

The manner in which the SZ instruction is utilized depends on the details of the information processing problem to be programmed. One example is as follows. Assume that the price of an item is 75¢ each for the first 100 units and 70¢ for each additional unit, but the quantity purchased may be more or less than 100 units. Programs for computing the total cost may be devised in any of several different ways by using the instruction repertory previously described, but each way appears to require an awkward jumping procedure and an objectionably large number of housekeeping instructions in view of the simplicity of the task. However, with the SZ instruction added to the repertory the total cost of N units can be computed with the following reasonably short program containing no jumps. B and C are any conveniently available addresses, and a notation such as $L(N)$ means the address ("location") at which N is stored.

Address	Word	
B	RE	Determines the number of units in excess of 100— zero if $N \leq 100$
$B + 1$	AD $L(N)$	
$B + 2$	SZ $L(100)$	
$B + 3$	MU $L(0.05)$	Multiplies excess by 5¢ and stores result in C
$B + 4$	ST C	
$B + 5$	RE	
$B + 6$	AD $L(N)$	Multiplies N by 70¢
$B + 7$	MU $L(0.70)$	
$B + 8$	AD C	Adds the two quantities obtained above to obtain total cost and stores result in C
$B + 9$	ST C	

Programming Languages

The instruction repertory of a physical computer, no matter how cleverly the computer may be designed and no matter how efficient the resultant programs may be, has proved to be an extremely awkward "language" to use in expressing the details of the steps to be followed in the solution of most problems of consequence.

Accordingly, much effort has been expended on the development of "computer languages" in which a given problem to be solved can be expressed not only explicitly and unambiguously (as in a machine-language program) but also conveniently and easily. Convenience and ease cannot be accurately measured, but generally these objectives are realized by causing the resultant language to be as similar as practical to conventional written and spoken languages and by causing the number of instructions (generally expanded to "statements") to be as few as possible in any given program. Unfortunately, conventional written and spoken languages are poor vehicles for expressing information with the degree of explicitness and unambiguousness required for computer action. Not the least of the problems is that some single words have many different meanings or shades of meaning which can be known only when considering the context in which the words appear, and even then different people will often interpret sentences differently.

Of course, when a problem is expressed in some language other than the machine language of some computer, the program thus written must subsequently be translated to the machine language before the computer can execute the program. This translation task, although extensive, is routine and can itself be reduced to a computer program generally called a "translator" or a "compiler." For each combination of a specific language and a specific computer upon which problems written in that language are to be solved, a separate translator is required.

The different computer languages that have been developed now number in the hundreds. Of course, not all of these are in widespread usage, and many of them are for highly specialized types of problems and are often called "problem-oriented languages" or "POL's." An example would be a language specifically intended to express problems related to the design of road bridges and similar structures.

On the other hand, a few languages have enjoyed widespread usage and have become more or less standard languages within the computer industry, although countless variations or "dialects" have arisen to meet the peculiarities of specific computer designs or to meet the specialized needs of various problem types. Of the major languages, one known as ALGOL is probably at the peak of mathematical elegance and sophistication. While widely used, ALGOL is not a particularly easy language to learn for people who are not trained in notational abstractions as encountered in mathematics. A language called FORTRAN is older and less sophisticated, but in the range of applications for which it is intended (general computations), it has proved to be highly effective. This language has been revised from time to time, and the current version is called FORTRAN IV, but the older versions are still useful and used. At the opposite extreme from ALGOL, a language called COBOL has been developed for persons with no interest whatsoever in mathematical abstractions but whose primary concern is the writing of programs for business applications where the computations are elementary and perhaps little more than incidental to the over-all accounting problem.

Any one of most of the computer languages can be completely defined in perhaps twenty pages or so of text. However, when examples and explanatory material is included in sufficient detail to enable a reader to become adept with the language, an average-sized book is required to present the language. Many books, including at least 12 on ALGOL, at least 30 on FORTRAN, at least 10 on COBOL, and at least 10 on a newer language called PL-1, plus several books on other programming languages, have been published—not to mention countless instruction manuals prepared by computer hardware and software manufacturers. For this reason, the subject matter of this book will not be extended further into programming languages. For readers who desire a more meaningful introduction to this subject but who are not required to gain a full working knowledge of any particular language, reference is made to Chapter 5 of Richard's book *Electronic Digital Systems* (1966).

The General Purpose Computer in a System—Input-Output Devices

The general purpose stored-program computer with a single input device and a single output device as described earlier in this chapter would constitute a complete and useful system for many applications. However, in most applications the intended end result is achieved much more expeditiously by including much more input-output equipment in the system and by incorporating various design elaborations within the computer so that the input-output devices can be utilized efficiently.

For one thing, instead of calling an input device into play by means of an instruction in the elementary manner previously described, each input device is caused to interupt the computer when that device is ready with a word (or a group of words) to be entered into the computer. Upon being interrupted, the computer accepts this word and stores it at an appropriate address for use when needed. This procedure avoids the relatively long waiting times otherwise required when the computer must stand idle while the mechanical and electromechanical portions of the input device undergo physical motion. Of course, to make this procedure advantageous, the program must be written in such a manner that the computer can be doing useful work during the periods of time when it would otherwise be waiting for input words. Again, reference is made to the author's *Electronic Digital Systems* for more details. For understanding the following systems, it is sufficient that the reader be aware of the fact that the computer can be caused to accept input information at any time such information is available from an input device, as opposed to the requirement that an input device must wait until an INPUT instruction pertaining to that device is encountered in the execution of the program.

Figure 9-4 illustrates in a rudimentary way the manner by which a general purpose computer might be used in a conventional computing or accounting system. The block labeled "general purpose computer" corresponds to the equipment portrayed in Figure 9-1 except for the input and output devices which are designated more specifically in Figure 9-4. Input information may be obtained from a punched card reader,

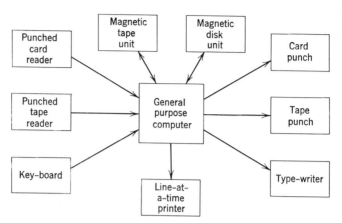

FIGURE 9-4. Conventional computing or accounting system.

a punched paper tape reader or a manually operated keyboard. The computed results may be recorded by means of a card punch, a paper tape punch, an electromagnetically operated typewriter, or a line-at-a-time printer. Of course, the above listing is not necessarily complete, and more than one of any or all of the input-output devices may be included.

The magnetic disk storage unit shown in Figure 9-4 may be used as an input-output device, and such usage has become common in recent years. However, more commonly the magnetic disk unit is visualized as a large-capacity storage unit to augment the digital storage unit within the computer itself. The computer could actually be made to work with a magnetic disk unit as the main digital storage unit of the computer, and some computers have in fact been designed in essentially this manner (although the surface geometry is perhaps more likely to be in the form of a drum than a disk). Such storage allows a large capacity to be provided at relatively low cost, but the time required to gain access to a given stored word tends to be so great that the preferred design approach is to employ magnetic cores within the computer and to transmit large blocks of words between the computer's internal storage unit and the magnetic disk storage unit in a manner (the details of which depend on the problem) by which both the speed and capacity advantages are realized.

Similarly, the magnetic tape unit in Figure 9-4 can be employed either as an input-output device or as an auxiliary large-capacity storage unit. The capacity of a reel of tape is even greater than that of a disk assembly of reasonable size, but the access time problem is substantially more severe. When either the tape unit or the disk unit is used in the role of input-output, the tape reels or the disk assemblies are designed to be physically removable so that they may be replaced by other reels or disk assemblies, as the case may be.

Another General Purpose Computer Application

Figure 9-5 illustrates a quite different application of a general purpose computer. Input and output is primarily from and to a number of "desk sets" each of which is operated by a person. The desk sets may number in the thousands and may be located at widely scattered points, each of which is in communication with the computer by some means, usually digitally operated telephone lines. In addition to the desk sets, suitable supervisory input-output devices are provided to control the course of the information processing (to enter the initial program, if for nothing else).

Applications for which a system as illustrated in Figure 9-5 is useful include reservations (as for airline seats, theater or athletic event seats, motels, and others), price quotations and other market activity (notably for corporation stocks but also poultry and countless other things), point-of-sale accounting in retail merchandising, the time-shared use of a computer by persons wishing to perform miscellaneous manually programmed calculations. Of course, the detailed design of all aspects of the system would be strongly dependent upon the application. However, in any application the supervisory input-output devices can be designed so that, with the aid of suitable internal programming of the computer, all man-

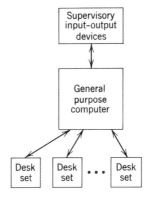

FIGURE 9-5. System for essentially simultaneous use by many people.

ner of detailed reports of computer activity may be prepared with records made of individual items (high-low-closing prices of individual corporation stocks for each day, to cite one example) or of statistical summaries of the information processed (such as total number of shares traded). Other things, such as bills to be charged the user of each desk set, can also be prepared automatically by means of the supervisory input-output devices.

The General Purpose Computer as Used in a Message Switching System

Figure 9-6 illustrates the use of a general purpose computer in a message switching system. In this application, messages are to be transmitted between any pair of teletypewriter terminals. Apart from the fact that providing separate communication links between each pair of terminals would be impracticably expensive, a basic problem arises from the fact that no one teletypewriter can be receiving messages simultaneously from two different sources. To avoid the "line is busy" situations as encountered with telephone communication, the messages are transmitted to a central switching terminal where they are stored until they can be transmitted to their intended destination. In a large system, many switching terminals may be established, one for each general area, and to reach its final destination a message may need to be transmitted through two or more such terminals with a temporary storage taking place at each one while waiting for a suitable transmission path to become available.

A general purpose stored-program digital computer is capable of providing all of the functions needed for message switching. Each message

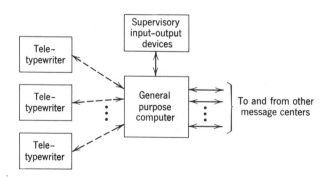

FIGURE 9-6. Message switching system.

is accompanied by a "heading" which, among other things, designates the intended destination of that message. By programming, the computer can be caused to "remember" that it has a message for a given destination and can be caused to transmit that message on the appropriate output line when that line is available. Moreover, a given message can be transmitted to two or more different destinations, with each transmission occurring at a different time as the respective lines become available in accordance with the other messages currently being handled by the system. The heading of each message can be further expanded to include a priority, and the selection of a particular message to be transmitted on a given line at a given time can be based on the relative priorities of the messages currently awaiting transmission on that line. Moreover, if a low priority message has been delayed for more than a specified length of time, its priority can be automatically increased by suitable programming. Still other message handling functions such as the acknowledgement of receipt of messages and the correction of certain types of errors can also be handled by elaborations in the computer program.

Although not shown in Figure 9-6, various large-capacity storage devices in the form of magnetic tape units or magnetic disk units would be necessary in most message switching systems of consequential size. The supervisory input-output devices may be used for all manner of control functions and for preparing reports of system activity.

The General Purpose Computer as Used in a Telephone Switching System

A telephone switching system, as illustrated in Figure 9-7, can also be based on a general purpose stored-program computer, but the characteristics of the resulting system are quite different from those required for message switching. The storing of messages, even in analog form, is not acceptable in a telephone system. Instead, the system must be capable of establishing a communication path between any one telephone and any other telephone. In some systems the number of such paths simultaneously operative may be as great as half the number of telephones, but more commonly the system need provide for only a much smaller number of simultaneous conversations. Regardless of the requirements in this respect, a digital system is not of itself capable of making the connections necessary for the transmission of analog signals as encountered in telephones.

Therefore, to perform telephone switching, the computer must have a major auxiliary unit to establish the actual interconnections among the various telephone lines. This auxiliary unit can be based on the Strowger electromechanical stepping switching as found in some dial

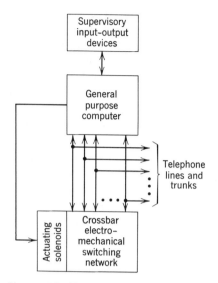

FIGURE 9-7. Telephone switching system.

systems or it can be based on the currently more popular crossbar electromechanical switching arrays. Other switch types such as those formed with magnetic reed elements or gas tubes may alternatively be used. Although any of these switch types perform the basic function of making and breaking circuits intended for analog-type signals, the switches are actuated by "on-off" types of signals that are essentially of digital character. In a telephone switching system the primary function of the computer is to determine (from the "dial pulses" transmitted at the beginning of each call) the particular elements to be actuated in the network, with the crossbar switching array being assumed in Figure 9-7. With the dial pulses as input signals, the computer generates output signals which actuate the array.

In telephone applications that include two or more interconnected systems of the form indicated in Figure 9-7, another important function of the system is to "remember" and regenerate the dial pulses when a transmission path must pass through two or more such systems before reaching the called telephone. The first system establishes a connection from the calling telephone to a "trunk" line to the second system. The regenerated dial pulses are then sent along this trunk to the second system where the pulses produce essentially the same result as would pulses arriving from a local telephone line at that system. The second system then determines the connections to be made there (which may be to a local line or to a second trunk to a third system), and the process continues until the called telephone is reached.

Thus, in Figure 9-7 each line or trunk connected to that system is connected to both the general purpose computer and the crossbar switching array. Dial pulses may be received from any number of the lines and trunks simultaneously, but because the duration between successive pulses on any one line is a substantial fraction of a second, the computer may be presumed fast enough to accept the dialing information from all lines (or at least from a substantial number of them) simultaneously. When the computer has the information needed to establish a connection, it transmits signals to the appropriate actuating solenoids in the crossbar array. The figure shows only a single line for this purpose, but as many physical wires as needed are implied. As explained in the previous paragraph, if the connection is to a trunk instead of a local line, the computer additionally regenerates the dial pulses and transmits them as output signals on the corresponding calling telephone line, which is now connected to the selected trunk.

The crossbar array is so designed that the connections established by the computer remain established without further attention by the computer. A d-c signal exists on the telephone line as long as the line is in use for a call, but when the call is completed the "hanging up" of a telephone set causes the d-c signal to be terminated, at which time the connections in the crossbar array are automatically opened. When the call is being transmitted along trunks through two or more switching systems, all crossbar arrays are similarly deactuated at the completion of a call. The termination signal is not needed by the computer in performing its primary function, but when telephone charges are based on the duration of the call, the termination signal is utilized by the computer in computing the charges.

The same computer as used for determining the connections may also be programmed to compute the charges and to perform other accounting tasks. Bills and other documents may then be prepared on the supervisory input-output devices indicated in Figure 9-7.

Digital Control or "Real Time" Systems

Figure 9-8 represents still another quite different system in which the general purpose stored-program digital computer is applicable. Here the computer is used to control the operation of some other system, which is essentially of an analog character. This other system may be an analog computer, a chemical manufacturing plant, a simulated aircraft for training, a spacecraft, a missile, or any of many other things.

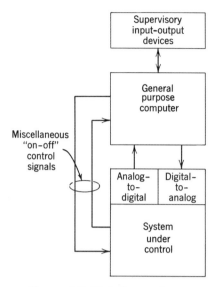

FIGURE 9-8. Digital control system.

Applications of this form are generally said to be "real time" applications because, in the equations pertaining to the controlled system, changes in the time variable t correspond to the progress of real time. Changes in other variables must be computed with sufficient speed to cause the computer output signals to be within the required accuracy at each step or point in time. (The case where the system under control is an analog computer could be regarded as an exception to the "real time" definition because t may be only an abstraction in the problem being solved by the analog computer—although in the equations expressing the analog computer set-up itself, t would be real. By judicious scaling of the problem parameters the programmer can, in principle, cause the analog computer to proceed at any desired speed, although practical limitations are encountered at both the low and high ends of the speed range.)

In Figure 9-8 the major control signals are transmitted to the controlled system in digital form, and the digital-to-analog converters are used to convert the signals to whatever form is needed within the controlled system. The major signals which indicate the status or progress of the system are generally in analog form, and these signals are converted to digital form by analog-to-digital converters before being transmitted to the computer. In addition to these major signals, numerous "on-off" signals may be needed to perform miscellaneous control functions. Similarly, many signals generated within the controlled system may be essentially of "on-off" form, and these signals may be utilized as conventional digital computer input signals.

In the case where the controlled system is an analog computer, the usual purpose is to have the analog computer perform certain parts of the computations for which analog computers are especially well adapted. The digital computer transmits "set-up" information and initial values of the variables to the analog computer. The "set-up" information establishes the desired interconnections among the integrators, multipliers, and other units forming the analog computer. The initial values are, in effect, converted to analog form by means of the digitally actuated potentiometers and other devices within the analog computer. A miscellaneous "on-off" control signal may be used to start the computations in the analog computer, and from time to time the digital computer will accept the results of the analog computations as transmitted through the analog-to-digital converters. All of these actions are under the ultimate control of the program in the digital computer.

In a chemical manufacturing plant, the devices under control may be things such as valves, heating elements, variable-speed conveyors, and so on. These devices are basically analog but can be actuated digitally by suitable mechanisms which, in effect, constitute the digital-to-analog converters. The signals indicating the progress of the plant may be generated by flow meters, thermometers, pressure gauges, and numerous other such instruments. In nearly all cases the signals generated by these devices are converted to electrical analog signals by mechanisms which are parts of the instruments, and then analog-to-digital converters are used to generate the digital signals needed by the controlling computer.

In the case of a simulated aircraft for training, a cockpit with all controls and indicators is provided. The controls, which are operated by the pilot being trained, generate analog signals which after conversion to digital form are transmitted to the digital computer. The computer determines the motion that a real aircraft would undergo in the event that the same actuation of the controls had been applied to a real aircraft. This motion is represented by signals which are, after conversion to analog form, used to acutate indicators which indicate speed, altitude, direction, and other information for the pilot. Thus, the pilot would constitute the "system under control" as the block diagram in Figure 9-8 appears, although of course the pilot is really controlling a simulated aircraft represented by the computer.

When the arrangement in Figure 9-8 is used to control something such as a spacecraft or a missile, the signal lines between the general purpose computer and the system under control may contain radio transmitters and receivers operated in the form of digital information transmission systems. Alternatively, the general purpose computer may be on board the spacecraft or missile in which case the line between the computer and the supervisory input-output devices may contain digitally operated radio transmission paths. Also, the system under control may include one or more radar installations operating in the following manner. The spacecraft or missile need not, of itself, generate any signals at all to be sent to the general purpose computer. Instead, the radar installations may be used to sense the location, direction, and speed of the spacecraft or missile. The analog signals thus generated may then be converted to digital form at the radar sight and transmitted to the computer, which may be remote from both the radar and the spacecraft or missile. After computing the necessary controlling information, signals are then transmitted from the computer to the spacecraft or missile (the system under control) in the manner portrayed by Figure 9-8.

Time-Sharing

One other concept pertaining to the general purpose stored-program computer will be mentioned briefly, although in some respects it is an extensive subject. The cost per computation tends to decrease, often quite dramatically, as the computer size (storage capacity), component speed, and sophistication of design is increased. However, physical design limitations of the input-output devices are often such that, for some applications, input-output information cannot be practically transmitted to and from the computer fast enough to keep a large, high speed, sophisticated computer busy. In other applications, the information processing requirements simply may not be great enough to justify the powerful computer.

Therefore, to achieve the cost advantages of a powerful computer, a mode of operation known as "time-sharing" may be necessary. With time sharing, two or more sets of input-output devices, which need not be similar to each other, are connected to a single computer which solves the various problems more or less simultaneously (as it appears to the several users of the computer). Actually, the computer works on one problem for a while, then another, and another, and so on, in whatever sequence is appropriate to perform all computations within prescribed time limits. In some applications, low priority problems may be delayed while the computer is busy with high priority problems. All of the above is controlled by suitable programs, which may involve equipment modifications within the computer.

In some instances the economies of time sharing can be so great as to justify the transmission of input-output information long distances (thousands of miles) to a central powerful computer. On the other hand, the added programming complexity (cost), added equipment complexity, and the information transmission systems are often so substantial that the potential advantages of time sharing are not really realized. Accordingly, medium sized computeres and small computers of many descriptions have found and are expected to continue to find wide application.

The extent to which time sharing should be employed in a give application or group of applications will probably always be a matter of great concern in the design of an information processing system, with compromises among many complicated and specialized economic factors being required.

Exercises

1. Write out in detail the program variation suggested at the bottom of page 536.
2. Assume that x and y are stored at addresses 0034 and 0035, respectively. Write a program that will interchange the locations at which x and y are stored.
3. For the flow chart in Figure 9-3, determine the addresses at which the instructions for each step are stored, as the program is given on page 540.
4. Alter the program on page 540 as necessary to cause it to store x, y, and z in numerical sequence at addresses 0239, 0240, and 0241.
5. For the computer as described, write a loop-type program which will enter C numbers as input words and will store them at consecutively numbered addresses with H as the first address.
6. Write the instructions for handling N and M in the alternative manner suggested on page 545.
7. For the interpretive program on page 549, prepare a subprogram to execute an interpreted JUMP instruction. (Arbitrarily store the first instruction of this subprogram at address 0380.) Also prepare a subprogram for an interpreted JUMP IF NEGATIVE instruction, where the jump condition is dependent on the sign of the number at address 9999, the imaginary accumulator. (Arbitrarily store the first instruction of this subprogram at address 0420.)
8. With the instruction repertory given near the beginning of the chapter, write a program for the following problem. Employees in a company may participate in a stock purchase plan to the extent of 5% of their salary, but with a maximum of $50.00 per month. Compute the maximum monthly participation amount for an employee having a monthly salary M.
9. Write a different program for the same problem, but use the SUBTRACT-TO-ZERO (SZ) instruction to eliminate all jump instructions in the program.
10. With the instruction repertory given in the text, write a program that will extract the square root of whatever number is stored at address 2070.

INDEX

Accumulator, binary, 282–284, 324
Accuracy, in multiplication, 321
Addition, binary, 277–308, 318–319, 365–367
  decimal, 385–390, 399–401
Address, defined, 528
ALGOL, 560
Algorithm, defined, 436
Analog computer, used with digital computer, 569
Angalog-to-digital converters, 267
AND function, defined, 4–5, 9
AND–INVERT (A–I or NAND) modules, 21, 78–103
AND–OR networks, 24–53, 68
AND–OR–AND networks, 56–58
AND–OR–INVERT (A–O–I) modules, 80–81, 103–105, 107–108, 110
Associative storage unit, to replace switching network, 116–118
Asynchronous circuits, 132–134, 138, 141, 144, 150, 157–166, 300–308, 325–326, 533
Automaton, defined, 122

Basic term, defined, 29
  possible number of, 41
Binary-to-decimal conversion, 436–437, 448–459
Binary numbers and arithmetic, 272–374
Binit, defined, 179
Biquinary code, 383–385, 522

Bit, defined, 120, 179–181, 379
Block codes, 187–267
Boolean algebra, 1–22
Borrow, defined, 308, 390
Bose-Chaudhuri codes, 233–234
Burst-error control, 190–192, 210, 213, 236–238
Buzzing, 133
Bypass, carry, 289–300, 302–303, 323, 387–388, 523
  for improving comparison speed, 463

Canonical form of switching function, defined, 28
Capacity, of information bits, 182
  of a channel, 183–185
Carry, binary, 274, 278–281
  bypass, 289–300, 302–303, 323, 387–388, 523
  decimal, 385–390
  end-around, 317, 393
"Carry save" technique, 323, 325–326, 333, 442
Casting out 9's, 522
Catalogs of switching networks, 19–21
Channel, defined, 182
  capacity, 183–185
Check digits, defined, 192
  codes using, 197–215
Clock pulses and clocks, 131, 172–176, 532
COBOL, 560
Codes, cryptographic, 473–477

573

574 INDEX

decimal, 376–385
  for digit errors in arithmetic operations, 516–523
  for digit errors in storage and transmission, 187–251, 267–269, 382–385
  for synchronization errors, 251–266
Coherent digit detection, 186
Combinatorial networks, 24–118
"Commas," for control of synchronization errors, 253–257
Comparison, 460–463
Complements, binary, defined, 312
  in multiplication, 338–341
  subtraction by addition of, 315–318, 392–396
  9's, 379, 397–399
Computer, general purpose, 526–571
Computer-aided design, 21–22, 42–46
Content, information, in bits, 181–183
Control systems, 567–570
Convolutional codes, 268
Counter circuits, 151–157
Cryptographic operations, 473–477
Current-mode (ECL) circuits, 2, 68, 77, 107–110
Cyclic codes, 215, 221–251

D flop-flops, 143–151
Decimal-to-binary conversion, 436–448, 458–459
Decimal numbers and arithmetic, 376–427, 521–522
Decoding, 194–195, 198–200, 203–204, 209–210, 213–215, 241–251, 268–270
De Morgan's laws, 10
Diagrams, for simplifying switching networks, 48–56
Digital differential analyzers (DDA), 506–516
Diode circuits, 5–6, 24–25
"Distance" of a code, 187–190
Division, absence from Boolean algebra, 15
  binary, 341–364, 368
  decimal, 417–427
  in a DDA, 508–509
Documentation of programs, 547
"Don't care" conditions, 40, 46–48, 55, 67, 95–97, 104, 116, 145–148, 398, 413–414

Doublers, as used in binary-decimal conversion, 447, 451–452
Doubling, binary, 276, 336
  decimal, 404–407
DTL circuits, 79, 81, 104
Dynamic switching elements, 2

ECL circuits, 2, 68, 77, 107–110
Elemental term, defined, 28
Encoding, 193–194, 198–200, 238–241, 268
End-around carry, 317, 393
Equality, Boolean, defined, 9
Errors, arithmetic operations, 516–523
  in asynchronous circuits, 160
  bit (or digit), 187–251, 267–269, 382–385, 516–523
  codes for control of, 187–269, 382–385, 516–523
  effect on information content of a bit, 181–183
  as a function of channel parameters, 185–187
  programming, 552–554
  synchronization, 251–266, 269
Essential term, defined, 31
Excess-weighted codes, 380–382
"Exclusive-OR" function, 6–7, 199–200, 215

Factoring, of Boolean expressions, 12–13, 56–58, 93–95
Fan-in and fan-out, 21, 80, 82, 98–101
Feedback, digital, 103, 120, 176, 397
Fire codes, 238
Flip-flop, complementing, 134–138, 146–151
  D, 143–151
  J-K, 139–142, 146–156, 282–284
  master-slave, 142
  pseudo-J-K, 142–143
  R-S, 122–124, 146–151
  R-S-T, 135, 146–151
Floating point, binary-decimal conversions, 458–459
  general, 279–280, 321, 349, 364–374, 427
  miscellaneous functions, 473
  square row, 468–469
Flow charts, 537–539
FORTRAN, 560
"Fractional" computer, defined, 321
Fractions, binary, 272–273
Functions, Boolean, 13–14

Gates, 72, 130–131, 134–136, 146

INDEX 575

General purpose computer, 526–571
Generator polynomial, defined, 222–223
Gray-codes, 267, 429–436
Group codes, 187

Halving, 276
Hamming codes, 179, 200–206
Hazard, defined, 132
Housekeeping instructions, 554–558

Incoherent digit detection, defined, 186
Incremental computer, defined, 506
Index registers, 555–558
Information, meaure of, 179–181, 377–378
Input-output, 131, 529, 535, 561–564
Instruction repertory, 533–535
"Integral" computer, defined, 320
Integration and integrators, 502–505
Interleaving, 214–215
Interpretive programs, 547–552
Inversion, defined, 7–9
Inverters, minimization of number of, 77–78
Irreducible polynomials, 226–228
Iterative procedures, binary division, 362–364
  decimal division, 426
  square root, 470–471

J–K flip-flops, 139–142, 146–156, 282–284
"Jump" instructions, 530, 534, 537–539, 545, 551

Karnaugh maps, 48

Languages, programming, 559–561
Large scale integration (LSI), 115–116, 130, 172, 210, 323, 361, 388, 390, 440, 443, 470
Levels of switching, defined, 25
Loops, program, 539–543
  switching network, 103, 120–176, 397

Magnetic core switching circuits, 2, 101
Majority circuits, 3
Maps, Karnaugh, 48
Master-slave flip-flops, 142
Matrices, 68–74
Mechanization, of network design, 21–22, 42–46

Message, defined, 179, 184
Message switching, 564–565
Minimization, A–I and O–I modules, 82–103
  AND–OR, 28–38
  multiplexers, 113–114
  in a sequential circuit, 166–168
  using output of one flip-flop to actuate another, 156–157
  using R–S–T flip-flop characteristics, 145
Modulo, 474, 481, 519
Modulo–2 arithmetic, 215–221
Multidimensional codes, 206–213
Multi-output networks, 59–68, 102–103
Multiplexers, 110–114
Multiplication, binary, 319–341, 368
  decimal, 401–417
  in a DDA, 509–511
Multi-state switching networks, 124–130, 170

NAND modules, 21, 78–103
Negative numbers, binary, 312–315, 338–341, 361–362
"No carry" signal, defined, 301
Nonblock codes, 267–269
Nonsystematic block codes, 192–196, 206, 222, 247
Normalized numbers, 349, 364, 369–370
NOR modules, 21, 78–103
NOT function, defined, 7, 9

Octal digit, defined, 333
OR function, defined, 6, 9
OR–to–AND networks, 53–56, 68
OR–AND–INVERT (O–A–I) modules, 105–108, 110
OR–AND–OR networks, 56–58
OR–INVERT (O–I) or NOR) modules, 21, 78–103
Output devices, 131, 529, 535, 561–564
Overflow, 315, 320, 347–348, 366, 368–369, 393–396, 418

Parametrons, 2, 3
Parity check digits, defined, 192
PL–1, 560
Point, binary, 275–276, 314, 348–349, 370
  floating, see "Floating point"
Polynomial, as a set of mod 2 digits, 216
Prefix, for control of synchronization errors, 253–257

576 INDEX

Prime implicants, defined, 29
Primitive polynomials, 226–228
Programs and programming, 527, 530, 535–561
Pseudo–J–K flip-flops, 142–143
Pseudo-random numbers, 479–483
Pulse-type signals, 2
Pyramid matrix, 69–71

Quintupling, 407–408
  as used in decimal-binary conversion, 438–440, 454–456

Races, 133
Random numbers, 477–483, 502
Read-only storage (ROM), as a replacement for a switching network, 15, 115–118, 363, 470–473
"Real time" operation, 567–570
Recurrent codes, 268
Redundancy digits, defined, 192
Reflected binary codes, 267, 429–436
Repertory of instructions, 533–535, 547–548, 551
Residue, 517–520
ROM, as a replacement for a switching network, 15, 115–118, 363, 470–473
Round-off, 277, 320–321, 342, 346–347, 366–367, 372–373, 426–427
R–S and R–S–T flip-flops, 122–124, 135, 146–151
RTL circuits, 101, 285

Scaling, 502–503, 505, 515
Self-complementing codes, 379–381
Separation of variables, 41
Sequential circuits, 121, 130–134, 157–176, 422, 478–479, 481
Sequential decoding, 268
Serial operation, binary addition-substraction, 318–319
  binary division, 362
  binary multiplication, 330–333
  decimal, 399–401
  in a DDA, 513–515
Shifting codes, 382
Shift register, 160–162
Shortened codes, 206, 235–236, 238, 262–266
Significance of digits, 320–321, 369–374, 468

Simplification, *see* "Minimization"
Simulation, 547
Simultaneous addition, 287–289, 293–294
Sliver signals, 168
Snake-in-the-box codes, 266–267
Sorting, address calculation, 487
  balanced merge, 493–495
  counting, 491–492
  digit, 483–485
  exchanging, 487–488
  general, 483–502
  insertion, 485–486
  internal, 484–493, 498–499
  merging, 484–485, 488–489, 493–502
  partitioning, 492–493
  replacement-selection, 491
  selection, 490–491
  tape, 493–502
  unbalanced merge, 495–497
Square root, 463–471
Storage, formed with switching networks, 120–130
  associative, 116–118
Storage unit, as a replacement for a switching network, 15, 115–118, 363, 470–473
"Store address" instruction, 543
Stored-program computer, 526–571
Subprograms, 543–547
Subtraction, absence from Boolean algebra, 15
  binary, 289, 308–319, 365–367
  decimal, 390–401
"Subtract-to-zero" instruction, 558–559
Superfluous terms and variables, 26–28
Synchronization errors, 251–266, 269
Synchronous circuits, 130–132
Syndrome, defined, 243, 249
Systematic block codes, defined, 192
  decoding method, 194

Telephone switching, 565–567
Term, defined, 26
Term tables, 29–48, 62–66
Testing, for superfluous terms and variables, 26–28
Three-state networks, 124–126, 170
Threshold circuits, 3
Time-dependent signals, 120–176
Time-sharing, 570–571
Tracing programs, 552–554

Trapezoidal integration, 505
Tree codes, 268
Tree matrix, 69
Trigger circuit, 135
Truth tables, 14–15
TTL circuits, 81
Tunnel diode circuits, 23, 101
Two-dimensional codes, 206–213

Two-out-of-five decimal codes, 382–383

Universal switching modules, 112

Variable, binary, defined, 4

Weighted codes, 379–382

Zero, representation of, 318, 367, 393, 396